A First Course in the Design of Experiments

A Linear Models Approach

A First Course in the Design of Experiments

A Linear Models Approach

Donald C. Weber
John H. Skillings
Miami University
Oxford, Ohio

CRC Press
Boca Raton London New York Washington, D.C.

Library of Congress Cataloging-in-Publication Data

Catalog record is available from the Library of Congress.

International Standard Book Number 0-8493-9671-9
Printed in the United States of America 1 2 3 4 5 6 7 8 9 0
Printed on acid-free paper

PREFACE

For majors in many areas, such as mathematics, statistics, economics, engineering, and science, the first course in statistics is a course in mathematical statistics. Such a course includes much of the theory underlying the subject matter in addition to some applications. After that first course, students traditionally proceed into mainstream courses such as regression and experimental designs. These courses are typically taught with an emphasis on only methods or with an emphasis on only theory. Courses emphasizing methods largely neglect the distribution theory, which was so central to the beginning course. Theoretical courses neglect the important applications in regression and design of experiments.

This text provides a blend of the methods and theory for the important topic of design of experiments. By utilizing the linear models approach, we are able to integrate the theory with the methods and present a unified strategy for analyzing all designs. As a by-product of the linear models approach, one can obtain the traditional analysis of variance, adjusted and unadjusted sum of squares, and the f-tests. An advantage of the linear model approach is that, unlike the analysis of variance approach, the procedures are essentially the same whether the data results from designed, undesigned, unbalanced or missing cells experiments.

Features of Text

As indicated previously, the analysis of various designs is integrated in this text through the general theory of linear models. Computer usage is encouraged and illustrated throughout the text. SAS, one of the most commonly used statistical programs, is demonstrated extensively. In addition, this textbook illustrates how any regression program can also be used to carry out most analyses.

To aid in understanding the subject matter almost every section contains examples to illustrate the concepts. There are also nearly 400 exercises divided into two groups, application exercises and theoretical exercises, each with a wide range of difficulty. Those who want a more methods-oriented course may emphasize the application exercises, while the theoretical exercises may be emphasized for a more theoretical course. Data used in the exercises generally come from problems that the authors have encountered over the years in actual practice. While the data sets have often been reduced in size or otherwise modified, the reader will have the experience of working on data from real problems.

The authors have aimed to use accurate notation. For example, parameters, estimators and estimates are carefully distinguished from each other through the use of different symbols. Further, every effort is made to distinguish random variables from their realized values.

Prerequisites

Readers need to have completed a course in mathematical statistics. To fully understand and appreciate the entirety of this textbook the student should also have had exposure to matrix algebra, enough to feel comfortable with the ideas and operations. For those who have not had such an exposure to matrix algebra, there is an introduction to the topic in Appendix A of this book. Appendix A also introduces some topics that are important for this book, but are topics that are often not found in an elementary matrix algebra course. A background in regression analysis, though desirable, is not absolutely necessary. Principles of regression that are needed in this book are introduced in Chapter 6. Those who have had an introduction to regression will want to skip most of this chapter, while those without a regression background will need to cover the material. The suggested background for a reader implies that this book is intended primarily as a text for an upper-level undergraduate course or a beginning graduate course on the design of experiments.

Suggested Use

Based upon classroom testing, the authors believe that this text contains enough material to serve as at least a four semester hour course in experimental designs. There are several approaches that may be used. For a course that emphasizes both theory and applications one should first cover the linear models material in Chapters 1- 5. While most of the material in these chapters should be covered, many of the derivations should be handled outside of class to allow sufficient time to cover the applied chapters. Chapter 6 should be covered for those who have not had a course in regression. Most of the applications material in Chapters 7 - 16 may be covered, along with a few selections from Chapters 17 and 18. A course emphasizing the theory of linear models should concentrate on Chapters 1 - 5 by completing all or most of the derivations. Parts of Chapters 6 - 18 may be covered to allow for some illustrations of the linear models material. A course emphasizing applications would skim the material in Chapters 1 - 5 and then emphasize the applications material in Chapters 6 - 18. A two semester sequence emphasizing linear models in the first course (Chapters 1 - 5) and applications in the second course (Chapters 6 - 18) provides another possible use of the book.

Contents in Text

Chapter 1 provides a general overview of the topic of the design of experiments using verbal descriptions and illustrations. It presents ideas, concepts and nomenclature that the student will encounter in the remainder of the text.

In Chapters 2 through 5 the linear model is introduced with some examples. Theoretical results for the linear model, which are needed for the analysis of experimental designs, are then presented. Many examples, including numerical ones, are given throughout these chapters to illustrate the usefulness of the obtained results. Regression models and the one-way classification model are used as the primary examples in these chapters. Chapters 3 - 5 are the most theoretical in the text, while the remaining chapters are much more application orientated.

The multiple regression model and the inference formulas for this model are presented in Chapter 6. The chapter does not intend to provide a complete treatment of regression analysis, but rather it introduces the regression material that is needed later in the book.

The completely randomized design is discussed in great detail in Chapters 7 - 9. This material illustrates how the general theory for the linear model is applied to experimental design models. Special topics such as checking model assumptions, planned comparisons (Chapter 8) and multiple comparisons (Chapter 9) are introduced in this setting.

Using a format similar to that used in Chapter 7 on the completely randomized design, two other basic designs, randomized block and Latin square, are introduced and discussed in Chapters 10 - 12. The presentation flows rather easily as the theoretical foundations and analysis approach have been outlined and illustrated in earlier chapters. The ideas of adjusted versus unadjusted sum of squares, and orthogonality are introduced in this setting.

Factorial experiments, replicated in previously studied designs, are the subject matter of Chapters 13 and 14. The idea of interaction is introduced, and a brief introduction to fractional factorials closes the discussion on analyzing fixed effects factorial experiments.

In Chapter 15 the analysis of covariance is presented as a method of adjusting for the effect of extraneous factors on the response variable. Adjusted and least squares means are introduced.

Up through Chapter 15, only fixed effects experimental design models have been presented. Chapter 16 introduces the student to random and mixed effects models, the nested classification and variance components. In the final two chapters (17 and 18), these ideas and concepts are expanded upon as they are applied to special settings, such as repeated measures designs, three-stage nesting, and split-plot designs. An introduction to response surface methodology and comments on design selection conclude the work.

Acknowledgments

We wish to express our appreciation to those who have aided in the completion of this book. Our colleagues at Miami University, Drs. Robert Schaefer and Kyoungah See, and reviewers, Drs. John Stufken (Iowa State University) and Sudhir Gupta (Northern Illinois University), have provided many helpful comments that have led to an improved presentation. Ms. Jean Cavalieri has been extremely helpful with her high quality, efficient work in the preparation of this manuscript. Finally, we appreciate the patience and support provided by our spouses, Elaine and Sue, over the years.

CONTENTS

Chapter 6. Inference for Multiple Regression Models, 139

Chapter 7. The Completely Randomized Design, 185

Chapter 8. Planned Comparisons, 253

Chapter 9. Multiple Comparisons, 281

Chapter 10. Randomized Complete Block Design, 307

CHAPTER 1

INTRODUCTION TO THE DESIGN OF EXPERIMENTS

1.1 Designing Experiments

Experiments

Experiments are performed and observations (data) are generated in almost every discipline. Engineers perform experiments to improve the quality of products. Environmental scientists perform experiments and collect data to determine the effect of toxic substances on animals. Social scientists perform experiments to help determine the causes of certain behavior patterns.

There are many purposes for experiments. One purpose is to determine how changing the value of one variable or measurement affects another variable. For example, in making steel rods we can determine the effect of changes in the amount of a chemical A (first variable) on the tensile strength of the rod (second variable). Alternatively, we can determine the effect of increasing levels of copper in water on the lifetime of some aquatic life.

Another purpose of experiments is to compare the differences between the mean of a variable for various groups. These groups are often formed by the different values of another variable. For example, in a sex discrimination study for accountants we might want to compare the mean salary (first variable) for men and women (values of the variable sex). As another example, we could have three groups of students who were taught to read using different methods, and we might want to compare their performance on a standardized test.

Designed Experiments

A **designed experiment** is an experiment in which the experimenter plans the structure of the experiment. For example, the experimenter determines the variables to be measured, the settings for the variables, the order in which multiple trials are run, and in general sets up the experiment so that data can be obtained to help answer appropriate questions.

There are several steps that are required to set up an experiment. For most problems these steps include the ones listed below. We illustrate these various steps with the following example. A company produces metal sheets and wants to improve the strength of the sheets. The company determines that the strength of the sheets is mainly determined by two key chemicals A and B that are used in making the metal. The goal of the experiment is to study the relationship between the amounts of these two chemicals and the strength of the metal sheet.

Steps for Setting Up an Experiment

- *(Goal).* The first step is to determine the goals for the experiment. If possible, a research hypothesis should be established. For the example, the stated goal is to determine the relationship between the amount of chemical A and the amount of chemical B with the strength of the metal.

- *(Response Variable).* Next, the **response variable** (dependent variable) for the problem should be determined. The response variable is the main variable of interest. It is the variable that is affected by the other variables in the problem. For the example, the response variable is a measurement of the strength of the metal. We are assuming that the other variables (i.e., amount of chemical A and the amount of chemical B) affect the strength.

- *(Factors).* The **independent variables** for the problem need to be determined. Independent variables are ones that potentially affect the response variable. These variables will either be **controlled**, **uncontrolled** or remain **constant** in the experiment. Variables are called "controlled" if they are intentionally varied by the experimenter. These variables constitute the **factors** for the experiment. In the example, the factors are the amount of chemical A and the amount of chemical B. Since these amounts will be determined by the experimenter, these are called **controlled** variables.

Variables that the experimenter does not control are called **uncontrolled** variables. In the example there may be several uncontrolled variables. Possibilities include temperature, humidity and other factors that might affect the strength of the metal but are not controlled by the experimenter. A diagram of this situation is given in Figure 1.1.1.

In some experiments, one or more variables are held **constant** throughout the experiment. A possibility in our example is the amount of another chemical, say C, that is to remain the same, regardless of the amount of chemical A or B.

- *(Levels).* For each factor the experimenter must choose the different levels to be used. The **levels** of the factor are the different values that the factor assumes during the experiment. In some cases the levels are obvious and no selection needs to be made. For example, if the factor is sex, then there are clearly only two levels: female and male.

In our metal sheet example, the experimenter would need to determine how many different levels would be used for chemicals A and B. This requires knowing the "feasible" range for the chemical. In many cases two or three different levels of A will be used. For example, a high level and a low level of A might be used. Similar decisions would be required for chemical B.

If chemical A has two levels (high and low) and chemical B has three levels (high, medium and low), then the distinct test runs possible for the experiment would be (H = high, M = medium, L = low):

Test Runs

Level of chemical A:	H	H	H	L	L	L
Level of chemical B:	H	M	L	H	M	L

Controlled Variables

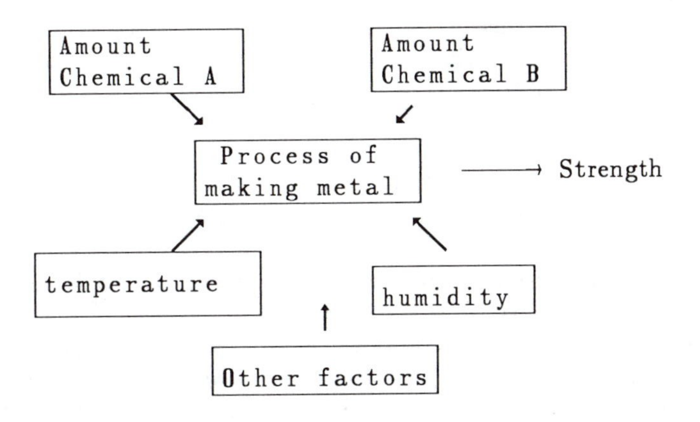

Uncontrolled Variables

Figure 1.1.1 Experiment with Controlled and Uncontrolled Variables

• *(Design).* The experimenter now needs to choose a design for the experiment. Many decisions need to be made here including:

—Which of the test runs will be used? (All, half — which half, etc.)

—Should any test runs be duplicated? If so, how many times?

—In what order should we make the test runs? Which should be done first, last?

• *(Experiment).* After all of this careful preparation, the experiment is run and the data is collected.

• *(Data Analysis and Conclusion).* The data obtained from the designed experiment is now analyzed with the intent of answering the questions raised in the goals part. Conclusions are now presented with support from the data.

An article by Coleman and Montgomery (1993) provides an interesting overview of many of the issues involved in setting up an experiment.

As another example consider a study with the goal of comparing several drugs that are supposed to reduce blood pressure in humans. The response variable would be blood pressure or perhaps the amount it is reduced by the drug. Some potential independent variables would be: type drug used, subject, sex of subject, age of subject, physical condition of subject, blood pressure at the beginning of the study, other drugs a subject is taking, and many others. Important decisions that an experimenter would need to make include deciding which of these potential independent variables will be controlled as factors and which will be left uncontrolled.

Need for Good Designs

There are a number of reasons for needing a good design. Some of these are:

• We want to be sure that the data we collect is sufficient to answer the questions asked in the goal.

• The response variable varies from run to run — partly due to the different values of the controlled variables and partly due to the uncontrolled variables. We need to set up our design in such a way that the variability in response due to the uncontrolled variables (sometimes called experimental error) is not so great that it masks the effects of the controlled variables. For example, in the metal sheet problem the variation in the strength of the metal due to temperature and humidity (uncontrolled variables) could be so large that the effect on the strength due to the amounts of chemical A and B are undetectable.

• We want to design experiments that are efficient, that is, designs where we can answer the questions of interest with a minimal amount of data because of the expense associated with data collection.

1.2 Types of Designs

There are many different types of designs for experiments. Designs are classified by their characteristics. Among the ones used to classify designs are:

- The number of factors,
- The types of factors,
  - categorical versus numerical variable
  - number of levels
  - fixed versus random
- Type of randomization used, i.e., how the assignment of factor levels is made,
- Replication structure, i.e., how many times each type of test run is repeated,
- Any relationships between the factors.

To observe that more than one design is possible for a particular problem consider the following example.

Example: 1.2.1 Consider an environmental study in which the objective is to determine the level of copper in water that affects aquatic life. Daphnia, which are small aquatic animals, will be used for the study. The response variable will be the lifetime of the daphnia. The factor of interest will be the amount of copper in the water. The levels for copper will be: 0 (control group), 20 μg / liter, and 40 μg / liter. Note that a **control group** is a level of a factor that is essentially used as a base condition or a status quo condition. It is typically used for comparisons with other levels of the factor. Some uncontrolled factors include sex, physical condition of the animals, and several others. Thirty daphnia of the same age are available for the experiment. These animals will be the **experimental units**. Note that the subjects or objects on which the measurements are made are called the **experimental units**.

One possible design for this experiment is to have thirty separate chambers, one for each daphnia. We randomly assign ten daphnia to each level of copper. The resulting data structure is given in the Table 1.2.1. (Note that each * represents a single observation, i.e., a lifetime of a daphnia.)

Table 1.2.1 Experiment with Thirty Separate Chambers
Data — Lifetime

Copper Level										
0	*	*	*	*	*	*	*	*	*	*
20	*	*	*	*	*	*	*	*	*	*
40	*	*	*	*	*	*	*	*	*	*

Another possible design for this problem could be based upon having only six separate chambers. We randomly choose two chambers for each copper level, and the thirty daphnia are randomly assigned to the six chambers subject to having five per chamber. The resulting data structure would correspond to Table 1.2.2.

Table 1.2.2 Experiment with Six Separate Chambers

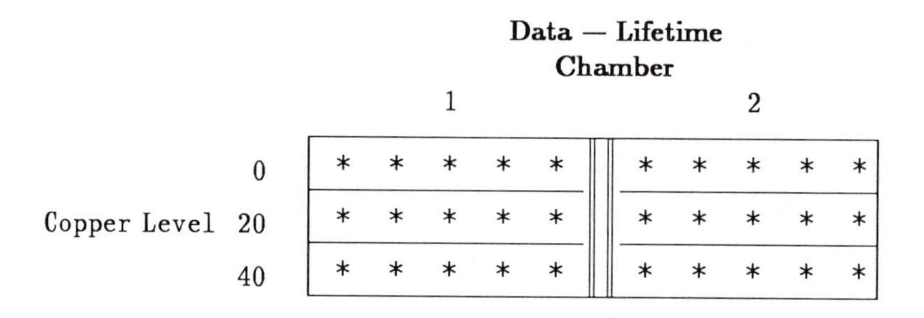

	Data — Lifetime Chamber 1					Chamber 2				
Copper Level 0	*	*	*	*	*	*	*	*	*	*
Copper Level 20	*	*	*	*	*	*	*	*	*	*
Copper Level 40	*	*	*	*	*	*	*	*	*	*

We can observe that the two designs are different. They have different structures for the data — even though the goal for the experiment remains unchanged. An experimenter will need to choose the design that best meets the needs for the experiment.

In some cases the design that is used depends upon the types of questions and analysis that are to be considered in the data analysis phase. Some typical questions or goals that could be of interest in a problem are:

- Determine whether there is a difference between the mean of the response variable for the levels of a factor.
- Determine which factor influences the response variable the most.
- Determine a functional relationship between the factors (independent variables) and the response variable.
- Determine how the factors relate to each other.
- Determine the settings for the factors that maximize or minimize the response variable.

1.3 Topics in Text

Our major focus in this text will be to discuss the issues involved in designing an experiment and analyzing the data that results from these experiments. The more commonly used designs will be discussed in great detail as will the techniques generally used in the analysis of the given data.

For each design we will present the properties of the design, including advantages and disadvantages; look at applications involving the design; and consider how to analyze data from the design. For the first design, the completely randomized design, we will introduce a number of ideas including:

- Models for designs,
- Forming tests and confidence intervals,
- Comparing levels of factors, and
- Checking assumptions.

In order to facilitate the analysis of data we will rely heavily upon models. The model featured in this text is known as the linear model. The first part of the course will emphasize the theory of linear models, which will be used in the later chapters of the book. Chapter 2 introduces the linear model, while the properties of linear models are presented in Chapters 3, 4 and 5. In Chapter 6 we focus on the types of linear models used in problems involving numerical independent variables called regression models. The remainder of the text (Chapter 7 onward) forms the core of the course in which the various designs are considered in detail. By the nature of the presentation the text is a bit more theoretical in the mathematical sense at the beginning and much more applied thereafter.

CHAPTER 2

LINEAR MODELS

2.1 Definition of a Linear Model

As noted in Chapter 1, the analysis of data obtained in designed experiments is facilitated by the use of a statistical model. Statistical models are used to describe an assumed structure for both the underlying population and the sample. In experimental designs the models we use are special cases of a very general model called the linear statistical model, or, simply, the linear model.

Definition 2.1.1 A **linear statistical model** is a model that can be represented in the form

$$Y = \beta_0 x_0 + \beta_1 x_1 + \beta_2 x_2 + \cdots + \beta_k x_k + \varepsilon \tag{2.1.1}$$

where Y is an observable random variable, $x_0, x_1, \ldots, x_k$ are known mathematical (nonrandom) variables, $\beta_0, \beta_1, \ldots, \beta_k$ are parameters and ε is an unobservable random variable.

This model is used to represent a random variable Y as a linear combination of $\beta_0, \beta_1, \ldots, \beta_k$ plus a random component. Since we are utilizing linear combinations, it is natural to call this model a linear model. The parameters can be thought of as unknown constants that need to be estimated using sample data. The random variable Y is called the response variable or dependent variable and $x_0, x_1, \ldots, x_k$ are referred to as the independent variables.

Linear models are widely used for applications in many disciplines. For example, in agriculture we can think of determining the yield of wheat, Y, based on various values of independent variables which could include the variety of wheat, the amount of fertilizer, and the type of soil. In social science we can think of the problem of predicting a student's academic ability as measured by grade point average using independent variables such as IQ, social adjustment, and prior grades. Since one can seldom predict the value of a response variable perfectly based only on the independent variables, and since the response variable does not always yield the same value for identical independent variable values, one accounts for this anomaly by including the random "error" term, ε , in the model.

If we have several observations $Y_1, Y_2, \ldots, Y_n$ on the random variable Y, we can express each observation as a function of the independent variables in a linear model structure. Of course, the random errors and the values of the independent variables may change from observation to observation. Accordingly, the linear model representation of these n observations consists of n equations of the form

$$Y_i = \beta_0 x_{i0} + \beta_1 x_{i1} + \cdots + \beta_k x_{ik} + \varepsilon_i, \quad i = 1,2,\ldots,n. \tag{2.1.2}$$

Matrix Representation

It is convenient to express the linear model for n observations in matrix form. Although this may seem cumbersome at first, it will greatly simplify our efforts as we proceed through the study of linear models. A review of matrix algebra is given in Appendix A for those who desire such a review.

Suppose we let $\mathbf{Y}$ and $\boldsymbol{\varepsilon}$ be $n \times 1$ vectors, $\boldsymbol{\beta}$ be a $(k + 1) \times 1$ vector and $\mathbf{X}$ be an $n \times (k + 1)$ matrix where each defined as follows:

$$\mathbf{Y} = \begin{bmatrix} Y_1 \\ Y_2 \\ \cdot \\ \cdot \\ \cdot \\ Y_n \end{bmatrix}, \quad \mathbf{X} = \begin{bmatrix} x_{10} & x_{11} & \cdots & x_{1k} \\ x_{20} & x_{21} & \cdots & x_{2k} \\ \cdot & \cdot & & \cdot \\ \cdot & \cdot & & \cdot \\ \cdot & \cdot & & \cdot \\ x_{n0} & x_{n1} & \cdots & x_{nk} \end{bmatrix}, \quad \boldsymbol{\beta} = \begin{bmatrix} \beta_0 \\ \beta_1 \\ \cdot \\ \cdot \\ \cdot \\ \beta_k \end{bmatrix}, \quad \boldsymbol{\varepsilon} = \begin{bmatrix} \varepsilon_1 \\ \varepsilon_2 \\ \cdot \\ \cdot \\ \cdot \\ \varepsilon_n \end{bmatrix} \tag{2.1.3}$$

Using properties of matrix algebra, we see that the linear model for n observations in equation (2.1.2) can be written as

$$\mathbf{Y} = \mathbf{X}\boldsymbol{\beta} + \boldsymbol{\varepsilon} . \tag{2.1.4}$$

It is this form of the linear model for n observations that we will utilize a great deal.

In most of the cases encountered in this text we will use the linear model in which we set $x_{i0} = 1$ for all i This leads to the following linear model for n observations:

$$Y_i = \beta_0 + \beta_1 x_{i1} + \cdots + \beta_k x_{ik} + \varepsilon_i , \quad i = 1,2,\ldots,n . \tag{2.1.5}$$

We call this the **intercept form** of the linear model. Observe that in the intercept case there are only k independent variables. There are many special forms of the linear model used in various sorts of applications. In the remaining sections of this chapter we will present some examples which will be useful in this text.

2.2 Simple Linear Regression

One of the most commonly used forms of the linear model is the simple linear regression model. In this model we have the dependent variable Y and a single independent variable x. The **simple linear regression model** for n observations is

$$Y_i = \beta_0 + \beta_1 x_i + \varepsilon_i, \quad i = 1,2,\ldots,n. \tag{2.2.1}$$

This model is often called the linear regression model since it essentially describes a linear relationship between the variables x and Y in a geometric sense. If we momentarily ignore the random error term, ε_i, the dependent variable will be a nonrandom variable. In this case the modified model becomes

$$y_i = \beta_0 + \beta_1 x_i ,$$

which is the equation of a line. (Notationally, we have used a lower case y here since y is a nonrandom variable.) This line has slope β_1 and y–intercept β_0; therefore, it is natural to call β_1 the **slope parameter** and β_0 the **y–intercept parameter.**

The type of situation in which the simple linear regression model is appropriate is one in which we expect Y to generally change in a linear fashion when x changes. To illustrate this, consider the following example.

Example 2.2.1 A social scientist is interested in determining how the educational level (measured by the number of years of education) of a person is related to the amount of television watched in the evening. Ten people are selected with varying educational levels, and the number of hours spent watching television after 6:00 p.m. is recorded for each person during a selected seven–day period. The educational level and the number of hours watching TV per day (seven–day average) is recorded for each person in Table 2.2.1.

Table 2.2.1 Television Study

Person	1	2	3	4	5	6	7	8	9	10
Education (years)	9	10	12	12	12	14	14	16	16	18
TV watching (hours)	3.1	2.8	2.4	2.3	2.5	1.9	2.2	1.7	1.6	1.4

In this example we can think of education level as the independent variable and the average number of hours watching TV as the dependent variable. A graph of these data is given in Figure 2.2.1

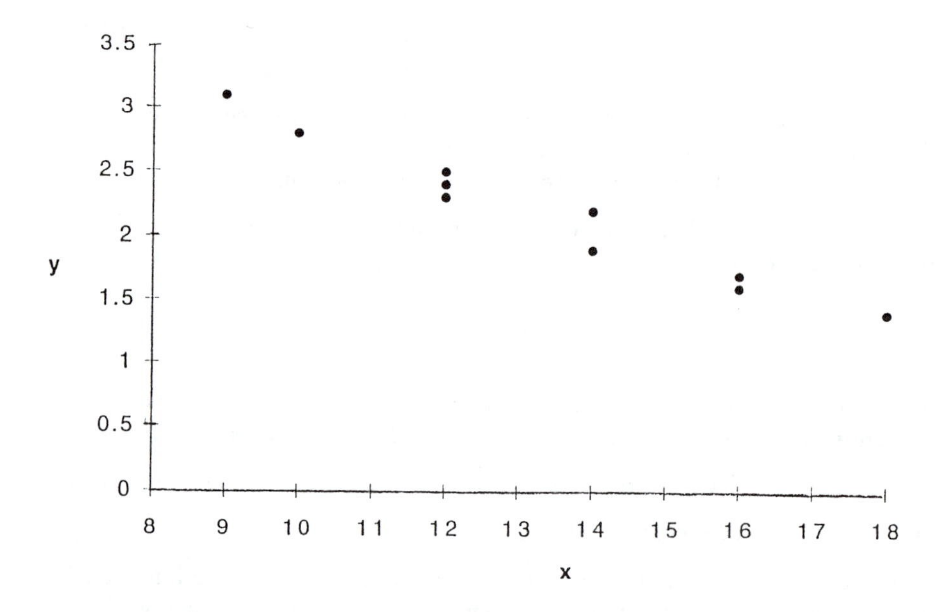

Figure 2.2.1 Scatter diagram of data points in television study

While the data points in Figure 2.2.1 do not all fall on a straight line, there is a line that can be found which "closely fits" the data values. Thus it seems reasonable to use the simple linear regression model for the television study problem.

In general, when data are adequately modeled by simple linear regression, a line can be found and used to describe the inherent linear relationship between x and Y. This allows one to predict values of Y for given values of x. For example, in the TV study we may wish to predict the number of hours watching TV for a person with 11 years of education. In the next section we develop a procedure for finding the equation of a line that will enable us to do this.

Matrix Representation

If we let

$$\mathbf{Y} = \begin{bmatrix} Y_1 \\ Y_2 \\ \vdots \\ Y_n \end{bmatrix}, \ \mathbf{X} = \begin{bmatrix} 1 & x_1 \\ 1 & x_2 \\ \vdots & \vdots \\ 1 & x_n \end{bmatrix}, \ \boldsymbol{\beta} = \begin{bmatrix} \beta_0 \\ \beta_1 \end{bmatrix}, \text{ and } \boldsymbol{\varepsilon} = \begin{bmatrix} \varepsilon_1 \\ \varepsilon_2 \\ \vdots \\ \varepsilon_n \end{bmatrix},$$

we see that model (2.2.1) can be expressed in the matrix form (2.1.4), that is, $\mathbf{Y} = \mathbf{X}\boldsymbol{\beta} + \boldsymbol{\varepsilon}$. Note that this representation is (2.1.3) with $k = 1$. Since in this case the columns of the $\mathbf{X}$ matrix are linearly independent, $\mathbf{X}$ has rank 2. To illustrate, for the TV data given in Table 2.2.1 of the previous example, $\mathbf{X}$ is the 10×2 matrix

$$\mathbf{X} = \begin{bmatrix} 1 & 9 \\ 1 & 10 \\ 1 & 12 \\ 1 & 12 \\ 1 & 12 \\ 1 & 14 \\ 1 & 14 \\ 1 & 16 \\ 1 & 16 \\ 1 & 18 \end{bmatrix}$$

Clearly this is a matrix of rank 2.

When the rank of $\mathbf{X}$ equals the number of columns in the matrix, then $\mathbf{Y} = \mathbf{X}\boldsymbol{\beta} + \boldsymbol{\varepsilon}$ is said to be a **full rank** linear model. Thus the simple linear regression model is a full rank model. As we shall see later, the concept of rank will be very important to us as we proceed through this text. (Readers unfamiliar with the concepts of rank and linear independence should refer to the review of matrix algebra provided in Appendix A.)

2.3 Least Squares Criterion

In the simple linear regression model

$$Y = \beta_0 + \beta_1 x + \varepsilon ,$$

the parameters β_0 and β_1 are fixed constants. However, typically we do not know the value of these constants. Furthermore, ε is a random variable and cannot be easily predicted. This presents us with a basic problem, namely, how can we estimate a value of Y for given values of the independent variable when β_0, β_1 and ε are all unknown? For example, in the TV problem of the previous section, how can we estimate the number of hours of TV watching (Y) if we know the educational level (x) of the person? The standard approach is to use sample data to estimate unknown parameters. Before describing how to do this, however, we need to develop some necessary notation.

Some Notation

We shall adopt the usual convention in statistics of using capital letters for random variables and lowercase letters for the actual observed values of these variables. The one exception to this, again by convention, is the symbol ε to represent a random variable. Thus $y_1, y_2, \ldots, y_n$ denotes n actual observed values (numbers) for the response variable Y. In matrix notation the actual observation vector is simply $\mathbf{y}$. We can think of using $\mathbf{y}$ to obtain values b_0 and b_1 (estimates for β_0 and β_1) so that

$$\hat{y}_i = b_0 + b_1 x_i$$

yields an estimate for y_i based on the simple linear regression model. We refer to $\hat{y}_i$ as the **predicted value** for the response variable when the independent variable has the value x_i. The observed value, y_i, may differ from the predicted value, $\hat{y}_i$. This difference is called the **residual**. That is, if e_i denotes the residual for the ith observation, then

$$e_i = y_i - \hat{y}_i .$$

Since there is a residual corresponding to each observation, we write

$$y_i = b_0 + b_1 x_i + e_i , \quad i = 1,2,\ldots,n , \tag{2.3.1}$$

which we call the **sample** or **observed** form of the simple linear regression model.

To illustrate our notation, let us again consider the TV data given in Example 2.2.1. In Figure 2.3.1, we have plotted the x (educational level) and y (hours watching TV) values, and in addition, we have graphed the line $\hat{y} = b_0 + b_1x$. Notice that for each data point (x_i,y_i), the vertical distance from the point to the line is the residual $e_i = y_i - \hat{y}_i$ where $\hat{y}_i = b_0 + b_1x_i$ yields ordinate values on the line.

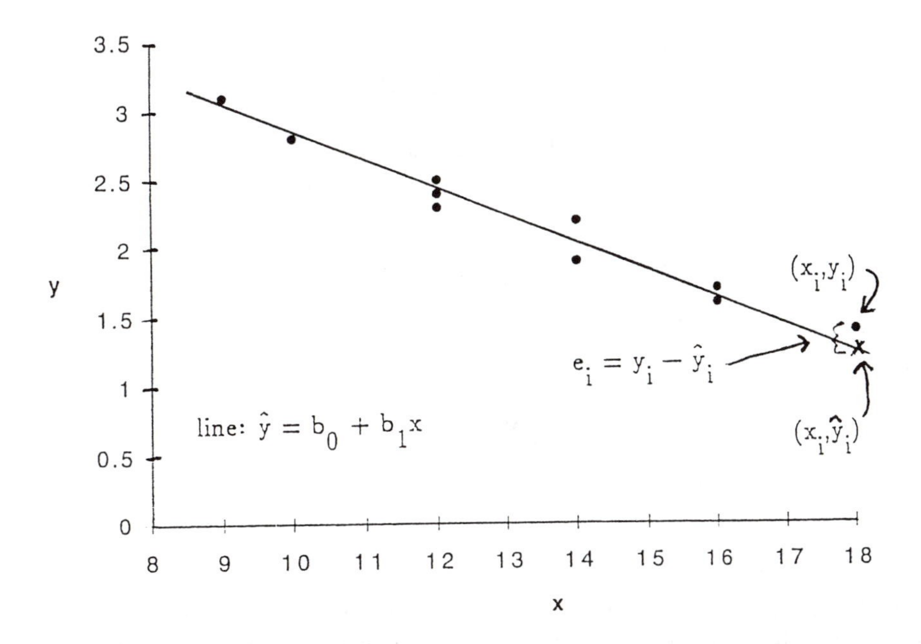

Figure 2.3.1 Residuals and Prediction Line

Least Squares Line

It is clear that we want to choose b_0 and b_1 so that the residuals $e_1,e_2,\cdots,e_n$ are as small as possible. This principle is the basis on which we find appropriate values for b_0 and b_1. We use the **method of least squares** to obtain the values for b_0 and b_1. In the least squares method we select the values b_0 and b_1 which minimize the sum of squares

$$\sum_{i=1}^{n} (y_i - \hat{y}_i)^2, \text{ where } \hat{y}_i = b_0 + b_1x_i .$$

Note that this sum of squares can also be written as $\sum_{i=1}^{n} e_i^2$; thus in the least squares method we minimize the sum of the squared residuals.

To find the proper values for b_0 and b_1, we need to minimize

$$\sum_{i=1}^{n} e_i^2 = \sum_{i=1}^{n} (y_i - \hat{y}_i)^2, = \sum_{i=1}^{n} (y_i - b_0 + b_1 x_i)^2$$

with respect to b_0 and b_1. Taking partial derivatives we get

$$\frac{\partial \sum_{i=1}^{n} e_i^2}{\partial b_0} = -2 \sum_{i=1}^{n} (y_i - b_0 - b_1 x_i)$$

and

$$\frac{\partial \sum_{i=1}^{n} e_i^2}{\partial b_1} = -2 \sum_{i=1}^{n} x_i (y_i - b_0 - b_1 x_i) .$$

After setting these partials equal to zero, distributing the summation signs, and algebraically rearranging terms, we arrive at the equations

$$n b_0 + b_1 \sum_{i=1}^{n} x_i = \sum_{i=1}^{n} y_i$$

and (2.3.2)

$$b_0 \sum_{i=1}^{n} x_i + b_1 \sum_{i=1}^{n} x_i^2 = \sum_{i=1}^{n} x_i y_i$$

These last two equations are called the **normal equations**. Solving these equations yields the unique solution

$$b_1 = \frac{n \sum_{i=1}^{n} x_i y_i - \sum_{i=1}^{n} x_i \sum_{i=1}^{n} y_i}{n \sum_{i=1}^{n} x_i^2 - (\sum_{i=1}^{n} x_1)^2}$$

and (2.3.3)

$$b_0 = \bar{y} - b_1 \bar{x},$$

where $\bar{y} = \sum_{i=1}^{n} y_i/n$ and $\bar{x} = \sum_{i=1}^{n} x_i/n$.

It can be shown that these solutions do, in fact, minimize the sum of the squared residuals.

We call the values of b_0 and b_1 given in (2.3.3) the **least squares estimates** for β_0 and β_1. The value $\hat{y}$, obtained from $\hat{y} = b_0 + b_1 x$, is called the least squares estimate for Y. The line $\hat{y} = b_0 + b_1 x$ is called the **least squares line**, or **regression line**, or **line of best fit**.

Example 2.3.1 Refer to the TV data of Example 2.2.1. As before we let x represent the educational level and y represent the number of hours watching TV. We proceed to find the least squares estimates and the corrssponding least squares line for these data. Performing some summary calculations we obtain

$$\sum_{i=1}^{10} {}_i = 21.9 , \qquad \sum_{i=1}^{10} x_i = 133 ,$$

$$\sum_{i=1}^{10} x_i^2 = 1841 , \quad \text{and} \quad \sum_{i=1}^{10} x_i y_i = 277.7 .$$

Using (2.3.3) formulas, we find the least square estimates

$$b_1 = \frac{10(277.7) - (133)(21.9)}{10(1841) - (133)^2} = -0.1882$$

$$b_0 = \frac{21.9}{10} - (-0.1882)\left(\frac{133}{10}\right) = 4.6932.$$

Hence the least squares line is

$$\hat{y} = 4.6932 - 0.1882x.$$

The actual data and the least squares line are plotted in Figure 2.3.2. From this plot we see that this least squares line "fits" the data quite well.

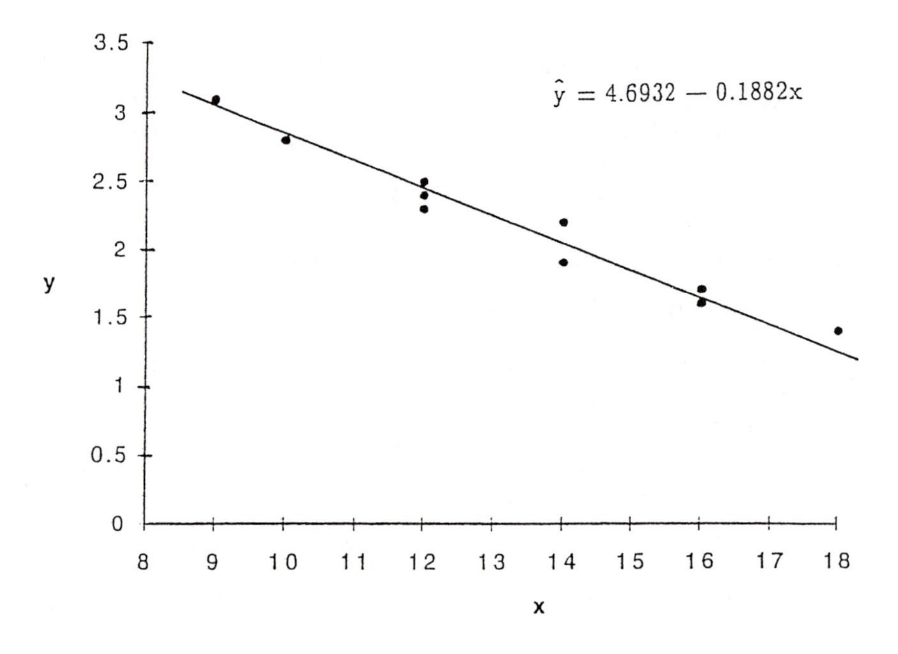

Figure 2.3.2 Television Least Squares Line

We can use the least squares line, alternately called the regression line, in several ways. To illustrate, in the television study example, the slope is negative. It therefore follows that hours watching TV tends to decrease as the educational level increases. Secondly, the regression line can be used to predict values of y for given values of x. For example, if the educational level of a person is $x = 11$ years, we predict that his or her evening TV watching averages $\hat{y} = 4.6932 - 0.1882(11) = 2.6$ hours.

Matrix Representation

We close this section with a note concerning the matrix representation in simple linear regression. The sample form (2.3.1) can be written as

$$\mathbf{y} = \mathbf{X}\mathbf{b} + \mathbf{e} \ ,$$

where

$$\mathbf{b} = \begin{bmatrix} b_0 \\ b_1 \end{bmatrix} .$$

In the future, for space reasons, we will sometimes detail a column vector by its **transpose**, namely, the corresponding row vector using the "prime" symbol. For example, in the simple linear regression case we would write $\mathbf{b}' = (b_0, b_1)$.

In the exercises, it is shown that the normal equations (2.3.2) can be written as

$$\mathbf{X}'\mathbf{X}\mathbf{b} = \mathbf{X}'\mathbf{y}. \qquad (2.3.4)$$

In keeping with the symbolism introduced in the previous paragraph, here $\mathbf{X}'$, a $2 \times n$ matrix, is the transpose of the $n \times 2$ matrix $\mathbf{X}$.) Furthermore, the solution to the normal equations is given by

$$\mathbf{b} = (\mathbf{X}'\mathbf{X})^{-1}\mathbf{X}'\mathbf{y}, \qquad (2.3.5)$$

where $(\mathbf{X}'\mathbf{X})^{-1}$ is the inverse of $\mathbf{X}'\mathbf{X}$. Equations (2.3.4) and (2.3.5) will play an important role later in this text.

2.4 Multiple Regression

In the simple linear regression model we assume a relationship between a dependent variable Y and a single independent variable x. In many situations it is unrealistic to try to predict Y based on only one variable. For example, if our dependent variable Y represents the crime rate in a city, it is clear that Y is related to a considerable number of other variables such as city population, percentage of poor in city, and population density, to name a few. Thus, in general, it seems reasonable to attempt to relate Y to $k \geq 2$ independent variables $x_1, x_2, \cdots, x_k$.

If we extend the simple linear regression model to k variables, the most natural extension is the model

$$Y = \beta_0 + \beta_1 x_1 + \cdots + \beta_k x_k + \varepsilon\ . \tag{2.4.1}$$

If we have n observations $Y_1, Y_2, \cdots, Y_n$, then we write

$$Y_i = \beta_0 + \beta_1 x_{i1} + \cdots + \beta_k x_{ik} + \varepsilon_i, \quad i = 1,2,\cdots, n. \tag{2.4.2}$$

A **multiple regression model** is an example of a model having this general functional form. Hence, the multiple regression problem is one in which we relate one dependent variable to $k \geq 2$ independent variables.

It is easy to see that the multiple regression model for n observations (2.4.2) is functionally the same as the intercept—form of the linear model given in equations (2.1.5). As a consequence, we can write the multiple regression model in matrix form

$$\mathbf{Y} = \mathbf{X}\boldsymbol{\beta} + \boldsymbol{\varepsilon}\ ,$$

where $\mathbf{Y}$, $\boldsymbol{\beta}$, and $\boldsymbol{\varepsilon}$ are as given in (2.1.3). Here the $\mathbf{X}$ matrix is identical to the given $\mathbf{X}$ except the first column is a column of ones.

In multiple regression problems the columns of the $\mathbf{X}$ matrix are assumed to be linearly independent. Thus, the rank of the matrix $\mathbf{X}$ equals the number of columns in $\mathbf{X}$, that is, the rank of $\mathbf{X}$ is $k + 1$. It follows that the multiple regression model is a full rank linear model. This is the characterization of the multiple regression model that we will use most; that is, we will refer to a full rank linear model as a regression model. In Section 6 of this chapter we encounter our first less than full rank linear model. Next we illustrate the multiple regression model via a numerical example.

A Multiple Regression Example

Example 2.4.1 In an energy usage survey, homes that are heated with electricity are surveyed in several areas for a one—week period. The following quantities were measured during January for ten homes:

y = cost of electric use,

x_1 = average outside temperature,

x_2 = usual thermostat setting for the home,

x_3 = square footage of the home.

The results are given in Table 2.4.1.

Table 2.4.1 Electric Costs for Ten Homes

y	35.09	54.59	43.82	38.61	30.62	30.25	47.15	66.59	31.41	50.24
x_1	29.0	20.0	31.5	22.0	36.5	37.5	18.0	13.5	29.5	17.5
x_2	65	67	68	67	66	67	65	68	68	66
x_3	2125	2350	2325	1825	2550	2100	2400	1325	2050	1575

In this study the objective is to determine how x_1, x_2, and x_3 affect y. Therefore, it is natural to think of y as the dependent variable and x_1, x_2, x_3 as the independent variables. This is a typical multiple regression problem with $k = 3$. Thus the model we use is

$$Y_i = \beta_0 + \beta_1 x_{i1} + \beta_2 x_{i2} + \beta_3 x_{i3} + \varepsilon_i, \quad i = 1,2,\cdots,10,$$

which we can write in matrix notation $\mathbf{Y} = \mathbf{X}\boldsymbol{\beta} + \boldsymbol{\varepsilon}$. For the sample form $\mathbf{y} = \mathbf{X}\mathbf{b} + \mathbf{e}$,

$$\mathbf{y} = \begin{bmatrix} 35.09 \\ 54.59 \\ 43.82 \\ 38.61 \\ 30.62 \\ 30.25 \\ 47.15 \\ 66.59 \\ 31.41 \\ 50.24 \end{bmatrix} \quad \mathbf{X} = \begin{bmatrix} 1 & 29.0 & 65 & 2125 \\ 1 & 20.0 & 67 & 2350 \\ 1 & 31.5 & 68 & 2325 \\ 1 & 22.0 & 67 & 1825 \\ 1 & 36.5 & 66 & 2550 \\ 1 & 37.5 & 67 & 2100 \\ 1 & 18.0 & 65 & 2400 \\ 1 & 13.5 & 68 & 1325 \\ 1 & 29.5 & 68 & 2050 \\ 1 & 17.5 & 66 & 1575 \end{bmatrix}$$

and **b** and **e** are vectors of 4 and 10 constants, respectively, yet to be determined. Note that the **X** matrix here has full rank 4 which categorizes this as a multiple regression problem. Later in this text we will demonstrate how to find the least squares estimates for β_0, β_1, β_2, and β_3. We can then obtain the prediction model

$$\hat{y} = b_0 + b_1 x_1 + b_2 x_2 + b_3 x_3 ,$$

which can be used to predict electric usage costs for given values of x_1, x_2, and x_3.

We close this section by noting that the simple linear regression model is the multiple regression model with $k = 1$. This fact will be utilized in the following manner. Whenever we obtain results for the multiple regression model, we have also obtained results for the simple case.

2.5 Polynomial Regression

When using the simple linear regression model, we assumed that a dependent variable Y and a single independent variable x had a linear relationship. While this is a reasonable assumption in many problems, there are situations where the relationship between Y and x is not linear in nature. The following example is an illustration of such a situation.

Example 2.5.1. A large department store chain has twelve stores of approximately equal size in a district. In past years some stores stocked more Christmas trees than they could sell, and as a consequence, these stores had a lower profit on tree sales than some other stores. A study was undertaken to determine the optimal number of trees that a store should stock. For the study, each store stocked a different number of trees and the profit obtained from Christmas tree sales was recorded. The results are given in Table 2.5.1 and these data are plotted in Figure 2.5.1.

Table 2.5.1 Christmas Tree Profit Study

Stores	1	2	3	4	5	6	7	8	9	10	11	12
Trees (in hundreds)	1.2	1.4	1.6	1.8	2.0	2.2	2.4	2.6	2.8	3.0	3.2	3.4
Profit (in $100s)	4.50	12.59	13.32	25.17	26.77	22.21	27.99	29.61	27.02	17.91	18.45	7.23

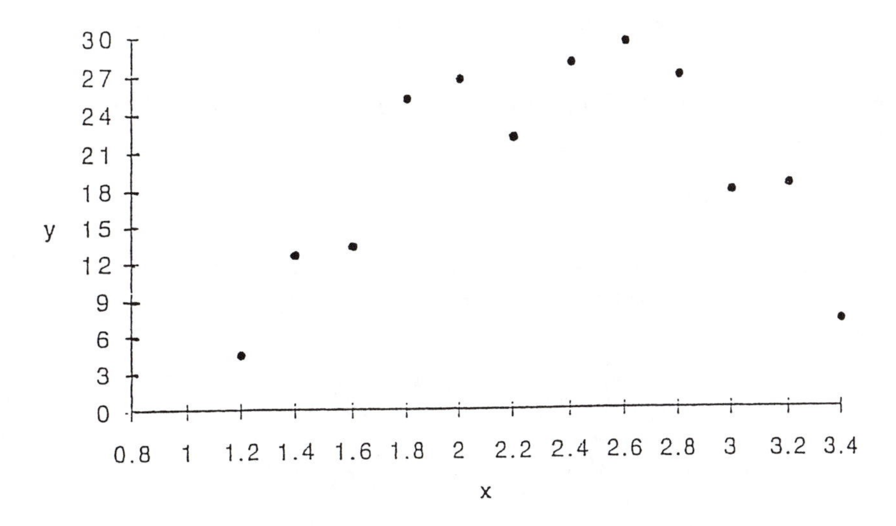

Figure 2.5.1 Number of Christmas Trees Stocked

It is clear that we cannot expect a straight line to fit the data in Figure 2.5.1 very well. Instead we need to assume some curvilinear relationship.

A common, easy—to—use, curvilinear relationship between Y and x is a polynomial. Recall that a polynomial of degree k has the form

$$y = a_0 + a_1 x + a_2 x^2 + \cdots + a_k x^k .$$

Incorporating this form in a model for n observations $Y_1, Y_2, \cdots, Y_n$, we write

$$Y_i = \beta_0 + \beta_1 x_i + \beta_2 x_i^2 + \cdots + \beta_k x_i^k + \varepsilon_i, \quad i = 1,2,\cdots,n. \tag{2.5.1}$$

This model is called the **polynomial regression model of degree k** and is the kind of model we can use to describe many curvilinear relationships between Y and x.

At first glance this polynomial regression model seems like an entirely different model. However, upon closer inspection we can see that it is really a multiple regression model and, hence, also a linear model. If we make the identification $x_j = x^j$ for $j = 1,2,\cdots,k$, then the model (2.5.1) can be written as

$$Y_i = \beta_0 + \beta_1 x_{i1} + \beta_2 x_{i2} + \cdots + \beta_k x_{ik} + \varepsilon_i, \quad i = 1,2,\cdots,n,$$

which we recognize as the multiple regression model discussed in the previous section.

Matrix Representation

Since the polynomial regression model is a linear model, we can express it in the matrix form $\mathbf{Y} = \mathbf{X}\boldsymbol{\beta} + \boldsymbol{\varepsilon}$. In this case the $\mathbf{X}$ matrix, written in general, is

$$\mathbf{X} = \begin{bmatrix} 1 & x_1 & x_1^2 & \cdots & x_1^k \\ 1 & x_2 & x_2^2 & \cdots & x_2^k \\ \cdot & \cdot & \cdot & & \cdot \\ \cdot & \cdot & \cdot & & \cdot \\ \cdot & \cdot & \cdot & & \cdot \\ 1 & x_n & x_n^2 & \cdots & x_n^k \end{bmatrix}.$$

Again we see that the columns of $\mathbf{X}$ will be linearly independent so that the polynomial regression model is another full rank linear model.

Let us return to the Christmas tree example of this section. Based on Figure 2.5.1, it seems reasonable to use a second degree polynomial regression model to fit these data. The sample form of this model is

$$y_i = b_0 + b_1 x_i + b_2 x_i^2 + e_i, \quad i = 1,2,\cdots,n,$$

where y_i is the profit and x_i is the number of trees for store i. In matrix form we have

$$\mathbf{y} = \mathbf{X}\mathbf{b} + \mathbf{e},$$

where, for the data in Table 2.5.1, the $\mathbf{y}$ vector and $\mathbf{X}$ matrix are

$$\mathbf{y} = \begin{bmatrix} 4.50 \\ 12.59 \\ 13.32 \\ 25.17 \\ 26.77 \\ 22.21 \\ 27.99 \\ 29.61 \\ 27.02 \\ 17.91 \\ 18.45 \\ 7.23 \end{bmatrix} \qquad \mathbf{X} = \begin{bmatrix} 1 & 1.2 & 1.44 \\ 1 & 1.4 & 1.96 \\ 1 & 1.6 & 2.56 \\ 1 & 1.8 & 3.24 \\ 1 & 2.0 & 4.00 \\ 1 & 2.2 & 4.84 \\ 1 & 2.4 & 5.76 \\ 1 & 2.6 & 6.76 \\ 1 & 2.8 & 7.84 \\ 1 & 3.0 & 9.00 \\ 1 & 3.2 & 10.24 \\ 1 & 3.4 & 11.56 \end{bmatrix}$$

In the first section of the next chapter the three element vector **b** will be determined using the method of least squares.

Because the polynomial regression model is a special case of the multiple regression model, we recognize that whenever mathematical results are obtained for the multiple regression model, we also are obtaining results for the polynomial regression model. It should be noted, however, that interpretation and use of polynomial regression is sometimes different than that for "usual" multiple regression. For example, in polynomial regression we often focus on the leading coefficient, β_k, while in multiple regression one is usually interested in most, if not all, coefficients.

2.6 One—Way Classification

In experimental design problems we commonly are interested in comparing several means. For our last example of a linear model in this chapter we consider a model which is useful for such comparisons. The model is one associated with the one—way classification problem.

In the one—way classification setting we consider p populations (sometimes called **treatment** groups) which have means $\mu_1, \mu_2, \cdots, \mu_p$, respectively. Independent random samples are taken from each population yielding observations $Y_{11}, Y_{12}, \cdots, Y_{1n_1}$, $Y_{21}, Y_{22}, \cdots, Y_{2n_2}, \cdots$, $Y_{p1}, Y_{p2}, \cdots, Y_{pn_p}$. Here Y_{ij} represents the jth observation from the ith sample and n_i represents the sample size for the ith sample. Our usual goal in a one—way classification is to test the null hypothesis

$$H_0\colon \mu_1 = \mu_2 = \cdots = \mu_p. \tag{2.6.1}$$

The following example in an illustration of the one—way classification.

Example 2.6.1. A secchi disk is a metal object that is used to measure water clarity. The secchi disk is lowered in the water until it can no longer be seen and the attained depth (in centimeters) is recorded. In an artificial lake the water clarity is compared at three locations of the lake. Location 1 is near the intake of the lake, location 2 is near the deepest part of the lake, and location 3 is near the dam. At each location five secchi disk readings are taken. The results are given in Table 2.6.1.

Table 2.6.1 Secchi Disk Reading

Location		
1	2	3
39	46	48
37	45	46
40	46	48
39	45	46
39	47	49

It is clear that this is a one—way classification problem. The locations are the populations (or treatments) so $p = 3$. Our interest in this problem is the comparison of the mean secchi disk readings for the three locations.

The Model

To model the one—way classification we take note of the following facts. Since Y_{ij} has mean μ_i we can express Y_{ij} as

$$Y_{ij} = \mu_i + \varepsilon_{ij}, \quad i = 1,2,\cdots,p\ ;\ \ j = 1,2,\cdots,n_i, \tag{2.6.2}$$

where ε_{ij} is a random variable with mean zero. The ε_{ij} term in our model accounts for the fact that observations within a population are not all the same. By tradition, dating back to R. A. Fisher's work in the 1920s, the population means are assumed to have the structure

$$\mu_i = \mu + \tau_i \tag{2.6.3}$$

where τ_i is referred to as the **effect** of the ith "treatment." This is in deference to the historical experimental setting where conceptually observations in one population differ fundamentally from the observations in another population due to the fact that the units in the respective populations were "treated" differently. In this sense, the parameter μ represents an "overall" mean from which the population means, the μ_i, deviate. The magnitudes of these deviations are represented by the parameters, τ_i .

By using (2.6.3) in equation (2.6.2) we rewrite the one—way classification model as

$$Y_{ij} = \mu + \tau_i + \varepsilon_{ij}, \quad i = 1,2,\cdots,p; \; j = 1,2,\cdots,n_i . \tag{2.6.4}$$

Henceforth we refer to (2.6.4) as the **traditional** one—way classification model, and it is the form that we will usually use. Next we illustrate that model (2.6.4) is a linear model.

Example 2.6.2. Consider the following observations taken from three populations:

Population 1	Population 2	Population 3
Y_{11}	Y_{21}	Y_{31}
Y_{12}	Y_{22}	Y_{32}
Y_{13}	Y_{23}	
Y_{14}		

Let us model these nine observations using the one—way classification model (2.6.4).

$$Y_{ij} = \mu + \tau_i + \varepsilon_{ij}, \quad i = 1,2,3; \; j = 1,\cdots,n_i. \tag{2.6.5}$$

Here Y_{ij} represents the jth observation taken from the ith population and $n_1 = 4$, $n_2 = 3$, and $n_3 = 2$. In matrix notation, (2.6.5) is written as

$$\mathbf{Y} = \mathbf{X}\boldsymbol{\beta} + \boldsymbol{\varepsilon}$$

where

$$\mathbf{Y} = \begin{bmatrix} Y_{11} \\ Y_{12} \\ Y_{13} \\ Y_{14} \\ Y_{21} \\ Y_{22} \\ Y_{23} \\ Y_{31} \\ Y_{32} \end{bmatrix}, \quad \mathbf{X} = \begin{bmatrix} 1 & 1 & 0 & 0 \\ 1 & 1 & 0 & 0 \\ 1 & 1 & 0 & 0 \\ 1 & 1 & 0 & 0 \\ 1 & 0 & 1 & 0 \\ 1 & 0 & 1 & 0 \\ 1 & 0 & 1 & 0 \\ 1 & 0 & 0 & 1 \\ 1 & 0 & 0 & 1 \end{bmatrix}, \quad \boldsymbol{\beta} = \begin{bmatrix} \mu \\ \tau_1 \\ \tau_2 \\ \tau_3 \end{bmatrix}, \quad \boldsymbol{\varepsilon} = \begin{bmatrix} \varepsilon_{11} \\ \varepsilon_{12} \\ \varepsilon_{13} \\ \varepsilon_{14} \\ \varepsilon_{21} \\ \varepsilon_{22} \\ \varepsilon_{23} \\ \varepsilon_{31} \\ \varepsilon_{32} \end{bmatrix}.$$

This illustrates that for this example the one—way model (2.6.5) is a linear model.

It is easy to generalize that the one—way model (2.6.4) can be written as $\mathbf{Y} = \mathbf{X}\boldsymbol{\beta} + \boldsymbol{\varepsilon}$ for any $p \geq 2$ which verifies that (2.6.4) is, in fact, always a linear model. There are several interesting things to note about this one—way model. First, from the previous example we see that the $\mathbf{X}$ matrix is composed of just zeros and ones. This is the typical situation for an experimental design type model. Second, the first column, being a column of ones, corresponds to the μ term in the model, the second column in $\mathbf{X}$ corresponds to τ_1, and so forth. Third, the $\mathbf{X}$ matrix in Example 2.6.2 has rank 3 since the sum of the columns two through four equals column 1 using vector addition. Since the rank of $\mathbf{X}$ is less than the number of columns, we call this **a less than full rank model**. Historically, the **traditional** representation for experimental design models are of the less than full rank type.

We will use the one—way model as an illustration in succeeding chapters and, as we do, we will be learning useful facts that will be utilized extensively throughout this text. The importance of this model will be fully appreciated in Chapter 7 where we find that the one—way model is used to analyze data obtained in an experiment utilizing a **completely randomized design**.

Our objective in this chapter has been to introduce the linear model and to illustrate its versatility in explaining the underlying structure of a data set. At the same time, we have briefly described several important versions of the linear model that we will be using in our study of experimental design methodology. In the next few chapters we will study various properties of the linear model and illustrate these properties using regression models and the one—way model. Our ultimate goal is to use our knowledge of linear models to analyze experimental design data. It is important to keep this goal in mind as we proceed through the chapters to follow.

Chapter 2 Exercises

Application Exercises

2–1 Which of the following models are linear statistical models? Where ε appears in the model, assume it to be a random variable and assume any x's in the model to be mathematical (nonrandom) variables. Consider α and β's to be parameters.

(a) $Y = \alpha + \beta x$

(b) $Y = \alpha + \beta x + \varepsilon$

(c) $Y = \alpha + \beta x\varepsilon$

(d) $Y = \alpha + \frac{\beta}{x} + \varepsilon$

(e) $Y = \alpha e^{\beta x}\varepsilon$, where e is the base of natural logarithms.

(f) $Y = \beta_0 + \beta_1 x + \beta_2 x^2 + \varepsilon$

(g) $Y = \beta_0 + \beta_1 x^{\beta_2} + \varepsilon$

(h) $Y = \alpha\beta^x + \varepsilon$

(i) $Y = \alpha\beta^x\varepsilon$

(j) $Y = \beta_0 + \beta_1 x_1 + \beta_2 x_2 + \beta_3 x_1 x_2 + \varepsilon$

(k) $Y = \beta_0 + \beta_1 x_1 + \beta_2 x_2 + \beta_3 x_1^2 + \beta_4 x_2^2 + \beta_5 x_1 x_2 + \varepsilon$

(l) $Y = \alpha x^{\beta}\varepsilon$

(m) $1/Y = \alpha + \beta x + \varepsilon$

(n) $Y = \beta x + \varepsilon$

2–2 Consider the following data:

Y	0	1	2	2	4	9
x	1	1	1	2	2	5

(a) Plot these data.

(b) Find the regression line $\hat{y} = b_0 + b_1 x$ for these data.

(c) Find the residual for each observation.

(d) Calculate the sum of the squared residuals

$$\sum_{i=1}^{6} (y_i - \hat{y}_i)^2.$$

(e) Consider an alternative linear equation given by $\hat{\hat{y}} = -1 + 1.5x$.

Find the quantity $y_i - \hat{\hat{y}}_i$ for each observation.

(f) Find $\sum_{i=1}^{6} (y_i - \hat{\hat{y}}_i)^2$ and compare your answer with that obtained in part d.

2–3 A study was undertaken to determine the water temperature in Acton Lake at Hueston Woods State Park in Ohio. As part of this study the water temperature was measured on a warm summer day at the deepest part of the lake. Temperature readings were recorded at several depths. The results are as follows:

Depth (meters)	0	1	2	3	4	5	6	7	8	9	10
Temperature (oC)	29	29	28	27	25	25	24	23	21	20	20

(a) Plot these data.

(b) Find the regression line using temperature as the dependent variable.

(c) Estimate the temperature at a depth of 7.5 meters.

2–4 Trees in a forest are surveyed to determine lumber potential. For a sample of maple trees growing in the forest, the tree height and age are recorded. The results are given as follows:

Age (years)	20	18	32	17	25	36	40	23	33	28	27	31
Height (feet)	15.7	12.2	18.9	12.7	12.6	19.7	22.8	10.9	21.6	15.2	17.2	18.4

(a) Plot these data.
(b) Find the regression line using height as the dependent variable.
(c) Estimate the height when the age is twenty–five years.

2–5 An experiment is conducted to study the yield of a chemical process at two levels of temperature and two levels of pressure. Consider the model

$$Y = \beta_0 + \beta_1 x_1 + \beta_2 x_2 + \beta_3 x_1 x_2 + \varepsilon ,$$

where y is the percent of yield, x_1 is the temperature, and x_2 is the pressure. The levels of x_1 and x_2 are arbitrarily coded (–1, +1) with the following results.

y	30	50	40	60	20	30	40	50
x_1	–1	–1	1	1	–1	–1	1	1
x_2	–1	1	–1	1	–1	1	–1	1

(a) Show that the model is a multiple regression model.
(b) Identify the quantities **y**, **X** and **b** in the sample form **y** = **Xb** + **e**.
(c) What is the rank of **X**?

2–6 Consider the following data:

y	6.2	3.4	3.3	7.1	8.4	9.0	7.5
x	1	2	3	4	5	6	7

(a) Give the equation for a third degree polynomial regression model.
(b) Identify the quantities **y**, **X** and **b** for the sample form **y** = **Xb** + **e** for this model.
(c) What is the rank of **X** ?

2–7 Consider the secchi disk data in Table 2.6.1.
(a) What is the appropriate model for these data?

(b) Write the model in part a in the form $\mathbf{Y} = \mathbf{X}\beta + \varepsilon$.
Be sure to identify the components in each vector or matrix.

(c) What is the rank of $\mathbf{X}$?

Theoretical Exercises

2–8 A model is called an **intrinsically linear model** if a transformation of both sides of the equation puts the equation in the linear statistical model form. For example, the model $Y = \alpha e^{\beta x}\varepsilon$ can be placed in linear model form by taking natural logs of both sides of the equation and using $\log Y$ as the response variable. Show that model (i) and (ℓ) in Exercise 2–1 are intrinsically linear models.

2–9 Verify that equation (2.3.3) does provide a solution to the system of equations in (2.3.2).

2–10 Using second derivatives, verify that the solution given to the normal equations in (2.3.3) does **minimize** the sum of the squared residuals.

2–11 Show that the normal equations (2.3.2) for the simple linear regression model can be written as $\mathbf{X'Xb} = \mathbf{X'y}$ (equation 2.3.4).

2–12 Show that for the simple linear regression model, $\mathbf{b} = (\mathbf{X'X})^{-1}\mathbf{X'y}$ (equation 2.3.5) yields the solution to the normal equations given by (2.3.3).

2–13 Suppose that we define the form of the simple linear regression model as

$$Y_i = \alpha + \beta(x_i - \bar{x}) + \varepsilon_i\ , \quad i = 1, \cdots, n.$$

(a) Using derivatives, find the least squares estimates for α and β.

(b) Use these estimates to find the regression line for the data of Exercise 2–2. Compare the resulting line with the one obtained in Exercise 2–2.

(c) Show that the regression line obtained with the model in this problem always equals the regression line obtained by using the traditional simple linear regression model $Y = \beta_0 + \beta_1 x + \varepsilon$.

2–14 The no–intercept model for the simple linear regression problem is

$$Y_i = \beta x_i + \varepsilon_i, \quad i = 1,\cdots,n.$$

(a) By using a derivative, find the least squares estimate for β.

(b) Show that the estimate in part a can be found by calculating

$$b = (X'X)^{-1}X'y.$$

2–15 Consider a one–way classification with $p = 3$ treatments and sample sizes given by $n_1 = 4$, $n_2 = 3$ and $n_3 = 2$. The model

$$Y_{ij} = \mu_i + \varepsilon_{ij}, \quad i = 1,2,3; \quad j = 1,\cdots,n_i$$

is often called the means model.

(a) Form **Y**, **X**, $\boldsymbol{\beta}$ and $\boldsymbol{\varepsilon}$ for this problem to show that the model can be expressed in linear model form.

(b) What is the rank of **X**? Is this a full rank model?

CHAPTER 3

LEAST SQUARES ESTIMATION AND NORMAL EQUATIONS

3.1 Least Squares Estimation

In the previous chapter we discovered how to estimate the unknown parameters in the simple linear regression model using the least squares method. In this section we show how the least squares method can be used for the linear model, in general.

For the linear model $\mathbf{Y} = \mathbf{X}\boldsymbol{\beta} + \boldsymbol{\varepsilon}$, the unknown parameters are the components of the vector $\boldsymbol{\beta}' = (\beta_0, \beta_1, \cdots, \beta_k)$. Our objective is to obtain a vector $\mathbf{b}' = (b_0, b_1, \cdots, b_k)$ which is the estimate for $\boldsymbol{\beta}'$. As before, we will let $y_1, y_2, \cdots, y_n$ be the realized observations and $\hat{y}_1, \hat{y}_2, \cdots, \hat{y}_n$ will denote the predicted, where $\hat{y}_i = b_0 + b_1 x_{i1} + \cdots + b_k x_{ik}$ for $i = 1,2,\cdots,n$. In matrix representation, $\mathbf{y}' = (y_1, y_2, \cdots, y_n)$, $\hat{\mathbf{y}} = (\hat{y}_1, \hat{y}_2, \cdots, \hat{y}_n)$ and $\hat{\mathbf{y}} = \mathbf{Xb}$. The residuals $e_1, e_2, \cdots, e_n$ defined by $e_i = y_i - \hat{y}_i$ are denoted in matrix form by the vector $\mathbf{e}' = (e_1, e_2, \cdots, e_n)$. We can now express the vector $\mathbf{y}$ as

$$\mathbf{y} = \mathbf{Xb} + \mathbf{e}$$

which we call the **sample form** of the linear model.

We now proceed to find the vector $\mathbf{b}$ using the least squares method. Recall that in the least squares method the vector $\mathbf{b}$ is chosen so that $\sum_{i=1}^{n} e_i^2$ is minimal. Note that

$$\sum_{i=1}^{n} e_i^2 = \mathbf{e}'\mathbf{e} = (\mathbf{y} - \hat{\mathbf{y}})'(\mathbf{y} - \hat{\mathbf{y}}) = (\mathbf{y} - \mathbf{Xb})'(\mathbf{y} - \mathbf{Xb}).$$

Thus we need to find the vector $\mathbf{b}$ so that $(\mathbf{y} - \mathbf{Xb})'(\mathbf{y} - \mathbf{Xb})$ is minimized with respect to its elements. When we expand and collect like terms, the above expression becomes

$$(\mathbf{y} - \mathbf{Xb})'(\mathbf{y} - \mathbf{Xb}) = \mathbf{y}'\mathbf{y} - 2\mathbf{b}'\mathbf{X}'\mathbf{y} + \mathbf{b}'\mathbf{X}'\mathbf{Xb}. \tag{3.1.1}$$

To minimize (3.1.1) we use the rules of vector differentiation found in Appendix A which yields

$$\frac{\partial}{\partial \mathbf{b}}(\mathbf{y}'\mathbf{y} - 2\mathbf{b}'\mathbf{X}'\mathbf{y} + \mathbf{b}'\mathbf{X}'\mathbf{X}\mathbf{b}) = -2\mathbf{X}'\mathbf{y} + 2\mathbf{X}'\mathbf{X}\mathbf{b}.$$

It easily follows that the derivative is zero whenever

$$\mathbf{X}'\mathbf{X}\mathbf{b} = \mathbf{X}'\mathbf{y}\ . \tag{3.1.2}$$

The expression (3.1.2) represents a set of $k + 1$ equations in the $k + 1$ unknowns $b_0, b_1 \cdots, b_k$. These equations are referred to as the **normal equations.** There may be one or even infinitely many solution sets to these normal equations. It can be shown, however, that **any** solution $\mathbf{b}$ to these normal equations does indeed minimize $(\mathbf{y} - \hat{\mathbf{y}})'(\mathbf{y} - \hat{\mathbf{y}})$. Thus to find $\mathbf{b}$ using the least squares method we need only find a solution to equation (3.1.2). We summarize this in the following theorem which we state without proof. Those interested will find a proof on page 73 in Seber (1977).

Theorem 3.1.1 Let $\mathbf{y} = \mathbf{X}\mathbf{b} + \mathbf{e}$ be the sample form of a linear model. Any solution $\mathbf{b}$ of the normal equations $\mathbf{X}'\mathbf{X}\mathbf{b} = \mathbf{X}'\mathbf{y}$ will minimize $(\mathbf{y} - \hat{\mathbf{y}})'(\mathbf{y} - \hat{\mathbf{y}})$ where $\hat{\mathbf{y}} = \mathbf{X}\mathbf{b}$.

Next we consider the problem of obtaining a solution to the normal equations in the full rank and less than full rank cases.

Full Rank Case

If $\mathbf{X}$ is an $n \times (k + 1)$ matrix with $n > k + 1$, we say the linear model $\mathbf{Y} = \mathbf{X}\boldsymbol{\beta} + \boldsymbol{\varepsilon}$ is a full rank model whenever the rank of $\mathbf{X}$ equals $k + 1$. If $\mathbf{X}$ has rank $k + 1$, then the $(k+1) \times (k+1)$ matrix $\mathbf{X}'\mathbf{X}$ also has rank $k + 1$. Therefore $\mathbf{X}'\mathbf{X}$ is a nonsingular matrix having a unique inverse $(\mathbf{X}'\mathbf{X})^{-1}$. It follows that in the full rank case the normal equations $\mathbf{X}'\mathbf{X}\mathbf{b} = \mathbf{X}'\mathbf{y}$ have a unique solution given by

$$\mathbf{b} = (\mathbf{X}'\mathbf{X})^{-1}\mathbf{X}'\mathbf{y}. \tag{3.1.3}$$

The vector $\mathbf{b}$ is called the unique **least squares estimate** for the parameter vector $\boldsymbol{\beta}$ in the full rank model $\mathbf{Y} = \mathbf{X}\boldsymbol{\beta} + \boldsymbol{\varepsilon}$.

As noted in Chapter 2, a multiple regression model is an example of a linear model with a full rank matrix $\mathbf{X}$. Recall that in a multiple regression problem we have a dependent variable Y related to k independent variables $x_1, x_2, \cdots, x_k$ as follows:

$$Y_i = \beta_0 + \beta_1 x_{i1} + \cdots + \beta_k x_{ik} + \varepsilon_i, \quad i = 1,2,\cdots,n.$$

This is a linear model where

$$\beta = \begin{bmatrix} \beta_0 \\ \beta_1 \\ \cdot \\ \cdot \\ \cdot \\ \beta_k \end{bmatrix} \quad \text{and} \quad \mathbf{X} = \begin{bmatrix} 1 & x_{11} & \cdots & x_{1k} \\ 1 & x_{21} & \cdots & x_{2k} \\ \cdot & \cdot & & \cdot \\ \cdot & \cdot & & \cdot \\ \cdot & \cdot & & \cdot \\ 1 & x_{n1} & \cdots & x_{nk} \end{bmatrix}.$$

If the columns of $\mathbf{X}$ are linearly independent, which is the case in most multiple regression settings then $\mathbf{X}$ has rank $k + 1$ and $\mathbf{X'X}$ is full rank. Thus in a multiple regression problem there is a unique least squares estimate $\mathbf{b}$ and it is given by formula (3.1.3). Note that the matrix $\mathbf{X'X}$ and the vector $\mathbf{X'y}$, which are used to solve the normal equations, are relatively easy to calculate. In this case we have

$$\mathbf{X'X} = \begin{bmatrix} n & \sum_{i=1}^{n} x_{i1} & \cdots & \sum_{i=1}^{n} x_{ik} \\ \sum_{i=1}^{n} x_{i1} & \sum_{i=1}^{n} x_{i1}^2 & \cdots & \sum_{i=1}^{n} x_{i1}x_{ik} \\ \cdot & \cdot & & \cdot \\ \cdot & \cdot & & \cdot \\ \cdot & \cdot & & \cdot \\ \sum_{i=1}^{n} x_{ik} & \sum_{i=1}^{n} x_{ik}x_{i1} & \cdots & \sum_{i=1}^{n} x_{ik}^2 \end{bmatrix}$$

$$\text{and} \quad \mathbf{X'y} = \begin{bmatrix} \sum_{i=1}^{n} y_i \\ \sum_{i=1}^{n} x_{i1}y_i \\ \vdots \\ \sum_{i=1}^{n} x_{ik}y_i \end{bmatrix}.$$

Since the simple linear regression model is the special case of the multiple regression model (i.e., k = 1), there are unique estimates for the y—intercept and the slope in this model. Note that we have previously derived the solution $\mathbf{b} = (\mathbf{X'X})^{-1}\mathbf{X'y}$ in the simple linear regression setting, displayed as equation (2.3.5).

Example 3.1.1 To illustrate the use of the solution to the normal equations, we turn to the Christmas tree data given in Table 2.5.1. In the discussion that followed, it was suggested that these data be modeled by the second degree polynomial model

$$Y_i = \beta_0 + \beta_1 x_i + \beta_2 x_i^2 + \varepsilon_i, \; i = 1,2,\cdots,n.$$

Since the polynomial regression model can be treated as a multiple regression model, i.e., a linear model of full rank, the least squares estimate for $\boldsymbol{\beta}' = (\beta_0, \beta_1, \beta_2)$ is given by (3.1.3). Using the values for $\mathbf{y}$ and $\mathbf{X}$ given in Section 2.5 we obtain

$$\mathbf{X'X} = \begin{bmatrix} 12 & 27.6 & 69.2 \\ 27.6 & 69.2 & 185.472 \\ 69.2 & 185.472 & 522.224 \end{bmatrix} \quad \text{and} \quad \mathbf{X'y} = \begin{bmatrix} 232.77 \\ 549.216 \\ 1368.3024 \end{bmatrix}$$

Matrix inversion techniques yields

$$(\mathbf{X'X})^{-1} = \begin{bmatrix} 11.85739261 & -10.77047952 & 2.25399600 \\ -10.77047952 & 10.083666332 & -2.15409590 \\ 2.25399600 & -2.15409590 & 0.46828172 \end{bmatrix}$$

from which we get

$$\mathbf{b} = (\mathbf{X}'\mathbf{X})^{-1}\mathbf{X}'\mathbf{y} = \begin{bmatrix} -71.12626 \\ 83.61178 \\ -17.65029 \end{bmatrix} .$$

We now can find predicted values for y using

$$\hat{y} = -71.12626 + 83.61178x - 17.65029x^2 ,$$

which is the equation of a parabola. The graph of this parabola as well as the data values are given in Figure 3.1.1. From the figure we see that this parabola "fits" the data quite well.

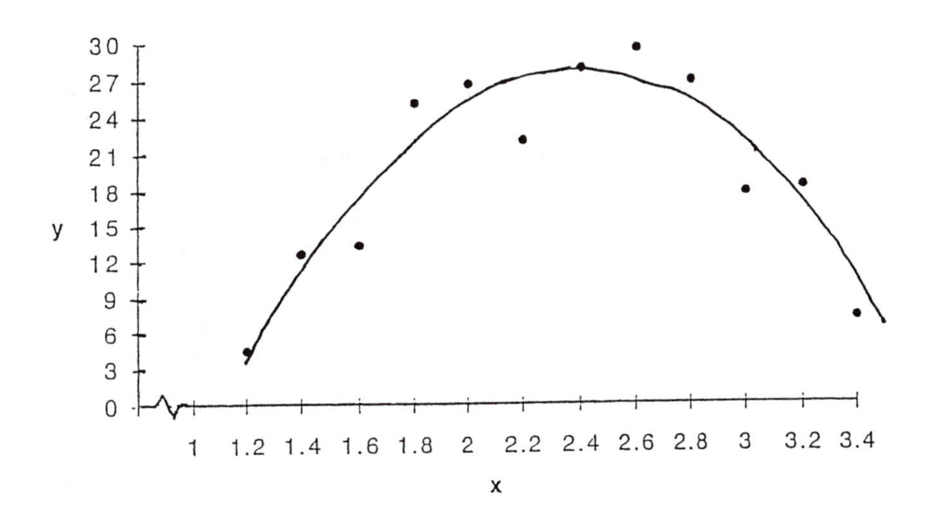

Figure 3.1.1 Christmas Tree Data Fitted by a Parabola

Less Than Full Rank Case

Let us again consider the linear model $\mathbf{Y} = \mathbf{X}\boldsymbol{\beta} + \boldsymbol{\varepsilon}$. If the rank of the $n \times (k + 1)$ matrix $\mathbf{X}$ is r where $r < k + 1 < n$, the rank of the $(k + 1) \times (k + 1)$ matrix $\mathbf{X}'\mathbf{X}$ will be r so that $\mathbf{X}'\mathbf{X}$ does not have an inverse since $r < k + 1$. We call this setting the less than full rank case. When the matrix $\mathbf{X}'\mathbf{X}$ does not have an inverse, we cannot solve the normal equations $\mathbf{X}'\mathbf{X}\mathbf{b} = \mathbf{X}'\mathbf{y}$ by finding $\mathbf{b} = (\mathbf{X}'\mathbf{X})^{-1}\mathbf{X}'\mathbf{y}$. In the exercises we show that the normal equations are consistent even in the less than full rank case. This implies there is at least one solution. Unfortunately, when $r < k + 1$ there are infinitely many different sets of solutions to the normal equations. Because of this we should not think of a solution $\mathbf{b}$ to the normal equations as an estimate for the parameter vector $\boldsymbol{\beta}$. Instead, in this case we simply refer to $\mathbf{b}$ as a solution to the normal equations.

Some obvious problems arise in the less than full rank case. First, we need to determine how to find solutions to the normal equations. Second, we need to determine whether it matters which solution we use. Third, since $\mathbf{b}$ is not an estimate of $\boldsymbol{\beta}$, we need to determine what, if anything, can be estimated. We shall address each of these questions later.

More Notation and a Less Than Full Rank Example

We close this section with a less than full rank example. Before doing so, however, it is necessary to introduce additional notation beyond that already used in our illustration of the one—way classification model in Section 2.6. Recall that the general one—way classification model (2.6.4) was given as

$$Y_{ij} = \mu + \tau_i + \varepsilon_{ij}, \quad i = 1,2,\cdots,p\ ;\ j = 1,2,\cdots, n_i.$$

Therefore, in the $\mathbf{Y} = \mathbf{X}\boldsymbol{\beta} + \boldsymbol{\varepsilon}$ matrix representation, the elements of the parameter vector are given by $\boldsymbol{\beta}' = (\mu, \tau_1\ \tau_2, \cdots, \tau_p)$. Corresponding to this vector, we denote vector $\mathbf{b}$ by $\mathbf{b}' = (m, t_1,t_2,\cdots,t_p)$ so that the sample form of the general one—way classification model will be denoted by

$$y_{ij} = m + t_i + e_{ij}, \ i = 1, 2,\cdots,p;\ j = 1,2,\cdots,n_i \ . \tag{3.1.4}$$

In the next example we use the "dot notation" for the first time, a convenience for compactly indicating sums throughout this text. The sums of sets of observations associated with (3.1.4) will be denoted as follows:

$$\sum_{i=1}^{p} \sum_{j=1}^{n_i} y_{ij} = y_{..} \ \text{ and } \ \sum_{j=1}^{n_i} y_{ij} = y_{i.} \ . \tag{3.1.5}$$

That is, $y_{\cdot\cdot}$ is the sum of all of the observations and $y_{i\cdot}$ is the sum of the observations in sample i.

Similarly, if $\sum_{i=1}^{p} n_i = n$, we denote averages of observations by

$$\sum_{i=1}^{p} \sum_{j=1}^{n_i} y_{ij}/n = \bar{y}_{\cdot\cdot} \qquad \sum_{j=1}^{n_i} y_{ij}/n_i = \bar{y}_{i\cdot} \quad . \tag{3.1.6}$$

Example 3.1.2 Refer to Example 2.6.2. Here we used the model

$$Y_{ij} = \mu + \tau_i + \varepsilon_{ij}$$

$$i = 1,2,3; \quad j = 1,\cdots,n_i$$

where $n_1 = 4$, $n_2 = 3$, and $n_3 = 2$. Using the **Y** vector and the **X** matrix displayed in Example 2.6.2 together with the notation introduced in the two previous paragraphs, it follows that the normal equations $\mathbf{X'Xb} = \mathbf{X'y}$ are

$$\begin{bmatrix} 9 & 4 & 3 & 2 \\ 4 & 4 & 0 & 0 \\ 3 & 0 & 3 & 0 \\ 2 & 0 & 0 & 2 \end{bmatrix} \begin{bmatrix} m \\ t_1 \\ t_2 \\ t_3 \end{bmatrix} = \begin{bmatrix} y_{\cdot\cdot} \\ y_{1\cdot} \\ y_{2\cdot} \\ y_{3\cdot} \end{bmatrix}$$

where $y_{\cdot\cdot}$ and $y_{i\cdot}$ for i = 1,2,3 are defined by (3.1.5). Previously it was noted that the rank of the 9 × 4 matrix X is 3. Observe that the rank of the $\mathbf{X'X}$ matrix is also 3 since the first column of $\mathbf{X'X}$ is the vector–wise sum of the other three columns. Multiplication reveals the normal equations to be

$$\begin{aligned} 9m + 4t_1 + 3t_2 + 2t_3 &= y_{\cdot\cdot} \\ 4m + 4t_1 &= y_{1\cdot} \\ 3m + 3t_2 &= y_{2\cdot} \\ 2m + 2t_3 &= y_{3\cdot} \quad . \end{aligned}$$

Defining $\bar{y}_{\cdot\cdot}$, $\bar{y}_{1\cdot}$, $\bar{y}_{2\cdot}$, and $\bar{y}_{3\cdot}$ by (3.1.6), is easy to verify that $m = \bar{y}_{\cdot\cdot}$, $t_1 = \bar{y}_{1\cdot} - \bar{y}_{\cdot\cdot}$, $t_2 = \bar{y}_{2\cdot} - \bar{y}_{\cdot\cdot}$, and $t_3 = \bar{y}_{3\cdot} - \bar{y}_{\cdot\cdot}$ constitute a solution to these normal equations. Another set of solutions is $m = \bar{y}_{3\cdot}$, $t_1 = \bar{y}_{1\cdot} - \bar{y}_{3\cdot}$, $t_2 = \bar{y}_{2\cdot} - \bar{y}_{3\cdot}$, and $t_3 = 0$. In fact, there are infinitely many solutions to these normal equations, a problem that we consider later in this chapter.

Even though the two presented solutions to the normal equations in Example 3.1.2 are very different, we note that both yield the same $\hat{y}_{ij}$. That is, for both presented solutions, $\hat{y}_{ij} = m + t_i = \bar{y}_{i\cdot}$ for $i = 1,2,3$. We shall show later in this chapter that this is not an accident as $\hat{y}_{ij}$ is the same (**invariant**) for any solution of the normal equations.

3.2 Solutions to Normal Equations — Generalized Inverse Approach

In the previous section we discovered that when $\mathbf{X}'\mathbf{X}$ is less than full rank there can be many solutions to the normal equations $\mathbf{X}'\mathbf{X}\mathbf{b} = \mathbf{X}'\mathbf{y}$. Our objective in this section is to consider a procedure for finding these solutions. In order to do this we first introduce a new concept, namely, the idea of a generalized inverse.

Generalized Inverse

Definition 3.2.1 For any matrix $\mathbf{A}$, if a matrix $\mathbf{A}^-$ satisfies the relation

$$\mathbf{A}\mathbf{A}^-\mathbf{A} = \mathbf{A}, \tag{3.2.1}$$

then $\mathbf{A}^-$ is called a **generalized inverse** (g—inverse for short) of $\mathbf{A}$.

An illustration, take

$$\mathbf{A} = \begin{bmatrix} 2 & 0 \\ 0 & 0 \end{bmatrix}, \quad \text{then } \mathbf{A}^- = \begin{bmatrix} 1/2 & 0 \\ 0 & 0 \end{bmatrix}$$

is a g—inverse of $\mathbf{A}$ since

$$\begin{bmatrix} 2 & 0 \\ 0 & 0 \end{bmatrix} \begin{bmatrix} 1/2 & 0 \\ 0 & 0 \end{bmatrix} \begin{bmatrix} 2 & 0 \\ 0 & 0 \end{bmatrix} = \begin{bmatrix} 2 & 0 \\ 0 & 0 \end{bmatrix}.$$

It is easy to see that the matrices

$$\begin{bmatrix} 1/2 & 0 \\ 1 & 1 \end{bmatrix} \quad \text{and} \quad \begin{bmatrix} 1/2 & 0 \\ 4 & -3 \end{bmatrix}$$

are also g—inverses for $\mathbf{A}$. This illustrates that g—inverses are not necessarily unique. Hence when we write "$\mathbf{A}^- =$" we mean that the given matrix is a g—inverse for $\mathbf{A}$.

If $\mathbf{A}$ is a square full rank matrix, then $\mathbf{AA}^-\mathbf{A} = \mathbf{A}$ implies that $\mathbf{A}^{-1}\mathbf{AA}^-\mathbf{AA}^{-1} = \mathbf{A}^{-1}\mathbf{AA}^{-1}$ from which we find that $\mathbf{A}^- = \mathbf{A}^{-1}$. Hence, if $\mathbf{A}$ is a square full rank matrix the g—inverse and the "usual" inverse are the same. That is, there is a unique g—inverse which turns out to be the inverse matrix encountered in introductory matrix algebra.

Our interest in g—inverses is shown in the following two theorems.

Theorem 3.2.1 If $(\mathbf{X}'\mathbf{X})^-$ is a g—inverse of $\mathbf{X}'\mathbf{X}$, then $\mathbf{b} = (\mathbf{X}'\mathbf{X})^-\mathbf{X}'\mathbf{y}$ is a solution of the normal equations $\mathbf{X}'\mathbf{Xb} = \mathbf{X}'\mathbf{y}$.

Proof. To show that $(\mathbf{X}'\mathbf{X})^-\mathbf{X}'\mathbf{y}$ is a solution we need only verify that $\mathbf{X}'\mathbf{X}[(\mathbf{X}'\mathbf{X})^-\mathbf{X}'\mathbf{y}] = \mathbf{X}'\mathbf{y}$. The normal equations are known to have at least one solution. Let $\mathbf{b}^*$ be a solution. Then $\mathbf{X}'\mathbf{Xb}^* = \mathbf{X}'\mathbf{y}$. Using the fact that $\mathbf{X}'\mathbf{X} = \mathbf{X}'\mathbf{X}(\mathbf{X}'\mathbf{X})^-\mathbf{X}'\mathbf{X}$ by Definition 3.2.1, we obtain

$$\mathbf{X}'\mathbf{X}(\mathbf{X}'\mathbf{X})^-\mathbf{X}'\mathbf{Xb}^* = \mathbf{X}'\mathbf{y}.$$

Replacing $\mathbf{X}'\mathbf{Xb}^*$ by $\mathbf{X}'\mathbf{y}$, the previous equation becomes

$$\mathbf{X}'\mathbf{X}[(\mathbf{X}'\mathbf{X})^-\mathbf{X}'\mathbf{y}] = \mathbf{X}'\mathbf{y},$$

which is what we needed to show.

The next important theorem will be stated without proof. A proof can be found on page 76 of Seber (1977).

Theorem 3.2.2 Every solution of the normal equations $\mathbf{X}'\mathbf{Xb} = \mathbf{X}'\mathbf{y}$ can be expressed as $\mathbf{b} = (\mathbf{X}'\mathbf{X})^-\mathbf{X}'\mathbf{y}$ where $(\mathbf{X}'\mathbf{X})^-$ is some g—inverse of $\mathbf{X}'\mathbf{X}$.

The previous theorem indicates how to find solutions to the normal equations. Given $\mathbf{X}'\mathbf{X}$ we need to find **any** g—inverse of $\mathbf{X}'\mathbf{X}$, say $(\mathbf{X}'\mathbf{X})^-$. Then $(\mathbf{X}'\mathbf{X})^-\mathbf{X}'\mathbf{y}$ is a solution to the normal equations. Furthermore, by finding all possible g—inverses $(\mathbf{X}'\mathbf{X})^-$ and the associated $(\mathbf{X}'\mathbf{X})^-\mathbf{X}'\mathbf{y}$, we can generate all the solutions to the normal equations. Throughout the remainder of this text we shall denote any solution to the normal equations by the symbol

$$\mathbf{b} = (\mathbf{X}'\mathbf{X})^{-}\mathbf{X}'\mathbf{y}. \tag{3.2.2}$$

Properties of A^-

There is a procedure for finding a g–inverse of a matrix $\mathbf{A}$ that is particularly useful in the analysis of experimental designs. Suppose that matrix $\mathbf{A}$ is a square $(k + 1) \times (k + 1)$ matrix of rank r. Also suppose that $\mathbf{A}$ can be partitioned as follows:

$$\mathbf{A} = \begin{bmatrix} \mathbf{A}_{11} & \mathbf{A}_{12} \\ \mathbf{A}_{21} & \mathbf{A}_{22} \end{bmatrix}$$

where $\mathbf{A}_{11}$ is an $r \times r$ full rank matrix, $\mathbf{A}_{12}$ has dimensions $r \times (k + 1 - r)$, $\mathbf{A}_{21}$ is $(k + 1 - r) \times r$ and $\mathbf{A}_{22}$ is $(k + 1 - r) \times (k + 1 - r)$. Then we can easily show that a g–inverse for $\mathbf{A}$ is

$$\begin{bmatrix} \mathbf{A}_{11}^{-1} & \mathbf{0} \\ \mathbf{0} & \mathbf{0} \end{bmatrix},$$

where the dimensions of the null matrices, denoted by $\mathbf{0}$, are as given for $\mathbf{A}_{12}$, $\mathbf{A}_{21}$, and $\mathbf{A}_{22}$, respectively.

Some properties of g–inverses are presented in the next theorem. Since our interest in g–inverses will be for the matrix $\mathbf{X}'\mathbf{X}$ associated with the normal equations (3.1.2), we present our results in terms of this matrix.

Theorem 3.2.3 Consider $\mathbf{X}$ an $n \times (k + 1)$ matrix. Then

(1) $[(\mathbf{X}'\mathbf{X})^{-}]'$ is a g–inverse of $\mathbf{X}'\mathbf{X}$.
(2) $(\mathbf{X}'\mathbf{X})^{-}\mathbf{X}'$ is a g–inverse of $\mathbf{X}$. Thus $\mathbf{X}(\mathbf{X}'\mathbf{X})^{-}\mathbf{X}'\mathbf{X} = \mathbf{X}$. Also $\mathbf{X}'\mathbf{X}(\mathbf{X}'\mathbf{X})^{-}\mathbf{X}' = \mathbf{X}'$.
(3) $\mathbf{X}(\mathbf{X}'\mathbf{X})^{-}\mathbf{X}'$ is invariant to $(\mathbf{X}'\mathbf{X})^{-}$. That is, $\mathbf{X}(\mathbf{X}'\mathbf{X})^{-}\mathbf{X}'$ is the same for any g–inverse of $\mathbf{X}'\mathbf{X}$.
(4) $\mathbf{X}(\mathbf{X}'\mathbf{X})^{-}\mathbf{X}'$ is symmetric.

(5) $\mathbf{X(X'X)^- X'}$ is idempotent. (Recall $\mathbf{A}$ idempotent means $\mathbf{AA = A}$.)

(6) If $\mathbf{I}$ is the $n \times n$ identity matrix, then $\mathbf{I - X(X'X)^- X'}$ is symmetric, idempotent and invariant to $\mathbf{(X'X)^-}$.

Proof. We shall present the proofs of parts (1) and (3) of the theorem. The remaining parts will be left for exercises.

For (1) we recall that $\mathbf{X'X(X'X)^- X'X = X'X}$ (Definition 3.2.1). By taking transposes of both sides and recognizing that $\mathbf{(X'X)' = X'X}$ we obtain

$$\mathbf{[X'X(X'X)^- X'X]' = (X'X)'}\ , \text{ which is}$$
$$\mathbf{(X'X)'[(X'X)^-]'(X'X)' = (X'X)'}\ , \text{ or}$$
$$\mathbf{(X'X)[(X'X)^-]'(X'X) = X'X}\ ,$$

which implies that $\mathbf{[(X'X)^-]'}$ is a g–inverse for $\mathbf{X'X}$.

For part (3) we let $\mathbf{(X'X)_1^-}$ and $\mathbf{(X'X)_2^-}$ be two g–inverses for $\mathbf{X'X}$. **Let** $\mathbf{N = X(X'X)_1^- X' - X(X'X)_2^- X'}$. Then

$$\mathbf{N' = X[(X'X)_1^-]'X' - X[(X'X)_2^-]'X'} \quad \text{so that}$$

$$\mathbf{N'N = X[(X'X)_1^-]'X'X(X'X)_1^- X' - X[(X'X)_1^-]'X'X(X'X)_2^- X'}$$
$$\mathbf{- X[(X'X)_2^-]'X'X(X'X)_1^- X' + X[(X'X)_2^-]'X'X(X'X)_2^- X'}.$$

Now using part (2) of the theorem we find

$$\mathbf{N'N = X[(X'X)_1^-]X' - X[(X'X)_2^-]'X' - X[(X'X)_1^-]'X'}$$
$$\mathbf{+ X[(X'X)_2^-]'X' = 0}$$

where $\mathbf{0}$ is a null matrix. Since $\mathbf{N'N = 0}$, it follows that $\mathbf{N = 0}$. Hence $\mathbf{X(X'X)_1^- X' - X(X'X)_2^- X' = 0}$, so that

$$\mathbf{X(X'X)_1^- X' = X(X'X)_2^- X'}\ .$$

This implies $\mathbf{X(X'X)^- X'}$ is the same for any g–inverse of $\mathbf{X'X}$.

We close this section with an example which illustrates the use of a g–inverse in solving normal equations.

Example 3.2.1 Refer to Example 3.1.2 . In the previous section where

$$\mathbf{X}'\mathbf{X} = \begin{bmatrix} 9 & 4 & 3 & 2 \\ 4 & 4 & 0 & 0 \\ 3 & 0 & 3 & 0 \\ 2 & 0 & 0 & 2 \end{bmatrix}$$

is of rank 3 and is therefore less than full rank. One of the g—inverses for $\mathbf{X}'\mathbf{X}$ is

$$(\mathbf{X}'\mathbf{X})^{-} = \begin{bmatrix} \begin{bmatrix} 9 & 4 & 3 \\ 4 & 4 & 0 \\ 3 & 0 & 3 \end{bmatrix}^{-1} & \begin{matrix} 0 \\ 0 \\ 0 \end{matrix} \\ \begin{matrix} 0 & 0 & 0 \end{matrix} & 0 \end{bmatrix} = \begin{bmatrix} 1/2 & -1/2 & -1/2 & 0 \\ -1/2 & 3/4 & 1/2 & 0 \\ -1/2 & 1/2 & 5/6 & 0 \\ 0 & 0 & 0 & 0 \end{bmatrix}.$$

It is easily verified that $(\mathbf{X}'\mathbf{X})(\mathbf{X}'\mathbf{X})^{-}(\mathbf{X}'\mathbf{X}) = \mathbf{X}'\mathbf{X}$. The corresponding solution to the normal equations $\mathbf{b} = (\mathbf{X}'\mathbf{X})^{-}\mathbf{X}'\mathbf{y}$ is

$$\mathbf{b} = \begin{bmatrix} m \\ t_1 \\ t_2 \\ t_3 \end{bmatrix} = \begin{bmatrix} 1/2 & -1/2 & -1/2 & 0 \\ -1/2 & 3/4 & 1/2 & 0 \\ -1/2 & 1/2 & 5/6 & 0 \\ 0 & 0 & 0 & 0 \end{bmatrix} \begin{bmatrix} y_{\cdot\cdot} \\ y_{1\cdot} \\ y_{2\cdot} \\ y_{3\cdot} \end{bmatrix}.$$

Using the fact that $y_{\cdot\cdot} = \sum_{i=1}^{3} y_{i\cdot}$ and $\bar{y}_{i\cdot} = y_{i\cdot}/n_i$, the solution set simplifies to $m = \bar{y}_{3\cdot}$, $t_1 = \bar{y}_{1\cdot} - \bar{y}_{3\cdot}$, $t_2 = \bar{y}_{2\cdot} - \bar{y}_{3\cdot}$, and $t_3 = 0$. This is one of the solutions noted in Example 3.1.2.

3.3 Invariance Properties and Error Sum of Squares

We discovered in the previous section that there are potentially many solutions to the normal equations. It is necessary to ask, therefore, does it matter which solution we use? Or to state it another way, does it matter which g—inverse of $\mathbf{X}'\mathbf{X}$ we select to find $\mathbf{b} = (\mathbf{X}'\mathbf{X})^{-}\mathbf{X}'\mathbf{y}$? We present two results in this section which illustrate the fact that for some important measures it does not matter which g—inverse is selected.

Theorem 3.3.1. Let $\mathbf{b}$ represent a solution to the normal equations. The expression $\hat{\mathbf{y}} = \mathbf{Xb}$ is invariant to $\mathbf{b}$.

Proof. Every solution to the normal equations can be written as $\mathbf{b} = (\mathbf{X'X})^{-}\mathbf{X'y}$ for some generalized inverse. Hence

$$\hat{\mathbf{y}} = \mathbf{Xb} = \mathbf{X}(\mathbf{X'X})^{-}\mathbf{X'y}.$$

From Theorem 3.2.3, $\mathbf{X}(\mathbf{X'X})^{-}\mathbf{X'}$ is invariant to $(\mathbf{X'X})^{-}$ so that $\hat{\mathbf{y}}$ is invariant to $(\mathbf{X'X})^{-}$. Since $\mathbf{b}$ is determined by $(\mathbf{X'X})^{-}$, it follows that $\hat{\mathbf{y}}$ is invariant to $\mathbf{b}$.

Theorem 3.3.1 explains why both given solutions to the normal equations in Example 3.1.2 yield the same expression for $\hat{y}_{ij}$. It is now clear that **any** solution to the normal equations yields the same result for $\hat{y}_{ij}$.

As we proceed further in this text, we will find that certain sums of squares are important in the analysis of experimental design data. Next we define one such sum of squares which plays a central role in linear model inference procedures.

Definition 3.3.1. Let $\hat{\mathbf{y}} = \mathbf{Xb}$, where $\mathbf{b}$ is a solution to the normal equations. The **error sum of squares**, denoted by SSe, is defined by

$$\text{SSe} = (\mathbf{y} - \mathbf{Xb})'(\mathbf{y} - \mathbf{Xb}). \tag{3.3.1}$$

There are several alternative expressions for SSe. If we expand the right—hand side of (3.3.1) we get

$$\text{SSe} = \mathbf{y'y} - 2\mathbf{b'X'y} + \mathbf{b'X'Xb}\ .$$

But since $\mathbf{b}$ is a solution to the normal equations $\mathbf{X'Xb} = \mathbf{X'y}$, on substituting this relation in the above expression we obtain

$$\begin{aligned}\text{SSe} &= \mathbf{y'y} - 2\mathbf{b'X'y} + \mathbf{b'X'y} \\ &= \mathbf{y'y} - \mathbf{b'X'y}\ . \end{aligned} \tag{3.3.2}$$

We will refer to (3.3.2) as the computing form for SSe. Also, since $\hat{\mathbf{y}} = \mathbf{Xb}$, we can immediately write (3.3.1) as

$$\text{SSe} = (\mathbf{y} - \hat{\mathbf{y}})'(\mathbf{y} - \hat{\mathbf{y}})$$

$$= \sum_{i=1}^{n} (y_i - \hat{y}_i)^2 = \sum_{i=1}^{n} e_i^2 . \qquad (3.3.3)$$

It is (3.3.3) that explains why SSe is called the error sum of squares. Observe that $y_i - \hat{y}_i$ is the difference between the actual y_i and the **predicted** y_i, denoted by $\hat{y}_i$; hence, it is natural to call this difference an "error". Using this terminology, it follows that SSe is the "error sum of squares." It is clear that SSe is a measure of the accuracy of the prediction since, if prediction is our only major objective, then a small SSe is desirable.

As previously noted, the error sum of squares will prove to be a very important measure in the analysis of experimental design data. Since in Definition 3.3.1, **b** is used, it is appropriate to ask if SSe is invariant to the solution to the normal equations. The following theorem tells us that it is.

Theorem 3.3.2. Let **b** be a solution to the normal equations. Then SSe is invariant to **b**, that is, SSe is unique.

Proof. From equation (3.3.3), $\text{SSe} = (\mathbf{y} - \hat{\mathbf{y}})'(\mathbf{y} - \hat{\mathbf{y}})$. Since $\hat{\mathbf{y}}$ is invariant to **b** (Theorem 3.3.1), it follows that $(\mathbf{y} - \hat{\mathbf{y}})'(\mathbf{y} - \hat{\mathbf{y}})$ is invariant to **b**.

Two Examples

We close this section with two examples, the first illustrating the invariance of SSe.

Example 3.3.1. Refer to Example 3.1.2, a one–way classification problem with three populations (treatments). There we noted that two of the infinitely many solutions to the normal equations are

first solution: $m = \bar{y}_{\cdot\cdot}$, $t_1 = \bar{y}_{1\cdot} - \bar{y}_{\cdot\cdot}$, $t_2 = \bar{y}_{2\cdot} - \bar{y}_{\cdot\cdot}$, $t_3 = \bar{y}_{3\cdot} - \bar{y}_{\cdot\cdot}$

second solution: $m = \bar{y}_{3\cdot}$, $t_1 = \bar{y}_{1\cdot} - \bar{y}_{3\cdot}$, $t_2 = \bar{y}_{2\cdot} - \bar{y}_{3\cdot}$, $t_3 = 0$.

As noted before, $\hat{y}_{ij} = m + t_i$, for all i and j, we see that for both solutions, $\hat{y}_{ij} = \bar{y}_{i.}$. This is not unexpected since $\hat{y}_{ij}$ is invariant to the solution of the normal equations; hence, $\hat{y}_{ij} = \bar{y}_{i.}$ for **any** solution. The error sum of squares is

$$\text{SSe} = \sum_{i=1}^{3} \sum_{j=1}^{n_i} (y_{ij} - \hat{y}_{ij})^2$$

$$= \sum_{i=1}^{3} \sum_{j=1}^{n_i} (y_{ij} - \bar{y}_{i.})^2 ,$$

which is the same for the two solutions, or any other solution for that matter, since SSe is unique.

Example 3.2.2 To demonstrate the calculation of the error sum of squares in a multiple regression problem, we turn to information contained in Example 3.1.1. We use calculation formula (3.3.2),

$$\text{SSe} = \mathbf{y}'\mathbf{y} - \mathbf{b}'\mathbf{X}'\mathbf{y} .$$

Referring to the $\mathbf{y}$ vector for the data in Table 2.5.1, we get

$$\mathbf{y}'\mathbf{y} = \sum_{i=1}^{12} y_i^2 = (4.5)^2 + (12.59)^2 + \cdots + (7.23)^2 = 5303.3425 .$$

Also using vectors $\mathbf{b}$ and $\mathbf{X}'\mathbf{y}$ displayed in Example 3.1.1 we obtain

$$\mathbf{b}'\mathbf{X}\mathbf{y} = [-71.12626 \quad 83.61178 \quad -17.65029] \begin{bmatrix} 232.77 \\ 549.216 \\ 1368.3024 \end{bmatrix} = 5213.934$$

Therefore, SSe = 5303.3425 − 5213.934 = 89.408.

3.4 Solutions to Normal Equations — Side Conditions Approach

The work required to find a solution to the normal equations using a g–inverse is, at times, a bit cumbersome. When $\mathbf{X}'\mathbf{X}$ is a matrix with many rows and columns, it can be difficult to find a g–inverse without the aid of a computer. For example, in a one–way classification problem with six treatments, using the method of Section 3 in this chapter requires inverting a 6×6 matrix. Clearly, this is not the kind of exercise one wishes to do on a regular basis. Hence it is desirable to find another method for finding a solution to the normal equations. Remember we are not particular about which solution we find since $\hat{y}$ and SSe are the same for every solution.

If we examine the normal equations carefully, we see that there are $k + 1$ equations and $k + 1$ unknowns. The problem occurs when the rank of $\mathbf{X'X}$, which is r, is less than $k + 1$. In such a case, $q = (k + 1) - r$ equations are redundant. What is required to get a unique solution is to "add" $q = (k + 1) - r$ appropriate new equations.

To formalize this, suppose we consider a matrix $\mathbf{H}$ of constants having dimension $q \times (k + 1)$. Let us select $\mathbf{H}$ in such a way so that the rows of $\mathbf{H}$ are linearly independent of the rows of $\mathbf{X}$. Thus the rank of $\mathbf{H}$ is q and the rank of a new matrix

$$\mathbf{G} = \begin{bmatrix} \mathbf{X} \\ \mathbf{H} \end{bmatrix}$$

is $k + 1$. Note that $\mathbf{G}$ is an $(n + q) \times (k + 1)$ matrix. To form a system of equations we first form the **side condition** or constraints

$$\mathbf{H}\boldsymbol{\beta} = \mathbf{0} \tag{3.4.1}$$

on the actual parameters. In terms of $\mathbf{b}$, we impose

$$\mathbf{Hb} = \mathbf{0}. \tag{3.4.2}$$

Using the normal equations and side conditions (3.4.2), we have an augmented system of equations.

$$\mathbf{X'Xb} = \mathbf{X'y} \tag{3.4.3}$$

$$\mathbf{Hb} = \mathbf{0} \ .$$

In this system there are $k + 1$ unknowns, $b_0, b_1, \cdots, b_k$, and $q + k + 1$ equations. However, this set includes exactly $k + 1$ linearly independent equations, r from the first set of equations and $q = k + 1 - r$ from the second set. We should, therefore, be able to obtain a unique solution.

Justification

To find a solution we proceed as follows. Now

$$\mathbf{G'G} = [\mathbf{X'H'}]\begin{bmatrix}\mathbf{X}\\ \mathbf{H}\end{bmatrix} = \mathbf{X'X} + \mathbf{H'H}$$

which is a $(k+1) \times (k+1)$ matrix of rank $k+1$. Hence the inverse $(\mathbf{G'G})^{-1}$ exists. Now $\mathbf{Hb} = \mathbf{0}$ implies that $\mathbf{H'Hb} = \mathbf{0}$. Therefore any solution to the system of equations (3.4.3) must satisfy

$$\mathbf{X'Xb} = \mathbf{X'y} \tag{3.4.4}$$

$$\mathbf{H'Hb} = \mathbf{0}\ .$$

But if equations (3.4.4) hold, then by addition

$$\mathbf{X'Xb} + \mathbf{H'Hb} = \mathbf{X'y},$$

which can be written as

$$(\mathbf{X'X} + \mathbf{H'H})\mathbf{b} = \mathbf{X'y}$$

or

$$\mathbf{G'Gb} = \mathbf{X'y}\ . \tag{3.4.5}$$

Of course, equations (3.4.5) have a unique solution which is given by

$$\mathbf{b} = (\mathbf{G'G})^{-1}\mathbf{X'y}\ . \tag{3.4.6}$$

Since any solution to equations (3.4.3) must also be a solution to equations (3.4.5), and since each has a unique solution, it follows that (3.4.6) is a solution to equations (3.4.3). Naturally, if (3.4.6) is a solution to (3.4.3), it also is a solution to the normal equations. That is, $\mathbf{b} = (\mathbf{G'G})^{-1}\mathbf{X'y}$ is the unique solution to the normal equations that also satisfies the side conditions $\mathbf{Hb} = \mathbf{0}$. Proof in more detail can be found in Scheffé (1959) on page 17 and in Seber (1977) on page 74.

Earlier we showed that every solution of the normal equations can be expressed as $\mathbf{b} = (\mathbf{X'X})^{-}\mathbf{X'y}$ for some g–inverse $(\mathbf{X'X})^{-}$. Since $\mathbf{b} = (\mathbf{G'G})^{-1}\mathbf{X'y}$ is a solution to the normal equations, it follows that there is some g–inverse of $\mathbf{X'X}$ so that $(\mathbf{G'G})^{-1}\mathbf{X'y} = (\mathbf{X'X})^{-}\mathbf{X'y}$. In other words, the solution we find using side conditions can be found using some g–inverse. Example 3.2.1 illustrates this. The g–inverse used in this example to obtain the solution set $m = \bar{y}_{3\cdot}$, $t_1 = \bar{y}_{1\cdot} - \bar{y}_{3\cdot}$, $t_2 = \bar{y}_2 - \bar{y}_{3\cdot}$, $t_3 = 0$ is the solution to the normal equations with the imposed side condition $\tau_3 = 0$ (or $t_3 = 0$).

To illustrate how side conditions can facilitate our search for a solution to the normal equations, we consider the following example.

Example 3.4.1. In Example 3.1.2, the normal equations are

$$\begin{aligned}
9m + 4t_1 + 3t_2 + 2t_3 &= y_{\cdot\cdot} \\
4m + 4t_1 &= y_{1\cdot} \\
3m + 3t_2 &= y_{2\cdot} \\
2m + 2t_3 &= y_{3\cdot}
\end{aligned}$$

Since the corresponding $\mathbf{X}$ and $\mathbf{X'X}$ matrices have rank 3 and $k + 1 = 4$, we can get a solution to these normal equations by imposing one side condition equation. A side condition equation which yields a solution set in an easy manner is

$$4t_1 + 3t_2 + 2t_3 = 0.$$

From the first equation we immediately get $m = y_{\cdot\cdot}/9 = \bar{y}_{\cdot\cdot}$ and then using this value for m in the other three equations we obtain $t_1 = \bar{y}_{1\cdot} - \bar{y}_{\cdot\cdot}$, $t_2 = \bar{y}_{2\cdot} - \bar{y}_{\cdot\cdot}$, and $t_3 = \bar{y}_{3\cdot} - \bar{y}_{\cdot\cdot}$. It is easily seen that the above side condition can be written as the vector dot product

$$[0 \quad 4 \quad 3 \quad 2]\mathbf{b} = 0.$$

Thus the matrix $\mathbf{H}$ in (3.4.2) is $\mathbf{H} = [0 \quad 4 \quad 3 \quad 2]$ for this case.

This example illustrates our procedure for finding a solution to the normal equations by utilizing side conditions. We summarize this process here as a four-step procedure.

Step 1: Find the normal equations.
Step 2: Select $(k + 1) - r$ side conditions that simplify the algebra required to find a solution to the system of equations.
Step 3: By imposing the side conditions, we solve the normal equations using algebraic manipulations.

Step 4: One can use the obtained solution to find the unique $\hat{\mathbf{y}}$ and SSe.

Chapter 3 Exercises

Application Exercises

3–1 Consider the following data and the model

$$Y = \beta_0 + \beta_1 x_1 + \beta_2 x_2 + \varepsilon :$$

y	3	0	0	2	5
x_1	−2	−1	0	1	2
x_2	2	−1	−2	−1	2

(a) Find the least squares estimates for β_0, β_1 and β_2.

(b) Predict the value of y when $x_1 = 1$, $x_2 = 1$.

(c) Calculate SSe.

3–2 Refer to Example 3.1.1 in which a second degree polynomial model was fit to the Christmas tree data. Use the fitted model to estimate the number of trees that should be stocked to yield maximum profit.

3–3 Refer to Exercise 2–5 and the model given there.

(a) Find the least squares prediction equation $\hat{y}$.
(b) Predict the percent yield of the chemical process when the temperature is .5 and the pressure is .3.
(c) Find SSe.

3–4 The time to failure (in thousand of hours) for an electric motor is generally related to the speed (revolutions per minute) of the motor. To determine this relationship, a sample of 15 motors is treated by varying the motor speed and recording the time to failure. The results are given below.

Time to Failure for Electric Motors

Time to Failure	5.07	4.80	5.03	4.60	4.58	4
Speed	600	700	800	900	950	1

4.17	4.32	3.57	3.15	2.85	2.67	2.15	1.56
1050	1100	1150	1200	1250	1300	1350	1400

(a) Plot the data using y = time and x = speed.
(b) Find the least squares estimates for the second degree polynomial model.
(c) Predict the time to failure when the speed is 1020.
(d) Find SSe.

3–5 The object of an experiment is to compare the effects of two diets, A and B, on weight in rats. Eight rats of the same initial weight and age were chosen to participate in the experiment. Four of the rats were randomly selected to receive diet A, the other four were given diet B with the following results.

Weight gain, y	8	11	12	9	8	6	8	10
Diet	A	A	A	A	B	B	B	B

Here diet is the independent variable of the qualitative type.

(a) Fit the model $Y = \beta_0 + \beta_1 x + \varepsilon$,

$$\text{where } x = \begin{cases} 1 \ , & \text{if the rat is fed diet B} \\ 0 \ , & \text{if the rat is fed diet A} \end{cases}$$

(b) Use the result obtained in (a) to predict the amount of weight gained by similar rats fed on

(i) diet A, (ii) diet B.

(c) Calculate SSe.

(Note: An independent variable x which assumes only the values 0 and 1 is called a dummy or indicator variable. Such variables will be discussed in more detail later in this text.)

3–6 Consider the following data which consists of eight observations at four levels of the independent variable repeated two times.

y	1	3	−2	−4	−3	−5	6	4
x	1	1	2	2	3	3	4	4

Fit the third degree polynomial model

$$Y = \beta_0 + \beta_1 x + \beta_2 x^2 + \beta_3 x^3 + \varepsilon .$$

3–7 Verify that $\begin{bmatrix} 1/2 & 0 \\ 1 & 1 \end{bmatrix}$ and $\begin{bmatrix} 1/2 & 0 \\ 4 & -3 \end{bmatrix}$ are g–inverses for $\begin{bmatrix} 2 & 0 \\ 0 & 0 \end{bmatrix}$.

3–8 Find a g–inverse for each of the following matrices.

(a) $\begin{bmatrix} 2 & 1 & 3 \\ 4 & 0 & 5 \\ 4 & 2 & 6 \end{bmatrix}$ (b) $\begin{bmatrix} 5 & 2 & 1 & 3 \\ 4 & 3 & 4 & 1 \\ 5 & 2 & 1 & 3 \\ 4 & 3 & 4 & 1 \end{bmatrix}$.

3–9 Seber (1977) points out the following method for finding a generalized inverse for a $p \times p$ symmetric matrix $\mathbf{A}$ of rank r:

(1) Delete $p - r$ rows and the corresponding columns so as to leave an $r \times r$ matrix that is nonsingular.

(2) Invert the $r \times r$ matrix.

(3) $\mathbf{A}^-$ is obtained by taking the $r \times r$ matrix of step 2 and inserting zeros into the $p \times p$ matrix to correspond to the rows and columns originally deleted.

Use this procedure to find a generalized inverse for the following matrix of rank 2:

$$\begin{bmatrix} 4 & 2 & 12 & 8 \\ 2 & 1 & 6 & 4 \\ 12 & 6 & 36 & 24 \\ 8 & 4 & 24 & 12 \end{bmatrix}.$$

3–10 Suppose y_{ij} represents the observed yield of wheat obtained on the jth plot upon which fertilizer is applied. An experiment is conducted in which the responses to two types of fertilizers are studied. Fertilizer 1 is applied to 4 plots while fertilizer 2 is applied to 2 plots. The yields obtained in this experiment can be displayed as in the table below.

Fertilizer 1	Fertilizer 2
y_{11}	y_{21}
y_{12}	y_{22}
y_{13}	
y_{14}	

We assume the one–way classification model for this experiment:

$$Y_{ij} = \mu + \tau_i + \varepsilon_{ij}$$

$$i = 1,2; \quad j = 1,\cdots,n_i; \quad n_1 = 4,\ n_2 = 2.$$

(a) Obtain the normal equations using $\mathbf{b}' = (m, t_1, t_2)$.

(b) Show that the following are generalized inverses of $\mathbf{X}'\mathbf{X}$:

$$\text{i)} \begin{bmatrix} 0 & 0 & 0 \\ 0 & 1/4 & 0 \\ 0 & 0 & 1/2 \end{bmatrix} \quad \text{ii)} \begin{bmatrix} 1/2 & -1/2 & 0 \\ -1/2 & 3/4 & 0 \\ 0 & 0 & 0 \end{bmatrix} \quad \text{iii)} \begin{bmatrix} 1/4 & 0 & -1/4 \\ 0 & 0 & 0 \\ -1/4 & 0 & 3/4 \end{bmatrix}.$$

Note: **These g–inverses were obtained using the procedure outlined in Exercise 3–9.**

(c) Find the solution to the normal equations $\mathbf{b} = (\mathbf{X}'\mathbf{X})^- \mathbf{X}'\mathbf{y}$ using each of the generalized inverses in part b.

(d) For each solution in part c calculate $t_1 - t_2$. Notice that

$$t_1 - t_2 = \bar{y}_{1\cdot} - \bar{y}_{2\cdot} \quad \text{for each solution.}$$

(e) For each solution in part c calculate $t_1 + t_2$. Is $t_1 + t_2$ invariant to the solution of the normal equations?

(f) Find an expression for SSe.

3–11 Refer to the setting in Exercise 3–10.

(a) Find the solution to the normal equations corresponding to each of the following side conditions:

$$\text{i) } 4t_1 + 2t_2 = 0 \qquad \text{ii) } t_1 + t_2 = 0 \qquad \text{iii) } t_1 = 0.$$

(b) Find $\hat{y}_{ij}$ for each solution.

(c) Find $\mathbf{b'X'y}$ for each solution and notice that it is invariant to these solutions.

3–12 Calculate SSe for the data in Exercise 2–3.

3–13 Calculate SSe for the data in Exercise 2–4.

3–14 Refer to the secchi disk data given in Table 2.6.1 and discussed in Example 2.6.1.

(a) Write down an appropriate model for these data.
(b) Find the normal equations.
(c) Solve the normal equations using a generalized inverse.
(d) Find $\hat{y}_{ij}$.
(e) Find SSe.
(f) Solve the normal equations using the side condition

$$t_1 + t_2 + t_3 = 0.$$

(g) Which of the following quantities are invariant to the solutions in part c and part f?

i) $t_1 - 2t_2 + t_3$ ii) $t_1 - 2t_2$ iii) $t_1 - t_3$

3–15 Refer to Example 3.1.2.

(a) Verify that the given two solution sets are solutions of the normal equations.
(b) Verify that these two solutions yield the same values for

i) $t_2 - t_1$ ii) $t_1 - 2t_2 + t_3$.

3–16 Refer to Example 3.2.1. Verify that $\mathbf{X'X}$ is as given and $\mathbf{(X'X)^-}$ is a g–inverse. Then verify the following g–inverses yield the solutions

$$m = \bar{y}_{1\cdot}, \quad t_1 = 0, \quad t_2 = \bar{y}_{2\cdot} - \bar{y}_{1\cdot}, \quad t_3 = \bar{y}_{3\cdot} - \bar{y}_{1\cdot};$$

and

$$m = \bar{y}_{3\cdot}, \quad t_1 = \bar{y}_{1\cdot} - \bar{y}_{2\cdot}, \quad t_2 = 0, \quad t_3 = \bar{y}_{3\cdot} - \bar{y}_{2\cdot},$$

respectively.

$$\begin{bmatrix} 1/4 & 0 & -1/4 & -1/4 \\ 0 & 0 & 0 & 0 \\ -1/4 & 0 & 7/12 & 1/4 \\ -1/4 & 0 & 1/4 & 3/4 \end{bmatrix} \qquad \begin{bmatrix} 1/3 & -1/3 & 0 & -1/3 \\ -1/3 & 7/12 & 0 & 1/3 \\ 0 & 0 & 0 & 0 \\ -1/3 & 1/3 & 0 & 5/6 \end{bmatrix}$$

3–17 Refer to Example 3.4.1. Using the side condition $\mathbf{Hb} = 0$ where $\mathbf{H} = [1 \ 0 \ 0 \ 0]$, obtain yet another solution set: $m = 0$, $t_1 = \bar{y}_{1\cdot}$, $t_2 = \bar{y}_{2\cdot}$, and $t_3 = \bar{y}_{3\cdot}$.

3–18 Consider a one–way classification with two treatments and n_1 observations in treatment one and n_2 in treatment two.

(a) Write down an appropriate model for these data.
(b) Find the normal equations in terms of n_1, n_2, m, t_1, t_2.
(c) Find a solution to these normal equations using

(i) a generalized inverse,

(ii) the side condition $t_1 + t_2 = 0$,

(iii) the side condition $t_1 = 0$.

3–19 The yield per plot in bushels of corn was observed on $n = 10$ plots which had been fertilized in varying degrees. Here the amount of fertilizer applied is the independent variable x. Let us code x by the formula $x_1 = x -$ average amount of fertilizer applied. The data with the two sets of fertilizer values follows:

Yield, y, in bushels	12	13	13	14	15	15	14	16	17	18
Fertilizer, x, in pounds	2	2	3	3	4	4	5	5	6	6
x_1, Coded x	–2	–2	–1	–1	0	0	1	1	2	2

(a) Obtain the least squares prediction equation y using the models

(i) $Y = \beta_0 + \beta_1 x + \varepsilon$ and (ii) $Y = \beta_0 + \beta_1 x_1 + \varepsilon$.

(b) Estimate the **average** yield per plot when the amount of fertilizer applied is x = 3 pounds using the results obtained in (a) for **both** models.

(c) Calculate SSe for both models (i) and (ii) in (a).

3–20 At an elementary school a study is aimed at determining what affects a student's reading ability. In a preliminary part of the study, the effect of IQ level is studied. The IQ level is used to divide the students into three groups: low IQ (less than 90), medium IQ (90 – 110) and high IQ (over 110). A sample of six students is chosen from each IQ group, and each student is given a reading test. The scores on the test range from 0–75. The results are given as follows:

Reading Test Scores

IQ Group

Low	Medium	High
39	49	61
47	38	58
26	50	60
36	49	57
39	40	60
50	58	63

(a) Write down an appropriate model for these data.
(b) Find the normal equations.
(c) Find a solution to the normal equations.
(d) Find SSe.

Theoretic Exercises

3–21 Show that the normal equations are consistent. Hint: Recall that the system of equations $\mathbf{X'Xb} = \mathbf{X'y}$ are consistent if rank $(\mathbf{X'X}, \mathbf{X'y}) =$ rank $(\mathbf{X'X})$.

3–22 In the full rank case use a second partial derivative matrix to show that $\mathbf{b} = (\mathbf{X'X})^{-1}\mathbf{X'y}$ minimizes $\mathbf{e'e}$.

3–23 Suppose that matrix $\mathbf{A}$ is a square $(k + 1) \times (k + 1)$ matrix of rank r. Also suppose that $\mathbf{A}$ can be partitioned as follows:

$$A = \begin{bmatrix} A_{11} & A_{12} \\ A_{21} & A_{22} \end{bmatrix},$$

where $\mathbf{A}_{11}$ is an $r \times r$ full rank matrix, $\mathbf{A}_{12}$ has dimensions $r \times (k + 1 - r)$, $\mathbf{A}_{21}$ is $(k + 1 - r) \times r$ and $\mathbf{A}_{22}$ is $(k + 1 - r) \times (k + 1 - r)$. Consider the matrix

$$\mathbf{B} = \begin{bmatrix} A_{11}^{-1} & 0 \\ 0 & 0 \end{bmatrix}$$

where the dimensions of the null matrices, denoted by $\mathbf{0}$, are as given for $\mathbf{A}_{12}$, $\mathbf{A}_{21}$, and $\mathbf{A}_{22}$, respectively. Show that $\mathbf{B}$ is a g–inverse for $\mathbf{A}$.

3–24 Consider a matrix $\mathbf{A}$ of the form

$$\mathbf{A} = \begin{bmatrix} \mathbf{B} & \mathbf{0} \\ \mathbf{0} & \mathbf{C} \end{bmatrix}. \text{ Show that } \begin{bmatrix} \mathbf{B}^{-} & \mathbf{0} \\ \mathbf{0} & \mathbf{C}^{-} \end{bmatrix}$$

is a generalized inverse for $\mathbf{A}$.

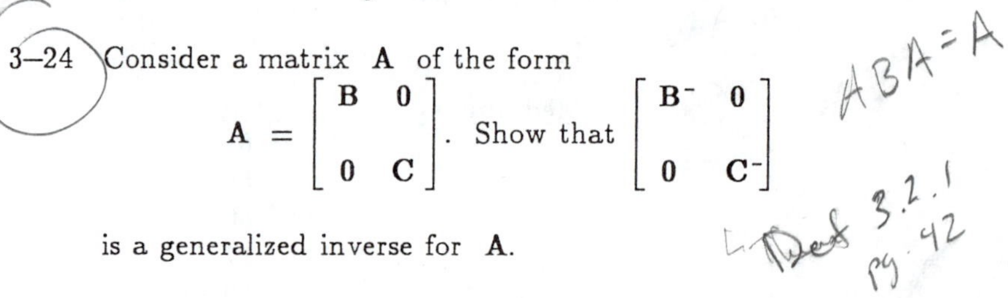

3–25 Let $\hat{\mathbf{y}} = \mathbf{Xb}$, where $\mathbf{b}$ is any solution to the normal equations based on the linear model (2.1.5). Show that $\sum_{i=1}^{n} (y_i - \hat{y}_i) = 0$, that is, the sum of the residuals is zero.

3–26 Prove parts (2), (4), (5) and (6) of Theorem 3.2.3. Hint: For part (2) let

$$\mathbf{M} = \mathbf{X} - \mathbf{X}(\mathbf{X}'\mathbf{X})^{-}\mathbf{X}'\mathbf{X} \text{ and consider } \mathbf{M}'\mathbf{M}.$$

3–27 Prove that $\mathbf{b}'\mathbf{X}'\mathbf{y}$ is invariant to the solution of the normal equations.

3–28 Prove the following:

(a) Rank $\mathbf{X}^- \geq$ Rank $\mathbf{X}$.
(b) Rank $(\mathbf{X}^-\mathbf{X})$ = Rank $(\mathbf{X}\mathbf{X}^-)$ = Rank $\mathbf{X}$.
(c) Rank $[(\mathbf{X}'\mathbf{X})(\mathbf{X}'\mathbf{X})^-]$ = Rank $\mathbf{X}$.

3–29 Prove or disprove the following: $(\mathbf{A}^-)^- = \mathbf{A}$.

3–30 Consider Exercise 2–15 and the model given there.

(a) Find the solution to the normal equations for this model.
(b) Find SSe based on this model.
(c) Compare the formula for SSe obtained in part (b) with the one obtained in Example 3.3.1.

3–31 Consider the no–intercept model $Y = \beta x + \varepsilon$ introduced in Exercise 2–14.

(a) Derive the least squares formula for the estimate of the parameter β,

$$b = \frac{\sum_{i=1}^{n} x_i y_i}{\sum_{i=1}^{n} x_i^2} .$$

(b) Show that $SSe = \sum_{i=1}^{n} y_i^2 - b \sum_{i=1}^{n} x_i y_i$.

3–32 Consider the linear model and the error sum of squares SSe. Determine how SSe is affected by linear transformations on the dependent variable. First show how SSe is affected by the transformation $W_i = a\, Y_i$, where a is a non–zero constant. Second, consider the effect of the transformation $W_i = a\, Y_i + c$, where a and c are non–zero constants.

3–33 Repeat Exercise 3–32 for the less than full rank linear model.

3–34 Consider a two–factor experimental design model

$$Y_{ijk} = \mu + \tau_i + \alpha_j + \varepsilon_{ijk}\,, \quad i = 1,2; \quad j = 1,2; \quad k = 1,2.$$

The pattern of observations is given in the following table.

	α_1	α_2
τ_1	y_{111} y_{112}	y_{121} y_{122}
τ_2	y_{211} y_{212}	y_{221} y_{222}

(a) Show that this model is a linear model.
Use $\boldsymbol{\beta}' = (\mu, \tau_1, \alpha_1, \tau_2, \alpha_2)$.

(b) Find the normal equations using $\mathbf{b}' = (m, t_1, a_1, t_2, a_2)$.

(c) Solve the normal equations using the two side conditions

$$t_1 + t_2 = 0, \qquad a_1 + a_2 = 0.$$

(d) Find $\hat{y}_{ijk}$.

CHAPTER 4

LINEAR MODEL DISTRIBUTION THEORY

4.1 Usual Linear Model Assumptions

Thus far in our discussion of the linear model

$$Y_i = \beta_0 + \beta_1 x_{i1} + \cdots + \beta_k x_{ik} + \varepsilon_i \ , \quad i = 1,2,\cdots,n \ ,$$

we have not made any assumptions about the distributions of the random variables Y_i and ε_i. As a consequence, all the results presented prior to this are true for any distributional form on the Y_i and ε_i random variables. For example, all our results on least squares and on invariance to the solution of the normal equations were verified without the need of making distributional assumptions for components of the linear model. To proceed further in the study of linear models, however, certain assumptions will be required.

First we will make assumptions concerning the moments of ε_i and Y_i as this will enable us to determine some properties of the least squares estimates. Later in the chapter we will add the assumption of normality which is needed for the purpose of making statistical inferences via test and estimation procedures. It is customary to make the assumptions on ε_i, recognizing that this then yields corresponding distributional properties for Y_i.

Assumptions in Univariate Form

The usual assumptions for the linear model

$$Y_i = \beta_0 + \beta_1 x_{i1} + \cdots + \beta_k x_{ik} + \varepsilon_i \ , \quad i = 1,2,...,n$$

are as follows:

(a) $E(\varepsilon_i) = 0 \ , \quad i = 1,2,...,n.$

(b) $Var(\varepsilon_i) = \sigma^2 \ , \quad i = 1,2,...,n.$ (4.1.1)

(c) $Cov(\varepsilon_i,\varepsilon_j) = 0$ for $i \neq j$.

The linear model having these assumptions is sometimes referred to as the **general** or **classical** linear model. Assumption (b), $Var(\varepsilon_i) = \sigma^2$ for all i, is called the **homoscedasticity** assumption.

The three assumptions of (4.1.1) determine the **moments** for the Y_i random variables. From Chapter 2, recall that the β_i's are fixed parameters and the x_{ij}'s are known constants. As a result, it follows from (4.1.1) that the Y_i random variables satisfy the following:

(a) $E(Y_i) = \beta_0 + \beta_1 x_{i1} + \cdots + \beta_k x_{ik}$, i = 1,2,...,n.

(b) $Var(Y_i) = \sigma^2$, i = 1,2,...,n. (4.1.2)

(c) $Cov(Y_i, Y_j) = 0$ for $i \neq j$.

Assumptions in Matrix Form

For convenience, it is often desirable to express assumptions (4.1.1) and (4.1.2) in matrix form. Before doing so we need some notation and terminology. Consider a vector of random variables $\mathbf{Y}' = (Y_1, Y_2, \cdots, Y_n)$. The mean vector, denoted by $E(\mathbf{Y})$, and the covariance of the vector, denoted by $Cov(\mathbf{Y})$, are defined as follows:

Definition 4.1.1. For an $n \times 1$ random vector $\mathbf{Y}$, the **mean vector** $\boldsymbol{\mu}$ or $E(\mathbf{Y})$ is given by

$$\boldsymbol{\mu} = E(\mathbf{Y}) = \begin{bmatrix} E(Y_1) \\ E(Y_2) \\ \cdot \\ \cdot \\ \cdot \\ E(Y_n) \end{bmatrix},$$

and the **covariance matrix** (variance—covariance matrix) $Cov(\mathbf{Y})$ is given by

$$Cov(\mathbf{Y}) = E[(\mathbf{Y} - \boldsymbol{\mu})(\mathbf{Y} - \boldsymbol{\mu})'].$$

If we let $\sigma_{ij} = Cov(Y_i, Y_j) = E[\{Y_i - E(Y_i)\}\{Y_j - E(Y_j)\}]$, and

$\sigma_{ii} = \sigma_i^2 = \mathrm{Var}(Y_i) = E[\{Y_i - E(Y_i)\}^2]$, then

$$\mathrm{Cov}(\mathbf{Y}) = \begin{bmatrix} \sigma_1^2 & \sigma_{12} & \cdots & \sigma_{1n} \\ \sigma_{21} & \sigma_2^2 & \cdots & \sigma_{2n} \\ \vdots & \vdots & & \vdots \\ \sigma_{n1} & \sigma_{n2} & \cdots & \sigma_n^2 \end{bmatrix}.$$

Thus we see that $\mathrm{Cov}(\mathbf{Y})$ is an $n \times n$ matrix which contains the variances of the components of $\mathbf{Y}$ on the diagonal while the off–diagonal elements are the covariances between the components of $\mathbf{Y}$. Since $\mathrm{Cov}(Y_i, Y_j) = \mathrm{Cov}(Y_j, Y_i)$, it follows that $\mathrm{Cov}(\mathbf{Y})$ is a symmetric $n \times n$ matrix.

Covariance matrices will be very important in our development. Observe that when the components of the random vector Y have pairwise zero covariance, $\mathrm{Cov}(\mathbf{Y})$ is a diagonal matrix. Such is the case when (4.1.2c) holds and is a characteristic of many of the models used in experimental design analysis. We can now restate the assumptions associated with the general linear model in matrix notation. For the linear model $\mathbf{Y} = X\boldsymbol{\beta} + \varepsilon$, the three assumptions of (4.1.1) can be condensed into the following two statements:

(a) $E(\varepsilon) = \mathbf{0}$,

(b) $\mathrm{Cov}(\varepsilon) = \sigma^2\mathbf{I}$ (4.1.3)

where $\mathbf{I}$ is the $n \times n$ identity matrix.

It is obvious that the mean of vector ε is a null vector and the covariance of the ε vector is a diagonal matrix with common diagonal elements, σ^2. That is,

$$\text{Cov}(\varepsilon) = \begin{bmatrix} \sigma^2 & 0 & \cdots & 0 \\ 0 & \sigma^2 & \cdots & 0 \\ \cdot & \cdot & & \cdot \\ \cdot & \cdot & & \cdot \\ \cdot & \cdot & & \cdot \\ 0 & 0 & & \sigma^2 \end{bmatrix}.$$

Unless otherwise stated, assumptions (4.1.3) or equivalently, (4.1.1) alternately (4.1.2), will be used throughout the remainder of the text. In the next sections of this chapter, we discover how these assumptions can help us determine desirable estimation properties for linear models. In order to develop these properties, however, it is necessary to digress for more notation so that we may clearly distinguish between vectors of random variables and vectors of constants.

More Notation

In the previous chapter we represented a solution to the normal equations as $\mathbf{b} = (\mathbf{X}'\mathbf{X})^- \mathbf{X}'\mathbf{y}$. In this formula, $\mathbf{y}$ denotes a realization of the random vector $\mathbf{Y}$ and hence $\mathbf{b}$ is a vector of constants resulting from a particular set of values of $\mathbf{Y}$. In other words, $\mathbf{b}$ is a realization of a random vector that is a function of the random vector $\mathbf{Y}$. As a consequence, we now introduce the symbol $\hat{\boldsymbol{\beta}}$ to represent the **random vector** form of $\mathbf{b}$. To clarify, if $\mathbf{b} = (\mathbf{X}'\mathbf{X})^- \mathbf{X}'\mathbf{y}$, then $\hat{\boldsymbol{\beta}} = (\mathbf{X}'\mathbf{X})^- \mathbf{X}'\mathbf{Y}$. Later we will learn the exact distribution of the random vector $\hat{\boldsymbol{\beta}}$ in the normal general linear model setting.

4.2 Moments of Response and Solutions Vector

Incorporating the assumptions of the last section we can now write the general linear model with assumptions as

$$\mathbf{Y} = \mathbf{X}\boldsymbol{\beta} + \varepsilon\,, \quad \mathrm{E}(\varepsilon) = \mathbf{0}\,, \quad \mathrm{Cov}(\varepsilon) = \sigma^2\mathrm{I}. \tag{4.2.1}$$

Using (4.2.1) we can obtain the moments of $\mathbf{Y}$ and $\hat{\boldsymbol{\beta}} = (\mathbf{X}'\mathbf{X})^- \mathbf{X}'\mathbf{Y}$. Before doing so, however, we need to consider some properties involving mathematical expectation, mean vectors, and covariance matrices. In the last section we learned that the expected value of a random vector (or random matrix) is the vector (or matrix) of the expected values of its elements. (See Definition 4.1.1.) It follows that if $\mathbf{A}$ and $\mathbf{C}$ are constant matrices and $\mathbf{W}$ and $\mathbf{Z}$ are random matrices with finite moments, then

$$\begin{aligned} &(a) \quad E(\mathbf{A}) = \mathbf{A} \\ &(b) \quad E(\mathbf{AWC}) = \mathbf{A}E(\mathbf{W})\mathbf{C} \\ &(c) \quad E(\mathbf{W} + \mathbf{Z}) = E(\mathbf{W}) + E(\mathbf{Z}) \ . \end{aligned} \tag{4.2.2}$$

The reader will recognize these properties of the expected value operator to be analogous to those in the univariate case. Of course, the (4.2.2) statements tacitly assume the dimensions of the matrices to be compatible with the indicated operations and they also apply to vectors as special matrices.

Now let us consider an $n \times 1$ random vector $\mathbf{Y}$, a $q \times 1$ vector of constants $\mathbf{c}$, and an $n \times 1$ vector of constants $\mathbf{a}$, and a $q \times n$ matrix of constants $\mathbf{B}$. Two expressions that will prove useful to us are $\mathbf{a}'\mathbf{Y}$ and $\mathbf{BY} + \mathbf{c}$. Note that $\mathbf{a}'\mathbf{Y}$ is a linear combination of the elements of the random vector $\mathbf{Y}$, namely, $a_1Y_1 + a_2Y_2 + \cdots + a_nY_n$. On the other hand, $\mathbf{BY} + \mathbf{c}$ is a $q \times 1$ random vector. The next theorem lists the means and the variances of these entities. The proofs of these results will be left to the student in the exercises. Here Definition 4.1.1 and statements (4.2.2) are applicable.

Theorem 4.2.1 Consider a random vector $\mathbf{Y}$, constant vectors $\mathbf{a}$ and $\mathbf{c}$, and constant matrix $\mathbf{B}$ with dimensions as given above. Let $E(\mathbf{Y}) = \boldsymbol{\mu}$ and $\mathrm{Cov}(\mathbf{Y}) = \mathbf{V}$. Then

$$\begin{aligned} &(a) \quad E(\mathbf{BY} + \mathbf{c}) = \mathbf{B}\boldsymbol{\mu} + \mathbf{c}, \\ &(b) \quad E(\mathbf{a}'\mathbf{Y}) = \mathbf{a}'\boldsymbol{\mu}, \\ &(c) \quad \mathrm{Cov}(\mathbf{BY} + \mathbf{c}) = \mathbf{BVB}', \\ &(d) \quad \mathrm{Var}(\mathbf{a}'\mathbf{Y}) = \mathbf{a}'\mathbf{Va} \ . \end{aligned}$$

Using Theorem 4.2.1, we can now find the mean vector and covariance matrix for $\mathbf{Y}$. Since $\mathbf{X}\boldsymbol{\beta}$ is in fact a constant vector and $\boldsymbol{\varepsilon}$ is a random vector, we find that

$$\begin{aligned} E(\mathbf{Y}) &= E(\mathbf{X}\boldsymbol{\beta} + \boldsymbol{\varepsilon}) \\ &= \mathbf{X}\boldsymbol{\beta} + E(\boldsymbol{\varepsilon}) \\ &= \mathbf{X}\boldsymbol{\beta} \ , \end{aligned} \tag{4.2.3}$$

and

$$\begin{aligned} \mathrm{Cov}(\mathbf{Y}) &= \mathrm{Cov}(\mathbf{X}\boldsymbol{\beta} + \boldsymbol{\varepsilon}) \\ &= \mathrm{Cov}(\boldsymbol{\varepsilon}) \\ &= \sigma^2\mathbf{I}. \end{aligned} \tag{4.2.4}$$

Next we turn our attention to $\hat{\beta}$, the random form of the normal equations solutions vector. Its mean vector and covariance matrix are given in the next theorem.

Theorem 4.2.2 If $\mathbf{Y} = \mathbf{X}\beta + \varepsilon$ where $E(\varepsilon) = \mathbf{0}$ and $\text{Cov}(\varepsilon) = \sigma^2\mathbf{I}$, then for $\hat{\beta} = (\mathbf{X}'\mathbf{X})^-\mathbf{X}'\mathbf{Y}$ we have

(a) $E(\hat{\beta}) = (\mathbf{X}'\mathbf{X})^-\mathbf{X}'\mathbf{X}\beta$,

(b) $\text{Cov}(\hat{\beta}) = \sigma^2(\mathbf{X}'\mathbf{X})^-\mathbf{X}'\mathbf{X}[(\mathbf{X}'\mathbf{X})^-]'$.

Proof. Note that $(\mathbf{X}'\mathbf{X})^-\mathbf{X}'$ in the expression $\hat{\beta} = (\mathbf{X}'\mathbf{X})^-\mathbf{X}'\mathbf{Y}$ is a constant matrix. Using Theorem 4.2.1 together with results (4.2.3) and (4.2.4) we obtain

$$
\begin{aligned}
E(\hat{\beta}) &= E[(\mathbf{X}'\mathbf{X})^-\mathbf{X}'\mathbf{Y}] \\
&= (\mathbf{X}'\mathbf{X})^-\mathbf{X}'E(\mathbf{Y}) \quad \text{from (4.2.2b)} \\
&= (\mathbf{X}'\mathbf{X})^-\mathbf{X}'\mathbf{X}\beta ,
\end{aligned}
$$

and

$$
\begin{aligned}
\text{Cov}(\hat{\beta}) &= \text{Cov}[(\mathbf{X}'\mathbf{X})^-\mathbf{X}'\mathbf{Y}] \\
&= (\mathbf{X}'\mathbf{X})^-\mathbf{X}' \ \text{Cov}(\mathbf{Y})[(\mathbf{X}'\mathbf{X})^-\mathbf{X}']' \\
&= (\mathbf{X}'\mathbf{X})^-\mathbf{X}'\sigma^2\mathbf{I}\mathbf{X}[(\mathbf{X}'\mathbf{X})^-]' \\
&= \sigma^2(\mathbf{X}'\mathbf{X})^-\mathbf{X}'\mathbf{X}[(\mathbf{X}'\mathbf{X})^-]' \ .
\end{aligned}
$$

The reader will recall the definition of an unbiased estimator from a previous course. We define an unbiased random vector in like manner.

Definition 4.2.1 The random vector $\hat{\beta}$ is said to be **unbiased** for the constant vector β if $E(\hat{\beta}) = \beta$.

The previous theorem indicates that $\hat{\beta}$ is not necessarily unbiased for the parameter vector in the general linear model (4.2.1). In general, then, we cannot think of $\hat{\beta}$ as being an estimator of β . But if $\mathbf{X}$ is a full rank matrix, then $\mathbf{X}'\mathbf{X}$ is also full rank so the ordinary inverse,

$(X'X)^{-1}$ exists. This means that in Theorem 4.2.2, $(X'X)^- \equiv (X'X)^{-1}$. As a result,

$$E(\hat{\beta}) = (X'X)^{-1}X'X\beta = \beta$$

so that in the full rank case $\hat{\beta} = (X'X)^{-1}X'Y$ is an unbiased estimator of β. Also $(X'X)^{-1}$ is symmetric since $X'X$ is symmetric. It follows that $[(X'X)^{-1}]' = (X'X)^{-1}$, and therefore, the Theorem 4.2.2(b) result reduces to

$$\begin{aligned}\mathrm{Cov}(\hat{\beta}) &= \sigma^2(X'X)^{-1}X'X(X'X)^{-1} \\ &= \sigma^2(X'X)^{-1} \ .\end{aligned}$$

To summarize, we present a corollary.

Corollary 4.2.2 Let $Y = X\beta + \varepsilon$ where $E(\varepsilon) = \mathbf{0}$ and $\mathrm{Cov}(\varepsilon) = \sigma^2 I$. If X is full rank, then for $\hat{\beta} = (X'X)^{-1}X'Y$

(a) $E(\hat{\beta}) = \beta$,

(b) $\mathrm{Cov}(\hat{\beta}) = \sigma^2(X'X)^{-1}$.

A Multiple Regression Example

As a result of Corollary 4.2.2, we know that in the multiple regression setting, $\hat{\beta}$ is unbiased for β and $\mathrm{Cov}(\hat{\beta}) = \sigma^2(X'X)^{-1}$. We illustrate with an example

Example 4.2.1 In Example 2.4.1 we posed the problem of relating energy usage to three prediction variables, namely, average outside temperature (x_1), thermostat setting (x_2), and square footage (x_3). The proposed model for the ten observations was

$$Y_i = \beta_0 + \beta_1 x_{i1} + \beta_2 x_{i2} + \beta_3 x_{i3} + \varepsilon_i, \quad i = 1,2,\ldots,10,$$

where the dependent variable is the cost of electric use. For the given data, the least squares function (rounded to three significant digits) turns out to be

$$\hat{y} = -112 - 1.32x_1 + 2.72x_2 + 0.00311x_3.$$

Therefore the unbiased estimates for β_0, β_1, β_2 and β_3 are -112, -1.32, 2.72 and 0.00311, respectively. To interpret, we can say that for a particular home, we estimate every one degree increase in average outside temperature decreases the January electricity cost by \$1.32 on the average. (Notice that x_2 and x_3 are fixed in this situation.) Similarly, we estimate that increasing the thermostat setting by one degree will result in an average increased January cost of \$2.72 for the home. Information concerning the variances and covariances of the estimators is contained in the $(\mathbf{X}'\mathbf{X})^{-1}$ matrix. Here to four significant digits

$$(\mathbf{X}'\mathbf{X})^{-1} = \begin{bmatrix} 4.759\times10^{2} & 3.166\times10^{-1} & -6.880\times10^{0} & -1.211\times10^{-2} \\ 3.166\times10^{-1} & 2.587\times10^{-3} & -4.588\times10^{-3} & -3.709\times10^{-5} \\ -6.880\times10^{0} & -4.588\times10^{-3} & 1.001\times10^{-1} & 1.538\times10^{-4} \\ -1.211\times10^{-2} & -3.709\times10^{-5} & 1.538\times10^{-4} & 1.359\times10^{-6} \end{bmatrix}$$

From Corollary 4.2.2(b), the variances of the estimators are $\text{Var}(\hat{\beta}_0) = 475.9\sigma^2$, $\text{Var}(\hat{\beta}_1) = 0.002587\sigma^2$, $\text{Var}(\hat{\beta}_2) = 0.1001\sigma^2$, and $\text{Var}(\hat{\beta}_3) = 0.000001359\sigma^2$. Choosing two of the six covariances we observe that $\text{Cov}(\hat{\beta}_0, \hat{\beta}_1) = 0.3166\sigma^2$ and $\text{Cov}(\hat{\beta}_1, \hat{\beta}_2) = -0.004588\sigma^2$. In the next chapter we will learn how to estimate the numerical value of the unknown parameter σ^2.

Less Than Full Rank Example

In the previous example we saw that in the full rank linear models case the least squares estimates are unique. This is not so in the less than full rank case. To illustrate, we again turn to a one—way classification problem.

Example 4.2.2. In Example 3.1.2, we considered the model

$$Y_{ij} = \mu + \tau_i + \varepsilon_{ij}, \quad i = 1,2,3; \quad j = 1,2,\ldots,n_i,$$

where $n_1 = 4$, $n_2 = 3$ and $n_3 = 2$. With the additional specifications that $E(\varepsilon) = \mathbf{0}$ and $Cov(\varepsilon) = \sigma^2 I$, it follows that $E(Y_{ij}) = \mu + \tau_i$ and $Var(Y_{ij}) = \sigma^2$. Thus in the one–way classification model we assume that all three populations (treatment groups) have the same variance. However, the population means are $\mu + \tau_1$, $\mu + \tau_2$ and $\mu + \tau_3$ which can differ according to the values of τ_1, τ_2 and τ_3. In Example 3.1.2, two of the infinitely many solutions to the normal equations were given. In random vector form, they are

$$\hat{\beta}_1' = (\bar{Y}_{..}, \bar{Y}_{1.} - \bar{Y}_{..}, \bar{Y}_{2.} - \bar{Y}_{..}, \bar{Y}_{3.} - \bar{Y}_{..})$$

and

$$\hat{\beta}_2' = (\bar{Y}_{3.}, \bar{Y}_{1.} - \bar{Y}_{3.}, \bar{Y}_{2.} - \bar{Y}_{3.}, 0).$$

Next, we find $E(\hat{\beta}_1)$ and $E(\hat{\beta}_2)$. First we note that

$$E(\bar{Y}_{..}) = \sum_{i=1}^{3} \sum_{j=1}^{n_i} E(Y_{ij})/9$$

$$= \frac{1}{9} \sum_{i=1}^{3} \sum_{j=1}^{n_i} (\mu + \tau_i)$$

$$= \mu + (4\tau_1 + 3\tau_2 + 2\tau_3)/9.$$

Also

$$E(\bar{Y}_{i.}) = \sum_{j=1}^{n_i} E(Y_{ij})/n_i$$

$$= \frac{1}{n_i} \sum_{j=1}^{n_i} (\mu + \tau_i)$$

$$= \mu + \tau_i \; .$$

Then

$$E(\hat{\beta}_1) = E\begin{bmatrix} \bar{Y}_{\cdot\cdot} \\ \bar{Y}_{1\cdot} - \bar{Y}_{\cdot\cdot} \\ \bar{Y}_{2\cdot} - \bar{Y}_{\cdot\cdot} \\ \bar{Y}_{3\cdot} - \bar{Y}_{\cdot\cdot} \end{bmatrix} = \begin{bmatrix} \mu + (4\tau_1 + 3\tau_2 + 2\tau_3)/9 \\ \tau_1 - (4\tau_1 + 3\tau_2 + 2\tau_3)/9 \\ \tau_2 - (4\tau_1 + 3\tau_2 + 2\tau_3)/9 \\ \tau_3 - (4\tau_1 + 3\tau_2 + 2\tau_3)/9 \end{bmatrix},$$

and

$$E(\hat{\beta}_2) = E\begin{bmatrix} \bar{Y}_{3\cdot} \\ \bar{Y}_{1\cdot} - \bar{Y}_{3\cdot} \\ \bar{Y}_{2\cdot} - \bar{Y}_{3\cdot} \\ 0 \end{bmatrix} = \begin{bmatrix} \mu + \tau_3 \\ \tau_1 - \tau_2 \\ \tau_2 - \tau_3 \\ 0 \end{bmatrix}.$$

Note that with $\beta' = (\mu, \tau_1, \tau_2, \tau_3)$, $E(\hat{\beta}_1) \neq \beta$ and $E(\hat{\beta}_2) \neq \beta$. That is, neither $\hat{\beta}_1$ and $\hat{\beta}_2$ are unbiased estimators for β. In fact, this example illustrates that in the less than full rank case $E(\hat{\beta})$ varies according to the solution of the normal equations and can be quite different from β.

In this section we have seen that the solutions vector $\hat{\beta}$ is an unbiased estimator of the parameter vector β in the full rank case but not in the less than full rank case. This leads to a natural question, namely, can anything be estimated in the latter case? If so, what? In the next section we obtain answers to these questions.

4.3 Estimable Functions

Our objective in this section is to discover what expressions involving β can be estimated in the less than full rank case. As a by—product, we will develop a class of functions of β that are very useful in statistical inference. To proceed, we first need a fundamental definition.

Definition 4.3.1 Consider the linear model $\mathbf{Y} = \mathbf{X}\beta + \varepsilon$ with $E(\varepsilon) = \mathbf{0}$. Let $\boldsymbol{\ell}$ be a $(k + 1) \times 1$ vector of constants and $\psi = \boldsymbol{\ell}'\beta$ be a linear combination of the parameters. We say that ψ is an **estimable function** if there exists a vector $\mathbf{c}$ of constants such that $E(\mathbf{c}'\mathbf{Y}) = \psi$ for all possible values of β.

We see that a linear function of the parameters, $\psi = \boldsymbol{\ell}'\beta$ is estimable if there is a linear combination of the observations Y, say $c'Y$, so that $c'Y$ is unbiased for ψ. In the case of several linear functions, we can extend the above definition as follows: If L is a $q \times (k + 1)$ matrix of constants, then $L\beta$ is a set of q estimable functions if there is a matrix C of constants so that $E(CY) = L\beta$ for all possible values of β.

To help discover what is estimable in the general linear model framework, we utilize the following useful three results.

Theorem 4.3.1 Consider the linear model $Y = X\beta + \varepsilon$ with $E(\varepsilon) = 0$. Then $\psi = \boldsymbol{\ell}'\beta$ is an estimable function if and only if there is a vector c of constants such that $\boldsymbol{\ell}' = c'X$. (Note that $\boldsymbol{\ell}' = c'X$ implies that $\boldsymbol{\ell}$ is a linear combination of the rows of X.) Moreover, for a $q \times (k + 1)$ matrix L, $L\beta$ is a set of estimable functions if and only if there is a matrix C of constants such that $L = CX$.

Proof. First suppose that $\psi = \boldsymbol{\ell}'\beta$ is estimable. Then by Definition 4.3.1 there is a vector of constants c so that $E(c'Y) = \boldsymbol{\ell}'\beta$. But $E(c'Y) = c'E(Y) = c'X\beta, = \boldsymbol{\ell}'\beta$. Since this last equation is true for all β, it follows that $c'X = \boldsymbol{\ell}'$. Next suppose that $\boldsymbol{\ell}' = c'X$. Then $\boldsymbol{\ell}'\beta = c'X\beta = E(c'Y)$ and by Definition 4.3.1, $\boldsymbol{\ell}'\beta$ is estimable. Similarly, if $E(CY) = L\beta$, then $CX\beta = L\beta$ for all β and thus $CX = L$. Likewise, if $L = CX$, then $L\beta = CX\beta = E(CY)$ and hence $L\beta$ is estimable. This completes the proof.

Corollary 4.3.1 In the linear model $Y = X\beta + \varepsilon$ with $E(\varepsilon) = 0$, $X\beta$ is a set of estimable functions.

Proof. From the previous theorem, $L\beta$ is a set of estimable functions if and only if $L = CX$. By taking C to be an identity matrix, we get $L = X$, and as a consequence $X\beta$ is estimable.

Theorem 4.3.2 Any linear combination of estimable functions is estimable, i.e., if $\psi_1, \psi_2, \cdots, \psi_q$ are estimable functions, then

$\psi = \sum_{i=1}^{q} h_i\psi_i$ is estimable where $h_1, h_2, \cdots, h_q$ are constants.

Proof. Let $\psi_i = \boldsymbol{\ell}_i'\beta$ for $i = 1, 2, \cdots, q$, then $\psi = \sum_{i=1}^{q} h_i\boldsymbol{\ell}_i'\beta$. By Definition 4.3.1, there is a vector c_i such that $E(c_i' Y) = \boldsymbol{\ell}_i'\beta$ for

$i = 1,2,\cdots,q$. Then $E[(\sum_{i=1}^{q} h_i \mathbf{c}_i')Y] = \sum_{i=1}^{q} h_i E(\mathbf{c}_i' Y) = \sum_{i=1}^{q} h_i \boldsymbol{\ell}_i' \beta = \psi$.

Hence, by Definition 4.3.1, ψ is estimable.

Finding Estimable Functions

Using the previous results immediately above, we can now present a simple procedure for finding all estimable functions. First find $\mathbf{X}\beta$. Each row of this vector is an estimable function. (Note that some rows may contain the same estimable function.) Then any linear combination of the rows of $\mathbf{X}\beta$ is also an estimable function. It can be shown (see for example Chapter 13 in Graybill (1976)) that all estimable functions can be expressed as a linear combination of the entries in $\mathbf{X}\beta$. Thus by finding $\mathbf{X}\beta$, and all linear combinations of the entries in $\mathbf{X}\beta$, we have found all possible estimable functions. We illustrate the procedure in the next example.

Example 4.3.1 Consider the one–way classification of Example 2.6.2 with three populations (treatments) and $n_1 = 4$, $n_2 = 3$, $n_3 = 2$. Let us find the estimable functions in this case. First we find $\mathbf{X}\beta$, namely,

$$\mathbf{X}\beta = \begin{bmatrix} \mu + \tau_1 \\ \mu + \tau_1 \\ \mu + \tau_1 \\ \mu + \tau_1 \\ \mu + \tau_2 \\ \mu + \tau_2 \\ \mu + \tau_2 \\ \mu + \tau_3 \\ \mu + \tau_3 \end{bmatrix}.$$

From this we identify the three basic estimable functions $\mu + \tau_1$, $\mu + \tau_2$ and $\mu + \tau_3$. Any linear combination of these three is estimable and any estimable function can be expressed as a linear combination of these three. Hence, for example, $\tau_1 - \tau_2$, $\tau_1 - \frac{1}{2}(\tau_2 + \tau_3)$ and $3\mu + 2\tau_1 + \tau_2$ all are estimable since they can each be expressed as a linear combination of $\mu+\tau_1$, $\mu+\tau_2$ and $\mu+\tau_3$. On the other hand, $\mu+\tau_1+\tau_2$ and $\tau_1+\tau_2$ are not estimable since they cannot be expressed as a linear combination of the three basic estimable functions. In particular, notice that μ, τ_1, τ_2 and τ_3 individually are not estimable.

To this point it is not totally clear why estimable functions are important. In the previous one–way classification example, it is strange that none of μ, τ_1, τ_2, τ_3 is estimable, yet certain functions of these parameters are estimable. Of what consequence is this? To answer the question, we will subsequently show that estimable functions are important in that they are functions that can be unbiasedly estimated using **any** solution to the normal equations. First we define some notation.

Definition 4.3.2 Let $\psi = \boldsymbol{\ell}'\boldsymbol{\beta}$ be an estimable function and let $\mathbf{b}$ be any solution to the normal equations. We call $\boldsymbol{\ell}'\mathbf{b}$ the **least squares estimate** for ψ since it is derived from the least squares process. The symbol $\hat{\psi} = \boldsymbol{\ell}'\hat{\boldsymbol{\beta}}$ is called the **least squares estimator** for ψ, where $\hat{\boldsymbol{\beta}}$ is the random form of $\mathbf{b}$.

In the above definition, we implied that $\boldsymbol{\ell}'\mathbf{b}$ is the one and only least squares estimate for $\boldsymbol{\ell}'\boldsymbol{\beta}$. This is not at all obvious since in some cases there can be infinitely many solutions $\mathbf{b}$ to the normal equations. The next theorem verifies that indeed the least squares estimate $\boldsymbol{\ell}'\mathbf{b}$ is unique.

Theorem 4.3.3 Let $\psi = \boldsymbol{\ell}'\boldsymbol{\beta}$ be an estimable function. The least squares estimate $\boldsymbol{\ell}'\mathbf{b}$ is invariant to $\mathbf{b}$, that is, $\boldsymbol{\ell}'\mathbf{b}$ has the same value for every solution $\mathbf{b}$ to the normal equations.

Proof. From Theorem 4.3.1 we have $\boldsymbol{\ell}' = \mathbf{c}'\mathbf{X}$ and from Theorem 3.2.2, $\mathbf{b} = (\mathbf{X}'\mathbf{X})^{-}\mathbf{X}'\mathbf{y}$. Using these results we get

$$\boldsymbol{\ell}'\mathbf{b} = \mathbf{c}'\mathbf{X}(\mathbf{X}'\mathbf{X})^{-}\mathbf{X}'\mathbf{y}\,.$$

Since $\mathbf{X}(\mathbf{X}'\mathbf{X})^{-}\mathbf{X}'$ is invariant to $(\mathbf{X}'\mathbf{X})^{-}$ by Theorem 3.2.3, it follows that $\mathbf{c}'\mathbf{X}(\mathbf{X}'\mathbf{X})^{-}\mathbf{X}'\mathbf{y}$ has the same value for all $(\mathbf{X}'\mathbf{X})^{-}$, and hence, for all $\mathbf{b}$. Thus $\boldsymbol{\ell}'\mathbf{b}$ is invariant to $(\mathbf{X}'\mathbf{X})^{-}$, and as a consequence, it is invariant to $\mathbf{b}$.

Next we show that the least squares estimator for an estimable function is unbiased and obtain its variance.

Theorem 4.3.4 Let $\psi = \boldsymbol{\ell}'\beta$ be an estimable function and let $\hat{\psi} = \boldsymbol{\ell}'\hat{\beta}$ be its least squares estimator. Then $\hat{\psi}$ is unbiased for ψ. Moreover, if $\text{Cov}(\varepsilon) = \sigma^2 I$ then $\text{Var}(\hat{\psi}) = \sigma^2 \boldsymbol{\ell}'(X'X)^- \boldsymbol{\ell}$, which is invariant to the choice of the g—inverse.

Proof. Since $\hat{\beta} = (X'X)^- X'Y$ and $\boldsymbol{\ell}' = c'X$ we have

$$
\begin{aligned}
E(\boldsymbol{\ell}'\hat{\beta}) &= \boldsymbol{\ell}' E(\hat{\beta}) \\
&= \boldsymbol{\ell}'(X'X)^- X'X\beta \qquad \text{(Theorem 4.2.2)} \\
&= c'X(X'X)^- X'X\beta \\
&= c'X\beta \qquad \text{(Theorem 3.2.3)} \\
&= \boldsymbol{\ell}'\beta
\end{aligned}
$$

This verifies that $\boldsymbol{\ell}'\hat{\beta}$ is unbiased for $\boldsymbol{\ell}'\beta$. Next, since $\text{Cov}(Y) = \sigma^2 I$,

$$
\begin{aligned}
\text{Var}(\boldsymbol{\ell}'\hat{\beta}) &= \text{Var}(\boldsymbol{\ell}'(X'X)^- X'Y) \\
&= \boldsymbol{\ell}'(X'X)^- X'\sigma^2 IX[(X'X)^-]'\boldsymbol{\ell} \qquad \text{(Theorem 4.2.1)} \\
&= \sigma^2 c'X(X'X)^- X'X[(X'X)^-]'X'c \\
&= \sigma^2 c'X[(X'X)^-]'X'c \qquad \text{(Theorem 3.2.3)} \\
&= \sigma^2 c'X(X'X)^- X'c \qquad \text{(Theorem 3.2.3)} \\
&= \sigma^2 \boldsymbol{\ell}'(X'X)^- \boldsymbol{\ell}.
\end{aligned}
$$

The above theorem is most important. This follows from the fact that $\boldsymbol{\ell}'\hat{\beta}$ is not always unbiased for an arbitrary function $\boldsymbol{\ell}'\beta$. Recall from Theorem 4.2.2, in general

$$E(\boldsymbol{\ell}'\hat{\beta}) = \boldsymbol{\ell}' E(\hat{\beta}) = \boldsymbol{\ell}'(X'X)^- X'X\beta$$

which does not necessarily equal $\boldsymbol{\ell}'\beta$. Thus the case when $\boldsymbol{\ell}'\beta$ is estimable, and hence when $\boldsymbol{\ell}'\hat{\beta}$ is unbiased, is a very special case.

It should now be clear why estimable functions are important. If a function $\psi = \boldsymbol{\ell}'\beta$ is an estimable function, then the least squares estimator $\hat{\psi} = \ell'\hat{\beta}$ is unique and it is unbiased for ψ. Thus to obtain an estimable function we can find any solution b to the normal equations and be assured that ℓ'b provides an unbiased point estimate for the function. Further, the value of ℓ'b is the same (invariant) for all solutions to the normal equations. One should realize that when a function $\psi = \ell'\beta$ is not estimable, then $\ell' b$ is not the same for all solutions b, nor is $\ell'\hat{\beta}$ unbiased. The next example illustrates these points.

Example 4.3.2 Consider the one—way classification with three treatments and $n_1 = 4$, $n_2 = 3$, $n_3 = 2$. In Example 4.3.1 we showed that, among other things, $\mu + \tau_1$ and $\tau_1 - \tau_2$ are estimable and that τ_1 is not estimable. In Example 3.1.2 we noted that two solutions of the normal equations are:

Solution 1: $m = \bar{y}_{..}$, $t_1 = \bar{y}_{1.} - \bar{y}_{..}$, $t_2 = \bar{y}_{2.} - \bar{y}_{..}$,

$t_3 = \bar{y}_{3.} - \bar{y}_{..}$

Solution 2: $m = \bar{y}_{3.}$, $t_1 = \bar{y}_{1.} - \bar{y}_{3.}$, $t_2 = \bar{y}_{2.} - \bar{y}_{3.}$, $t_3 = 0$.

First note that for the non—estimable function τ_1, the value of τ_1 is different for the two solutions. However, for the estimable function $\mu + \tau_1$, the least squares estimate $m + t_1$ equals $\bar{y}_{1.}$ for both solutions. Similarly, for $\tau_1 - \tau_2$ the least squares estimate $t_1 - t_2$ equals $\bar{y}_{1.} - \bar{y}_{2.}$ for both solutions. We know, of course, that since $\mu + \tau_1$ is estimable, the quantity $m + t_1$ will equal $\bar{y}_{1.}$ for any solution. Similarly, $t_1 - t_2$ will always equal $\bar{y}_{1.} - \bar{y}_{2.}$. Thus the unbiased least squares estimator for $\mu + \tau_1$, denoted by $\widehat{\mu + \tau_1}$, is $\bar{Y}_{1.}$, and the unbiased least squares estimator for $\tau_1 - \tau_2$ is $\widehat{\tau_1 - \tau_2} = \bar{Y}_{1.} - \bar{Y}_{2.}$.

Full Rank Case

In the full rank case, finding estimable functions of the parameters does not present a problem. To see why, let us recall that in the full rank case, $\hat{\beta} = (X'X)^{-1}X'Y$ and from Corollary 4.2.2, $E[(X'X)^{-1}X'Y] = \beta$. By letting $C = (X'X)^{-1}X'$, we see that $E(CY) = \beta$. This implies that β is a vector of estimable functions. Thus, every element in β is estimable. Further, since $\ell'\beta$ is a linear combination of the elements of β, it follows that $\ell'\beta$ is estimable for all constant vectors ℓ. We formally state this as a theorem.

Theorem 4.3.5 Consider the linear model $Y = X\beta + \varepsilon$ with $E(\varepsilon) = 0$. If the model is full rank (e.g., a multiple regression model), then every element of β is estimable and $\ell'\beta$ is estimable for any vector of constants ℓ.

Summary

In this section we have introduced a class of functions, namely the estimable functions, for which the least squares estimator is unbiased. Equally important is the fact that we can use any solution of the normal equations to find the unique value of the least squares estimate. In this sense we have found a class of functions which can be easily estimated.

While unbiasedness is generally a desirable property for an estimator, it does not guarantee that it is the "best" estimator. Is the least squares estimator the best estimator in some sense? In the next section we address this issue.

4.4. Gauss—Markoff Theorem

In the previous section we discovered that the least squares estimator $\hat{\psi} = \ell'\hat{\beta}$ is an unbiased estimator for an estimable function $\psi = \ell'\beta$. We now proceed to show that this least squares estimator is the best estimator for ψ in a certain sense.

Definition 4.4.1 An estimator is said to be a **linear estimator** if it is a linear combination of the observations, $\mathbf{Y}$. That is, a linear estimator is of the form $\mathbf{a}'Y$ where $\mathbf{a}$ is a vector of constants.

BLUE Estimators

We now show that for any estimable function, the least squares estimator is the linear estimator that is unbiased and has the smallest variance among all linear unbiased estimators. For this reason we call the least squares estimator, $\hat{\psi}$, **the best linear unbiased estimator (BLUE)** for ψ. This result, which we now state and prove, is called the Gauss—Markoff Theorem.

Theorem 4.4.1 (Gauss—Markoff) Consider the linear model $Y = X\beta + \varepsilon$, where $E(\varepsilon) = 0$ and $Cov(\varepsilon) = \sigma^2 I$. Let $\psi = \ell'\beta$ be an estimable function. In the class of all linear unbiased estimators of ψ, the least squares estimator $\hat{\psi} = \ell'\hat{\beta}$ has minimum variance and it is the only linear unbiased estimator with this minimal variance.

Proof. Suppose that $m'Y$ is some other linear unbiased estimator for ψ. Then $E(m'Y) = \ell'\beta$, and $E(m'Y) = m'E(Y) = m'X\beta$ for all β. Hence $\ell'\beta = m'X\beta$ for all β and, as a consequence, $\ell' = m'X$. Next, we proceed to show that $Var(\ell'\hat{\beta}) \leq Var(m'Y)$.

$$Cov(\ell'\hat{\beta}, m'Y) = Cov[\ell'(X'X)^- X'Y, m'Y]$$

$$= E\{[\ell'(X'X)^- X'Y - \ell(X'X)^- X'X\beta]\,[m'Y - m'X\beta]'\}$$

$$= \ell'(X'X)^- X'E[(Y - X\beta)(Y - X\beta)']m$$

$$= \ell'(X'X)^- X' Cov(Y)\, m$$

$$= \sigma^2 \ell'(X'X)^- X'm$$

$$= \sigma^2 \ell'(X'X)^- \ell$$

$$= Var(\ell'\hat{\beta}) . \qquad \text{(from Theorem 4.3.4)}$$

Now,

$$Var(\ell'\hat{\beta} - m'Y) = Var(\ell'\hat{\beta}) + Var(m'Y) - 2\,Cov(\ell'\hat{\beta}, m'Y)$$

$$= Var(m'Y) - Var(\ell'\hat{\beta}).$$

Since $Var(\ell'\hat{\beta} - m'Y) \geq 0$, it follows that

$$Var(m'Y) \geq Var(\ell'\hat{\beta}).$$

This verifies that $\boldsymbol{\ell}'\hat{\beta}$ has minimum variance among all linear unbiased estimators. To verify that $\boldsymbol{\ell}'\hat{\beta}$ is the only linear unbiased estimator with minimum variance, we suppose that $\text{Var}(\mathbf{m}'\mathbf{Y}) = \text{Var}(\boldsymbol{\ell}'\hat{\beta})$. Then $\text{Var}(\boldsymbol{\ell}'\hat{\beta} - \mathbf{m}'\mathbf{Y}) = 0$, and hence $\boldsymbol{\ell}'\beta - \mathbf{m}'\mathbf{Y}$ is a constant, say c. Since $E(\boldsymbol{\ell}'\hat{\beta} - \mathbf{m}'\mathbf{Y}) = 0$, it follows that $c = 0$ and $\boldsymbol{\ell}'\hat{\beta} = \mathbf{m}'\mathbf{Y}$. Thus any linear unbiased estimator for $\boldsymbol{\ell}'\beta$ which has minimum variance must equal $\boldsymbol{\ell}'\hat{\beta}$. This completes the proof.

Now we know how to find best (BLUE) estimators for less than full rank linear models such as experimental design models. For any estimable function $\psi = \boldsymbol{\ell}'\beta$, we find any solution to the normal equations, the random form of which is denoted by $\hat{\beta}$. The least squares estimator $\hat{\psi} = \boldsymbol{\ell}'\hat{\beta}$ will have the following properties.

(1) $\hat{\psi}$ is the same for all solutions.

(2) $\hat{\psi}$ is BLUE for $\boldsymbol{\ell}'\beta$.

(3) $\text{Var}(\hat{\psi}) = \sigma^2\boldsymbol{\ell}'(\mathbf{X}'\mathbf{X})^-\boldsymbol{\ell}$, invariant to choice of $(\mathbf{X}'\mathbf{X})^-$. We say, of course, that $\boldsymbol{\ell}'\mathbf{b}$ is the best linear unbiased **estimate** for $\psi = \boldsymbol{\ell}'\beta$. The next two examples illustrate the usefulness of the Gauss–Markoff Theorem.

Example 4.4.1 Refer to the secchi disk data given in Example 2.6.1 which is an example of a one–way classification with three populations, sometimes called treatments. Our model for this example is

$$Y_{ij} = \mu + \tau_i + \varepsilon_{ij}, \ i = 1, 2, 3; \ \ j = 1, 2, 3, 4, 5,$$

which can be written as a linear model $\mathbf{Y} = \mathbf{X}\beta + \boldsymbol{\varepsilon}$, where $\beta' = (\mu, \tau_1, \tau_2, \tau_3)$. We again assume that $E(\boldsymbol{\varepsilon}) = \mathbf{0}$ and $\text{Cov}(\boldsymbol{\varepsilon}) = \sigma^2\mathbf{I}$. To find the estimable functions, we first find the entries in $\mathbf{X}\beta$. Since $E(Y_{ij}) = \mu + \tau_i$ for $i = 1, 2, 3$, it follows that the three distinct entries (basic estimable functions) are $\mu + \tau_1$, $\mu + \tau_2$ and $\mu + \tau_3$. As a consequence $\tau_3 - \tau_1 = (\mu + \tau_3) - (\mu + \tau_1)$ is estimable. We shall find the BLUE for the functions $\mu + \tau_1$ and $\tau_3 - \tau_1$.

In the exercises we verify that the generalized inverse

$$(\mathbf{X}'\mathbf{X})^{-} = \begin{bmatrix} 1/5 & -1/5 & -1/5 & 0 \\ -1/5 & 2/5 & 1/5 & 0 \\ -1/5 & 1/5 & 2/5 & 0 \\ 0 & 0 & 0 & 0 \end{bmatrix}$$

yields the solution set

$$\mathbf{b} = \begin{bmatrix} m \\ t_1 \\ t_2 \\ t_3 \end{bmatrix} = \begin{bmatrix} \bar{y}_{3\cdot} \\ \bar{y}_{1\cdot} - \bar{y}_{3\cdot} \\ \bar{y}_{2\cdot} - \bar{y}_{3\cdot} \\ 0 \end{bmatrix} .$$

It is now clear that the least squares estimators for $\mu + \tau_1$, and $\tau_3 - \tau_1$ are, respectively,

$$\widehat{\mu + \tau_1} = \bar{Y}_{3\cdot} + \bar{Y}_{1\cdot} - \bar{Y}_{3\cdot} = \bar{Y}_{1\cdot} \; .$$

and

$$\widehat{\tau_3 - \tau_1} = 0 - (\bar{Y}_{1\cdot} - \bar{Y}_{3\cdot}) = \bar{Y}_{3\cdot} - \bar{Y}_{1\cdot} \; .$$

According to the Gauss—Markoff Theorem, these are the best linear unbiased estimators for the two functions. By referring to the actual data values, we can make several calculations. First we find that $\bar{y}_{1\cdot} = 38.8$, $\bar{y}_{2\cdot} = 45.8$ and $\bar{y}_{3\cdot} = 47.4$. Then the numerical value of the BLUE for $\mu + \tau_1$ is 38.8 which is an estimate for the true mean secchi disk reading for location 1 at the lake. Similarly, the numerical BLUE for $\tau_3 - \tau_1$ is $47.4 - 38.8 = 8.6$ and is an estimate for the difference in true mean secchi disk readings at locations 3 and 1. From the formula $\text{Var}(\hat{\psi}) = \boldsymbol{\ell}'(\mathbf{X}'\mathbf{X})^{-}\boldsymbol{\ell}\sigma^2$, the reader can verify that

$\text{Var}(\widehat{\mu + \tau_1}) = \sigma^2/5$ using $\boldsymbol{\ell}' = (1\ 1\ 0\ 0)$, and

$\text{Var}(\widehat{\tau_3 - \tau_1}) = 2\sigma^2/5$ using $\boldsymbol{\ell}' = (0\ {-1}\ 0\ 1)$.

Notice that $\text{Var}(\widehat{\mu + \tau_1}) = \text{Var}(\bar{Y}_{1\cdot})$ and $\text{Var}(\widehat{\tau_3 - \tau_1}) = \text{Var}(\bar{Y}_{3\cdot} - \bar{Y}_{1\cdot})$ where $\text{Cov}(\bar{Y}_{1\cdot}, \bar{Y}_{3\cdot}) = 0$ so these last two results are not unexpected. Later we will learn that an unbiased estimate for the parameter σ^2 is 37/30.

Computer Example for Less than Full Rank Case

There are several commonly used statistical packages that can be used to help analyze experimental data. One program we will refer to in this text is SAS, the acronym for Statistical Analysis System. In Table 4.4.1 we have listed the SAS program commands and the resulting printout for the secchi disk problem. The program used, called GLM, is a general linear model program and uses a generalized inverse to solve the normal equations. In this regard, see the note at the end of the printout.

Table 4.4.1 SAS Commands and Printout for Secchi Disk Data

SAS Commands

```
OPTIONS LS = 80
DATA SECCHI;
INPUT LOCATION READING
CARDS;
1 39 1 37 1 40 1 39 1 39
2 46 2 45 2 46 2 45 2 47
3 48 3 46 3 48 3 46 3 49
PROC PRINT;
PROC GLM;
CLASS LOCATION;
MODEL READING = LOCATION/I SOLUTION;
MEANS LOCATION;
```

SAS Printout

OBS	LOCATION	READING
1	1	39
2	1	37
3	1	40
4	1	39
5	1	39
6	2	46
7	2	45
8	2	46
9	2	45
10	2	47
11	3	48
12	3	46
13	3	48
14	3	46
15	3	49

General Linear Models Procedure
Class Level Information

Class	Levels	Values
LOCATION	3	1 2 3

Number of observations in data set = 15

General Linear Models Procedure

X'X Generalized Inverse (g2)

	INTERCEPT	LOCATION 1	LOCATION 2	LOCATION 3	READING
INTERCEPT	0.2	-0.2	-0.2	0	47.4
LOCATION 1	-0.2	0.4	0.2	0	-8.6
LOCATION 2	-0.2	0.2	0.4	0	-1.6
LOCATION 3	0	0	0	0	0
READING	47.4	-8.6	-1.6	0	14.8

```
                  General Linear Models Procedure

Dependent Variable: READING
                                      Sum of            Mean
Source                   DF          Squares          Square   F Value     Pr > F

Model                     2     209.20000000    104.60000000     84.81     0.0001

Error                    12      14.80000000      1.23333333

Corrected Total          14     224.00000000

               R-Square             C.V.        Root MSE           READING Mean

               0.933929         2.523990       1.1105554              44.000000


Source                   DF        Type I SS     Mean Square   F Value     Pr > F

LOCATION                  2     209.20000000    104.60000000     84.81     0.0001

Source                   DF      Type III SS     Mean Square   F Value     Pr > F

LOCATION                  2     209.20000000    104.60000000     84.81     0.0001

                                         T for H0:    Pr > |T|    Std Error of
Parameter                 Estimate     Parameter=0                  Estimate

INTERCEPT            47.40000000 B           95.44      0.0001      0.49665548
LOCATION    1        -8.60000000 B          -12.24      0.0001      0.70237692
            2        -1.60000000 B           -2.28      0.0418      0.70237692
            3         0.00000000 B             .          .           .

NOTE: The X'X matrix has been found to be singular and a generalized inverse
      was used to solve the normal equations.   Estimates followed by the
      letter 'B' are biased, and are not unique estimators of the parameters.


               General Linear Models Procedure

         Level of          -----------READING-----------
         LOCATION     N        Mean                SD

         1            5     38.8000000          1.09544512
         2            5     45.8000000          0.83666003
         3            5     47.4000000          1.34164079
```

The estimate part of the printout displays the solution to the normal equations as

$$b' = (m, t_1, t_2, t_3) = (47.4, -8.6, -1.6, 0).$$

Essentially this is the solution one gets if we impose the side condition $t_3 = 0$ on the normal equations and is identical to the solution set arrived at in Example 4.1.1. None of the values in **b** are unbiased point estimates for the parameters in $\boldsymbol{\beta}$ because none of μ, τ_1, τ_2 or τ_3 is estimable. For that reason the letter B appears next to each entry in **b** on the printout. Note, however, that the least squares estimate for $\mu + \tau_1$ is $m + t_1 = 47.4 - 8.6 = 38.8$ and the least squares estimate for $\tau_3 - \tau_1$ is $t_3 - t_1 = 0 - (-8.6) = 8.6$. These values coincide with the values obtained previously since $\mu + \tau_1$ and $\tau_3 - \tau_1$ are estimable quantities and their estimates are invariant to the solution to the normal equations. We close this section with the SAS printout for a multiple regression problem, the full rank case.

Full Rank Case Computer Example

Refer to the energy usage survey data for homes presented in Example 2.4.1 and the initial analysis of these data given in Example 4.2.1. Recall that we used the multiple regression model

$$Y_i = \beta_0 + \beta_1 x_{i1} + \beta_2 x_{i2} + \beta_3 x_{i3} + \varepsilon_i,\ i = 1,2,\ldots,10,$$

where Y represents the electric costs for a home, x_1 is the outside temperature, x_2 is the thermostat setting and x_3 is home square footage. To analyze these data with the above observation structure and assumptions $E(\boldsymbol{\varepsilon}) = \mathbf{0}$ and $Cov(\boldsymbol{\varepsilon}) = \sigma^2\mathbf{I}$, we use the SAS statistical software package. Table 4.4.2 gives the computer commands and the corresponding printout for the analysis. On the printout there is no message about using a generalized inverse or about estimates being biased as in our previous (less than full rank) SAS illustration. The reason for this is the fact that the usual (unique) inverse of $\mathbf{X'X}$ is utilized to solve the normal equations. Among the displays in the printout are the $(\mathbf{X'X})^{-1}$ matrix and the solutions vector **b** given earlier, after rounding, in Example 4.2.1.

Table 4.4.2 Commands and Printout for Home Energy Data

SAS Commands

```
OPTIONS LS = 80;
DATA ENERGY;
INPUT ELCOST AVTEMP THERM SQFT;
CARDS;
35.09  29.0  65  2125
54.59  20.0  67  2350
43.82  31.5  68  2325
38.61  22.0  67  1825
30.62  36.5  66  2550
30.25  37.5  67  2100
47.15  18.0  65  2400
66.59  13.5  68  1325
31.41  29.5  68  2050
50.24  17.5  66  1575
PROC PRINT;
PROC GLM;
MODEL ELCOST = AVTEMP THERM SQFT/I XPX P;
```

SAS Printout

OBS	ELCOST	AVTEMP	THERM	SQFT
1	35.09	29.0	65	2125
2	54.59	20.0	67	2350
3	43.82	31.5	68	2325
4	38.61	22.0	67	1825
5	30.62	36.5	66	2550
6	30.25	37.5	67	2100
7	47.15	18.0	65	2400
8	66.59	13.5	68	1325
9	31.41	29.5	68	2050
10	50.24	17.5	66	1575

General Linear Models Procedure

The X'X Matrix

	INTERCEPT	AVTEMP	THERM	SQFT	ELCOST
INTERCEPT	10	255	667	20625	428.37
AVTEMP	255	7138.5	17011.5	542962.5	10144.625
THERM	667	17011.5	44501	1374400	28597.27
SQFT	20625	542962.5	1374400	43885625	861713.75
ELCOST	428.37	10144.625	28597.27	861713.75	19642.9439

General Linear Models Procedure

X'X Inverse Matrix

	INTERCEPT	AVTEMP	THERM	SQFT	ELCOST
INTERCEPT	475.88547999	0.3165926566	-6.879711323	-0.012112411	-111.6164051
AVTEMP	0.3165926566	0.0025869609	-0.004588483	-0.000037095	-1.320774863
THERM	-6.879711323	-0.004588483	0.1001435968	0.0001537647	2.7242702347
SQFT	-0.012112411	-0.000037095	0.0001537647	1.3586549E-6	0.0031148313
ELCOST	-111.6164051	-1.320774863	2.7242702347	0.0031148313	264.04467617

General Linear Models Procedure

Dependent Variable: ELCOST

Source	DF	Sum of Squares	Mean Square	F Value	Pr > F
Model	3	1028.8135338	342.9378446	7.79	0.0171
Error	6	264.0446762	44.0074460		
Corrected Total	9	1292.8582100			

R-Square	C.V.	Root MSE	ELCOST Mean
0.795767	15.48617	6.6338108	42.837000

Source	DF	Type I SS	Mean Square	F Value	Pr > F
AVTEMP	1	953.68713223	953.68713223	21.67	0.0035
THERM	1	67.98538755	67.98538755	1.54	0.2603
SQFT	1	7.14101405	7.14101405	0.16	0.7010

Source	DF	Type III SS	Mean Square	F Value	Pr > F
AVTEMP	1	674.32261720	674.32261720	15.32	0.0079
THERM	1	74.11006343	74.11006343	1.68	0.2420
SQFT	1	7.14101405	7.14101405	0.16	0.7010

Parameter	Estimate	T for H0: Parameter=0	Pr > |T|	Std Error of Estimate
INTERCEPT	-111.6164051	-0.77	0.4698	144.7152534
AVTEMP	-1.3207749	-3.91	0.0079	0.3374100
THERM	2.7242702	1.30	0.2420	2.0993008
SQFT	0.0031148	0.40	0.7010	0.0077325

Observation	Observed Value	Predicted Value	Residual
1	35.09000000	33.77770553	1.31229447
2	54.59000000	51.81405681	2.77594319
3	43.82000000	39.27154533	4.54845467
4	38.61000000	47.53722067	-8.92722067
5	30.62000000	27.91996758	2.70003242
6	30.25000000	27.92178889	2.32821111
7	47.15000000	49.16280763	-2.01280763
8	66.59000000	59.93066161	6.65933839
9	31.41000000	41.05651646	-9.64651646
10	50.24000000	49.97772950	0.26227050

Sum of Residuals	-0.00000000
Sum of Squared Residuals	264.04467617
Sum of Squared Residuals - Error SS	0.00000000
First Order Autocorrelation	-0.48102888
Durbin-Watson D	2.95527518

Example 4.4.2 Referring to the previous SAS printout, based on the data set, the best linear unbiased estimates for β_0, β_1, β_2 and β_3 in the energy usage regression model are $b_0 = -111.6164$, $b_1 = -1.3208$, $b_2 = 2.7243$ and $b_3 = 0.0031$, respectively. With these values, we are now in a position to find the BLUE for the mean value of Y, $E(Y) = \boldsymbol{\ell}'\boldsymbol{\beta}$, when x_1, x_2 and x_3 are given. If we set $x_1 = 32$, $x_2 = 70$ and $x_3 = 2000$, then

$$E(Y) = \beta_0 + 32\beta_1 + 70\beta_2 + 2000\beta_3 .$$

The BLUE for this expression is $b_0 + 32b_1 + 70b_2 + 2000b_3$ which equals $-111.6164 + 32(-1.3208) + 70(2.7243) + 2000(0.0031) = 43.02$.

Hence our best estimate for the mean January electric costs for 2000 square foot homes with a 70 degrees thermostat setting and average outside temperature of 32 degrees is \$43.02 per week. In the next chapter we will learn how to obtain an interval estimate for E(Y). This requires first finding an estimate for $\text{Var}(\boldsymbol{\ell}'\hat{\boldsymbol{\beta}}) = \sigma^2\boldsymbol{\ell}'(\mathbf{X}'\mathbf{X})^{-1}\boldsymbol{\ell}$ where, for this problem, $\boldsymbol{\ell}' = (1, 32, 70, 2000)$. Here $\sigma^2\boldsymbol{\ell}'(\mathbf{X}'\mathbf{X})^{-1}\boldsymbol{\ell} = 1.08\sigma^2$.

4.5 The Multivariate Normal Distribution

In order to proceed further in statistical inference for linear models, we need to become acquainted with the multivariate normal distribution and several basic theorems associated with this distribution. First we define what it means for a random vector $\mathbf{Y}$ to have a multivariate normal distribution.

Definition 4.5.1 Let $\mathbf{Y}$ be an $n \times 1$ random vector. We say that $\mathbf{Y}$ has a (non–singular) **multivariate normal distribution** with mean vector $\boldsymbol{\mu}$ and covariance matrix $\mathbf{V}$ if its elements have joint density given by

$$f(y_1, y_2, \ldots, y_n) = \frac{1}{(2\pi)^{n/2}|V|^{1/2}} e^{-\frac{1}{2}(y-\boldsymbol{\mu})'V^{-1}(y-\boldsymbol{\mu})}, \qquad (4.5.1)$$

where $|\mathbf{V}|$ denotes the determinant of matrix $\mathbf{V}$.

Note that the multivariate normal distribution for the $n \times 1$ random vector $\mathbf{Y}$ depends only on $\boldsymbol{\mu}$ and $\mathbf{V}$. We will use the notation $MN(\boldsymbol{\mu},\mathbf{V})$ to represent the multivariate normal distribution (4.5.1). Since $\mathbf{V}$ is a covariance matrix, it must be a positive definite symmetric matrix. Next we state some properties of the multivariate normal that will be useful for us.

Let us consider the $n \times 1$ random vector $\mathbf{Y}$ which is $MN(\boldsymbol{\mu},\mathbf{V})$. First, since $\mathbf{Y}$ is multivariate normal, each component of $\mathbf{Y}' = (Y_1,Y_2,\ldots,Y_n)$ is univariate normal. That is, each Y_i is univariate normal with mean μ_i, the ith component of $\boldsymbol{\mu}$, and variance σ_i^2, the ith diagonal element of $\mathbf{V}$. We denote this by Y_i is $N(\mu_i,\sigma_i^2)$. Here the symbol N represents univariate normal. Second, the moment generating function for $\mathbf{Y}$ is denoted by $M_{\mathbf{Y}}(\mathbf{t})$ where $\mathbf{t}' = (t_1,t_2,\ldots,t_n)$ and is defined to be $E(e^{\mathbf{t}'\mathbf{Y}})$. In most standard multivariate statistics textbooks, such as Anderson (1984), it is shown that for the multivariate normal (4.5.1)

$$M_{\mathbf{Y}}(\mathbf{t}) = e^{\mathbf{t}'\boldsymbol{\mu} + \frac{1}{2}\mathbf{t}'\mathbf{V}\mathbf{t}} . \tag{4.5.2}$$

Example 4.5.1 Recall from a previous course in mathematical statistics, the theoretical correlation coefficient for two random variables, Y_1 and Y_2, is defined as

$$\rho = \frac{\mathrm{Cov}(Y_1,Y_2)}{\sigma_1\sigma_2} \quad \text{so that} \quad \mathrm{Cov}(Y_1,Y_2) = \rho\sigma_1\sigma_2 .$$

Hence, for the **bivariate** normal distribution,

$$\mathbf{y} - \boldsymbol{\mu} = \begin{bmatrix} y_1 - \mu_1 \\ y_2 - \mu_2 \end{bmatrix} \quad \text{and} \quad \mathbf{V} = \begin{bmatrix} \sigma_1^2 & \rho\sigma_1\sigma_2 \\ \rho\sigma_1\sigma_2 & \sigma_2^2 \end{bmatrix}$$

As a result, in the bivariate normal case

$$|\mathbf{V}|^{1/2} = \sqrt{|\mathbf{V}|} = \sigma_1\sigma_2\sqrt{1-\rho^2} ,$$

and the expression $(\mathbf{y}-\boldsymbol{\mu})'\mathbf{V}^{-1}(\mathbf{y}-\boldsymbol{\mu})$ becomes

$$\frac{1}{1-\rho^2}\left\{\left(\frac{y_1-\mu_1}{\sigma_1}\right)^2 - 2\rho\left(\frac{y_1-\mu_1}{\sigma_1}\right)\left(\frac{y_2-\mu_2}{\sigma_2}\right) + \left(\frac{y_2-\mu_2}{\sigma_2}\right)^2\right\}.$$

The reader should work out the details.

Linear Transformation Results

The next result pertains to linear transformations of the multivariate normal random vector $\mathbf{Y}$. From Theorem 4.2.1, we already know that if $\mathbf{B}$ is a constant matrix and $\mathbf{c}$ is a constant vector, then the random vector $\mathbf{BY}+\mathbf{c}$ has mean vector $\mathbf{B}\boldsymbol{\mu}+\mathbf{c}$ and covariance matrix $\mathbf{BVB}'$ when $E(\mathbf{Y})=\boldsymbol{\mu}$ and $\text{Cov}(\mathbf{Y})=\mathbf{V}$. We now discover that if, in addition, $\mathbf{Y}$ is multivariate normal, the transformation $\mathbf{BY}+\mathbf{c}$ is also multivariate normal.

Theorem 4.5.1 Let $\mathbf{Y}$ have the multivariate normal distribution given by (4.5.1). Let $\mathbf{B}$ be a $q \times n$ matrix of constants with rank q and $\mathbf{c}$ be a $q \times 1$ vector of constants. Then $\mathbf{BY}+\mathbf{c}$ has the q—variable multivariate normal distribution $MN(\mathbf{B}\boldsymbol{\mu}+\mathbf{c},\ \mathbf{BVB}')$.

Proof. We use the moment generating function technique to derive our result. The moment generating function of $\mathbf{BY}+\mathbf{c}$ is given by

$$M_{\mathbf{BY+c}}(\mathbf{t}) = E[e^{\mathbf{t}'(\mathbf{BY+c})}] = E[e^{\mathbf{t}'\mathbf{BY}+\mathbf{t}'\mathbf{c}}]$$

$$= e^{\mathbf{t}'\mathbf{c}}E[e^{(\mathbf{B}'\mathbf{t})'\mathbf{Y}}] = e^{\mathbf{t}'\mathbf{c}}M_{\mathbf{Y}}(\mathbf{B}'\mathbf{t}).$$

Now, since $\mathbf{Y}$ is $MN(\boldsymbol{\mu},\mathbf{V})$,

$$M_{\mathbf{BY+c}}(\mathbf{t}) = e^{\mathbf{t}'\mathbf{c}} \cdot e^{(\mathbf{B}'\mathbf{t})'\boldsymbol{\mu} + \frac{1}{2}(\mathbf{B}'\mathbf{t})'\mathbf{VB}'\mathbf{t}}$$

$$= e^{\mathbf{t}'(\mathbf{B}\boldsymbol{\mu}+\mathbf{c}) + \frac{1}{2}\mathbf{t}'(\mathbf{BVB}')\mathbf{t}},$$

which we recognize as the moment generating function of an $MN(\mathbf{B}\boldsymbol{\mu}+\mathbf{c},\ \mathbf{BVB}')$ distribution from (4.5.2). Therefore, $\mathbf{BY}+\mathbf{c}$ is distributed as $MN(\mathbf{B}\boldsymbol{\mu}+\mathbf{c},\ \mathbf{BVB}')$. This completes the proof.

By letting $\mathbf{B}$ in the above theorem be a row vector $\mathbf{a}'$ and $\mathbf{c}$ be the number zero, we immediately get the following corollary.

Corollary 4.5.1 Let $\mathbf{Y}$ have the multivariate normal distribution (4.5.1) and let $\mathbf{a}$ be an $n \times 1$ vector of constants. If $\mathbf{a} \neq \mathbf{0}$, then $\mathbf{a}'\mathbf{Y}$ is $N(\mathbf{a}'\boldsymbol{\mu}, \mathbf{a}'\mathbf{V}\mathbf{a})$.

This important corollary tells us that if $\mathbf{Y}$ is multivariate normal, then any linear combination $a_1Y_1 + a_2Y_2 + \cdots + a_nY_n$ of the elements of $\mathbf{Y}$ is **univariate** normal.

As a final note, we have the following fact. If $\mathbf{Y}$ is $MN(\boldsymbol{\mu},\mathbf{V})$, then two components of $\mathbf{Y}$, say Y_i and Y_j, are independent if and only if $Cov(Y_i, Y_j) = 0$. Furthermore, the random variables $Y_1, Y_2, \ldots, Y_n$, are mutually independent if and only if $\mathbf{V}$ is a diagonal matrix. A proof of these facts can be found in Anderson (1984). From this, it follows that independence and zero covariance (or correlation) are equivalent for random variables that have a multivariate normal distribution. This fact will be useful for applications throughout this text.

4.6 The Normal Linear Model

As mentioned in the first section of this chapter, the assumption of normality will facilitate the making of tests and the construction of confidence intervals for linear model parameters. For the linear model

$$Y_i = \beta_0 + \beta_1 x_{i1} + \cdots + \beta_k x_{ik} + \varepsilon_i\ ,\ i = 1,2,\ldots,n,$$

we usually assume that $\varepsilon_1, \varepsilon_2, \ldots, \varepsilon_n$ are independent and identically distributed normal random variables with mean zero and variance σ^2. This is equivalent to stating that the random vector $\boldsymbol{\varepsilon}' = (\varepsilon_1, \varepsilon_2, \ldots, \varepsilon_n)$ is distributed as $MN(\mathbf{0}, \sigma^2\mathbf{I})$ where $\mathbf{0}$ is the zero vector. Notice that this assumption incorporates our previous assumption (4.1.3), namely, $E(\boldsymbol{\varepsilon}) = \mathbf{0}$ and $Cov(\boldsymbol{\varepsilon}) = \sigma^2\mathbf{I}$. For later work, we now formally give a label to a linear model having these basic characteristics.

Definition 4.6.1 The **normal linear model** is defined to be the linear model $\mathbf{Y} = \mathbf{X}\boldsymbol{\beta} + \boldsymbol{\varepsilon}$ with the additional specification that $\boldsymbol{\varepsilon}$ is $MN(\mathbf{0}, \sigma^2\mathbf{I})$.

An immediate consequence of the above specification on ε is that the random vector $\mathbf{Y}$ in a normal linear model also has a multivariate normal distribution. This fact is summarized in the next theorem with the proof left to the student.

Theorem 4.6.1 In the normal linear model, the following statements are true.

(1) $\mathbf{Y} = \mathbf{X}\boldsymbol{\beta} + \boldsymbol{\varepsilon}$ is $\text{MN}(\mathbf{X}\boldsymbol{\beta}, \sigma^2\mathbf{I})$.
(2) $Y_1, Y_2, \ldots, Y_n$ are mutually independent.
(3) Y_i is $N(\beta_0 + \beta_1 x_{i1} + \cdots + \beta_k x_{ik}, \sigma^2)$ for $i = 1,2,\ldots,n$.

Next, we consider the ramifications of the normality assumption on the distribution of least squares estimators.

Theorem 4.6.2 Consider the normal linear model given in Definition 4.6.1 and an estimable function $\boldsymbol{\ell}'\boldsymbol{\beta}$. The least squares estimator $\boldsymbol{\ell}'\hat{\boldsymbol{\beta}}$ is $N(\boldsymbol{\ell}'\boldsymbol{\beta}, \sigma^2\, \boldsymbol{\ell}'(\mathbf{X}'\mathbf{X})^-\boldsymbol{\ell})$.

Proof. Since $\mathbf{Y}$ is multivariate normal and since $\boldsymbol{\ell}'\hat{\boldsymbol{\beta}} = \boldsymbol{\ell}'(\mathbf{X}'\mathbf{X})^-\mathbf{X}'\mathbf{Y}$ is a linear combination of the elements in $\mathbf{Y}$, it follows from Corollary 4.5.1 that $\boldsymbol{\ell}'\hat{\boldsymbol{\beta}}$ is univariate normally distributed. In Theorem 4.3.4 we learned that $E(\boldsymbol{\ell}'\hat{\boldsymbol{\beta}}) = \boldsymbol{\ell}'\boldsymbol{\beta}$ and $\text{Var}(\boldsymbol{\ell}'\hat{\boldsymbol{\beta}}) = \sigma^2\boldsymbol{\ell}'(\mathbf{X}'\mathbf{X})^-\boldsymbol{\ell}$. Hence $\boldsymbol{\ell}'\hat{\boldsymbol{\beta}}$ is $N(\boldsymbol{\ell}'\boldsymbol{\beta}, \sigma^2\boldsymbol{\ell}'(\mathbf{X}'\mathbf{X})^-\boldsymbol{\ell})$.

For the normal linear model we can also find the distribution of $\hat{\boldsymbol{\beta}}$. In the full rank case $\hat{\boldsymbol{\beta}} = (\mathbf{X}'\mathbf{X})^{-1}\mathbf{X}'\mathbf{Y}$, and since $\mathbf{Y}$ is multivariate normal, we can apply Theorem 4.5.1 to find that

$$\hat{\boldsymbol{\beta}} \text{ is } \text{MN}(\boldsymbol{\beta}, \sigma^2(\mathbf{X}'\mathbf{X})^{-1}). \tag{4.6.1}$$

If $\mathbf{X}$ is not full rank, then the distribution of $\hat{\boldsymbol{\beta}}$ is still multivariate normal; however, it is a singular multivariate normal. In a singular multivariate normal distribution the covariance matrix is less than full rank. Since we will not need to use the singular multivariate normal we will not discuss it further. A more detailed discussion is contained in Anderson (1984).

Summary

In this chapter we have learned a great deal about estimable functions. To summarize, we know that if $\psi = \boldsymbol{\ell}'\boldsymbol{\beta}$ is an estimable function, its least squares estimator $\hat{\psi} = \boldsymbol{\ell}'\hat{\boldsymbol{\beta}}$ has the following properties:

(a) $\boldsymbol{\ell}'\hat{\boldsymbol{\beta}}$ is the same for every $\hat{\boldsymbol{\beta}}$ corresponding to a solution of the normal equations.

(b) $\boldsymbol{\ell}'\hat{\boldsymbol{\beta}}$ is the best linear unbiased estimator (BLUE) for $\boldsymbol{\ell}'\boldsymbol{\beta}$.

(c) $\boldsymbol{\ell}'\hat{\boldsymbol{\beta}}$ has variance $\sigma^2\boldsymbol{\ell}'(\mathbf{X}'\mathbf{X})^-\boldsymbol{\ell}$. (4.6.2)

(d) In the case of the normal linear model, $\boldsymbol{\ell}'\hat{\boldsymbol{\beta}}$ is normally distributed.

Armed with these useful results, we are now in a position to further our discussion of statistical inference in the analysis of experimental designs in the next and subsequent chapters.

Chapter 4 Exercises

Application Exercises

4–1 Refer to the data given in Exercise 2–2 and the simple linear regression model

$$Y_i = \beta_0 + \beta_1 x_i + \varepsilon_i \ , \ i = 1,\ldots,6 \ .$$

(a) Find expressions for $\mathrm{Var}(\hat{\beta}_0)$, $\mathrm{Var}(\hat{\beta}_1)$ and $\mathrm{Cov}(\hat{\beta}_0, \hat{\beta}_1)$.

(b) Find the numerical value of the BLUE for $\beta_0 + 2.5\beta_1$.

4–2 Consider the one–way classification problem given in Example 4.3.2

(a) What is $\hat{Y}_{ij}$?

(b) Find an expression for $\mathrm{Var}(\hat{Y}_{ij})$.

(c) If $\psi = \tau_1 - \tau_2$, then find $\text{Var}(\hat{\psi})$.

4–3 Refer to the data given in Exercise 3–20.

(a) Find the numerical value of the BLUE for $\psi_1 = \tau_i - 2\tau_2 + \tau_3$.

(b) Find the numerical value of the BLUE for $\psi_2 = \mu + \tau_3$.

(c) Find expressions for $\text{Var}(\hat{\psi}_1)$ and $\text{Var}(\hat{\psi}_2)$.

(d) If we assume a normal linear model then what is the distribution of $\hat{\psi}_2$?

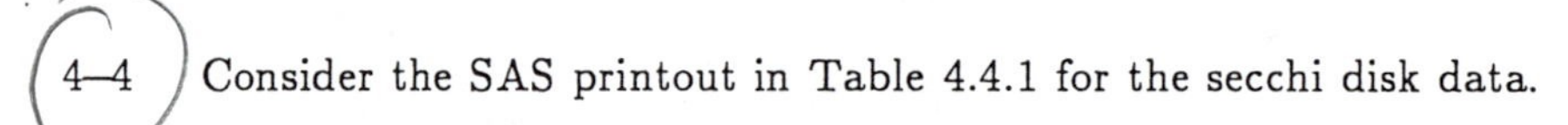

4–4 Consider the SAS printout in Table 4.4.1 for the secchi disk data.

(a) Use the solution to the normal equations given on the printout to find estimates for $\mu + \tau_1$ and $\tau_3 - \tau_1$. Compare your answers to those found in Example 4.4.1.

(b) What do the values 47.4, –8.6 and –1.6 given under "estimate" label on the printout really estimate for this problem?

(c) Consider $\psi = \tau_1 - 2\tau_2 + \tau_3$. Find the estimate for ψ.

(d) Find $\text{Var}(\hat{\psi})$ in two ways.

(i) Use $\text{Var}(\hat{\psi}) = \sigma^2 \boldsymbol{\ell}'(\mathbf{X}'\mathbf{X})^- \boldsymbol{\ell}$ and the $(\mathbf{X}'\mathbf{X})^-$ matrix given on the printout.

(ii) Recognize that $\hat{\psi} = \bar{Y}_{1\cdot} - 2\bar{Y}_{2\cdot} + \bar{Y}_{3\cdot}$ and find $\text{Var}(\bar{Y}_{1\cdot} - 2\bar{Y}_{2\cdot} + \bar{Y}_{3\cdot})$.

4–5 For the data and multiple regression model given in Exercise 3–1 find an expression for $\text{Cov}(\hat{\boldsymbol{\beta}})$.

4–6 An experiment is designed to relate the strength of paper, y, to the levels of three ingredients x_1, x_2, and x_3 that are used in making the paper. Two levels of each ingredient are chosen and we code the values as 1 for the high level and –1 for the low level. Eight different combinations of the levels of the ingredients are considered and the strength of the paper is measured for each.

The results are given as follows:

y	16.8	13.6	14.7	11.9	14.1	12.5	13.9	10.7
x_1	1	1	1	1	–1	–1	–1	–1
x_2	1	1	–1	–1	1	1	–1	–1
x_3	1	–1	1	–1	1	–1	1	–1

Consider the multiple regression model

$Y = \beta_0 + \beta_1 x_1 + \beta_2 x_2 + \beta_3 x_3 + \varepsilon$.

(a) Find the least squares estimates for β_0, β_1, β_2 and β_3.

(b) Find $Var(\hat{\beta}_j)$ for $j = 1,2,3$.

(c) Find Cov $(\hat{\beta}_j, \hat{\beta}_t)$ for $j \neq t$.

(d) Let $\psi = \beta_0 + \beta_1 - \beta_2 + \beta_3$. What does ψ represent?

(e) Find the estimate for ψ.

(f) Find an expression for $Var(\hat{\psi})$.

4–7 Refer to the multiple regression data given in Exercise 2–5 and analyzed in Exercise 3–3.

(a) Find an expression for $Var(\hat{\beta}_1)$.

(b) Find an estimate for $\psi = \beta_0 + .5\beta_1 + .3\beta_2 + .15\beta_3$. What does the value represent?

(c) Find an expression for $Var(\hat{\psi})$.

(d) If we assume a normal linear model, then what is the distribution of $\hat{\beta}_1$?

4–8 Refer to the computer printout given in Table 4.4.2 for the multiple regression problem described in Example 2.4.1 and in Example 4.4.2. Use the printout to answer the following.

(a) Find the estimated mean electric cost when $x_1 = 30$, $x_2 = 68$ and $x_3 = 1900$.

(b) Estimate $Var(\hat{\beta}_j)$ for $j = 0, 1, 2, 3$.

(c) Find the variance for the estimator associated with part a.

4–9 Let $Y' = (Y_1, Y_2, Y_3)$ have a trivariate normal distribution (i.e., Y is $MN(\mu, V)$) with

$$\mu = \begin{bmatrix} 10 \\ 2 \\ 5 \end{bmatrix} \quad \text{and} \quad V = \begin{bmatrix} 2 & 0 & -1 \\ 0 & 5 & 1 \\ -1 & 1 & 4 \end{bmatrix} .$$

(a) Are Y_1 and Y_2 independent?

(b) Find the distribution of $Y_1 + 2Y_2 + Y_3$. Hint: Let $\mathbf{a}' = (1\ 2\ 1)$ and consider $\mathbf{a}'\mathbf{Y}$.

(c) Let $\mathbf{B} = \begin{bmatrix} 2 & 0 & 0 \\ 0 & 1 & 2 \\ 0 & 1 & 3 \end{bmatrix}$. Find the distribution of **BY**.

d) Find the distribution of $\mathbf{W} = \begin{bmatrix} Y_1 - Y_2 \\ Y_1 - Y_3 \end{bmatrix}$.

Theoretical Exercises

4–10 Consider the one–way classification model $Y_{ij} = \mu + \tau_i + \varepsilon_{ij}$, $i = 1,...,p;\ \ j = 1,...,n_i$.

(a) Write assumptions (4.1.2) in terms of this model.
(b) If the normal linear model (4.6.1) is assumed, then what additions can be made to part a?

4–11 Verify the properties for expectations given in Equation 4.2.2.

4–12 Prove Theorem 4.2.1.

4–13 Consider the simple linear regression model $Y_i = \beta_0 + \beta_1 x_1 + \varepsilon_i$, $i = 1,...,n$ and the least squares estimators for β_0 and β_1. Find expressions for $Var(\hat{\beta}_0)$, and $Var(\hat{\beta}_1)$ and $Cov(\hat{\beta}_0, \hat{\beta}_1)$.

4–14 Consider the form of the simple linear regression model given in Exercise 2–13, namely, $Y_i = \alpha + \beta(x_i - \bar{x}) + \varepsilon_i$, $i = 1,...,n$. Find expressions for $\text{Var}(\hat{\alpha})$, $\text{Var}(\hat{\beta})$ and $\text{Cov}(\hat{\alpha},\hat{\beta})$.

4–15 Given the simple linear regression model $E(Y) = \beta_0 + \beta_1 x$. Let $\widehat{E(Y_0)} = \hat{\beta}_0 + \hat{\beta}_1 x_0$ be an estimator of $E(Y_0)$ for a given $x = x_0$. If $\hat{\beta}_0$ and $\hat{\beta}_1$ are the least squares estimators of β_0 and β_1, respectively, show that $\widehat{E(Y_0)}$ is an unbiased estimator and

$$\text{Var}[\widehat{E(Y_0)}] = \left[\frac{1}{n} + \frac{(x_0 - \bar{x})^2}{\sum_{i=1}^{n} (x_i - \bar{x})^2}\right] \sigma^2$$

4–16 Consider the simple no–intercept linear model

$$Y_i = \beta x_i + \varepsilon_i \ , \quad i = 1,2,\ldots,n \ ,$$

where $E(\varepsilon_i) = 0$ for all i and $\text{Var}(\varepsilon_i) = \sigma^2$ for all i.

(a) Show that

$$\hat{\beta}_1 = \sum_{i=1}^{n} Y_i \Big/ \sum_{i=1}^{n} x_i$$

is an unbiased estimator of β.

(b) Show that the least squares estimator of β is

$$\hat{\beta}_2 = \sum_{i=1}^{n} x_i Y_i \Big/ \sum_{i=1}^{n} x_i^2 . \qquad \text{(See Exercise 2–14).}$$

(c) Is $\hat{\beta}_2$ an unbiased estimator of β? Why?

(d) Find $\text{Var}(\hat{\beta}_1)$ and $\text{Var}(\hat{\beta}_2)$ and show that

$$\frac{\text{Var}(\hat{\beta}_1)}{\text{Var}(\hat{\beta}_2)} \geq 1 . \quad \text{Hint: "Expand" } n \sum_{i=1}^{n} (x_i - \bar{x})^2 .$$

(e) Use Lagrange multipliers to show that $\hat{\beta}_2$ is BLUE.

4–17 (BIAS.) Suppose we postulate the model $E(Y) = \mathbf{X}_1\boldsymbol{\beta}_1$ for which the least squares estimator of $\boldsymbol{\beta}_1$ is $\hat{\boldsymbol{\beta}}_1 = (\mathbf{X}_1'\mathbf{X}_1)^{-1}\mathbf{X}_1'\mathbf{Y}$. If the true model is, in fact,

$$E(Y) = \mathbf{X}_1\boldsymbol{\beta}_1 + \mathbf{X}_2\boldsymbol{\beta}_2 ,$$

show that $E(\hat{\boldsymbol{\beta}}_1) = \boldsymbol{\beta}_1 + (\mathbf{X}_1'\mathbf{X}_1)^{-1}\mathbf{X}_1'\mathbf{X}_2\boldsymbol{\beta}_2$, i.e., $\hat{\boldsymbol{\beta}}_1$ is a biased estimator of $\boldsymbol{\beta}_1$ when the incorrect model is used.

4–18 Refer to Exercise 4–17. Suppose we postulate the model $E(Y) = \beta_0 + \beta_1 x$ when the true model is $E(Y) = \beta_0 + \beta_1 x + \beta_2 x^2$. If the observations of Y are taken at $x = 0, 1, 2, 3, 4$, use the result of Exercise 4–17 to show that $E(\hat{\beta}_0) = \beta_0 - 2\beta_2$ and $E(\hat{\beta}_1) = \beta_1 + 4\beta_2$.

4–19 Show that for the linear regression model,

$$E(\mathbf{Y}'\mathbf{Y}) = \boldsymbol{\beta}'\mathbf{X}'\mathbf{X}\boldsymbol{\beta} + n\sigma^2 .$$

4–20 Show that the maximum number of **linearly independent** estimable functions equals the rank of X.

4–21 If $\boldsymbol{\ell}'\boldsymbol{\beta}$ is an estimable function, then show that $\mathrm{Var}(\boldsymbol{\ell}'\hat{\boldsymbol{\beta}}) = \sigma^2\boldsymbol{\ell}'(\mathbf{X}'\mathbf{X})^-\boldsymbol{\ell}$ invariant to $(\mathbf{X}'\mathbf{X})^-$.

4–22 Show that $\mathbf{X}'\mathbf{X}\boldsymbol{\beta}$ is a set of estimable functions. (Note: Since the normal equations are $\mathbf{X}'\mathbf{X}\mathbf{b} = \mathbf{X}'\mathbf{y}$ we see that the left side of the normal equations written in terms of $\boldsymbol{\beta}$ constitutes a set of estimable functions. The right side provides the value for the BLUE.)

4–23 Consider the one–way classification problem with $p = 3$, $n_1 = 4$, $n_2 = 3$ and $n_3 = 2$ as discussed in Example 4.2.2. Consider the model $Y_{ij} = \mu_i + \varepsilon_{ij}$ with $E(\boldsymbol{\varepsilon}) = \mathbf{0}$ and $\mathrm{Cov}(\boldsymbol{\varepsilon}) = \sigma^2\mathbf{I}$.

(a) Show that $\mathbf{b}' = (\bar{y}_{1\cdot}, \bar{y}_{2\cdot}, \bar{y}_{3\cdot})$ is a solution to the normal equations, where $\boldsymbol{\beta}' = (\mu_1, \mu_2, \mu_3)$.

(b) Is $\hat{\boldsymbol{\beta}}' = (\bar{Y}_{1\cdot}, \bar{Y}_{2\cdot}, \bar{Y}_{3\cdot})$ unbiased for $\boldsymbol{\beta}$? Why?

(c) Describe the estimable functions for this problem.

(d) Find the BLUE for $\psi = \mu_1 - \mu_2$.

(e) Find VAR($\hat{\psi}$).

4–24 Refer to the setting of Exercise 3–34 and the model given there. Assume that $E(\boldsymbol{\varepsilon}) = \mathbf{0}$ and $Cov(\boldsymbol{\varepsilon}) = \sigma^2\mathbf{I}$.

(a) Which of the following are estimable functions?

i) $\mu + \tau_1 + \alpha_1$ ii) $\mu + \alpha_1$ iii) μ

iv) α_1 v) $\alpha_1 - \alpha_2$ vi) $\tau_1 - \tau_2$.

(b) Find the BLUE for the functions that are estimable in part a. (Simplify your answers.)

(c) Let $\psi_1 = \mu + \tau_1 + \alpha_1$ and $\psi_2 = \alpha_1 - \alpha_2$. Find Var($\hat{\psi}_1$) and Var($\hat{\psi}_2$).

(d) Show that $\bar{Y}_{11\cdot}$ is unbiased for $\mu + \tau_1 + \alpha_{1\cdot}$.

(e) Compare Var($\bar{Y}_{11\cdot}$) and Var($\hat{\psi}_1$). Comment.

4–25 Consider the model: $Y_{ij} = (i-1)\theta + \tau_j + \varepsilon_{ij}$, $i = 1,2$; $j = 1,2$, where $E(\boldsymbol{\varepsilon}) = \mathbf{0}$ and $Cov(\boldsymbol{\varepsilon}) = \sigma^2\mathbf{I}$.

(a) Write the model in matrix notation $\mathbf{Y} = \mathbf{X}\boldsymbol{\beta} + \boldsymbol{\varepsilon}$ where $\mathbf{Y}' = [Y_{11}\ Y_{12}\ Y_{21}\ Y_{22}]$ and $\boldsymbol{\beta}' = [\theta\ \tau_1\ \tau_2]$.

(b) Are the functions
i) θ ii) $\tau_1 - \tau_2$ linearly estimable? Why?

(c) Write down the normal equations $\mathbf{X}'\mathbf{X}\mathbf{b} = \mathbf{X}'\mathbf{y}$ where $\mathbf{b}' = [a\ t_1\ t_2]$ and $\mathbf{X}'\mathbf{y}$ is a vector whose elements are functions of $y_{11}, y_{12}, y_{21}, y_{22}$.

(d) Use the normal equations in (c) to obtain the BLUE for $\tau_1 - \tau_2$ in terms of the y_{ij}.

(e) By definition $\ell'\beta$ is linearly estimable if there exists a vector c' such that $E(c'Y) = \ell'\beta$. Refer to your answer in (d).

(i) If $\ell'\beta = \tau_1 - \tau_2$ and $Y' = [Y_{11}\ Y_{12}\ Y_{21}\ Y_{22}]$, what is the vector c' which produces the BLUE for $\tau_1 - \tau_2$?

(ii) If $\widehat{\tau_1 - \tau_2}$ denotes the best linear unbiased estimator for $\tau_1 - \tau_2$, calculate $\mathrm{Var}(\widehat{\tau_1 - \tau_2})$.

Hint: What is $\widehat{\tau_1 - \tau_2}$ in terms of Y_{ij}?

(f) Refer to (e).

(i) Show that $Y_{11} - Y_{12}$ is another unbiased estimator for $\tau_1 - \tau_2$.

(ii) For the estimator in (i) of this part, what is vector c' in $E(c'Y) = \tau_1 - \tau_2$.

(iii) Is $Y_{11} - Y_{12}$ a BLUE for $\tau_1 - \tau_2$? Why or why not?

(g) Show that a generalized inverse of $X'X$ is

$$(X'X)^- = \begin{bmatrix} 1 & -\frac{1}{2} & -\frac{1}{2} \\ -\frac{1}{2} & \frac{3}{4} & \frac{1}{4} \\ -\frac{1}{2} & \frac{1}{4} & \frac{3}{4} \end{bmatrix}.$$

Suppose the four observations being modeled have values $y_{11} = 5$, $y_{12} = 5$, $y_{21} = 7$, $y_{22} = 3$. Using $(X'X)^-$, calculate the numerical values of the BLUES for

(i) θ

(ii) $2\tau_1 + 4\tau_2$.

(h) Refer to (g). If $\widehat{2\tau_1 + 4\tau_2}$ denotes the best linear unbiased estimator for $2\tau_1 + 4\tau_2$, calculate

$\widehat{\text{Var}(2\tau_1 + 4\tau_2)}$.

4–26 If $\bar{X}_1$ and $\bar{X}_2$ are means of independent random samples of size n_1 and n_2 from a population with mean μ and variance σ^2, show that the variance of the unbiased estimator $W = a\bar{X}_1 + (1 - a)\bar{X}_2$ is minimum when $a = n_1/(n_1 + n_2)$.

4–27 Let $Y_1, Y_2, \ldots, Y_n$ constitute a set of independent random variables with $E(Y_i) = \mu$ for each $i = 1,2,\ldots,n$. Consider the linear form

$$Z = c_1Y_1 + c_2Y_2 + \ldots + c_nY_n.$$

Using Lagrange multipliers, show that

(a) $Z = \frac{1}{n}\sum_{i=1}^{n} Y_i$ is a best linear unbiased estimator of μ if $\text{Var}(Y_i) = \sigma^2$ for each i.

(b) $Z = \dfrac{\sum_{i=1}^{n} (Y_i/\sigma_i^2)}{\sum_{i=1}^{n} (1/\sigma_i^2)}$ is a best linear unbiased estimator of μ. $\text{Var}(Y_i) = \sigma_i^2$.

4–28 Show that if the $p \times 1$ random vector **Y** is MN(**0**,**I**) then the joint density function of $Y_1, Y_2, \ldots, Y_p$ is

$$f(y_1, y_2, \ldots, y_p) = \frac{1}{(2\pi)^{p/2}} e^{-\frac{1}{2}\sum_{i=1}^{p} y_i^2}, \quad -\infty < y_i < \infty.$$

4–29 Suppose the $n \times 1$ random vector **Y** is $MN(\boldsymbol{\mu}, \sigma^2\mathbf{I})$ where $E(Y_i) = \mu$ for each i. Find the distribution of

$$Y = \frac{1}{n} \sum_{i=1}^{n} Y_i .$$

4–30 Find the moment generating function for the bivariate normal.

4–31 For a multivariate normal density the quadratic form is the negative of the exponent of e in formula (4.5.1) for the density. That is, the quadratic form is $\frac{1}{2}(\mathbf{y} - \boldsymbol{\mu})'\mathbf{V}^{-1}(\mathbf{y} - \boldsymbol{\mu})$. For the bivariate normal distribution identify $\mu_1, \mu_2, \sigma_1, \sigma_2$ and ρ if the quadratic form is as follows:

(a) $\frac{1}{2}[(y_1 - 5)^2 + (y_2 + 3)^2)]$

(b) $\frac{1}{2}[y_1^2 + y_2^2 + 2y_1 - 6y_2 + 10]$

(c) $\frac{1}{6}[4(y_1 - 5)^2 - 2(y_1 - 5)(y_2 - 8) + (y_2 - 8)^2]$.

4–32 $\mathbf{Y}' = [Y_1, Y_2, Y_3]$ is trivariate $MN(\mathbf{0},\mathbf{V})$ with quadratic form $\mathbf{Q} = 2Y_2^2 + 5Y_2^2 + Y_3^2 - 6Y_1Y_2$.

(a) Show that Q is positive definite.

(b) Find V and determine the correlation coefficients ρ_{12}, ρ_{13}, and ρ_{23}.

(c) Find the distribution of $Y_1 - 2Y_2 + 3Y_3$.

4–33 Consider a normal experiment design model. Let $\hat{\boldsymbol{\beta}} = (\mathbf{X}'\mathbf{X})^{-}\mathbf{X}'\mathbf{Y}$ correspond to any solution of the normal equations and let $\mathbf{L}\boldsymbol{\beta}$ be a set of linearly independent estimable functions where $\mathbf{L}$ is a $q \times (k + 1)$ matrix of rank $q \leq k + 1$. Show that $\mathbf{L}\hat{\boldsymbol{\beta}}$ is $MN(\mathbf{L}\boldsymbol{\beta}, \mathbf{L}(\mathbf{X}'\mathbf{X})^{-}\mathbf{L}'\sigma^2)$.

Hint:
$$\begin{aligned}
\mathrm{Cov}(\mathbf{L}\hat{\boldsymbol{\beta}}) &= \mathbf{L}(\mathbf{X}'\mathbf{X})^{-}\mathbf{X}'\mathbf{X}[(\mathbf{X}'\mathbf{X})^{-}]'\mathbf{L}'\sigma^2 \\
&= \mathbf{L}(\mathbf{X}'\mathbf{X})^{-}\mathbf{X}'\mathbf{X}[(\mathbf{X}'\mathbf{X})^{-}]'\mathbf{X}'\mathbf{C}'\sigma^2 \\
&= \mathbf{L}(\mathbf{X}'\mathbf{X})^{-}\mathbf{L}'\sigma^2.
\end{aligned}$$

4–34 Verify that the generalized inverse given in Example 4.4.1 leads to the solution set given in the example.

4–35 Prove Theorem 4.6.1.

CHAPTER 5

DISTRIBUTION THEORY FOR STATISTICAL INFERENCE

5.1 Distribution of Quadratic Forms

Among the major activities in the analysis of experimental data is the performance of tests of hypotheses and the calculation of confidence intervals. In order to proceed in that direction, we need to discover distributions that permit us to carry out these types of inference procedures. As in beginning courses in statistics, in this textbook we will be using test statistics that usually have t or F distributions. As we shall see, both of these distributions incorporate quadratic forms as "building blocks" in their construction.

Definition 5.1.1 Let $\mathbf{A}$ be an $n \times n$ symmetric matrix of constants and let $\mathbf{Y}$ be an $n \times 1$ vector. Then the expression $\mathbf{Y}'\mathbf{A}\mathbf{Y}$ is called a **quadratic form** in $\mathbf{Y}$. The matrix $\mathbf{A}$ is called the **matrix of the quadratic form**.

To obtain an idea about the structure of a quadratic form, we consider two simple examples. First, if $\mathbf{Y}$ is an $n \times 1$ vector and $\mathbf{A} = \mathbf{I}$, $\mathbf{I}$ being the $n \times n$ identity matrix, then the quadratic form $\mathbf{Y}'\mathbf{I}\mathbf{Y} = \mathbf{Y}'\mathbf{Y} = \sum_{i=1}^{n} Y_i^2$, which is a simple sum of squares. Second, if we take a symmetric matrix $\mathbf{A} = (a_{ij})$, then the resulting quadratic form is

$$\mathbf{Y}'\mathbf{A}\mathbf{Y} = \sum_{i=1}^{n} \sum_{j=1}^{n} a_{ij} Y_i Y_j$$

$$= \sum_{i=1}^{n} a_{ii} Y_i^2 + 2 \sum_{i<j}^{n} \sum a_{ij} Y_i Y_j .$$

Thus we see that when $\mathbf{A}$ is symmetric, the resulting quadratic form is composed of the sum of squares terms and twice the cross–product terms. It is important to note that a quadratic form is a scalar and not a matrix quantity.

Although we specified matrix $\mathbf{A}$ to be symmetric in Definition 5.1.1, $\mathbf{Y}'\mathbf{A}\mathbf{Y}$ is a quadratic form whether or not $\mathbf{A}$ is symmetric. In the exercises the student is asked to show that $\mathbf{A}$ can always be chosen to be symmetric. That is, any expression $\mathbf{Y}'\mathbf{B}\mathbf{Y}$ with $\mathbf{B}$ nonsymmetric can be written as $\mathbf{Y}'\mathbf{A}\mathbf{Y}$ with $\mathbf{A}$ symmetric. As a consequence, we shall always assume that the matrix of a quadratic form to be symmetric in this text. The following theorem tells us the distribution of a quadratic form is an important special case.

Theorem 5.1.1 Let $\mathbf{Y}$ be an $n \times 1$ random vector where $\mathbf{Y}$ is $MN(\mathbf{0}, \sigma^2\mathbf{I})$ and let $\mathbf{A}$ be a symmetric matrix of constants. If $\mathbf{A}$ is idempotent of rank m, then $\mathbf{Y}'\mathbf{A}\mathbf{Y}/\sigma^2$ has the chi–square distribution with m degrees of freedom.

Proof. For the proof, we appeal to some facts from matrix algebra presented in Appendix A. Since $\mathbf{A}$ is symmetric, there exists an $n \times n$ **orthogonal** matrix $\mathbf{P}$ such that $\mathbf{P}'\mathbf{A}\mathbf{P}$ is a diagonal matrix where the diagonal elements are the eigenvalues of $\mathbf{A}$. But since $\mathbf{A}$ is idempotent of rank m, m of the eigenvalues are ones and the remainder $n - m$ are zeros. Therefore, without loss of generality, the diagonalization of $\mathbf{A}$ can be represented by

$$\mathbf{P}'\mathbf{A}\mathbf{P} = \begin{bmatrix} \mathbf{I}_m & \mathbf{0} \\ \mathbf{0} & \mathbf{0} \end{bmatrix},$$

where $\mathbf{I}_m$ is the $m \times m$ identity matrix. Let $\mathbf{W} = \mathbf{P}'\mathbf{Y}$, and since $\mathbf{P}$ is orthogonal, $\mathbf{P}'\mathbf{P} = \mathbf{P}\mathbf{P}' = \mathbf{I}$, the $n \times n$ identity matrix. To solve for $\mathbf{Y}$ we note that $\mathbf{P}\mathbf{W} = \mathbf{P}\mathbf{P}'\mathbf{Y}$ which yields $\mathbf{P}\mathbf{W} = \mathbf{Y}$. By Theorem 4.5.1, $\mathbf{W}$ is multivariate normal with mean vector $\mathbf{P}'\mathbf{0} = \mathbf{0}$ and covariance matrix $\mathbf{P}'\sigma^2\mathbf{I}\mathbf{P} = \sigma^2\mathbf{I}$. If W_i is the ith component of $\mathbf{W}$, then W_i is $N(0, \sigma^2)$ and the W_i's are independent. Now

$$\mathbf{Y}'\mathbf{A}\mathbf{Y}/\sigma^2 = \mathbf{W}'\mathbf{P}'\mathbf{A}\mathbf{P}\mathbf{W}/\sigma^2 = \mathbf{W}' \begin{bmatrix} \mathbf{I} & \mathbf{0} \\ \mathbf{0} & \mathbf{0} \end{bmatrix} \mathbf{W}/\sigma^2 = \sum_{i=1}^{m} W_i^2/\sigma^2 .$$

Since each $W_1^2/\sigma^2, W_2^2/\sigma^2, \ldots, W_m^2/\sigma^2$ are independently distributed chi–square variables with one degree of freedom, their sum is distributed as chi–square with m degrees of freedom.

To illustrate the use of the preceding theorem, we apply it to find the distributions of three familiar expressions encountered in a first course in mathematical statistics. We now discover that these expressions $\sum_{i=1}^{n} Y_i^2$, $n\bar{Y}^2$ where $\bar{Y} = \sum_{i=1}^{n} Y_i/n$, and $\sum_{i=1}^{n} (Y_i - \bar{Y})^2$ are, in fact, quadratic forms in $\mathbf{Y}$. Of course, if $\mathbf{Y}$ is an $n \times 1$ vector,

$\sum_{i=1}^{n} Y_i^2 = \mathbf{Y}'\mathbf{I}\mathbf{Y}$ where $\mathbf{I}$ is the $n \times n$ identity matrix. The student is asked to verify that $n\bar{Y}^2 = \mathbf{Y}'\mathbf{A}\mathbf{Y}$, where all entries in the $n \times n$ matrix $\mathbf{A}$ are $1/n$. Additionally, the student should verify that $\sum_{i=1}^{n} (Y_i - \bar{Y})^2 = \mathbf{Y}'(\mathbf{I} - \mathbf{A})\mathbf{Y}$ where $\mathbf{A}$ is the matrix of $n\bar{Y}^2$.

Example 5.1.1 Let $\mathbf{Y}$ be an $n \times n$ random vector which is $MN(\mathbf{0}, \sigma^2\mathbf{I})$. We investigate the distribution of the following

$$\text{(a)} \sum_{i=1}^{n} Y_i^2/\sigma^2, \ \text{(b)} \ n\bar{Y}^2/\sigma^2, \ \text{and} \ \text{(c)} \sum_{i=1}^{n} (Y_i - \bar{Y})^2/\sigma^2.$$

For part (a), we note that $\sum_{i=1}^{n} Y_i^2/\sigma^2 = \mathbf{Y}'\mathbf{I}\mathbf{Y}/\sigma^2$ where $\mathbf{I}$ is symmetric, idempotent and of rank n. Hence, from Theorem 5.1.1, it follows immediately that $\sum_{i=1}^{n} Y_i^2/\sigma^2$ is chi–square with n degrees of freedom. For part (b), $n\bar{Y}^2/\sigma^2 = \mathbf{Y}'\mathbf{A}\mathbf{Y}/\sigma^2$ where $\mathbf{A}$ is the $1/n$ all entries matrix noted immediately preceding this example. It is easy to verify that $\mathbf{A}$ is symmetric, idempotent and of rank one. Therefore $n\bar{Y}^2/\sigma^2$ is a chi–square random variable with one degree of freedom. Finally, $\sum_{i=1}^{n} (Y_i - \bar{Y})^2/\sigma^2 = \mathbf{Y}'(\mathbf{I} - \mathbf{A})\mathbf{Y}/\sigma^2$ where $\mathbf{I} - \mathbf{A}$ is an $n \times n$ matrix with all off–diagonal elements $-1/n$ and all diagonal elements $(n - 1)/n$. It is easily shown that $\mathbf{I} - \mathbf{A}$ is symmetric, idempotent and of rank $n - 1$. From Theorem 5.1.1, $\sum_{i=1}^{n} (Y_i - \bar{Y})^2/\sigma^2$ has a chi–square distribution with $n - 1$ degrees of freedom.

The expression $(\mathbf{y} - \boldsymbol{\mu})'\mathbf{V}^{-1}(\mathbf{y} - \boldsymbol{\mu})$ in the exponent of the multivariate normal distribution (4.5.1) is sometimes called the quadratic form of the multivariate normal. We now show that it too, in random variable form, has a chi–square distribution.

Theorem 5.1.2 If the $n \times 1$ random vector $\mathbf{Y}$ is $MN(\boldsymbol{\mu}, \mathbf{V})$, then the quadratic form $(\mathbf{Y} - \boldsymbol{\mu})'\mathbf{V}^{-1}(\mathbf{Y} - \boldsymbol{\mu})$ has the chi–square distribution with n degrees of freedom.

Proof. From Appendix A, we know that since $\mathbf{V}^{-1}$ is symmetric positive definite, there exists a **nonsingular** $n \times n$ matrix $\mathbf{C}$ such that $\mathbf{C}'\mathbf{V}^{-1}\mathbf{C} = \mathbf{I}$. Let $\mathbf{W} = \mathbf{C}^{-1}(\mathbf{Y} - \boldsymbol{\mu})$. Using Theorem 4.5.1, it follows that $\mathbf{Y} - \boldsymbol{\mu}$ is $MN(\mathbf{0}, \mathbf{V})$ and so $\mathbf{W}$ has a multivariate normal distribution with mean vector $\mathbf{C}^{-1}\mathbf{0} = \mathbf{0}$ and covariance matrix $\mathbf{C}^{-1}\mathbf{V}(\mathbf{C}^{-1})' = (\mathbf{C}'\mathbf{V}^{-1}\mathbf{C})^{-1} = \mathbf{I}$. Hence the $\mathbf{W}$ elements, W_1, $W_2, \ldots, W_n$ are mutually independent $N(0,1)$ random variables so that $\sum_{i=1}^{n} W_i^2$ is chi–square with n degrees of freedom. But,

$$(\mathbf{Y} - \boldsymbol{\mu})'\mathbf{V}^{-1}(\mathbf{Y} - \boldsymbol{\mu}) = (\mathbf{CW})'\mathbf{V}^{-1}(\mathbf{CW}) = \mathbf{W}'\mathbf{C}'\mathbf{V}^{-1}\mathbf{CW} = \mathbf{W}'\mathbf{IW} = \sum_{i=1}^{n} W_i^2$$

This completes the proof.

To illustrate the above result, we turn to the multiple regression setting where $\mathbf{X}$ is full rank.

Example 5.1.2 Suppose $\mathbf{Y} = \mathbf{X}\boldsymbol{\beta} + \boldsymbol{\varepsilon}$ is a normal linear regression intercept model with independent variables $x_1, x_2, \ldots, x_k$. Recall statement (4.6.1) which says that for this model the least squares estimator for $\boldsymbol{\beta}$, $\hat{\boldsymbol{\beta}} = (\mathbf{X}'\mathbf{X})^{-1}\mathbf{X}'\mathbf{Y}$, is $MN[\boldsymbol{\beta}, \sigma^2(\mathbf{X}'\mathbf{X})^{-1}]$. Here, $\mathbf{V}^{-1} = [\sigma^2(\mathbf{X}'\mathbf{X})^{-1}]^{-1} = \mathbf{X}'\mathbf{X}/\sigma^2$ so it immediately follows from Theorem 5.1.2 that $(\hat{\boldsymbol{\beta}} - \boldsymbol{\beta})'\mathbf{X}'\mathbf{X}(\hat{\boldsymbol{\beta}} - \boldsymbol{\beta})/\sigma^2$ has the chi–square distribution with $k + 1$ degrees of freedom.

Error Sum of Squares Random Variable

In Section 3.3 we introduced the quantity called the error sum of squares denoted and defined in formula (3.3.1) as $SSe = (\mathbf{y} - \mathbf{Xb})'(\mathbf{y} - \mathbf{Xb})$. In practice, the number SSe varies from sample to sample and so it is a value of a random variable. We next focus our attention on this random variable. In keeping with our effort to distinguish between a random variable and its value, we write the random variable form of the error sum of squares as

$$\text{SSE} = (\mathbf{Y} - \mathbf{X}\hat{\boldsymbol{\beta}})'(\mathbf{Y} - \mathbf{X}\hat{\boldsymbol{\beta}}). \tag{5.1.1}$$

We will use Theorem 5.1.1 to find the distribution of SSE but first we need to express SSE as a quadratic form whose matrix is symmetric and idempotent. In the next theorem we show that SSE can be represented as a quadratic form in $\mathbf{Y}$ and also in $\mathbf{Y} - \mathbf{X}\boldsymbol{\beta}$, both of which have matrix $\mathbf{I} - \mathbf{X}(\mathbf{X}'\mathbf{X})^{-}\mathbf{X}'$. From Theorem 3.2.3, we already know that $\mathbf{I} - \mathbf{X}(\mathbf{X}'\mathbf{X})^{-}\mathbf{X}'$ is symmetric and idempotent.

Theorem 5.1.3 Consider the linear model $\mathbf{Y} = \mathbf{X}\boldsymbol{\beta} + \boldsymbol{\varepsilon}$ and let $\hat{\boldsymbol{\beta}}$ correspond to any solution to the normal equations. Then SSE can be expressed as

(a) $\text{SSE} = \mathbf{Y}'[\mathbf{I} - \mathbf{X}(\mathbf{X}'\mathbf{X})^{-}\mathbf{X}']\mathbf{Y}$ and

(b) $\text{SSE} = (\mathbf{Y} - \mathbf{X}\boldsymbol{\beta})'[\mathbf{I} - \mathbf{X}(\mathbf{X}'\mathbf{X})^{-}\mathbf{X}'](\mathbf{Y} - \mathbf{X}\boldsymbol{\beta})$

where $\mathbf{I}$ is the $n \times n$ identity matrix.

Proof. Recall that $\hat{\boldsymbol{\beta}} = (\mathbf{X}'\mathbf{X})^{-}\mathbf{X}'\mathbf{Y}$ so

$$\begin{aligned}
\text{SSE} &= (\mathbf{Y} - \mathbf{X}\hat{\boldsymbol{\beta}})'(\mathbf{Y} - \mathbf{X}\hat{\boldsymbol{\beta}}) \\
&= [\mathbf{Y} - \mathbf{X}(\mathbf{X}'\mathbf{X})^{-}\mathbf{X}'\mathbf{Y}]'[\mathbf{Y} - \mathbf{X}(\mathbf{X}'\mathbf{X})^{-}\mathbf{X}'\mathbf{Y}] \\
&= \mathbf{Y}'[\mathbf{I} - \mathbf{X}\{(\mathbf{X}'\mathbf{X})^{-}\}'\mathbf{X}'][\mathbf{I} - \mathbf{X}(\mathbf{X}'\mathbf{X})^{-}\mathbf{X}']\mathbf{Y}.
\end{aligned}$$

But since $\{(\mathbf{X}'\mathbf{X})^{-}\}'$ also is a g–inverse of $\mathbf{X}'\mathbf{X}$, $\mathbf{X}(\mathbf{X}'\mathbf{X})^{-}\mathbf{X}'$ is invariant to $(\mathbf{X}'\mathbf{X})^{-}$ and $\mathbf{I} - \mathbf{X}(\mathbf{X}'\mathbf{X})^{-}\mathbf{X}'$ is idempotent (three of the results in Theorem 3.2.3), the last expression for SSE becomes

$$\text{SSE} = \mathbf{Y}'[\mathbf{I} - \mathbf{X}(\mathbf{X}'\mathbf{X})^{-}\mathbf{X}']\mathbf{Y}.$$

This verifies part (a). For part (b) we proceed as follows:

$$\begin{aligned}
&(\mathbf{Y} - \mathbf{X}\boldsymbol{\beta})'[\mathbf{I} - \mathbf{X}(\mathbf{X}'\mathbf{X})^{-}\mathbf{X}](\mathbf{Y} - \mathbf{X}\boldsymbol{\beta}) \\
&\quad = \mathbf{Y}'[\mathbf{I} - \mathbf{X}(\mathbf{X}'\mathbf{X})^{-}\mathbf{X}']\mathbf{Y} - 2\boldsymbol{\beta}'\mathbf{X}'[\mathbf{I} - \mathbf{X}(\mathbf{X}'\mathbf{X})^{-}\mathbf{X}']\mathbf{Y} \\
&\quad + \boldsymbol{\beta}'\mathbf{X}'[\mathbf{I} - \mathbf{X}(\mathbf{X}'\mathbf{X})^{-}\mathbf{X}']\,\mathbf{X}\boldsymbol{\beta}.
\end{aligned}$$

But $\mathbf{X}'[\mathbf{I} - \mathbf{X}(\mathbf{X}'\mathbf{X})^{-}\mathbf{X}'] = \mathbf{X}' - \mathbf{X}'\mathbf{X}(\mathbf{X}'\mathbf{X})^{-}\mathbf{X}'$ is the zero matrix by Theorem 3.2.3(2). Hence $(\mathbf{Y} - \mathbf{X}\boldsymbol{\beta})'[\mathbf{I} - \mathbf{X}(\mathbf{X}'\mathbf{X})^{-}\mathbf{X}'](\mathbf{Y} - \mathbf{X}\boldsymbol{\beta}) = \mathbf{Y}'[\mathbf{I} - \mathbf{X}(\mathbf{X}'\mathbf{X})^{-}\mathbf{X}']\mathbf{Y} = \text{SSE}$. This completes the proof of part (b).

We are now able to achieve the major objective of this section, namely, the distribution of SSE. In the next theorem we show that if we divide SSE by the parameter σ^2 in a normal linear model, the resulting random variable has a chi–square distribution.

Theorem 5.1.4 Consider the linear model $\mathbf{Y} = \mathbf{X}\boldsymbol{\beta} + \boldsymbol{\varepsilon}$ where $\boldsymbol{\varepsilon}$ is $\text{MN}(\mathbf{0}, \sigma^2\mathbf{I})$. If the rank of $\mathbf{X}$ is r, then SSE/σ^2 has a chi–square distribution with $n - r$ degrees of freedom.

Proof. By part (b) of the previous theorem

$$\text{SSE}/\sigma^2 = (\mathbf{Y} - \mathbf{X}\boldsymbol{\beta})'[\mathbf{I} - \mathbf{X}(\mathbf{X}'\mathbf{X})^{-}\mathbf{X}'](\mathbf{Y} - \mathbf{X}\boldsymbol{\beta}')/\sigma^2 .$$

Since $\mathbf{Y}$ is $\text{MN}(\mathbf{X}\boldsymbol{\beta}, \sigma^2\mathbf{I})$, we know from Theorem 4.5.1 that $\mathbf{Y} - \mathbf{X}\boldsymbol{\beta}$ is $\text{MN}(\mathbf{0}, \sigma^2\mathbf{I})$. Also, we know that $\mathbf{I} - \mathbf{X}(\mathbf{X}'\mathbf{X})^{-}\mathbf{X}'$ is symmetric and idempotent. The fact that SSE/σ^2 is chi–square then follows immediately from Theorem 5.1.1. To verify that the degrees of freedom is $n - r$, we show that $\mathbf{I} - \mathbf{X}(\mathbf{X}'\mathbf{X})^{-}\mathbf{X}'$ has rank $n - r$ using some facts from Appendix A. Since $\mathbf{I} - \mathbf{X}(\mathbf{X}'\mathbf{X})^{-}\mathbf{X}'$ is idempotent, its rank equals its trace. To begin, we note that

$$\text{trace}[\mathbf{I} - \mathbf{X}(\mathbf{X}'\mathbf{X})^{-}\mathbf{X}'] = \text{trace}(\mathbf{I}) - \text{trace}[\mathbf{X}(\mathbf{X}'\mathbf{X})^{-}\mathbf{X}'] = n - \text{trace}[\mathbf{X}(\mathbf{X}'\mathbf{X})^{-}\mathbf{X}'] .$$

But $\mathbf{X}(\mathbf{X}'\mathbf{X})^{-}\mathbf{X}'$ is idempotent so $\text{trace}[\mathbf{X}(\mathbf{X}'\mathbf{X})^{-}\mathbf{X}'] = \text{rank}[\mathbf{X}(\mathbf{X}'\mathbf{X})^{-}\mathbf{X}']$. Now $\text{rank}(\mathbf{X}) \geq \text{rank}[\mathbf{X}(\mathbf{X}'\mathbf{X})^{-}\mathbf{X}'] \geq \text{rank}[\mathbf{X}(\mathbf{X}'\mathbf{X})^{-}\mathbf{X}'\mathbf{X}] = \text{rank}(\mathbf{X})$, with the strict equality due to $(\mathbf{X}'\mathbf{X})^{-}\mathbf{X}'$ being a g–inverse for $\mathbf{X}$ (Theorem 3.2.3). Thus $\text{rank}[\mathbf{X}(\mathbf{X}'\mathbf{X})^{-}\mathbf{X}'] = \text{rank}(\mathbf{X}) = r$. Finally, $\text{rank}[\mathbf{I} - \mathbf{X}(\mathbf{X}'\mathbf{X})^{-}\mathbf{X}'] = n - r$. This completes the proof.

Error Mean Square

In this subsection we define a useful quantity that turns out to be an unbiased estimator for the unknown parameter σ^2 in the general linear model.

Definition 5.1.2 The **error mean square** (mean square error) is denoted by MSE or $\hat{\sigma}^2$ and defined by $\hat{\sigma}^2 = \text{MSE} = \text{SSE}/(n - r)$. A sample value of the random variable $\hat{\sigma}^2$ will be denoted s^2 (or MSe). That is,

$$s^2 = SSe/(n - r) \quad . \tag{5.1.2}$$

In Theorem 5.1.4 we found that in the case of the normal linear model, SSE/σ^2 is a chi–square random variable with $n - r$ degrees of freedom. Recalling that the mean of a chi–square random variable equals its degrees of freedom, we immediately get

$$E(SSE/\sigma^2) = n - r \quad \text{or} \quad E[SSE/(n - r)] = \sigma^2 \ .$$

Thus for the normal linear model, $\hat{\sigma}^2 \equiv MSE = SSE/(n - r)$ is unbiased for σ^2. In order to arrive at a more general result which does not depend upon the assumption of normality, we will need to find the expected value of a quadratic form. To aid in this endeavor we first present a theorem which will be useful now and later on.

Theorem 5.1.5 If $\mathbf{Y}$ is a random vector with $E(\mathbf{Y}) = \boldsymbol{\mu}$ and $Cov(\mathbf{Y}) = \sigma^2\mathbf{I}$, then for a symmetric matrix $\mathbf{A}$ we have

$$E(\mathbf{Y}'\mathbf{A}\mathbf{Y}) = \boldsymbol{\mu}'\mathbf{A}\boldsymbol{\mu} + \sigma^2 \text{trace}(\mathbf{A}) \ .$$

Proof.

$$\begin{aligned} E(\mathbf{Y}'\mathbf{A}\mathbf{Y}) &= E[\sum_{i=1}^{n} a_{ii} Y_i^2 + 2 \sum_{i<j}^{n} \sum a_{ij} Y_i Y_j] \\ &= \sum_{i=1}^{n} a_{ii} E(Y_i^2) + 2 \sum_{i<j}^{n} \sum a_{ij} E(Y_i Y_j) \\ &= \sum_{i=1}^{n} a_{ii} (\sigma^2 + \mu_i^2) + 2 \sum_{i<j}^{n} \sum a_{ij} (\sigma_{ij} + \mu_i \mu_j) \ . \end{aligned}$$

But $\sigma_{ij} = Cov(Y_i, Y_j) = 0$. Therefore

$$\begin{aligned} E(\mathbf{Y}'\mathbf{A}\mathbf{Y}) &= \sigma^2 \sum_{i=1}^{n} a_{ii} + \sum_{i=1}^{n} a_{ii} \mu_i^2 + 2 \sum_{i<j}^{n} \sum a_{ij} \mu_i \mu_j \\ &= \sigma^2 \text{trace}(\mathbf{A}) + \boldsymbol{\mu}'\mathbf{A}\boldsymbol{\mu} \ . \end{aligned}$$

Theorem 5.1.6 Consider the linear model $\mathbf{Y} = \mathbf{X}\boldsymbol{\beta} + \boldsymbol{\varepsilon}$ where $E(\boldsymbol{\varepsilon}) = \mathbf{0}$ and $Cov(\boldsymbol{\varepsilon}) = \sigma^2\mathbf{I}$. Then the error mean square $\hat{\sigma}^2$ is an unbiased estimator for σ^2, that is, $E(\hat{\sigma}^2) = \sigma^2$.

Proof. From Theorem 5.1.3(b) we have

$$SSE = (\mathbf{Y} - \mathbf{X}\boldsymbol{\beta})'[\mathbf{I} - \mathbf{X}(\mathbf{X}'\mathbf{X})^{-}\mathbf{X}'](\mathbf{Y} - \mathbf{X}\boldsymbol{\beta})$$

where $\mathbf{I} - \mathbf{X}(\mathbf{X}'\mathbf{X})^{-}\mathbf{X}'$, the matrix of this quadratic form in $\mathbf{Y} - \mathbf{X}\boldsymbol{\beta}$, is symmetric, idempotent and has rank $n - r$. Applying Theorem 5.1.5 with $\boldsymbol{\mu} = E(\mathbf{Y} - \mathbf{X}\boldsymbol{\beta}) = \mathbf{0}$, we get

$$\begin{aligned} E(SSE) &= \sigma^2 \text{trace}[\mathbf{I} - \mathbf{X}(\mathbf{X}'\mathbf{X})^{-}\mathbf{X}'] + 0 \\ &= \sigma^2 \text{rank}[\mathbf{I} - \mathbf{X}(\mathbf{X}'\mathbf{X})^{-}\mathbf{X}'] \\ &= \sigma^2 (n - r). \end{aligned}$$

Thus $E(\hat{\sigma}^2) = E[SSE/(n - r)] = \sigma^2$.

It is interesting to note that $\hat{\sigma}^2$ is an unbiased estimator for σ^2 for any vector $\boldsymbol{\beta}$ and for any distribution of the vector $\boldsymbol{\varepsilon}$ as long as $E(\boldsymbol{\varepsilon}) = \mathbf{0}$ and $Cov(\boldsymbol{\varepsilon}) = \sigma^2\mathbf{I}$. For example, in the secchi disk problem we observe from the SAS printout of Table 4.4.1 that $SSe = 14.8$ and $s^2 = MSe = SSe/12 = 1.2333$ (rounded to four decimal places). We now know that 1.2333 is an unbiased estimate for σ^2, and that this estimate is unbiased regardless of the truthfulness of the hypothesis $H_0 : \tau_1 = \tau_2 = \cdots = \tau_p$ and regardless of the distribution of $\boldsymbol{\varepsilon}$.

5.2 Independence of Quadratic Forms

At the beginning of this chapter it was implied that we will be using t and F distributions for the purpose of statistical inference. The reader should recall that in the derivation of t and F random variables independence between constituent random variables is required. To establish this independence, we will use two results. The first is presented with a proof; the second is proved in a similar manner but the details are left out here. They can be found in Graybill (1976) or Searle (1971).

Theorem 5.2.1 Let $\mathbf{C}$ be a $q \times n$ matrix of constants and $\mathbf{A}$ be a symmetric $n \times n$ matrix of constants. Suppose the $n \times 1$ random vector $\mathbf{Y}$ is $MN(\boldsymbol{\mu}, \sigma^2\mathbf{I})$. If $\mathbf{CA}$ is the zero matrix, then $\mathbf{Y}'\mathbf{AY}$ and $\mathbf{CY}$ are independently distributed.

Proof. Let m be the rank of $\mathbf{A}$. Since $\mathbf{A}$ is symmetric there is an orthogonal matrix $\mathbf{P}$ such that $\mathbf{P}'\mathbf{AP}$ is a diagonal matrix of the form

$$\mathbf{P}'\mathbf{AP} = \begin{bmatrix} \mathbf{D} & \mathbf{0} \\ \mathbf{0} & \mathbf{0} \end{bmatrix}$$

where $\mathbf{D}$ is an $m \times m$ full rank diagonal matrix. (Note that $m < n$, for if $m = n$, then $\mathbf{A}^{-1}$ exists which implies that $\mathbf{C}$ is the zero matrix.) Let $\mathbf{W} = \mathbf{P}'\mathbf{Y}$. Then by Theorem 4.5.1, $\mathbf{W}$ is $MN(\mathbf{P}'\boldsymbol{\mu}, \sigma^2\mathbf{I})$; therefore, the components of $\mathbf{W}$ are independent. Now since $\mathbf{P}'\mathbf{P} = \mathbf{I}$,

$$\mathbf{Y}'\mathbf{AY} = \mathbf{Y}'\mathbf{PP}'\mathbf{APP}'\mathbf{Y} = \mathbf{W}'\begin{bmatrix} \mathbf{D} & \mathbf{0} \\ \mathbf{0} & \mathbf{0} \end{bmatrix}\mathbf{W}\,.$$

Hence $\mathbf{Y}'\mathbf{AY}$ is a function of the first m components of $\mathbf{W}$. Next $\mathbf{CA} = \mathbf{0}$ and $\mathbf{PP}' = \mathbf{I}$ implies $\mathbf{CPP}'\mathbf{A} = \mathbf{CPP}'\mathbf{AP} = \mathbf{0}$. Thus $\mathbf{CP}\begin{bmatrix} \mathbf{D} & \mathbf{0} \\ \mathbf{0} & \mathbf{0} \end{bmatrix} = \mathbf{0}$. This in turn implies that the $q \times n$ matrix $\mathbf{CP}$ is of the form $\begin{bmatrix} \mathbf{0} & \mathbf{CP}_{12} \\ \mathbf{0} & \mathbf{CP}_{22} \end{bmatrix}$ where $\mathbf{CP}_{12}$ and $\mathbf{CP}_{22}$ are $n - m$ column matrices. Then $\mathbf{CY} = \mathbf{CPW} = \begin{bmatrix} \mathbf{0} & \mathbf{CP}_{12} \\ \mathbf{0} & \mathbf{CP}_{22} \end{bmatrix}\mathbf{W}$ which is a function of the last $n - m$ components in $\mathbf{W}$. Since $\mathbf{Y}'\mathbf{AY}$ is a function of the first m components in $\mathbf{W}$ and since the components in $\mathbf{W}$ are independent, it follows that $\mathbf{Y}'\mathbf{AY}$ and $\mathbf{CY}$ are independent.

Theorem 5.2.2 Let $\mathbf{A}$ and $\mathbf{B}$ be symmetric $n \times n$ matrices of constants and suppose $\mathbf{Y}$, an $n \times 1$ random vector, is $MN(\boldsymbol{\mu}, \sigma^2\mathbf{I})$. The quadratic forms $\mathbf{Y}'\mathbf{AY}$ and $\mathbf{Y}'\mathbf{BY}$ are independent if $\mathbf{BA}$ is the zero matrix.

These last two theorems can be generalized to the case when $\mathbf{Y}$ is $MN(\boldsymbol{\mu},\mathbf{V})$. In this case, independence occurs if and only if $\mathbf{CVA}$ is the zero matrix in Theorem 5.2.1. In Theorem 5.2.2, independence occurs if and only if $\mathbf{BVA}$ is the zero matrix. These more general results are developed in Searle (1971) but will not be needed in this text.

We can now use the first result, Theorem 5.2.1, to establish independence between the least squares estimator for an estimable function $\boldsymbol{\ell}'\boldsymbol{\beta}$ and SSE. This fact will be utilized later in this chapter in order to derive a statistic that can be used to obtain a confidence interval for an estimable function.

Theorem 5.2.3 Consider the linear model $\mathbf{Y} = \mathbf{X}\boldsymbol{\beta} + \boldsymbol{\varepsilon}$, where $\boldsymbol{\varepsilon}$ is $MN(\mathbf{0}, \sigma^2\mathbf{I})$. Let $\boldsymbol{\ell}'\boldsymbol{\beta}$ be an estimable function. Then the least squares estimator $\boldsymbol{\ell}'\hat{\boldsymbol{\beta}}$ and SSE are independent.

Proof. First we note that $\boldsymbol{\ell}'\hat{\boldsymbol{\beta}} = \boldsymbol{\ell}'(\mathbf{X}'\mathbf{X})^-\mathbf{X}'\mathbf{Y}$ which is of the form $\mathbf{CY}$ with $\mathbf{C} = \boldsymbol{\ell}'(\mathbf{X}'\mathbf{X})^-\mathbf{X}'$. Second, using Theorem 5.1.3, we express SSE as $SSE = \mathbf{Y}'[\mathbf{I} - \mathbf{X}(\mathbf{X}'\mathbf{X})^-\mathbf{X}']\mathbf{Y}$ which is of the form $\mathbf{Y}'\mathbf{AY}$ with $\mathbf{A} = \mathbf{I} - \mathbf{X}(\mathbf{X}'\mathbf{X})^-\mathbf{X}'$. According to Theorem 5.2.1, we can establish independence by verifying that $\mathbf{CA} = \mathbf{0}$. Here

$$
\begin{aligned}
\mathbf{CA} &= \boldsymbol{\ell}'(\mathbf{X}'\mathbf{X})^-\mathbf{X}'[\mathbf{I} - \mathbf{X}(\mathbf{X}'\mathbf{X})^-\mathbf{X}'] \\
&= \boldsymbol{\ell}'(\mathbf{X}'\mathbf{X})^-\mathbf{X}' - \boldsymbol{\ell}'(\mathbf{X}'\mathbf{X})^-\mathbf{X}'\mathbf{X}(\mathbf{X}'\mathbf{X})^-\mathbf{X}' \\
&= \boldsymbol{\ell}'(\mathbf{X}'\mathbf{X})^-\mathbf{X}' - \mathbf{c}'\mathbf{X}(\mathbf{X}'\mathbf{X})^-\mathbf{X}'\mathbf{X}(\mathbf{X}'\mathbf{X})^-\mathbf{X}' \quad \text{(Theorem 4.3.1)} \\
&= \boldsymbol{\ell}'(\mathbf{X}'\mathbf{X})^-\mathbf{X}' - \mathbf{c}'\mathbf{X}(\mathbf{X}'\mathbf{X})^-\mathbf{X}' \quad \text{(Theorem 3.2.3(2))} \\
&= \boldsymbol{\ell}'(\mathbf{X}'\mathbf{X})^-\mathbf{X}' - \boldsymbol{\ell}'(\mathbf{X}'\mathbf{X})^-\mathbf{X}' = \mathbf{0}.
\end{aligned}
$$

An Example of Independence

In a first course in mathematical statistics the reader learned that in sampling from a normal population, the sample mean and the sample variance are independently distributed. Because of its difficulty, generally this important fact is stated without proof. We now have the "tools" to establish this result in a relatively easy manner.

Example 5.2.1 Consider a random sample $Y_1, Y_2, \ldots, Y_n$ from a

$N(\mu,\sigma^2)$ distribution. We demonstrate that $\bar{Y} = \sum_{i=1}^{n} Y_i/n$ and $S^2 = \sum_{i=1}^{n} (Y_i - \bar{Y})^2/(n-1)$ are independent. To establish independence we will express $\bar{Y}$ in the form $\mathbf{CY}$ and S^2 in the form $\mathbf{Y'AY}$ where $\mathbf{Y}' = (Y_1,Y_2,\ldots,Y_n)$. Since the Y_i's are mutually independent and identically distributed $N(\mu,\sigma^2)$ random variables, it follows that $\mathbf{Y}$ is $MN(\boldsymbol{\mu},\sigma^2\mathbf{I})$. Hence independence will follow from Theorem 5.2.1 if $\mathbf{CA} = \mathbf{0}$. Obviously, $\bar{Y}$ can be expressed as $\mathbf{CY}$ if we let $\mathbf{C}$ be the $1 \times n$ matrix $[\frac{1}{n}, \frac{1}{n},\ldots,\frac{1}{n}]$. Further, in Example 5.1.1 we noted that

$\sum_{i=1}^{n} (Y_i - \bar{Y})^2$ can be expressed as a quadratic form with a matrix having off–diagonal entries $-1/n$ and diagonal elements $(n-1)/n$. Accordingly, it is easy to see that S^2 can be written as $\mathbf{Y'AY}$ where the entries in $\mathbf{A}$ are $1/n$ on the diagonal and $-1/n(n-1)$ off–diagonal. Therefore, each element of the $1 \times n$ matrix $\mathbf{CA}$ is calculated as $(1/n)[(1/n) + (n-1)\{-1/n(n-1)\}] = 0$. This establishes independence.

5.3 Interval Estimation for Estimable Functions

Our objective in this section is to obtain a confidence interval for an estimable function $\psi = \boldsymbol{\ell}'\boldsymbol{\beta}$. For example, in one–way classification we may be interested in comparing two population means, say $\mu_1 = \mu + \tau_1$ and $\mu_2 = \mu + \tau_2$. A confidence interval for the difference in these population means, namely, $(\mu + \tau_1) - (\mu + \tau_2) = \tau_1 - \tau_2$ would then be desirable.

In order to find a confidence interval for an estimable function $\boldsymbol{\ell}'\boldsymbol{\beta}$, it will be necessary to find an expression with a known distribution that involves this function and its estimator, $\boldsymbol{\ell}'\hat{\boldsymbol{\beta}}$. To be useful for our purpose, any other quantities in the expression must be accessible from available data. Expressions having these characteristics will also be the basis for the hypothesis testing procedures presented in the next section. Since most of our inferences will be based on expressions having t and F distributions, it is appropriate for us to review the genesis of these distributions.

t and F Random Variables

Recall from a previous course in mathematical statistics that a random variable has a t distribution with ν degrees of freedom if it has the form

$$T = \frac{Z}{\sqrt{W/\nu}}, \tag{5.3.1}$$

where Z is N(0,1), W is chi–square with ν degrees of freedom, and Z and W are independently distributed. Also recall that a random variable has an F distribution with numerator degrees of freedom ν_1 and denominator degrees of freedom ν_2 if it has the form

$$F = \frac{W_1/\nu_1}{W_2/\nu_2}, \tag{5.3.2}$$

where W_1 and W_2 are independent chi–square random variables with ν_1 and ν_2 degrees of freedom, respectively. We now obtain the distributional form needed for confidence intervals.

Theorem 5.3.1 If $\mathbf{Y} = \mathbf{X}\boldsymbol{\beta} + \boldsymbol{\varepsilon}$, where $\boldsymbol{\varepsilon}$ is $MN(\mathbf{0}, \sigma^2\mathbf{I})$, and if $\psi = \boldsymbol{\ell}'\boldsymbol{\beta}$ is an estimable function with least squares estimator $\hat{\psi} = \boldsymbol{\ell}'\hat{\boldsymbol{\beta}}$, then

a) $Z = (\boldsymbol{\ell}'\hat{\boldsymbol{\beta}} - \boldsymbol{\ell}'\boldsymbol{\beta})/\sqrt{\sigma^2\boldsymbol{\ell}'(\mathbf{X}'\mathbf{X})^-\boldsymbol{\ell}}$ is N(0,1),

b) $T = (\boldsymbol{\ell}'\hat{\boldsymbol{\beta}} - \boldsymbol{\ell}'\boldsymbol{\beta})/\sqrt{\hat{\sigma}^2\boldsymbol{\ell}'(\mathbf{X}'\mathbf{X})^-\boldsymbol{\ell}}$ has the t distribution with $n - r$ degrees of freedom, where r is the rank of the matrix $\mathbf{X}$.

Proof. From Theorem 4.6.2, we know that $\boldsymbol{\ell}'\hat{\boldsymbol{\beta}}$ is $N(\boldsymbol{\ell}'\boldsymbol{\beta}, \sigma^2\boldsymbol{\ell}'(\mathbf{X}'\mathbf{X})^-\boldsymbol{\ell})$. By standardizing this random variable, we immediately arrive at the part (a) result. For part (b) we use Theorem 5.1.4 which states that $SSE/\sigma^2 = (n - r)\hat{\sigma}^2/\sigma^2$ is chi–square with $n - r$ degrees of freedom where r is the rank of $\mathbf{X}$. We know that SSE and $\boldsymbol{\ell}'\hat{\boldsymbol{\beta}}$ are independent (Theorem 5.2.3). Consequently,

$Z = (\boldsymbol{\ell}'\hat{\boldsymbol{\beta}} - \boldsymbol{\ell}'\boldsymbol{\beta})/\sqrt{\sigma^2\boldsymbol{\ell}'(\mathbf{X}'\mathbf{X})^-\boldsymbol{\ell}}$ and $W = (n - r)\hat{\sigma}^2/\sigma^2$ are independent. Using the form of a t random variable, as given in (5.3.1), we obtain the expression

$$T = \frac{(\boldsymbol{\ell}'\hat{\boldsymbol{\beta}} - \boldsymbol{\ell}'\boldsymbol{\beta})/\sqrt{\sigma^2\boldsymbol{\ell}'(\mathbf{X}'\mathbf{X})^-\boldsymbol{\ell}}}{\sqrt{\dfrac{(n-r)\hat{\sigma}^2}{\sigma^2(n-r)}}} = \frac{\boldsymbol{\ell}'\hat{\boldsymbol{\beta}} - \boldsymbol{\ell}'\boldsymbol{\beta}}{\sqrt{\hat{\sigma}^2\boldsymbol{\ell}'(\mathbf{X}'\mathbf{X})^-\boldsymbol{\ell}}},$$

which has the t distribution with $n - r$ degrees of freedom.

We defer to the next section for the derivation of an important expression having an F distribution using formula (5.3.2). At this time, however, it is convenient to recall another fact from mathematical statistics which results in an F random variable. The square of a t random variable with ν degrees of freedom is an F random variable with 1 and ν degrees of freedom. Hence from part (b) of the preceding theorem

$$F = (\boldsymbol{\ell}'\hat{\boldsymbol{\beta}} - \boldsymbol{\ell}'\boldsymbol{\beta})^2/(\hat{\sigma}^2\boldsymbol{\ell}'(\mathbf{X}'\mathbf{X})^-\boldsymbol{\ell}) \tag{5.3.3}$$

has the F distribution with 1 and $n - r$ degrees of freedom. Expression (5.3.3) will be useful for us in a later chapter.

Standard Error and Some Notational Aspects

Since $E(\boldsymbol{\ell}'\hat{\boldsymbol{\beta}}) = \boldsymbol{\ell}'\boldsymbol{\beta}$ and $\mathrm{Var}(\boldsymbol{\ell}'\hat{\boldsymbol{\beta}}) = \sigma^2\boldsymbol{\ell}'(\mathbf{X}'\mathbf{X})^-\boldsymbol{\ell}$ for an estimable function, we note that the form of the standard normal variable in Theorem 5.3.1 is $[\boldsymbol{\ell}'\hat{\boldsymbol{\beta}} - E(\boldsymbol{\ell}'\hat{\boldsymbol{\beta}})]/\sqrt{\mathrm{Var}(\boldsymbol{\ell}'\hat{\boldsymbol{\beta}})}$. In most cases, the quantity $\mathrm{Var}(\boldsymbol{\ell}'\hat{\boldsymbol{\beta}})$ is unknown because σ^2 is unknown. Consequently, we use $\hat{\sigma}^2$ as the symbol for an estimator of σ^2 giving us the expression $\hat{\sigma}^2\boldsymbol{\ell}'(\mathbf{X}'\mathbf{X})^-\boldsymbol{\ell}$ for the unbiased estimator of $\mathrm{Var}(\boldsymbol{\ell}'\hat{\boldsymbol{\beta}})$. This latter expression we will denote by $\widehat{\mathrm{Var}}(\boldsymbol{\ell}'\hat{\boldsymbol{\beta}})$, that is,

$$\widehat{\mathrm{Var}}(\boldsymbol{\ell}'\hat{\boldsymbol{\beta}}) = \hat{\sigma}^2\boldsymbol{\ell}'(\mathbf{X}'\mathbf{X})^-\boldsymbol{\ell}. \tag{5.3.4}$$

The point estimate for $\operatorname{Var}(\boldsymbol{\ell}'\hat{\boldsymbol{\beta}})$ is $s^2\boldsymbol{\ell}'(\mathbf{X}'\mathbf{X})^-\boldsymbol{\ell}$ where s^2 is the observed value of the random variable $\hat{\sigma}^2$, and the positive square root of this point estimate is called the **standard error** of $\boldsymbol{\ell}'\hat{\boldsymbol{\beta}}$. It should be noted here that we are using the term "standard error" as a numerical value, which is the common practice among users. As a consequence, the standard error is not the same as $\sqrt{\operatorname{Var}(\boldsymbol{\ell}'\hat{\boldsymbol{\beta}})}$, which is the standard deviation of $\boldsymbol{\ell}'\hat{\boldsymbol{\beta}}$. Since standard errors are very useful in statistical inference procedures, the numerical values of the standard errors of various point estimates are often contained in the printout of computer programs. See, for example, the SAS output illustrations in Chapter 4. For that reason we will use the distinguishing symbol $\operatorname{Se}(\boldsymbol{\ell}'\hat{\boldsymbol{\beta}})$ to represent the standard error of $\boldsymbol{\ell}'\hat{\boldsymbol{\beta}}$; that is,

$$\operatorname{Se}(\boldsymbol{\ell}'\hat{\boldsymbol{\beta}}) = \sqrt{s^2\boldsymbol{\ell}'(\mathbf{X}'\mathbf{X})^-\boldsymbol{\ell}}\ . \tag{5.3.5}$$

It is important to note that the estimator and point estimate for $\operatorname{Var}(\boldsymbol{\ell}'\hat{\boldsymbol{\beta}})$ are found by replacing σ^2 by $\hat{\sigma}^2$ and s^2, respectively, in the expression for $\operatorname{Var}(\boldsymbol{\ell}'\hat{\boldsymbol{\beta}})$. We summarize in the form of a rule.

Rule 5.3.1 Let $\boldsymbol{\ell}'\boldsymbol{\beta}$ be an estimable function and $\boldsymbol{\ell}'\hat{\boldsymbol{\beta}}$ be its least squares estimator.

1. The unbiased estimator for $\operatorname{Var}(\boldsymbol{\ell}'\hat{\boldsymbol{\beta}})$, denoted by $\widehat{\operatorname{Var}}(\boldsymbol{\ell}'\hat{\boldsymbol{\beta}})$, is obtained by replacing σ^2 by $\hat{\sigma}^2$ in the expression for $\operatorname{Var}(\boldsymbol{\ell}'\hat{\boldsymbol{\beta}})$.
2. The point estimate for $\operatorname{Var}(\boldsymbol{\ell}'\hat{\boldsymbol{\beta}})$ is obtained by replacing σ^2 by s^2 in the expression for $\operatorname{Var}(\boldsymbol{\ell}'\hat{\boldsymbol{\beta}})$. The standard error for $\boldsymbol{\ell}'\hat{\boldsymbol{\beta}}$, denoted by $\operatorname{Se}(\boldsymbol{\ell}'\hat{\boldsymbol{\beta}})$, is the positive square root of the point estimate for $\operatorname{Var}(\boldsymbol{\ell}'\hat{\boldsymbol{\beta}})$.

Confidence Interval for $\boldsymbol{\ell}'\boldsymbol{\beta}$

We now find the form of a $100\,(1-\alpha)\%$ confidence interval for an estimable function $\boldsymbol{\ell}'\boldsymbol{\beta}$. The statistic we use for this is the one in part (b) of Theorem 5.3.1, namely,

$$T = \frac{\boldsymbol{\ell}'\hat{\boldsymbol{\beta}} - \boldsymbol{\ell}'\boldsymbol{\beta}}{\sqrt{\widehat{\mathrm{Var}}(\boldsymbol{\ell}'\hat{\boldsymbol{\beta}})}},$$

which has the t distribution with $n - r$ degrees of freedom.

Let $t_{\alpha/2,n-r}$ denote the upper $\alpha/2$ percentile point for a t random variable having $n - r$ degrees of freedom. Then

$$P[-t_{\alpha/2,n-r} < (\boldsymbol{\ell}'\hat{\boldsymbol{\beta}} - \boldsymbol{\ell}'\boldsymbol{\beta})/\sqrt{\widehat{\mathrm{Var}}(\boldsymbol{\ell}'\hat{\boldsymbol{\beta}})} < t_{\alpha/2,n-r}] = 1 - \alpha$$

and

$$P[\boldsymbol{\ell}'\hat{\boldsymbol{\beta}} - t_{\alpha/2,n-r}\sqrt{\widehat{\mathrm{Var}}(\boldsymbol{\ell}'\hat{\boldsymbol{\beta}})} < \boldsymbol{\ell}'\boldsymbol{\beta} < \boldsymbol{\ell}'\hat{\boldsymbol{\beta}} + t_{\alpha/2,n-r}\sqrt{\widehat{\mathrm{Var}}(\boldsymbol{\ell}'\hat{\boldsymbol{\beta}})}]$$
$$= 1 - \alpha .$$

This leads to the following $100(1 - \alpha)\%$ confidence interval for $\boldsymbol{\ell}'\boldsymbol{\beta}$.

> A $100(1 - \alpha)\%$ confidence interval for an estimable $\boldsymbol{\ell}'\boldsymbol{\beta}$ is given by
>
> $$(\boldsymbol{\ell}'\mathbf{b} - t_{\alpha/2,n-r}\sqrt{s^2\boldsymbol{\ell}'(\mathbf{X}'\mathbf{X})^-\boldsymbol{\ell}},\ \boldsymbol{\ell}'\mathbf{b} + t_{\alpha/2,n-r}\sqrt{s^2\boldsymbol{\ell}'(\mathbf{X}'\mathbf{X})^-\boldsymbol{\ell}}) \quad (5.3.6)$$
>
> which is of the form $\boldsymbol{\ell}'\mathbf{b} \pm t_{\alpha/2,n-r}\,\mathrm{Se}(\boldsymbol{\ell}'\hat{\boldsymbol{\beta}})$, where **b** is a solution to the least squares normal equations.

Notice that we are presenting the form of a confidence interval in terms of **values** of random quantities. Also note that confidence interval (5.3.6) is of the form

$$\text{point estimate} \pm t_{\alpha/2,n-r}\ (\text{standard error}). \qquad (5.3.7)$$

This form is common for confidence intervals and is commonly used throughout statistics in situations with small samples. Finally, the confidence interval (5.3.6) is based on the assumption of normally distributed observations. If the assumption of normality is not tenable, then (5.3.6) may not be appropriate for calculating a confidence interval for $\boldsymbol{\ell}'\boldsymbol{\beta}$.

We illustrate the use of confidence interval (5.3.6) with two examples.

Example 5.3.1 This example will be a follow—up to previous examples based on the analysis of the secchi disk data of Example 2.6.1. Using the one—way classification model,

$$Y_{ij} = \mu + \tau_i + \varepsilon_{ij} \ , \quad i = 1,2,3; \ \ j = 1,2,3,4,5,$$

in Example 4.4.1 we considered the problem of finding the BLUE for $\tau_3 - \tau_1$. There we assumed $E(\boldsymbol{\varepsilon}) = \mathbf{0}$ and $Cov(\boldsymbol{\varepsilon}) = \sigma^2\mathbf{I}$ for the random vector $\boldsymbol{\varepsilon}$ in the linear model form $\mathbf{Y} = \mathbf{X}\boldsymbol{\beta} + \boldsymbol{\varepsilon}$ and found that the numerical value of the point estimate for $\tau_3 - \tau_1$ is $t_3 - t_1$ $= 8.6$ with $Var(\widehat{\tau_3 - \tau_1}) = Var(\bar{Y}_{3\cdot} - \bar{Y}_{1\cdot}) = 2\sigma^2/5$. In Section 5.1 we discovered that $s^2 = 1.2333$. Now, using Rule 5.3.1, we obtain the standard error

$$Se(\widehat{\tau_3 - \tau_1}) = \sqrt{2s^2/5} = \sqrt{2(1.2333)/5} = 0.7024 \ .$$

If our goal is to find a 95% confidence interval for $\tau_3 - \tau_1$, we need to further assume that $\boldsymbol{\varepsilon}$ is $MN(\mathbf{0},\sigma^2\mathbf{I})$. Then

$$\boldsymbol{\ell}'\mathbf{b} \pm t_{\alpha/2,n-r} \, Se(\boldsymbol{\ell}'\hat{\boldsymbol{\beta}})$$

where $\boldsymbol{\ell}' = (0 \ \ -1 \ \ 0 \ \ 1)$, $\mathbf{b}' = (m \ t_1 \ t_2 \ t_3)$, $\boldsymbol{\ell}'\hat{\boldsymbol{\beta}} = \widehat{\tau_3 - \tau_1}$ and $t_{\alpha/2,n-r} = t_{.025,12} = 2.179$ becomes

$$8.6 \pm 2.179(0.7024) \quad \text{or} \quad (7.07, \ \ 10.13) \ .$$

This interval indicates with 95% confidence that the mean secchi disk reading at location 3 of the lake is between 7.07 and 10.13 units higher than the mean at location 1.

Example 5.3.2 Let us consider the television watching data in Example 2.2.1. For these data we used the simple linear model

$$Y_i = \beta_0 + \beta_1 x_i + \varepsilon_i\ ,\quad i = 1,2,...,10\ .$$

where Y represents the number of hours a person watches television and x represents the number of years of education. In Example 2.3.1 we found the least squares estimate $\mathbf{b}' = (b_0, b_1) = (4.69320, -0.18821)$. Our objective here is to calculate a 95% confidence interval for the slope β_1. We assume that $\boldsymbol{\varepsilon}$ is $MN(\mathbf{0}, \sigma^2\mathbf{I})$. Since the simple linear regression model is a full rank model, every function $\boldsymbol{\ell}'\boldsymbol{\beta}$ is estimable. We can then use (5.3.6) to find the desired confidence interval. That is, the confidence interval will have the form

$$\boldsymbol{\ell}'\mathbf{b} \pm t_{\alpha/2,n-r}\sqrt{s^2\boldsymbol{\ell}'(\mathbf{X}'\mathbf{X})^-\boldsymbol{\ell}}\ .$$

In this problem $\boldsymbol{\ell}' = (0,1)$ and $\boldsymbol{\ell}'\mathbf{b} = b_1 = -0.18821$. The rank of the $\mathbf{X}$ matrix is 2 so $n - r = 10 - 2 = 8$ and $t_{\alpha/2,n-r} = t_{.025,8} = 2.306$. Next we find $s^2 = SSe/(n - r)$. Basic calculations yield

$$\begin{aligned} SSe &= \mathbf{y}'\mathbf{y} - \mathbf{b}'\mathbf{X}'\mathbf{y} = 50.61 - [4.69320 \quad -0.18821]\begin{bmatrix} 21.9 \\ 277.7 \end{bmatrix} \\ &= 50.61 - 50.5152 = 0.0948 \end{aligned}$$

so that $s^2 = 0.0948/8 = 0.0119.$

To calculate $\boldsymbol{\ell}'(\mathbf{X}'\mathbf{X})^-\boldsymbol{\ell}$ we find that

$$\mathbf{X}'\mathbf{X} = \begin{bmatrix} 10 & 133 \\ 133 & 1841 \end{bmatrix} \text{ and } (\mathbf{X}'\mathbf{X})^- = (\mathbf{X}'\mathbf{X})^{-1} = \begin{bmatrix} 2.553398 & -0.184466 \\ -0.184466 & 0.013870 \end{bmatrix}.$$

Then $\boldsymbol{\ell}'(\mathbf{X}'\mathbf{X})^-\boldsymbol{\ell} = 0.01387$ so that

$$Se(\hat{\beta}_1) = \sqrt{s^2\boldsymbol{\ell}'(\mathbf{X}'\mathbf{X})^-\boldsymbol{\ell}} = \sqrt{(0.0119)\ (0.01387)} = 0.0128\ .$$

Using the foregoing results we obtain a 95% confidence interval for β_1 as

$$-0.18821 \pm 2.306(0.0128)$$

$$-0.18821 \pm 0.02952$$

which is (−0.2177, −0.1587). Thus we are 95% confident that the slope is between −0.2177 and −0.1587.

The reader should keep in mind that the 95% confidence interval in the immediate previous example does **not** mean that $P(-0.2177 < \beta_1 < -0.1587) = 0.95$. Since β_1, in fact, is a number, the preceding probability is either one or zero. Instead, 95% confidence means that the procedure we are using will yield a confidence interval containing the true value of β_1 95% of the time.

Finally, it should be pointed out that when a computer regression program is available, the tedious calculations detailed in Example 5.3.2 can be avoided. To illustrate, Table 5.3.1 is an excerpt from the SAS GLM printout of the regression analysis for the television watching data. There we find the crucial quantities $b_1 = -0.18821$ and $Se(\hat{\beta}_1) = 0.01283$ printed out for use in the construction of a particular $100(1 - \alpha)\%$ confidence interval for β_1. Using the information that $b_0 = 4.69320$ $Se(\hat{\beta}_0) = 0.17411$, the student is asked to calculate a 90% confidence interval for β_0 in the exercises.

Table 5.3.1 Excerpt from SAS Printout for Television Watching Data

PARAMETER	ESTIMATE	T for HO: PARAMETER = 0	PR > |T|	STD ERROR OF ESTIMATE
INTERCEPT	4.69320388	26.96	0.0001	0.17411192
YRED	-0.18821982	-14.67	0.0001	0.01283221

5.4 Testing Hypotheses

One of the major activities in the analysis of experimental design data is the performance of certain tests of hypotheses. For example, in the one-way classification problem, we may be interested in testing the equality of means hypothesis,

$$H_0: \mu_1 = \mu_2 = \cdots = \mu_p \text{ or equivalently } H_0: \tau_1 = \tau_2 = \cdots = \tau_p .$$

Our objective in this section is to develop a general procedure for performing tests of hypotheses that yield answers to questions that gave rise to the experiment in the first place.

General Linear Hypothesis

The hypothesis of interest will generally involve one or more estimable functions of the form

$$H_0: \boldsymbol{\ell}_1'\boldsymbol{\beta} = \gamma_1, \ \boldsymbol{\ell}_2'\boldsymbol{\beta} = \gamma_2, \ldots, \boldsymbol{\ell}_q'\boldsymbol{\beta} = \gamma_q . \tag{5.4.1}$$

Here $\boldsymbol{\ell}_1'\boldsymbol{\beta}$, $\boldsymbol{\ell}_2'\boldsymbol{\beta}, \ldots, \boldsymbol{\ell}_q'\boldsymbol{\beta}$ are q linearly independent estimable functions and $\gamma_1, \gamma_2, \ldots, \gamma_q$ are specified constants. As an example, consider the one—way classification with $p = 3$ treatments. The hypothesis $H_0: \tau_1 = \tau_2 = \tau_3$ can equivalently be written as

$$H_0: \tau_1 - \tau_2 = 0, \quad \tau_1 - \tau_3 = 0 .$$

Since $\tau_1 - \tau_2$ and $\tau_1 - \tau_3$ are linearly independent estimable functions we have demonstrated how the hypothesis of interest, $H_0: \tau_1 = \tau_2 = \tau_3$, can be expressed in the form of (5.4.1).

It is convenient to use matrix notation in the writing of hypothesis (5.4.1). Since the q estimable functions in (5.4.1) are linearly independent, the vectors $\boldsymbol{\ell}_1, \boldsymbol{\ell}_2, \ldots, \boldsymbol{\ell}_q$ must be linearly independent. If we let $\mathbf{L}$ be a $q \times (k + 1)$ matrix whose rows are $\boldsymbol{\ell}_1', \boldsymbol{\ell}_2', \ldots, \boldsymbol{\ell}_q'$ and $q \leq k + 1$, then $\mathbf{L}$ has rank q. Note that $\mathbf{L}\boldsymbol{\beta}$ is a set of q linearly independent estimable functions. Let $\boldsymbol{\gamma}$ be the $q \times 1$ vector of constants $\gamma_1, \gamma_2, \ldots, \gamma_q$ in hypothesis (5.4.1). Then (5.4.1) can be written as

$$H_0: \mathbf{L}\boldsymbol{\beta} = \boldsymbol{\gamma} . \tag{5.4.2}$$

We refer to hypothesis (5.4.2) as the **general linear hypothesis.** Our objective now will be to obtain a test for this hypothesis. First we consider two examples which illustrate the versatility of the general linear hypothesis.

Example 5.4.1 Consider the one—way classification with $p = 4$ treatments and model $Y_{ij} = \mu + \tau_i + \varepsilon_{ij}$, $i = 1,2,3,4$; $j = 1,2,\ldots,n_i$, where $\boldsymbol{\varepsilon}$ is $MN(\mathbf{0}, \sigma^2\mathbf{I})$. In matrix terms $\boldsymbol{\beta}' = (\mu, \tau_1, \tau_2, \tau_3, \tau_4)$. The hypothesis H_0: $\tau_1 = \tau_2 = \tau_3 = \tau_4$ is equivalent to H_0: $\tau_1 - \tau_2 = 0$, $\tau_1 - \tau_3 = 0$, $\tau_1 - \tau_4 = 0$. If we let

$$\mathbf{L} = \begin{bmatrix} 0 & 1 & -1 & 0 & 0 \\ 0 & 1 & 0 & -1 & 0 \\ 0 & 1 & 0 & 0 & -1 \end{bmatrix}, \text{ then } \mathbf{L}\boldsymbol{\beta} = \begin{bmatrix} \tau_1 - \tau_2 \\ \tau_1 - \tau_3 \\ \tau_1 - \tau_4 \end{bmatrix}$$

is a vector of three linearly independent estimable functions. Thus the hypothesis H_0: $\tau_1 = \tau_2 = \tau_3 = \tau_4$ is a general linear hypothesis since it can be written in the form (5.4.2) where in this case $\boldsymbol{\gamma}' = (0,0,0)$. Note that the 3×5 matrix $\mathbf{L}$ has rank three corresponding to the number of linearly independent estimable functions in $\mathbf{L}\boldsymbol{\beta}$.

Example 5.4.2 Consider a multiple regression model

$$Y_i = \beta_0 + \beta_1 x_{i1} + \beta_2 x_{i2} + \beta_3 x_{i3} + \varepsilon_i,\ i = 1,2,\ldots,n\ ,$$

where $\boldsymbol{\varepsilon}$ is $MN(\mathbf{0}, \sigma^2\mathbf{I})$. Suppose we wish to test the hypothesis H_0: $\beta_2 = \beta_3 = 0$. Here our $\boldsymbol{\beta}$ vector is given by $\boldsymbol{\beta}' = (\beta_0, \beta_1, \beta_2, \beta_3)$ and since the regression model is a full rank model, each β_i is estimable. If we let

$$\mathbf{L} = \begin{bmatrix} 0 & 0 & 1 & 0 \\ 0 & 0 & 0 & 1 \end{bmatrix} \text{ so that } \mathbf{L}\boldsymbol{\beta} = \begin{bmatrix} \beta_2 \\ \beta_3 \end{bmatrix},$$

then our hypothesis H_0 can be expressed as H_0: $\mathbf{L}\boldsymbol{\beta} = \boldsymbol{\gamma}'$ where $\boldsymbol{\gamma}' = (0,0)$. Thus H_0: $\beta_2 = \beta_3 = 0$ is a general linear hypothesis and we note that $\mathbf{L}$ has rank two corresponding to the number of linearly independent estimable functions in $\mathbf{L}\boldsymbol{\beta}$.

In order to perform the test for the general linear hypothesis (5.4.2), we need a suitable test statistic. The next result provides us with an F statistic which can be used for this purpose.

Theorem 5.4.1 Consider the normal linear model and let $\mathbf{L}$ be a $q \times (k+1)$ matrix of constants with rank $q \leq k+1$. Let $\mathbf{L}\boldsymbol{\beta}$ be a vector of q estimable functions and let $\mathbf{L}\hat{\boldsymbol{\beta}}$ be the vector of their least squares estimators. Then the random variable

$$\frac{(\mathbf{L}\hat{\boldsymbol{\beta}} - \mathbf{L}\boldsymbol{\beta})'[\mathbf{L}(\mathbf{X}'\mathbf{X})^{-}\mathbf{L}']^{-1}(\mathbf{L}\hat{\boldsymbol{\beta}} - \mathbf{L}\boldsymbol{\beta})}{q\hat{\sigma}^2} \tag{5.4.3}$$

has an F distribution with q and $n - r$ degrees of freedom, where r is the rank of the $\mathbf{X}$ matrix.

Proof. Since $\mathbf{Y}$ is $\text{MN}(\mathbf{X}\boldsymbol{\beta}, \sigma^2\mathbf{I})$ and $\hat{\boldsymbol{\beta}} = (\mathbf{X}'\mathbf{X})^{-}\mathbf{X}'\mathbf{Y}$, it follows from Theorem 4.5.1 that $\mathbf{L}\hat{\boldsymbol{\beta}}$ is q—variable multivariate normal. In the exercises it is shown that $E(\mathbf{L}\hat{\boldsymbol{\beta}}) = \mathbf{L}\boldsymbol{\beta}$ and $\text{Cov}(\mathbf{L}\hat{\boldsymbol{\beta}}) = \sigma^2\mathbf{L}(\mathbf{X}'\mathbf{X})^{-}\mathbf{L}'$. As a result, $\mathbf{L}\hat{\boldsymbol{\beta}}$ is $\text{MN}(\mathbf{L}\boldsymbol{\beta}, \sigma^2\mathbf{L}(\mathbf{X}'\mathbf{X})^{-}\mathbf{L}')$. Using Theorem 5.1.2, we immediately know that

$$(\mathbf{L}\hat{\boldsymbol{\beta}} - \mathbf{L}\boldsymbol{\beta})'[\mathbf{L}(\mathbf{X}'\mathbf{X})^{-}\mathbf{L}']^{-1}(\mathbf{L}\hat{\boldsymbol{\beta}} - \mathbf{L}\boldsymbol{\beta})/\sigma^2 \tag{5.4.4}$$

is distributed as chi—square with q degrees of freedom. Recall from Theorem 5.1.4 that SSE/σ^2, alternately written as

$$(n - r)\hat{\sigma}^2/\sigma^2 \tag{5.4.5}$$

has the chi—square distribution with $n - r$ degrees of freedom. It is easy to see that the expression for F in (5.4.3) is the ratio of expression (5.4.4) divided by q and expression (5.4.5) divided by $n - r$. That is, F in (5.4.3) is the ratio of two chi—square random variables divided by their respective degrees of freedom. Therefore from (5.3.2), if we show that these two chi—squares are independent, it will follow that F in (5.4.3) has an F distribution with q and $n - r$ degrees of freedom. To show independence we recall from Theorem 5.1.3 that since $(n - r)\hat{\sigma}^2 = \text{SSE}$, (5.4.5) can be written as a quadratic form in $\mathbf{Y} - \mathbf{X}\boldsymbol{\beta}$ with matrix $\mathbf{A} = \mathbf{I} - \mathbf{X}(\mathbf{X}'\mathbf{X})^{-}\mathbf{X}'$, while in the exercises we find that the numerator of (5.4.4) can be written as a quadratic form in $\mathbf{Y} - \mathbf{X}\boldsymbol{\beta}$

with matrix $\mathbf{B} = \mathbf{X}[(\mathbf{X}'\mathbf{X})^-]'\mathbf{L}'[\mathbf{L}(\mathbf{X}'\mathbf{X})^-\mathbf{L}']^{-1}\mathbf{L}(\mathbf{X}'\mathbf{X})^-\mathbf{X}'$.

Recognizing that $\mathbf{X}'\mathbf{X}(\mathbf{X}'\mathbf{X})^-\mathbf{X}' = \mathbf{X}'$ (Theorem 3.2.3), we see that $\mathbf{BA} = \mathbf{0}$ and the independence of (5.4.4) and (5.4.5) follows from Theorem 5.2.2. This completes the proof.

We can use the F statistic (5.4.3) to test the general linear hypothesis H_0: $\mathbf{L}\boldsymbol{\beta} = \boldsymbol{\gamma}$ versus H_1: $\mathbf{L}\boldsymbol{\beta} \neq \boldsymbol{\gamma}$. If $\mathbf{b}$ is a solution to the normal equations, then we reject H_0 when

$$f = \frac{(\mathbf{Lb} - \boldsymbol{\gamma})'[\mathbf{L}(\mathbf{X}'\mathbf{X})^-\mathbf{L}']^{-1}(\mathbf{Lb} - \boldsymbol{\gamma})}{qs^2} \geq f_{\alpha,q,n-r}, \tag{5.4.6}$$

where $f_{\alpha,q,n-r}$ is the upper α percentile point of the F distribution with q and $n - r$ degrees of freedom. It can be shown that this test is a likelihood ratio test for the general linear hypothesis. For example, see Graybill (1976).

Test for One Estimable Function

When there is only one estimable function $\boldsymbol{\ell}'\boldsymbol{\beta}$ of interest, the general linear hypothesis can be written as

$$H_0: \boldsymbol{\ell}'\boldsymbol{\beta} = \gamma \tag{5.4.7}$$

where γ is a constant. A test for this hypothesis is provided by the f–test in (5.4.6). However, the expression for f can be simplified considerably in this case. With only one estimable function $\boldsymbol{\ell}'\boldsymbol{\beta}$, we can replace q by 1, $\mathbf{L}$ by $\boldsymbol{\ell}'$ and $\boldsymbol{\gamma}$ by γ in (5.4.6). Then

$$\begin{aligned} f &= (\boldsymbol{\ell}'\mathbf{b} - \gamma)'[\boldsymbol{\ell}'(\mathbf{X}'\mathbf{X})^-\boldsymbol{\ell}]^{-1}(\boldsymbol{\ell}'\mathbf{b} - \gamma)/s^2 \\ &= (\boldsymbol{\ell}'\mathbf{b} - \gamma)^2/(s^2\boldsymbol{\ell}'(\mathbf{X}'\mathbf{X})^-\boldsymbol{\ell}) . \end{aligned}$$

Hence our test of H_0: $\boldsymbol{\ell}'\boldsymbol{\beta} = \gamma$ versus H_1: $\boldsymbol{\ell}'\boldsymbol{\beta} \neq \gamma$ rejects H_0 if

$$f = (\boldsymbol{\ell}'\mathbf{b} - \gamma)^2/(s^2\boldsymbol{\ell}'(\mathbf{X}'\mathbf{X})^-\boldsymbol{\ell}) \geq f_{\alpha,1,n-r} . \tag{5.4.8}$$

We know that the square of a t random variable with $n - r$ degrees of freedom has the F distribution with 1 and $n - r$ degrees of freedom. Hence (5.4.8) can be equivalently written as

$$t^2 = (\boldsymbol{\ell}'\mathbf{b} - \gamma)^2/(s^2\boldsymbol{\ell}'(\mathbf{X}'\mathbf{X})^-\boldsymbol{\ell}) \geq t^2_{\alpha/2,n-r},$$

or (5.4.9)

$$|t| = |\boldsymbol{\ell}'\mathbf{b} - \gamma|/\sqrt{s^2\boldsymbol{\ell}'(\mathbf{X}'\mathbf{X})^-\boldsymbol{\ell}} \geq t_{\alpha/2,n-r}.$$

This last expression provides us with a t–test for testing H_0: $\boldsymbol{\ell}'\boldsymbol{\beta} = \gamma$ versus suitable alternatives. We summarize this as follows.

t–test for H_0: $\boldsymbol{\ell}'\boldsymbol{\beta} = \gamma$ versus H_1: $\boldsymbol{\ell}'\boldsymbol{\beta} \neq \gamma$

Test statistic:

$$t = (\boldsymbol{\ell}'\mathbf{b} - \gamma)/\sqrt{s^2\boldsymbol{\ell}'(\mathbf{X}'\mathbf{X})^-\boldsymbol{\ell}} \qquad (5.4.10)$$

$$= (\boldsymbol{\ell}'\mathbf{b} - \gamma)/\mathrm{Se}(\boldsymbol{\ell}'\hat{\boldsymbol{\beta}})$$

Rejection region: Reject H_0 if $|t| \geq t_{\alpha/2,n-r}$.

Note: For one–sided tests

(i) H_0: $\boldsymbol{\ell}'\boldsymbol{\beta} \leq \gamma$ versus H_1: $\boldsymbol{\ell}'\boldsymbol{\beta} > \gamma$
Reject H_0 if $t \geq t_{\alpha,n-r}$.

(ii) H_0: $\boldsymbol{\ell}'\boldsymbol{\beta} \geq \gamma$ versus H_1: $\boldsymbol{\ell}'\boldsymbol{\beta} < \gamma$
Reject H_0 if $t \leq -t_{\alpha,n-r}$.

We illustrate this t–test with the following examples.

Example 5.4.3 Consider the secchi disk data given in Example 2.6.1 and later partially analyzed in Example 5.3.1. The one–way classification model for these data is

$$Y_{ij} = \mu + \tau_i + \varepsilon_{ij}, \quad i = 1,2,3; \; j = 1,2,3,4,5,$$

where the ε_{ij} are independently and identically distributed $N(0,\sigma^2)$ random variables. Suppose we wish to test the null hypothesis which states that the first and third population means are equal; that is, in symbols H_0: $\mu + \tau_1 = \mu + \tau_3$. We can write the hypothesis in terms of a single estimable function, namely, H_0: $\tau_3 - \tau_1 = 0$. In Example 5.3.1 we found the point estimate of $\boldsymbol{\ell}'\mathbf{b} = \tau_3 - \tau_1$ to be 8.6, with standard error $\text{Se}(\boldsymbol{\ell}'\hat{\boldsymbol{\beta}}) = \text{Se}(\widehat{\tau_3 - \tau_1}) = 0.7024$, $n - r = 12$. The test statistic (5.4.10) is

$$t = (\boldsymbol{\ell}'\mathbf{b} - \gamma)/\text{Se}(\boldsymbol{\ell}'\hat{\boldsymbol{\beta}})$$

$$= (8.6 - 0)/0.7024 = 12.24.$$

To perform an $\alpha = .05$ level test of H_0: $\tau_3 - \tau_1 = 0$ versus H_1: $\tau_3 - \tau_1 \neq 0$, we reject H_0 if $|t| \geq t_{.025,12} = 2.179$. Since $12.24 \geq 2.179$, we reject H_0 and conclude that the evidence is sufficient to declare the first and third population means different. Of course, we could have used the confidence interval obtained in Example 5.3.1 to test H_0: $\tau_3 - \tau_1 = 0$ versus H_1: $\tau_3 - \tau_1 \neq 0$. There we found the 95% confidence interval for $\mu_3 - \mu_1 = \tau_3 - \tau_1$ to be (7.07, 10.13). Since zero is not in this interval we reject H_0.

Example 5.4.4 Refer to the home energy data described in Example 2.4.1. These data were further analyzed in Example 4.2.1 and a computer printout to facilitate the analysis is contained in Table 4.4.2. The multiple regression model for this problem is

$$Y_i = \beta_0 + \beta_1 x_{i1} + \beta_2 x_{i2} + \beta_3 x_{i3} + \varepsilon_i,\ i = 1,2,\ldots,10\ ,$$

where, for inference purposes, the vector of the ε_i, $\boldsymbol{\varepsilon}$, is assumed to be $\text{MN}(\mathbf{0}, \sigma^2\mathbf{I})$. Let us test H_0: $\beta_1 = 0$ using the t—test of (5.4.10) with the aid of the printout in Table 4.4.2. From the estimate vector of the printout we find that the least squares estimate for β_1, the theoretical coefficient for average temperature (AVTEMP), is $b_1 = -1.321$. Note that b_1 is $\boldsymbol{\ell}'\mathbf{b}$ where $\boldsymbol{\ell}' = (0,1,0,0)$. The column headed STD

ERROR OF ESTIMATE reveals that $Se(\hat{\beta}_1) = \sqrt{s^2 \boldsymbol{\ell}'(\mathbf{X}'\mathbf{X})^- \boldsymbol{\ell}} =$ 0.3374. Hence the value of the test statistic is

$$t = b_1/Se(\hat{\beta}_1) = -1.321/0.3374 = -3.91.$$

This number also is given on the printout under the heading T FOR H0: PARAMETER = 0. Here we reject $H_0\colon \beta_1 = 0$ versus $H_1\colon \beta_1 \neq 0$ at the $\alpha = .05$ level since $|t| = 3.91 > t_{.025,6} = 2.447$.

Principle of Conditional Error

When the general linear hypothesis involves more than one estimable function, it is a tedious task to calculate the f—test (5.4.6). Therefore, it requires us to seek an alternative form for the expression of f , one that is computationally easier. The procedure that we use to obtain an alternative test form is called the **principle of conditional error.** We now describe this procedure.

Consider the normal linear model $\mathbf{Y} = \mathbf{X}\boldsymbol{\beta} + \boldsymbol{\varepsilon}$ and the general linear hypothesis $H_0\colon \mathbf{L}\boldsymbol{\beta} = \boldsymbol{\gamma}$. We call the model $\mathbf{Y} = \mathbf{X}\boldsymbol{\beta} + \boldsymbol{\varepsilon}$ the **full model** and we use SSe and s^2 to denote the associated error sum of squares and error mean square, respectively. If we impose the condition $\mathbf{L}\boldsymbol{\beta} = \boldsymbol{\gamma}$ on the full model we get a new model called the **reduced** (or **conditional** or **restricted**) **model**. That is, the reduced model is

$$\mathbf{Y} = \mathbf{X}\boldsymbol{\beta} + \boldsymbol{\varepsilon} \quad \text{with} \quad \mathbf{L}\boldsymbol{\beta} = \boldsymbol{\gamma} \,. \tag{5.4.11}$$

For this reduced model we can find the vector $\mathbf{b}$ which minimizes $(\mathbf{y} - \mathbf{X}\mathbf{b})'(\mathbf{y} - \mathbf{X}\mathbf{b})$ among vectors $\mathbf{b}$ satisfying $\mathbf{L}\mathbf{b} = \boldsymbol{\gamma}$. Let $\mathbf{b}^*$ denote such a vector. It is natural to name $(\mathbf{y} - \mathbf{X}\mathbf{b}^*)'(\mathbf{y} - \mathbf{X}\mathbf{b}^*)$ the error sum of squares for the reduced model and to denote it by SSe*. That is,

$$SSe^* = (\mathbf{y} - \mathbf{X}\mathbf{b}^*)'(\mathbf{y} - \mathbf{X}\mathbf{b}^*) \,. \tag{5.4.12}$$

Our next result shows how the quantities SSe, s^2 and SSe* can be used to obtain an equivalent expression for the f—test statistic (5.4.6).

Theorem 5.4.2 Consider the normal linear model with full model $\mathbf{Y} = \mathbf{X}\boldsymbol{\beta} + \boldsymbol{\varepsilon}$ which has error sum of squares SSe and error mean square s^2 . Consider the general linear hypothesis $H_0\colon \mathbf{L}\boldsymbol{\beta} = \boldsymbol{\gamma}$ where the $q \times (k + 1)$

matrix $\mathbf{L}$ has rank $q \leq k + 1$. Let SSe* be the error sum of squares for the reduced model $\mathbf{Y} = \mathbf{X}\boldsymbol{\beta} + \boldsymbol{\varepsilon}$ with $\mathbf{L}\boldsymbol{\beta} = \boldsymbol{\gamma}$. Then

(1) $\text{SSe}^* - \text{SSe} = (\mathbf{Lb} - \boldsymbol{\gamma})'[\mathbf{L}(\mathbf{X}'\mathbf{X})^- \mathbf{L}']^{-1}(\mathbf{Lb} - \boldsymbol{\gamma})$

and

(2) $f = \dfrac{(\text{SSe}^* - \text{SSe})/q}{s^2}$ equals the expression f in (5.4.6) and accordingly is a value of an F random variable with q and $n - r$ degrees of freedom.

Proof. First we find an expression for $\mathbf{b}^*$ in (5.4.12). This means we need to find the vector which minimizes $(\mathbf{y} - \mathbf{Xb})'(\mathbf{y} - \mathbf{Xb})$ subject to the constraint $\mathbf{Lb} = \boldsymbol{\gamma}$. To do this we use the Lagrange multiplier technique with multiplier $2\boldsymbol{\lambda}$. Proceeding with this technique we let

$$\begin{aligned} D &= (\mathbf{y} - \mathbf{Xb})'(\mathbf{y} - \mathbf{Xb}) + 2\boldsymbol{\lambda}'(\mathbf{Lb} - \boldsymbol{\gamma}) \\ &= \mathbf{y}'\mathbf{y} - 2\mathbf{b}'\mathbf{X}'\mathbf{y} + \mathbf{b}'\mathbf{X}'\mathbf{Xb} + 2\boldsymbol{\lambda}'(\mathbf{Lb} - \boldsymbol{\gamma}) . \end{aligned}$$

Using vector differentiation rules we obtain

$$\frac{\partial D}{\partial \mathbf{b}} = -2\mathbf{X}'\mathbf{y} + 2\mathbf{X}'\mathbf{Xb} + 2\mathbf{L}'\boldsymbol{\lambda} \qquad \text{and}$$

$$\frac{\partial D}{\partial \boldsymbol{\lambda}} = 2(\mathbf{Lb} - \boldsymbol{\gamma}) .$$

Setting these partials equal to the zero vector and denoting the solution for the system of equations by $\mathbf{b}^*$ we get

$$\begin{cases} \mathbf{X}'\mathbf{Xb}^* + \mathbf{L}'\boldsymbol{\lambda} = \mathbf{X}'\mathbf{y} \\ \mathbf{Lb}^* = \boldsymbol{\gamma} . \end{cases} \tag{5.4.13}$$

Now since $\mathbf{b} = (\mathbf{X}'\mathbf{X})^- \mathbf{X}'\mathbf{y}$ is a solution to $\mathbf{X}'\mathbf{Xb} = \mathbf{X}'\mathbf{y}$, $\mathbf{L} = \mathbf{CX}$ for some matrix $\mathbf{C}$ (Theorem 4.3.1), and $\mathbf{X}'\mathbf{X}(\mathbf{X}'\mathbf{X})^- \mathbf{X}' = \mathbf{X}'$ (Theorem 3.2.3), we find that $\mathbf{X}'\mathbf{X}(\mathbf{X}'\mathbf{X})^- \mathbf{X}'\mathbf{y} - \mathbf{X}'\mathbf{X}(\mathbf{X}'\mathbf{X})^- \mathbf{L}'\boldsymbol{\lambda} = \mathbf{X}'\mathbf{y} - \mathbf{L}'\boldsymbol{\lambda}$. It then follows that

$$\mathbf{b}^* = \mathbf{b} - (\mathbf{X}'\mathbf{X})^- \mathbf{L}'\boldsymbol{\lambda} \tag{5.4.14}$$

satisfies the first equation in (5.4.13).

Substituting this value for $\mathbf{b}^*$ into the second equation of (5.4.13) yields

$$\mathbf{Lb} - \mathbf{L}(\mathbf{X}'\mathbf{X})^{-}\mathbf{L}\boldsymbol{\lambda} = \boldsymbol{\gamma}\ .$$

Solving for $\boldsymbol{\lambda}$, we get

$$\boldsymbol{\lambda} = [\mathbf{L}(\mathbf{X}'\mathbf{X})^{-}\mathbf{L}']^{-1}(\mathbf{Lb} - \boldsymbol{\gamma})\ .$$

Using this value for $\boldsymbol{\lambda}$ in (5.4.14) gives us

$$\mathbf{b}^* = \mathbf{b} - (\mathbf{X}'\mathbf{X})^{-}\mathbf{L}'[\mathbf{L}(\mathbf{X}'\mathbf{X})^{-}\mathbf{L}']^{-1}(\mathbf{Lb} - \boldsymbol{\gamma}) \qquad \text{or}$$

$$\mathbf{b} - \mathbf{b}^* = (\mathbf{X}'\mathbf{X})^{-}\mathbf{L}'[\mathbf{L}(\mathbf{X}'\mathbf{X})^{-}\mathbf{L}']^{-1}(\mathbf{Lb} - \boldsymbol{\gamma})\ . \tag{5.4.15}$$

Now we find an expression for SSe*.

$$\begin{aligned}
\text{SSe}^* &= (\mathbf{y} - \mathbf{Xb}^*)'(\mathbf{y} - \mathbf{Xb}^*) \\
&= [(\mathbf{y} - \mathbf{Xb}) + \mathbf{X}(\mathbf{b} - \mathbf{b}^*)]'[(\mathbf{y} - \mathbf{Xb}) + \mathbf{X}(\mathbf{b} - \mathbf{b}^*)] \\
&= (\mathbf{y} - \mathbf{Xb})'(\mathbf{y} - \mathbf{Xb}) + (\mathbf{b} - \mathbf{b}^*)'\mathbf{X}'\mathbf{X}(\mathbf{b} - \mathbf{b}^*) \\
&\qquad + 2(\mathbf{b} - \mathbf{b}^*)'\mathbf{X}'(\mathbf{y} - \mathbf{Xb})
\end{aligned}$$

But $\mathbf{X}'(\mathbf{y} - \mathbf{Xb}) = \mathbf{X}'\mathbf{y} - \mathbf{X}'\mathbf{Xb} = \mathbf{0}$ and $(\mathbf{y} - \mathbf{Xb})'(\mathbf{y} - \mathbf{Xb}) = \text{SSe}$ so that $\text{SSe}^* = \text{SSe} + (\mathbf{b} - \mathbf{b}^*)'\mathbf{X}'\mathbf{X}(\mathbf{b} - \mathbf{b}^*)$.

Next, using result (5.4.15) and then simplifying by implementing Theorems 4.3.1 and 3.2.3, as before, we eventually arrive at the equality

$$(\mathbf{b} - \mathbf{b}^*)'\mathbf{X}'\mathbf{X}(\mathbf{b} - \mathbf{b}^*) = (\mathbf{Lb} - \boldsymbol{\gamma})'[\mathbf{L}(\mathbf{X}'\mathbf{X})^{-}\mathbf{L}']^{-1}(\mathbf{Lb} - \boldsymbol{\gamma}).$$

Finally,

$$\text{SSe}^* - \text{SSe} = (\mathbf{Lb} - \boldsymbol{\gamma})'[\mathbf{L}(\mathbf{X}'\mathbf{X})^{-}\mathbf{L}']^{-1}(\mathbf{Lb} - \boldsymbol{\gamma})$$

and therefore $\text{f} = \dfrac{(\text{SSe}^* - \text{SSe})/\text{q}}{\text{s}^2}$ coincides with that in expression (5.4.6). This completes the proof.

Our test statistic for the test of the general linear hypothesis, H_0: $\mathbf{L}\boldsymbol{\beta} = \boldsymbol{\gamma}$, can now be calculated using

$$\text{f} = (\text{SSe}^* - \text{SSe})/\text{qs}^2\ .$$

In most experimental design problems, this form of f is easier to calculate than the form in (5.4.6). The expression for f that we have just derived is called the **principle of conditional error** test statistic. The reason for the phrase "conditional error" is due to the fact that SSe* is the "conditional minimum" of $(y - Xb)'(y - Xb)$ with respect to b subject to the constraint $Lb = \gamma$. We summarize this approach to testing the general linear hypothesis.

Principle of Conditional Error Test Approach

Set up: Full model: $Y = X\beta + \varepsilon$, ε is $MN(0, \sigma^2 I)$, X has rank r.

General linear hypothesis: H_0: $L\beta = \gamma$

where L is a $q \times (k + 1)$ matrix of rank $q \leq k + 1$ and $L\beta$ constitutes q linearly independent estimable functions.

Steps:

1) For the full model calculate SSe and s^2.

2) Impose the condition $L\beta = \gamma$ on the full model to obtain the reduced model. Calculate the error sum of squares, SSe*, for this model.

3) For an α–level test, we reject H_0 if

$$f = \frac{(SSe^* - SSe/q}{s^2} \geq f_{\alpha,q,n-r}. \tag{5.4.16}$$

To close this topic, we illustrate this testing procedure with an example.

Example 5.4.5 Consider the secchi disk data introduced in Example 2.6.1 and seen again in Examples 5.3.1 and 5.4.3 of this chapter. The one–way classification model and, in this instance, the **full model**, is

$$Y_{ij} = \mu + \tau_1 + \varepsilon_{ij} , \quad i = 1,2,3; \; j = 1,2,3,4,5.$$

We assume that $\boldsymbol{\varepsilon}$ is $MN(\mathbf{0}, \sigma^2\mathbf{I})$. In Example 5.1.3 we found

$$SSe = 14.8, \; s^2 = 1.2333 \text{ and } n - r = 15 - 3 = 12 .$$

Our objective is to test H_0: $\tau_1 = \tau_2 = \tau_3$.

This can be written in general linear hypothesis form, H_0: $\mathbf{L}\boldsymbol{\beta} = \boldsymbol{\gamma}$ where

$$\mathbf{L} = \begin{bmatrix} 0 & 1 & -1 & 0 \\ 0 & 1 & 0 & -1 \end{bmatrix} , \boldsymbol{\beta}' = (\mu, \tau_1, \tau_2, \tau_3) , \boldsymbol{\gamma}' = (0,0)$$

and $\mathbf{L}$ has rank $q = 2$. To test H_0, we use the principle of conditional error. To find the reduced model, we note that when H_0 is true, the $\mu + \tau_i$'s are equal and can be written as $\mu + \tau_i = \mu^*$. Hence the **reduced model** in this case is

$$Y_{ij} = \mu^* + \varepsilon^*_{ij} , \quad i = 1,2,3; \; j = 1,2,3,4,5 .$$

In the exercises, it is shown that $\hat{y}_{ij} = \bar{y}_{..} = 44$ and the error sum of squares for the reduced model is

$$SSe^* = \sum_{i=1}^{3} \sum_{i=1}^{5} (y_{ij} - \bar{y}_{..})^2 = 224 .$$

Using $\alpha = .05$, we will reject H_0 if

$$f = \frac{(SSe^* - SSe)/2}{s^2} \geq f_{.05,2,12} = 3.89 .$$

Since $f = \dfrac{(224 - 14.8)/2}{1.2333} = 84.81$, we reject H_0 and conclude that at least one pair of means are unequal.

Chapter 5 Exercises

Application Exercises

5–1 Calculate MSe for the data in Exercise 3–1.

5–2 Calculate MSe for the electric motor data in Exercise 3–4.

5–3 Suppose $\mathbf{Y}' = (Y_1, Y_2)$ is $MN(\boldsymbol{\mu}, \sigma^2\mathbf{I})$ with $\boldsymbol{\mu}' = (1,3)$.

If $\mathbf{A} = \begin{bmatrix} 2 & 3 \\ 3 & 4 \end{bmatrix}$ find $E(\mathbf{Y}'\mathbf{AY})$.

5–4 Explain why the normality assumption is needed for the confidence interval given by formula (5.3.6). What is incorrect if the confidence interval is found and the observations are actually not normally distributed?

5–5 Refer to the data in Exercise 2–3. Using the value for SSe obtained in Exercise 3–12, find

(a) MSe.
(b) A 90% confidence interval for the average change in temperature per unit change in depth.
(c) A 90% confidence interval for the mean temperature when the depth is 7.5 meters.

5–6 Table 5.3.1 contains information on the simple linear regression model used for the television watching data. Use this information to find a 90% confidence interval for β_0.

5–7 Find a 95% confidence interval for β_1 using the diet data in Exercise 3–5. What does the confidence interval suggest about diets A and B?

5–8 Refer to the SAS printout in Table 4.4.2 for the home energy data. Use the information on the printout to find the following:

(a) A 95% confidence interval for the coefficient of AVTEMP.
(b) A 95% confidence interval for the mean electric cost, E(Y), when AVTEMP = 30, THERM = 67 and SQFT = 2000.

5–9 Consider the secchi disk data introduced in Chapter 2 and referred to in Examples 4.4.1, 5.3.1, 5.4.3 and 5.4.5. Find a 95%

confidence interval for $\psi = [(\tau_1 + \tau_2)/2] - \tau_3$. What does the estimate for ψ measure?

5–10 Refer to the maple tree height data of Exercise 2–4. In Exercise 3–13 the value of SSe was found for the simple linear regression model.

(a) Test $H_0:\beta_1 = 1$ vs. $H_1:\beta_1 \neq 1$ at the 0.05 level of significance.
(b) Calculate a 95% confidence interval for β_1.
(c) Calculate a 95% confidence interval for the mean E(Y) when $x = 24$.

5–11 Test the hypothesis $H_0:\beta_3 = 0$ using a .05 level of significance for the chemical process data given in Exercise 2–5. What does the outcome of this test indicate?

5–12 Refer to the fertilizer data of Exercise 3–19 and the model $Y = \beta_0 + \beta_1 x_1 + \varepsilon$.

(a) Calculate a 95% confidence interval for β_1.
(b) Test $H_0:\beta_1 = 0$ vs. $H_1:\beta_1 \neq 0$ at the .05 level using the result obtained in (a).
(c) Calculate a 90% confidence interval for the mean yield when 5 pounds of fertilizer is used.

5–13 Consider the reading scores data of Exercise 3–20 and the one–way classification model with three treatments.

(a) Find a 95% confidence interval for the mean reading score for the medium IQ group.
(b) Test the hypothesis $H_0:\tau_1 = \tau_2 = \tau_3$ using a .05 level of significance.
(c) Find a 95% confidence interval for $\tau_3 - \tau_2$. What does this interval suggest?

5–14 Fit the multiple regression model $Y = \beta_0 + \beta_1 x_1 + \beta_2 x_2 + \beta_3 x_3 + \varepsilon$ to the following data:

y	1	0	2	8	4	3	3
x_1	−3	−2	−1	0	1	2	3
x_2	2	−1	1	−4	1	−1	2
x_3	−1	1	1	0	−1	−1	1

Show that $\mathbf{b}' = [3 \frac{1}{2} \ -\frac{3}{4} \ -\frac{1}{2}]$ and $s^2 = 5.25$. At the 0.10 level, test the null hypothesis $H_0{:}\beta_2 = \beta_3$

(a) using the general linear hypothesis test statistic (5.4.6);

(b) using principle of conditional error test approach and statistic (5.4.16).

(c) Repeat parts (a) and (b) for the null hypothesis $H_0{:}\beta_2 = -1, \beta_3 = -1$.

5–15 Refer to the strength of paper data considered in Exercise 4–6 and the multiple regression model.

(a) Calculate the value of the test statistic used to test $H_0{:}\beta_3 = 0$ versus $H_0{:}\beta_3 \neq 0$ using (i) a t statistic and (ii) an F statistic. Observe that the .05 level test based on each statistic yields the same result.

(b) Using the .05 level, test the following null hypotheses:

(i) $H_0{:}\beta_1 = \beta_2 = \beta_3 = 0$,

(ii) $H_0{:}\beta_1 = \beta_2 = 0$,

(iii) $H_0{:}\beta_1 = \beta_2$.

Give an interpretation associated with the outcome of each test.

Theoretical Exercises

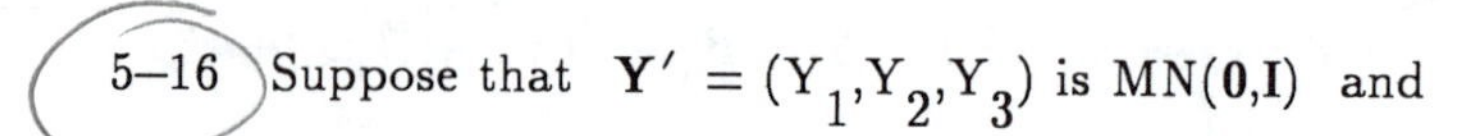

5–16 Suppose that $\mathbf{Y}' = (Y_1, Y_2, Y_3)$ is $MN(\mathbf{0}, \mathbf{I})$ and

$$\mathbf{A} = \begin{bmatrix} \frac{1}{3} & \frac{1}{3} & \frac{1}{3} \\ \frac{1}{3} & \frac{1}{3} & \frac{1}{3} \\ \frac{1}{3} & \frac{1}{3} & \frac{1}{3} \end{bmatrix} \qquad \mathbf{B} = \begin{bmatrix} \frac{2}{3} & -\frac{1}{3} & -\frac{1}{3} \\ -\frac{1}{3} & \frac{2}{3} & -\frac{1}{3} \\ -\frac{1}{3} & -\frac{1}{3} & \frac{2}{3} \end{bmatrix}$$

(a) Find the distribution of $Q_1 = \mathbf{Y}'\mathbf{AY}$.

(b) Find the distribution of $Q_2 = \mathbf{Y}'\mathbf{BY}$.

(c) Show that Q_1 and Q_2 are independent.

(d) What is the distribution of $F = 2Q_1/Q_2$?

(e) Verify that $Q_1 = 3\bar{Y}^2$ where $\bar{Y} = (Y_1 + Y_2 + Y_3)/3$.

(f) Verify that $Q_2 = \sum_{i=1}^{3} (Y_i - \bar{Y})^2$.

5–17 Show that the matrix $\mathbf{A}$ of a quadratic form can always be chosen to be symmetric. Hint: Consider $\mathbf{A}$ non–symmetric and compare $\mathbf{Y}'[(\mathbf{A} + \mathbf{A}')/2]\mathbf{Y}$ with $\mathbf{Y}'\mathbf{AY}$.

5–18 Suppose the $p \times 1$ random vector $\mathbf{Y}$ is $MN(\boldsymbol{\mu},\mathbf{I})$ and let $\mathbf{a}$ and $\mathbf{b}$ be $p \times 1$ constant vectors. Show that $\mathbf{a}'\mathbf{Y}$ and $\mathbf{b}'\mathbf{Y}$ are independent random variables if $\mathbf{a}'\mathbf{b} = 0$.

5–19 Suppose the $n \times 1$ vector $\mathbf{Y}$ is $MN(\mathbf{0},\mathbf{I})$ and $\mathbf{X}$ is an $n \times p$ constant matrix of rank p. Find the distribution of the quadratic form $Q = \mathbf{Y}'[\mathbf{I} - \mathbf{X}(\mathbf{X}'\mathbf{X})^{-1}\mathbf{X}']\mathbf{Y}$ and find $E(Q)$.

5–20 Let $\mathbf{Y}$ be an $n \times 1$ random vector which is $MN(\boldsymbol{\mu},\mathbf{I})$.

(a) Find the distribution of $Q_1 = \sum_{i=1}^{n} (Y_i - \mu_i)^2$.

(b) Find the distribution of $Q_2 = [\sum_{i=1}^{n} (Y_i - \mu_i)]^2/n$.

(c) Are Q_1 and Q_2 independent?

5–21 Consider the one–way classification problem of Example 2.6.2.

Explain why SSE and $\bar{Y}_{1\cdot}$ are independent.

5–22 Verify that $\widehat{\mathrm{Var}}(\boldsymbol{\ell}'\hat{\boldsymbol{\beta}}) = \hat{\sigma}^2\boldsymbol{\ell}'(\mathbf{X}'\mathbf{X})^-\boldsymbol{\ell}$ is unbiased for $\mathrm{Var}(\boldsymbol{\ell}'\hat{\boldsymbol{\beta}})$.

5–23 Use the confidence interval (5.3.6) to find a general expression for the confidence interval for β_0 and β_1 in a simple linear regression.

5–24 Consider the simple linear regression model. For a given value of x, say x_0, the mean of Y is $E(Y_0) = \beta_0 + \beta_1 x_0$. Find the form of a 100 $(1 - \alpha)$ % confidence interval for $E(Y_0)$.

5–25 For the simple linear regression model, find the form of the t–test (5.4.10) for testing $H_0 : \beta_1 = 0$.

5–26 Consider the multiple regression model. Use (5.3.6) to find the form of a confidence interval for the following:

(a) $\boldsymbol{\ell}'\boldsymbol{\beta}$, (b) β_j .

5–27 For the multiple regression model, find the form of the t–test (5.4.10) for testing $H_0 : \beta_j = 0$.

5–28 Consider the one–way classification problem of Example 2.6.2. Find the form of a 100 $(1 - \alpha)$ % confidence interval for $\tau_1 - \tau_2$.

5–29 For the one–way classification in example 5.4.5 verify that $\hat{y}_{ij} = \bar{y}_{\cdot\cdot}$ for the reduced model $Y_{ij} = \mu^* + \varepsilon^*_{ij}$.

5–30 Which of the following are general linear hypotheses?

(a) For a multiple regression problem with three independent variables, consider the hypotheses:

(i) $H_0 : \beta_1 = \beta_2 + 6$ (ii) $H_0 : \beta_1 = 2\beta_2$.

(b) For a one–way classification with $p = 3$ treatments consider the hypotheses:

(i) $H_0: \tau_1 = 0$ (ii) $H_0: \tau_1 = (\tau_2 + \tau_3)/2$

(iii) $H_0: \tau_1 = 2\tau_2$.

Identify the **L** matrix and its rank for the hypotheses that are general linear hypotheses.

5–31 Let $\boldsymbol{\ell}'\boldsymbol{\beta}$ be an estimable function. For testing $H_0: \boldsymbol{\ell}'\boldsymbol{\beta} = \gamma$ versus $H_1: \boldsymbol{\ell}'\boldsymbol{\beta} \neq \gamma$ the f–test of (5.4.9) and the t–test of (5.4.10) are equivalent. Are the two tests equivalent for testing $H_0: \boldsymbol{\ell}'\boldsymbol{\beta} \leq \gamma$ versus $H_1: \boldsymbol{\ell}'\boldsymbol{\beta} > \gamma$? Explain.

5–32 Consider a general linear hypothesis $H_0: \boldsymbol{\ell}'\boldsymbol{\beta} = \gamma$ and $H_1: \boldsymbol{\ell}'\boldsymbol{\beta} \neq \gamma$. How do the following two test procedures compare?

(i) The t–test of (5.4.10).
(ii) The test based on rejecting H_0 if a 100 $(1 - \alpha)$% confidence interval for $\boldsymbol{\ell}'\boldsymbol{\beta}$ does not contain γ.

5–33 Verify that the value for the f–test in (5.4.6) is invariant to the g–inverse used for $\mathbf{X}'\mathbf{X}$.

5–34 Show that $SSE^* - SSE$ and SSE are independent. (Note the proofs of Theorems 5.4.1 and 5.4.2.)

5–35 For the normal linear model, show that when the general linear hypothesis $H_0: \mathbf{L}\boldsymbol{\beta} = \boldsymbol{\gamma}$ holds, then $(SSE^* - SSE)/\sigma^2$ has a chi–square distribution with q degrees of freedom. (Note Theorem 5.4.2 and expression 5.4.4 in the proof of Theorem 5.4.1.)

5–36 Show that the numerator of the expression (5.4.4) can be written as a quadratic form in $\mathbf{Y} - \mathbf{X}\boldsymbol{\beta}$ with the matrix of the quadratic form given by

$$\mathbf{X}[(\mathbf{X}'\mathbf{X})^-]'\mathbf{L}'[\mathbf{L}(\mathbf{X}'\mathbf{X})^-\mathbf{L}']^{-1}\mathbf{L}(\mathbf{X}'\mathbf{X})^-\mathbf{X}'.$$

CHAPTER 6

INFERENCE FOR MULTIPLE REGRESSION MODELS

6.1 The Multiple Regression Model Revisited

Introduction

The purpose of this chapter is to present a selection of regression procedures which are useful in experimental design problems. Readers who desire a more comprehensive treatment of the various aspects and techniques of regression analysis together with other application settings are referred to one of the many multiple regression textbooks such as those written by Draper and Smith (1981), Neter, Wasserman and Kutner (1988) and Myers (1990).

In presenting a "catalog" of statistical methods associated with multiple regression models, originally introduced in Chapter 2, we will begin by restating fundamental regression results and then applying them to numerical examples. Later we develop the regression analysis of variance table using the principle of conditional error and discuss the concepts of adjusted and unadjusted sums of squares. Readers who have had a previous course in regression analysis should find the material in the first three or four sections a review, presented here for reference and self—containment purposes. The final sections deal with a topic perhaps unfamiliar to some readers called orthogonal polynomials. This subject is basic to the problem of determining the nature of the response in the analysis of experimental design data when the independent variables are of the quantitative type. Facts concerning multiple regression have been scattered throughout the previous chapters. At the risk of some repetition, we now bring them together in one place.

The Model

Because our interest is in the application of statistical inference procedures, we shall focus our attention on the **normal** multiple regression model. The intercept form of this model for n observation is

$$Y_i = \beta_0 + \beta_1 x_{i1} + \cdots + \beta_k x_{ik} + \varepsilon_i \, , \; i = 1, 2, \cdots, n, \tag{6.1.1}$$

where ε_i is $N(0,\sigma^2)$ for all i and $Cov(\varepsilon_i,\varepsilon_j) = 0$ for $i \neq j$. In matrix representation, this model is presented as

$$\mathbf{Y} = \mathbf{X}\boldsymbol{\beta} + \boldsymbol{\varepsilon} \text{ where } \boldsymbol{\varepsilon} \text{ is } MN(\mathbf{0},\sigma^2 I) \tag{6.1.2}$$

and

$$\mathbf{Y} = \begin{bmatrix} Y_1 \\ Y_2 \\ \vdots \\ Y_n \end{bmatrix}, \quad \mathbf{X} = \begin{bmatrix} 1 & x_{11} & \cdots & x_{1k} \\ 1 & x_{21} & \cdots & x_{2k} \\ \vdots & \vdots & & \\ 1 & x_{n1} & \cdots & x_{nk} \end{bmatrix}, \quad \boldsymbol{\beta} = \begin{bmatrix} \beta_0 \\ \beta_1 \\ \vdots \\ \beta_k \end{bmatrix}, \quad \boldsymbol{\varepsilon} = \begin{bmatrix} \varepsilon_1 \\ \varepsilon_2 \\ \vdots \\ \varepsilon_n \end{bmatrix}$$

$$n \times 1 \qquad n \times (k+1) \qquad (k+1) \times 1 \qquad n \times 1$$

The term regression implies that matrix $\mathbf{X}$ is full rank, that is, $\mathbf{X}$ has rank $k + 1$ where $n > k + 1$.

Normal Equations

To distinguish between random variables and their realization in a sample, we let

$$y_i = b_0 + b_1 x_{i1} + \cdots + b_k x_{ik} + e_i \ , \quad i = 1,2,\ldots,n,$$

denote the sample form corresponding to (6.1.1). In matrix notation, these n equations, as in (6.1.2), are written as

$$\mathbf{y} = \mathbf{Xb} + \mathbf{e}.$$

Least square estimation leads to the normal equations

$$\mathbf{X}'\mathbf{Xb} = \mathbf{X}'\mathbf{y} \ ,$$

which, in the regression case, has a unique solution, namely,

$$\mathbf{b} = (\mathbf{X}'\mathbf{X})^{-1} \mathbf{X}'\mathbf{y}. \tag{6.1.3}$$

Vector $\mathbf{b}$ is called the least squares estimate for the parameter vector $\boldsymbol{\beta}$; the least squares estimator for $\boldsymbol{\beta}$ is denoted by $\hat{\boldsymbol{\beta}} = (\mathbf{X}'\mathbf{X})^{-1}\mathbf{X}'\mathbf{Y}$.

Estimability

As noted in Theorem 4.3.5, a consequence of $\mathbf{X}$ being full rank is that all linear functions $\boldsymbol{\ell}'\boldsymbol{\beta}$ are estimable. In particular,

$$\hat{y} = b_0 + b_1 x_1 + \cdots + b_k x_k$$

is BLUE for the mean value of Y, $E(Y)$, for a fixed set of values of the independent variables $x_1, x_2, \ldots, x_k$, and the elements in $\mathbf{b}$ are the BLUEs for the corresponding elements in $\boldsymbol{\beta}$.

Confidence Intervals

From (5.3.6), a $100(1-\alpha)\%$ confidence interval for the linear function $\boldsymbol{\ell}'\boldsymbol{\beta}$ is

$$\boldsymbol{\ell}'\mathbf{b} \pm t_{\alpha/2,n-k-1}\sqrt{s^2\boldsymbol{\ell}'(\mathbf{X}'\mathbf{X})^{-1}\boldsymbol{\ell}} \qquad (6.1.4)$$

where, in this full rank case, $s^2 = SSe/(n-k-1)$ with $SSe = \sum_{i=1}^{n}(y_i-\hat{y}_i)^2$ $= \mathbf{y}'\mathbf{y} - \mathbf{b}'\mathbf{X}'\mathbf{y}$. In particular, a $100(1-\alpha)\%$ confidence interval for any β_j, $j = 0,1,\ldots,k$, is

$$b_j \pm t_{\alpha/2,n-k-1}\sqrt{s^2c_{jj}}\,, \qquad (6.1.5)$$

where c_{jj} is the jth diagonal element of $(\mathbf{X}'\mathbf{X})^{-1}$ starting the count with $j = 0$. As previously noted, the quantities $\sqrt{s^2\boldsymbol{\ell}'(\mathbf{X}'\mathbf{X})^{-1}\boldsymbol{\ell}}$ and $\sqrt{s^2c_{jj}}$ are called standard errors and are denoted by the symbols $Se(\boldsymbol{\ell}'\hat{\boldsymbol{\beta}})$ and $Se(\hat{\beta}_j)$, respectively.

t Tests

Expression (5.4.10) gives us the statistic for testing

$$H_0 : \boldsymbol{\ell}'\boldsymbol{\beta} = \gamma,$$

where γ is some specified constant. In a regression problem, the test statistic is

$$t = \frac{\boldsymbol{\ell}'\mathbf{b} - \gamma}{\sqrt{s^2\boldsymbol{\ell}'(\mathbf{X}'\mathbf{X})^{-1}\boldsymbol{\ell}}}\,. \qquad (6.1.6)$$

The rejection region is based upon the t distribution with $n-k-1$ degrees of freedom. In particular, for the null hypothesis

$$H_0 : \beta_j = 0,$$

the test statistic is

$$t = \frac{b_j}{\sqrt{s^2 c_{jj}}}, \tag{6.1.7}$$

where c_{jj} is the diagonal element defined earlier.

In the sections to follow, the formulas which we have just reintroduced will be used to assist in the analysis of data using regression models.

6.2. Computer—Aided Inference in Regression

In this section we use information contained in Statistical Analysis System (SAS) printouts in order to demonstrate basic inference procedures. As a bonus, we will learn more about the type of information "automatically" available to users of commercial regression programs as we trace through the results of certain programming commands.

Quantitative and Qualitative Type Variables

With the exception of Exercise 3—5, in all of the numerical regression illustrations and data exercises prior to this, the independent variables have been of the **quantitative** type where the values of the variables can be naturally ordered and placed on a number scale. (For instance, in Example 2.4.1, the independent variables outside temperature, thermostat setting and square footage are of this type.) In our first example in this section, the set of independent variables will contain a **qualitative** (or **nominal**) variable as well as two quantitive type variables. In this case, the qualitative variable is "sex" having two possible "values", namely, male and female, which cannot be naturally ordered on a number scale. In order to identify the two levels of the qualitative variable "sex" numerically, we make use of a (0,1) coding scheme. A variable consisting of the two values 0 and 1 is sometimes called a **dummy variable**. Perhaps a more appropriate name for such a variable is **indicator variable** since in practice the numbers 1 and 0 are used to indicate the presence or absence of a characteristic. This idea is not foreign to us. From the beginning, we have been using (0,1) coding in the construction of the **X** matrix in one—way classification problems where, for instance, in the second column (the first treatment column) an entry of 1 indicates the first treatment was applied, an entry of 0 tells us the first treatment was not applied.

Salary Discrimination Problem

In a study on salary structure, a particular company was

interested in the question of possible inequities between sexes. To investigate this possibility, data were obtained on 16 employees selected at random from several hundred employees listed as "white collar" workers. These data included information on education level and work experience, two prime determinants of salary level. (In practice, in a study such as this, a sample size much larger than 16 would normally be used. For this illustration, we have chosen $n = 16$ for the reason of economy in space.) To study the question of salary equity, we assume the functional form of the regression model to be

$$Y_i = \beta_0 + \beta_1 x_{i1} + \beta_2 x_{i2} + \beta_3 x_{i3} + \varepsilon_i \ , \ i = 1, 2, \ldots, 16, \qquad (6.2.1)$$

where

Y_i = annual salary of the ith employee expressed in \$1000 amounts,

$$x_{i1} = \begin{cases} 1, & \text{if ith employee is male} \\ 0, & \text{if ith employee is female,} \end{cases}$$

x_{i2} = the educational level attained by the ith employee in years,

x_{i3} = the experience of the ith employee in work years.

In model (6.2.1), we observe that for a **male** with education level x_{i2} and x_{i3} years of experience, the average salary is $E(Y_i) = \beta_0 + \beta_1 + \beta_2 x_{i2} + \beta_3 x_{i3}$, while for a **female** having the same education and experience levels, the average salary is $E(Y_i) = \beta_0 + \beta_2 x_{i2} + \beta_3 x_{i3}$. Therefore, the parameter β_1 represents the **actual** difference in average salary between male and female white collar workers in the company **adjusted** for level of education and amount of work experience. If $\beta_1 > 0$, males receive higher salaries (**on average**) than females with the **same** education and work experience; if $\beta_1 < 0$, females have higher average salaries than males at fixed levels of education and experience. To check for statistical evidence of salary discrimination, we need to test the null hypothesis

$$H_0 : \beta_1 = 0,$$

which, in words, is the hypothesis that no discrimination in salary exists between the sexes. For the proposed regression analysis using SAS, the SAS commands and the printout of the data set (resulting from the command PROC PRINT;) are given in Table 6.2.1.

Table 6.2.1 SAS Commands and Data Set for Salary Discrimination Problem

```
OPTIONS LS=80;
DATA REGPROB;
INPUT SALARY SEX EDUC WORKYRS @@ ;
CARDS;
30.6 1 12 27  34.0 0 16 15  33.0 1 18  5  26.0 1 10 38  24.6 0 12 14
50.6 1 16 38  24.4 1 14 10  29.4 0 12 21  44.2 1 20  8  23.4 0 16  2
26.6 1 13  9  26.8 1  9 34  32.0 0 12 19  37.4 0 16 20  45.0 1 18 20
20.6 1 12  4
PROC PRINT;
PROC GLM;
MODEL SALARY = SEX EDUC WORKYRS / I P CLM;
ESTIMATE 'FEMALE' INPTERCEPT 1 SEX 0 EDUC 16 WORKYRS 10 ;
ESTIMATE 'MALE' INTERCEPT 1 SEX 1 EDCU 16 WORKYRS 10 ;
```

OBS	SALARY	SEX	EDUC	WORKYRS
1	30.6	1	12	27
2	34.0	0	16	15
3	33.0	1	18	5
4	26.0	1	10	38
5	24.6	0	12	14
6	50.6	1	16	38
7	24.4	1	14	10
8	29.4	0	12	21
9	44.2	1	20	8
10	23.4	0	16	2
11	26.6	1	13	9
12	26.8	1	9	34
13	32.0	0	12	19
14	37.4	0	16	20
15	45.0	1	18	20
16	20.6	1	12	4

Before showing the results of the programmed commands, it is instructive to note that the sample figures do indeed **suggest** salary discrimination against females if one ignores the effects of education and experience on salary level. Simple calculations reveal that the average salary of male employees **in the sample** when rounded to \$100 is \$32,800 whereas the average for females is \$30,100.

To fit model (6.2.1) using the SAS General Linear Model program (command PROC GLM;) we have used the model command

$$\text{MODEL SALARY} = \text{SEX EDUC WORKYRS/I P CLM; .} \tag{6.2.2}$$

The results of this command appear in output form on the next and following pages. The three–line table of numbers at the top of Table 6.2.2 is the **regression analysis of variance table**. In the next section we will find out in detail what the numbers in this table represent; for the present we simply point out that the second line gives us the value of $s^2 = \text{MSe} = 10.53167$ with $n - k - 1 = 12$ degrees of freedom. Also, the line below the analysis of variance table includes $s = 3.24525$ labeled as ROOT MSE. The first entry on this line, identified as R–SQUARE, is the number 0.8866. This is a measure of the goodness of fit of the model to the data set **independent** of the measurement units used for the dependent variable. An $r^2 = 0.8866$ means that 88.66% of the variation in salaries is accounted for by the fitted model. The number r^2 is called the **coefficient of determination.** In the next section, the source of this quantity is developed.

Table 6.2.2 SAS GLM Printout of Salary Discrimination Regression Analysis

General Linear Models Procedure

Dependent Variable: SALARY

Source	DF	Sum of Squares	Mean Square	F Value	Pr > F
Model	3	988.21742760	329.40580920	31.28	0.0001
Error	12	126.38007240	10.53167270		
Corrected Total	15	1114.59750000			

R-Square	C.V.	Root MSE	SALARY Mean
0.886614	10.20921	3.2452539	31.787500

Source	DF	Type I SS	Mean Square	F Value	Pr > F
SEX	1	26.26816667	26.26816667	2.49	0.1403
EDUC	1	476.65270330	476.65270330	45.26	0.0001
WORKYRS	1	485.29655764	485.29655764	46.08	0.0001

Source	DF	Type III SS	Mean Square	F Value	Pr > F
SEX	1	0.12890875	0.12890875	0.01	0.9137
EDUC	1	864.87096247	864.87096247	82.12	0.0001
WORKYRS	1	485.29655764	485.29655764	46.08	0.0001

Parameter	Estimate	T for H0: Parameter=0	Pr > |T|	Std Error of Estimate
INTERCEPT	-16.47746667	-3.25	0.0070	5.07541528
SEX	-0.18983610	-0.11	0.9137	1.71587747
EDUC	2.72895468	9.06	0.0001	0.30114062
WORKYRS	0.55420447	6.79	0.0001	0.08164228

Observation	Observed	Predicted Residual	Lower 95% CLM Upper 95% CLM
1	30.60000000	31.04367416 -0.44367416	28.36403629 33.72331204
2	34.00000000	35.49887528 -1.49887528	32.33334207 38.66440848
3	33.00000000	35.22490379 -2.22490379	31.76363599 38.68617159
4	26.00000000	31.68201403 -5.68201403	27.75630338 35.60772467

Continuing on down in Table 6.2.2, we temporarily leave the TYPE I SS and TYPE III SS segments for later discussion. It is in the next set of data that we find the least squares regression equation

$$\hat{y} = b_0 + b_1x_1 + b_2x_2 + b_3x_3$$
$$= -16.477 - 0.190x_1 + 2.729x_2 + 0.554x_3, \qquad (6.2.3)$$

the coefficients given to three decimal places. The best linear unbiased estimate for β_1 is $b_1 = -0.190$. Considering the salary figures are given in \$1000 units, this means that for the individuals in the sample, females averaged \$190 more than males in annual salary after adjusting for education and experience, a major reversal from the unadjusted averages. To determine if the difference of \$190 is "real" or conceivably due to chance sampling fluctuations, we proceed to test the "research" hypothesis

$$H_0 : \beta_1 = 0 \text{ versus } H_1 : \beta_1 \neq 0$$

using test statistic (6.1.7). In this case the formula is given by

$$t = b_1/\sqrt{s^2 c_{11}} \ . \qquad (6.2.4)$$

The denominator of expression (6.2.4) is the standard error of the estimator for β_1, symbolically, $Se(\hat{\beta}_1) = \sqrt{s^2 c_{11}}$. Its numerical value is printed in Table 6.2.2 under the heading STD ERROR OF ESTIMATE where we find $\sqrt{s^2 c_{11}} = 1.715877$. The numerical value of the t ratio (6.2.4) is also given as $t = -0.11$ under the heading T FOR HO:PARAMETER = 0. On comparing the absolute value of this number with the tabulated $t_{.025,12} = 2.179$ value, we have insufficient evidence to reject $H_0 : \beta_1 = 0$ at the 0.05 level. In fact, we would not even reject the null hypothesis at an absurd significance level such as $\alpha = 0.90$ since the **observed significance level** is 0.9131, printed out under the heading PR > |T|. Technically, this means that the probability of obtaining a t ratio larger than 0.11 (in absolute value) when $\beta_1 = 0$ is true is 0.9137. This observed significance level is called the **probability value**, or simply, the **p–value** and is the smallest α risk level at which we can reject H_0. Therefore, any **fixed** α level less than 0.9137 leads to the non–rejection of $H_0:\beta_1 = 0$ in this problem implying the evidence is not sufficient to claim the existence of salary structure discrimination between the sexes.

Before we pose other inference problems, let us become further acquainted with information available to users of computer programs.

Recall that the standard error of $\hat{\beta}_1$, $\sqrt{s^2 c_{11}}$, can be obtained having knowledge of the matrix inverse $(X'X)^{-1}$. In command statement (6.2.2), the notation I following the slant sign causes SAS to print out the inverse of X′X. The excerpted result is given in Table 6.2.3.

Table 6.2.3 SAS Printout of $(X'X)^{-1}$ in Salary Discrimination Study

General Linear Models Procedure

X'X Inverse Matrix

	INTERCEPT	SEX	EDUC	WORKYRS	SALARY
INTERCEPT	2.4459400719	-0.038924307	-0.136478498	-0.024301611	-16.47746667
SEX	-0.038924307	0.279560102	-0.006062942	-0.002826011	-0.189836096
EDUC	-0.136478498	-0.006062942	0.0086107569	0.0010501913	2.7289546772
WORKYRS	-0.024301611	-0.002826011	0.0010501913	0.0006328967	0.5542044743
SALARY	-16.47746667	-0.189836096	2.7289546772	0.5542044743	126.3800724

The diagonal element in the second row and second column is c_{11} = 0.27956 and this, together with the already noted $s^2 = 10.53167$, yields

$$\sqrt{s^2 c_{11}} = \sqrt{10.53167(0.27956)} = 1.715877.$$

This is the same number we had seen earlier in Table 6.2.2.

Returning to Table 6.2.2, we note that education with a t value of 9.06 and work experience with $t = 6.79$ are highly significant predictors of salary level. (Both have observed significance levels of 0.0001.) Let us now demonstrate the use of formula (6.1.5) by calculating a 95% interval estimate for the average increase in annual salary attributable to each year of education as reflected in the sample data. Here $b_2 = 2.729$ translates into a point estimate of $2,729. With a standard error of 0.30114 and $t_{.025,12} = 2.179$, a 95% confidence interval for β_2 is

$$2.729 \pm 2.179(0.30114) \quad \text{or} \quad (2.073,\ 3.385).$$

In dollars, the 95% interval is $2,073 to $3,385.

The lines headed by the words FEMALE and MALE in Table 6.2.2 result from the use of Table 6.2.1 commands

ESTIMATE ‘FEMALE’ INTERCEPT 1 SEX 0 EDUC 16 WORKYRS 10;

ESTIMATE 'MALE' INTERCEPT 1 SEX 1 EDUC 16 WORKYRS 10; .

The ESTIMATE command evaluates the regression equation (6.2.3) for particular values of the independent variables, here for female and male employees having 16 years of education and 10 years of experience. The estimate for the mean annual salaries of these hypothetical individuals are $32,728 and $32,538, respectively. The STD ERROR OF ESTIMATE items are numerical evaluations of the expression $\sqrt{s^2 \boldsymbol{\ell}' (\mathbf{X}'\mathbf{X})^{-1} \boldsymbol{\ell}}$. Here for female, $\boldsymbol{\ell}' = (1, 0, 16, 10)$; for male, $\boldsymbol{\ell}' = (1, 1, 16, 10)$. With this information, it is a simple matter to calculate an interval estimate of the average salary for individuals with 16 years of education and 10 years of experience. Using (6.1.4), for males a 95% confidence interval for E(Y) is

$$32.728 \pm 2.179(1.24627).$$

In dollars, this works out to $30,012 to $35,444.

To complete our description of the items in Table 6.2.2 for the present,we turn our attention to the information concerning observations near the end of the printout. These numbers result from the instructions P CLM included at the end of model statement (6.2.2). The notation P yields for the ith observation the observed value, y_i , the predicted value $\hat{y}_i$ from (6.2.3), and the residual value $e_i = y_i - \hat{y}_i$. (Here, for reasons of space, we have shown these quantities for the first five observations only.) For example, associated with the first observation we have $y_1 = 30.6$, $\hat{y}_1 = 31.04367$, $e_1 = -0.44367$. Since the quantities "speak for themselves", no further comment will be made about them other than to point out that in model building and in checking assumptions, examination of residuals is an important activity. Readers who have had a "full" course in regression analysis previously will have had exposure to various diagnostic techniques involving the residuals.

The last column in the observation portion of the printout is a result of the instruction CLM in model statement (6.2.2). This column lists the 95% confidence intervals for the means, $E(Y_i)$, of the corresponding observations. For example, noting the characteristics of the first employee in the sample, the interval (28.364, 33.723) is the 95% interval estimate for the mean of those observations associated with a male employee having 12 years of education with 27 years of experience on the job.

Now that the reader has added familiarity with the type of information available in computer statistical software, in particular, with that accessible to users of the SAS GLM program, we will consider a second example for further illustration and for reference later in this chapter.

Grade Point Prediction Problem

Universities and colleges require freshman applicants to include in their admission materials the scores obtained in national aptitude tests such as the SAT and ACT. Another piece of information frequently considered important in the selection process is the applicant's percentile rank in his high school class. Of course, there are other criteria used by admission committees such as extracurricular and leadership activities, special talents and writing samples. These latter measures are largely qualitative in nature whereas aptitude scores and class rank are quantitative type variables. For our next numerical illustration, we will devise a simplistic model for predicting a student's success in college based on ACT score and high school class percentile rank only. To do this, we shall assume the data set in the next table represents the grade–point averages realized by twenty randomly selected students during their first year at a given university. (As in the salary discrimination study, in practice a larger sample normally would be used.) The data includes student composite ACT test scores and high school class percentile ranks.

To predict the level of college success, let us use the regression model

$$Y_i + \beta_0 + \beta_1 x_{i1} + \beta_2 x_{i1} + \varepsilon_i \ , \quad i = 1, 2, \ldots, 20,$$

where

Y_i = achieved grade–point average during freshman year by ith student,

x_{i1} = ith student's composite ACT score,

x_{i2} = percentile rank of the ith student in his/her high school class.

Table 6.2.4 contains the printout from SAS GLM for these data. The fitted least squares regression function is found to be

$$\hat{y} = -1.017 + 0.011x_1 + 0.044x_2. \tag{6.2.5}$$

An $r^2 = 0.9291$ (close to one) and $s^2 = 0.03366$ (close to zero) indicates that model (6.2.5) provides a good fit and that it explains almost 93% of the variation in the sample grade–point averages. However, we immediately observe that ACT score with $t = 0.99$ and a p–value of 0.3377 is not a significant predictor of grade–point average **in the presence of the other variable**, percentile rank. Does this mean that we can do about as well using only percentile rank as a predictor?

To investigate this question, we can run the simple linear regression model with x_2, percentile rank, the independent variable. The result appears in Table 6.2.5 where we find the regression line

$$\hat{y} = -1.029 + 0.048x_2.$$

Table 6.2.4 SAS Printout of Grade Point Prediction Problem Data and Regression Analysis

OBS	GPA	ACT	RANK
1	2.0	21	68
2	2.3	28	73
3	1.5	15	55
4	2.5	29	68
5	3.0	27	85
6	2.2	24	65
7	2.8	25	80
8	3.2	33	88
9	3.8	35	98
10	1.6	18	52
11	2.4	26	76
12	1.4	16	48
13	2.0	26	60
14	3.3	31	90
15	2.2	20	69
16	1.8	16	63
17	2.8	30	73
18	2.9	24	78
19	3.0	31	82
20	2.0	33	70

General Linear Models Procedure

Dependent Variable: GPA

Source	DF	Sum of Squares	Mean Square	F Value	Pr > F
Model	2	7.49370084	3.74685042	111.40	0.0001
Error	17	0.57179916	0.03363524		
Corrected Total	19	8.06550000			

R-Square	C.V.	Root MSE	GPA Mean
0.929106	7.531792	0.1833991	2.4350000

Source	DF	Type I SS	Mean Square	F Value	Pr > F
ACT	1	5.11377250	5.11377250	152.04	0.0001
RANK	1	2.37992835	2.37992835	70.76	0.0001

Source	DF	Type III SS	Mean Square	F Value	Pr > F
ACT	1	0.03273082	0.03273082	0.97	0.3377
RANK	1	2.37992835	2.37992835	70.76	0.0001

Parameter	Estimate	T for H0: Parameter=0	Pr > |T|	Std Error of Estimate
INTERCEPT	-1.016900452	-4.30	0.0005	0.23654377
ACT	0.011030752	0.99	0.3377	0.01118212
RANK	0.044021088	8.41	0.0001	0.00523331

Parameter	Estimate	T for H0: Parameter=0	Pr > |T|	Std Error of Estimate
INTERCEPT	-1.016900452	-4.30	0.0005	0.23654377
ACT	0.011030752	0.99	0.3377	0.01118212
RANK	0.044021088	8.41	0.0001	0.00523331

Table 6.2.5 SAS Printout of GPA vs. RANK Simple Regression

General Linear Models Procedure

Dependent Variable: GPA

Source	DF	Sum of Squares	Mean Square	F Value	Pr > F
Model	1	7.46097003	7.46097003	222.15	0.0001
Error	18	0.60452997	0.03358500		
Corrected Total	19	8.06550000			

R-Square	C.V.	Root MSE	GPA Mean
0.925047	7.526164	0.1832621	2.4350000

Source	DF	Type I SS	Mean Square	F Value	Pr > F
RANK	1	7.46097003	7.46097003	222.15	0.0001

Source	DF	Type III SS	Mean Square	F Value	Pr > F
RANK	1	7.46097003	7.46097003	222.15	0.0001

Parameter	Estimate	T for H0: Parameter=0	Pr > |T|	Std Error of Estimate
INTERCEPT	-1.029459707	-4.36	0.0004	0.23602441
RANK	0.048084104	14.90	0.0001	0.00322609

Note that here $r^2 = 0.9250$ and $s^2 = 0.03358$. On comparing these numbers with those obtained for the previous model (Table 6.2.4), it does appear that this simpler model has the same degree of utility for predicting success in college as the two—variable model (6.2.5). This does not mean that ACT score is useless as a predictor of early achievement level in college. Table 6.2.6 tells us otherwise. There the results of a simple regression with ACT score as the independent variable are given. In the test for the ACT parameter, a $t = 5.58$ with a p—value of 0.0001 clearly indicates a strong linear relationship between the variables GPA and ACT. However, to repeat, in this data set, the variable ACT score in the presence of the variable percentile rank does not contribute much more information toward estimating success in college. This anomaly is explained by a condition known as **collinearity**, a topic the reader may have encountered in a previous regression course. We will not discuss this subject in detail here; we will only note that multicollinearity occurs when independent variables are related (correlated) to each other. As a result, information concerning the dependent variable provided by an independent variable may be largely redundant in the presence of one or more variables. In this instance, it is obvious that x_1, aptitude, as measured by ACT score, and x_2, percentile rank in high school class, are positively correlated. The reader is invited to find its numerical value.

Table 6.2.6 SAS Printout of GPA vs. ACT Simple Regression

General Linear Models Procedure

Dependent Variable: GPA

Source	DF	Sum of Squares	Mean Square	F Value	Pr > F
Model	1	5.11377250	5.11377250	31.18	0.0001
Error	18	2.95172750	0.16398486		
Corrected Total	19	8.06550000			

R-Square	C.V.	Root MSE	GPA Mean
0.634030	16.63041	0.4049504	2.4350000

Source	DF	Type I SS	Mean Square	F Value	Pr > F
ACT	1	5.11377250	5.11377250	31.18	0.0001

Source	DF	Type III SS	Mean Square	F Value	Pr > F
ACT	1	5.11377250	5.11377250	31.18	0.0001

Parameter	Estimate	T for H0: Parameter=0	Pr > |T|	Std Error of Estimate
INTERCEPT	0.2744906621	0.69	0.4985	0.39734493
ACT	0.0850594228	5.58	0.0001	0.01523188

Type I and Type III Sum of Squares

Tables 6.2.4, 6.2.5 and 6.2.6 provide us with the opportunity to discover the difference between Type I and Type III sums of squares found in SAS printouts. We can characterize these two types of sums of squares by noting their properties. Turning to Table 6.2.4, we find that the Type I sums of squares for variables ACT and RANK (5.11377 and 2.37993, respectively) sum to the model sum of squares (7.49370) in the regression analysis of variance table. This model sum of squares is commonly known as the **regression sum of squares** in regression problems, a name we will use in this text. SAS Type I sums of squares, then, is a partitioning of the regression sum of squares in the following sense. The Type I sum of squares attributable to ACT in Table 6.2.4 is the regression sum of squares obtained when fitting the simple linear regression model with ACT as the independent variable. This can be seen to be the case in Table 6.2.6. Then the Type I sum of squares associated with percentile rank is the **addition** to the regression sum of squares obtained when fitting the multiple regression model having RANK as an independent variable in addition to variable ACT. The reader should be aware of the fact that the particular partitioning of the regression sum of squares as represented by TYPE I SS is dependent upon the order in which the independent variables are entered into the model. For example, if we were to run the multiple regression model with RANK as the first independent variable and ACT as the second independent variable, we would find the Type I sum of squares attributable to RANK to be 7.46097 (from Table 6.2.5) and the ACT additional sum of squares to be 7.49370 — 7.46097 = 0.03273, since the regression sum of squares, 7.49370, is a constant irrespective of order. Some authors have used the words "sequential F—tests" for the F values associated with these order dependent Type I sums of squares as a means of distinguishing between these and Type III F values.

Contrary to SAS Type I sums of squares, SAS Type III sums of squares are not dependent on order of entry of the independent variables and the associated F values are, in fact, sample values of F random variables (5.3.2). Focusing on the printed Type III F values in Table 6.2.4, it is easy to verify that, except for differences due to rounding, the numbers 0.97 and 70.76 are the squares of the t values 0.99 and 8.41 given for testing the hypotheses $H_0{:}\beta_1 = 0$ and $H_0{:}\beta_2 = 0$, respectively. Since the square of a t random variable is an F random variable with numerator degrees of freedom of one, Type III sums of squares are the numerator sums of squares in F statistics for the testing of individual coefficients to be zero. Further cursory evidence of this is seen when we observe that the accompanying p—values, 0.3377 and 0.0001, are identical to those given for the t—tests. (The reader will find these same relationships in the salary analysis printout, Table 6.2.2.)

To understand how the numbers in a SAS TYPE III SS table can be found, let us use the principle of conditional error to test $H_0{:}\beta_1 = 0$ where β_1 is the coefficient of the composite ACT score in model (6.2.4). The test statistic, here renumbered,

$$f = \frac{(SSe^* - SSe)/q}{s^2} \tag{6.2.6}$$

is the earlier equation (5.4.16). For this problem, H_0 can be expressed in the form $\mathbf{L}\boldsymbol{\beta} = \mathbf{0}$ by letting $\mathbf{L} = [0 \;\; 1 \;\; 0]$. Hence q, the rank of $\mathbf{L}$, is equal to 1. For the full model (6.2.5) we get SSe = 0.57180 and s^2 = MSe = 0.033635. (See the second line of the regression analysis of variance in Table 6.2.4.) For the reduced model $Y = \beta_0^* + \beta_2^* x_2 + \varepsilon^*$, we find in Table 6.2.5, $SSe^* = 0.60453$. Substituting these values into formula (6.2.6),

$$f = \frac{(0.60453 - 0.57180)/1}{0.033635} = \frac{0.03273}{0.033635} = 0.973 .$$

This agrees with the Type III F value for variable ACT (x_1) in Table 6.2.4.

To conclude this section, we show that the F value of 70.76 for independent variable RANK in Table 6.2.4 is the value of the principle of conditional error test statistic for testing $H_0{:}\beta_2 = 0$. Letting $\beta_2 = 0$, model (6.2.5) reduces to $Y = \beta_0^{**} + \beta_1^{**} x_1 + \epsilon^{**}$, the fit of which is given in Table 6.2.6. There we find the error sum of squares to be $SSe^{**} = 2.95173$. Accordingly, the value of (6.2.6) is

$$f = \frac{(2.95173 - 0.57180)/1}{0.033635} = \frac{2.37993}{0.033635} = 70.758.$$

6.3 Regression Analysis of Variance

In this section we study various measures concerned with the utility of the assumed regression model (6.1.1) in making estimates and predictions. Some of the material presented here was alluded to in earlier sections but a more complete treatment was deferred until now.

Test for $H_0 : \beta_1 = \beta_2 = \cdots = \beta_k = 0$

Given the general intercept multiple regression model (6.1.1) with k + 1 parameters $\beta_0, \beta_1, \beta_2, \ldots, \beta_k$, the regression analysis of variance table displays the f—test components for testing the null hypothesis

$$H_0 : \beta_1 = \beta_2 = \cdots = \beta_k = 0. \tag{6.3.1}$$

Should this null hypothesis be true, then none of the k independent variables theoretically affects the dependent variable Y and therefore they are of little use for estimation and prediction purposes. On the other hand, rejection of H_0 indicates that at least one of $\beta_1, \beta_2, \ldots, \beta_k$ does not equal zero. Tests for an individual β_j equal to zero were given and illustrated earlier in this chapter. To obtain the value of the F statistic (6.2.6) for testing $H_0 : \beta_j = 0$ in the last section, we used the principle of conditional error. We again turn to this approach to derive the form of the statistic for testing (6.3.1).

We first expressed the null hypothesis (6.3.1) in general linear hypothesis form $H_0 : \mathbf{L}\boldsymbol{\beta} = \mathbf{0}$. If we take $\mathbf{L}$ to be the $k \times (k+1)$ matrix

$$\mathbf{L} = \begin{bmatrix} 0 & 1 & 0 & 0 & \cdots & 0 \\ 0 & 0 & 1 & 0 & \cdots & 0 \\ 0 & 0 & 0 & 1 & \cdots & 0 \\ \vdots & \vdots & \vdots & \vdots & & \vdots \\ 0 & 0 & 0 & 0 & \cdots & 1 \end{bmatrix}$$

we get $H_0 : (\beta_1, \beta_2, \beta_3, \ldots \beta_k)' = (0, 0, 0, \ldots, 0)'$ which is equivalent to $H_0 : \beta_1 = \beta_2 = \beta_3 = \cdots = \beta_k = 0$. We note that the matrix $\mathbf{L}$ has rank k. Furthermore, $\mathbf{L}\boldsymbol{\beta}$ is a set of k linearly independent estimable functions since, in regression, each β_j is estimable.

Next, we recall the general principle of conditional error test statistic (6.2.6). In this case, $q = k$, making the statistic formula

$$f = \frac{(SSe^* - SSe)/k}{s^2}$$

where, from (3.3.2), $SSe = \mathbf{y}'\mathbf{y} - \mathbf{b}'\mathbf{X}'\mathbf{y}$ is the error sum of squares for the full model (6.1.1), $s^2 = SSe/(n-k-1)$ is called the error mean square also denoted by MSe, and SSe^* is the error sum of squares for the reduced model. Imposing the null hypothesis (6.3.1) on the full model results in the reduced model

$$Y_i = \beta_0^* + \varepsilon_i^*, \quad i = 1, 2, \ldots, n.$$

It is easy to show that $b_0^* = \bar{y}$ where $\bar{y} = \sum_{i=1}^{n} y_i/n$. (See Ex. 6 — 12). Therefore, the error sum of squares for the reduced model is

$$SSe^* = \sum_{i=1}^{n} (y_i - \hat{y}_i)^2 = \sum_{i=1}^{n} (y_i - \bar{y})^2 = \mathbf{y}'\mathbf{y} - n\bar{y}^2.$$

As a result,

$$\begin{aligned} SSe^* - SSe &= (\mathbf{y}'\mathbf{y} - n\bar{y}^2) - (\mathbf{y}'\mathbf{y} - \mathbf{b}'\mathbf{X}'\mathbf{y}) \\ &= \mathbf{b}'\mathbf{X}'\mathbf{y} - n\bar{y}^2. \end{aligned} \tag{6.3.2}$$

The above quantity (6.3.2) is called the **regression sum of squares**, which we denote by the symbol SSr. It follows that in testing the null hypothesis (6.3.1), the statistic (6.2.6) can be written as

$$f = \frac{SSr/k}{SSe/(n\text{-}k\text{-}1)} . \tag{6.3.3}$$

From Section 5.4, we know that the statistic (6.3.3) is a value of an F random variable with k and n—k—1 degrees of freedom. Accordingly, at the α level of risk, we reject $H_0{:}\beta_1 = \beta_2 = \cdots = \beta_k = 0$ if $f \geq f_{\alpha,k,n-k-1}$.

From the generic definition of an F random variable (5.3.2), we would suspect that SSr and SSe are numbers associated with chi—square random variables. This is indeed the case. Denoting their random variable forms by SSR and SSE, respectively, we already know that SSE/σ^2 is a chi—square random variable with n—k—1 degrees of freedom (Theorem 5.1.4). As a result, we say that the error sum of squares, SSe, has n—k—1 degrees of freedom. It can be shown that if $H_0{:}\beta_1 = \beta_2 = \cdots = \beta_k = 0$ is true, then SSR/σ^2 has a chi—square distribution with k degrees of freedom; hence we say that SSR has k degrees of freedom.

Analysis of Variance Table

In a multiple regression problem, the analysis of variance table is an arrangement of the steps required to calculate the F statistic (6.3.3). It is conventional to include a third sum of squares, called the **total sum of squares**, denoted by SSt. The formula for SSt is

$$SSt = \sum_{i=1}^{n} (y_i - \bar{y})^2 = \mathbf{y}'\mathbf{y} - n\bar{\mathbf{y}}^2.$$

In Example 5.1.1, we showed that $SST/\sigma^2 = \sum_{i=1}^{n} (Y_i - \bar{Y})^2/\sigma^2$ has a chi–square distribution with $n - 1$ degrees of freedom. Listing these three sums of squares we have

$$\begin{aligned}
&1. \quad \text{Regression sum of squares, } SSr = \mathbf{b}'\mathbf{X}'\mathbf{y} - n\bar{y}^2 \\
&\quad\quad \text{with } k \text{ degrees of freedom,} \\
&2. \quad \text{Error sum of squares, } SSe = \mathbf{y}'\mathbf{y} - \mathbf{b}'\mathbf{X}'\mathbf{y} \\
&\quad\quad \text{with } n-k-1 \text{ degrees of freedom,} \\
&3. \quad \text{Total sum of squares, } SSt = \mathbf{y}'\mathbf{y} - n\bar{y}^2 \\
&\quad\quad \text{with } n - 1 \text{ degrees of freedom.}
\end{aligned} \tag{6.3.4}$$

It is easy to see that

$$SSt = SSr + SSe$$

and that the associated degrees of freedom in the set add in the same manner. As we shall see, these are the relationships that form the basis of the analysis of variance table.

Returning to the F statistic (6.3.3),

$$f = \frac{SSr/k}{SSe/(n-k-1)},$$

we note that the form of f is the ratio of two sums of squares each divided by their respective degrees of freedom. We call a sum of squares divided by its degrees of freedom a **mean square**. We have already encountered this practice in Definition 5.1.2 where we defined and denoted the **error mean square** to be

$$s^2 = MSe = SSe/(n - k - 1).$$

In like manner, we denote the **regression mean square** by

$$MSr = SSr/k.$$

We have now reached the point where the above discussion can be conveniently summarized in a table called an **analysis of variance table**. The phrase "analysis of variance" will be abbreviated by ANOVA. The usual ANOVA table format for a multiple regression problem is shown in Table 6.3.1. When we refer to the set of computing formulas (6.3.4), it is obvious that the term $\mathbf{b}'\mathbf{X}'\mathbf{y}$ usually makes accurate computation of the ANOVA sums of squares rather forbidding without the aid of the computer whenever $k > 1$.

Table 6.3.1 Multiple Regression ANOVA Table

Source of Variation	Degrees of Freedom	Sum of Squares	Mean Squares	f
Regression	k	SSr	MSr = SSr/k	MSr/MSe
Error	n-k-1	SSe	MSe = SSe/(n-k-1)	
Total	n-1	SSt		

We note several things about this ANOVA table. First, in the remainder of this text, we shall abbreviate the column headings in the table as follows:

Source, df, SS, MS and f.

Second, we now know the reason why s^2, the estimate of σ^2, has been called error mean square. Third, in some texts, the total sum of squares, $\sum_{i=1}^{n} (y_i - \bar{y})^2$, is called the "total sum of squares corrected for the mean" in order to distinguish from the expression $\sum_{i=1}^{n} y_i^2$.

Example 6.3.1 In Table 6.2.2, we see the SAS GLM regression printout for the salary discrimination problem. The first three–line array of numbers appearing there is the ANOVA table for this problem. To be applicable in the analysis of all general linear model data sets, the first source of variation is titled "model". If the model is a regression model, as it is in this example, we have chosen to title this source "regression". Using the notation that we have settled on, the ANOVA for this problem is as follows.

Source	df	SS	MS	f
Regression	3	988.2174	329.4058	31.28
Error	12	126.3801	10.5317	
Total	15	1114.5975		

Here, we see SSr = 988.2174, SSe = 126.3801 and SSt = 1114.5975. Recall (6.2.1) which gives $Y = \beta_0 + \beta_1 x_1 + \beta_2 x_2 + \beta_3 x_3 + \varepsilon$ as the functional form of the model used in this analysis. Therefore, the resulting

$$f = \frac{MSr}{MSe} = \frac{329.4058}{10.5317} = 31.28$$

is the value of the statistic used in testing $H_0: \beta_1 = \beta_2 = \beta_3 = 0$. When we compare this value with $f_{.01,3,12} = 5.95$, we have reason to reject H_0 at the 0.01 significance level. In fact, according to the given p—value (PR > F), we would reject H_0 at all fixed risk levels greater than or equal to 0.0001.

Coefficient of Determination

The coefficient of determination, already encountered in the previous section, is another measure of the combined effect of the independent variables $x_1, x_2, \ldots, x_k$ on the independent variable Y. We now discover why r^2, the symbol for the coefficient of determination, is said to be the proportion of the variation in realized values of Y explained by the model.

By expressing the ANOVA sums of squares formulas in terms of the summation symbol, we see that $SSt = \sum_{i=1}^{n} (y_i - \bar{y})^2$, $SSr = \sum_{i=1}^{n} (\hat{y}_i - \bar{y})^2$, and $SSe = \sum_{i=1}^{n} (y_i - \hat{y}_i)^2$. Using the overall mean $\bar{y}$ as the point of reference, it is logical to say that SSt is a measure of the **total variation** in the y_i's and SSr and SSe measure **variation explained** and **unexplained** by the regression model, respectively. Since SSt = SSr + SSe, it follows that

$$\sum_{i=1}^{n} (y_i - \bar{y})^2 = \sum_{i=1}^{n} (\hat{y}_i - \bar{y})^2 + \sum_{i=1}^{n} (y_i - \hat{y}_i)^2$$

or, in words,

total variation = explained variation + unexplained variation.

The coefficient of determination is defined by the formula

$$r^2 = \frac{SSr}{SSt} = \frac{\text{explained variation}}{\text{total variation}} . \tag{6.3.5}$$

The reader should verify that this is the case in Tables 6.2.2 and 6.2.4. Since sums of squares are non—negative, $0 \leq SSr \leq SSt$, so that values of r^2 are confined to the range from 0 to 1.

6.4 SS( ) Notation and Adjusted Sums of Squares

We now introduce notation that provides us with a "shorthand" way of revealing the procedural origin of a sum of squares quantity. This notation will be especially helpful later in this text as a concise means of presenting ANOVA sums of squares "formulas" in the analysis of experimental design data.

As a basis for discussion, let us consider the multiple regression model used in the analysis of the salary discrimination data,

$$Y = \beta_0 + \beta_1 x_1 + \beta_2 x_2 + \beta_3 x_3 + \varepsilon .$$

On fitting the model to the data set in Table 6.2.1, we obtained $SSr = 988.21743$ and $SSe = 126.38007$ in the analysis of variance. (See Table 6.2.2.) In the SS( ) notational scheme, we denote the regression sum of squares for a fitted model by the expression $SS(\beta_0, \beta_1, \beta_2, \beta_3)$, where we identify within parentheses the parameters of the fitted model. A major objective in this problem was to test the hypothesis $H_0: \beta_1 = 0$. Suppose we perform this test using the principle of conditional error. This approach requires the calculation of $SSe^* - SSe$, where SSe^* is the error sum of squares for the reduced model $Y = \beta_0^* + \beta_2^* x_2 + \beta_3^* x_3 + \varepsilon^*$. For this reduced model we write $SSr^* = SS(\beta_0, \beta_2, \beta_3)$. (As an exercise, the reader can verify that $SSr^* = SS(\beta_0, \beta_2, \beta_3) = 988.08852$ and $SSe^* = 126.50898$.) Now, because of the identities $SSt = SSr + SSe$ and $SSt = SSr^* + SSe^*$, $SSe^* - SSe = SSr - SSr^* = SS(\beta_0, \beta_1, \beta_2, \beta_3) - SS(\beta_0, \beta_2, \beta_3) = 0.12891$. For conciseness, we denote the above difference in sums of squares by the single expression $SS(\beta_1 | \beta_0, \beta_2, \beta_3)$ which we read as "the sum of squares for β_1

adjusted for β_0, β_2 and β_3." The term "adjusted" is used here in a sense similar to our earlier usage when we referred to the difference in average salary, β_1, adjusted for education and work experience. It is instructive to note that the quantity $SS(\beta_1 | \beta_0, \beta_2, \beta_3)$ is the SAS Type III SS for the variable sex as found in Table 6.2.2, confirming our understanding of the meaning of this type of sum of squares.

In terms of regression sum of squares, we can think of $SS(\beta_1 | \beta_0, \beta_2, \beta_3)$ as the **increment** to the **regression sum of squares** when we include β_1 in the model in addition to the parameters β_0, β_2 and β_3. On the other hand, we can also view $SS(\beta_1 | \beta_0, \beta_2, \beta_3)$ as the **reduction** in **error sum of squares** due to the inclusion of β_1 in the model. As a result, the symbol $R(\beta_1 | \beta_0, \beta_2, \beta_3)$ is an alternative symbol sometimes found in the literature.

As an additional example using the same model $Y = \beta_0 + \beta_1 x_1 + \beta_2 x_2 + \beta_3 x_3 + \varepsilon$, let us consider the hypothesis $H_0: \beta_2 = \beta_3 = 0$. Here the reduced model is $Y = \beta_0^{**} + \beta_1^{**} x_1 + \varepsilon^{**}$ for which we denote the regression sum of squares by $SSr^{**} = SS(\beta_0, \beta_1)$. Using SS( ) notation, we write the difference $SS(\beta_0, \beta_1, \beta_2, \beta_3) - SS(\beta_0, \beta_1)$ as $SS(\beta_2, \beta_3 | \beta_0, \beta_1)$. This difference is said to be the sum of squares for β_2 and β_3 **adjusted** for β_0 and β_1.

More on SAS Type I and Type III Sums of Squares

The notation which we have introduced here enables us to label SAS Type I and Type III sums of squares and thereby distinguish between the two types in a precise manner.

Example 6.4.1 If we turn to the three tables associated with the grade point problem, we find $SS(\beta_0, \beta_1, \beta_2) = 7.49370$ (Table 6.2.4), $SS(\beta_0, \beta_1) = 5.11377$ (Table 6.2.6) and $SS(\beta_0, \beta_2) = 7.46097$. From our previous discussion of the meaning of Type I and Type III sums of squares, we can now label the output numbers in Table 6.2.4 as follows:

	Type I SS	Type III SS
ACT	$5.11377 = SS(\beta_0,\beta_1)$	$0.03273 = SS(\beta_1 \mid \beta_0,\beta_2)$
RANK	$2.37993 = SS(\beta_2 \mid \beta_0,\beta_1)$	$2.37993 = SS(\beta_2 \mid \beta_0,\beta_1)$.

Example 6.4.2 Consider Table 6.2.2 associated with the regression model $Y = \beta_0 + \beta_1 x_1 + \beta_2 x_2 + \beta_3 x_3 + \varepsilon$. By definition, Type I sums of squares is a **partitioning** of the regression sum of squares in the following manner:

$$SS(\beta_0,\beta_1,\beta_2,\beta_3) = SS(\beta_0,\beta_1) + SS(\beta_2 \mid \beta_0,\beta_1) + SS(\beta_3 \mid \beta_0,\beta_1,\beta_2)$$

$$= 26.26817 + 476.65270 + 485.29656 = 988.21743.$$

On the other hand, Type III sums of squares are "principle of conditional error sums of squares" used in testing the coefficients β_1, β_2 and β_3 **individually** equal to zero. Hence $SS(\beta_1 \mid \beta_0,\beta_2,\beta_3) =$ 0.12891, $SS(\beta_2 \mid \beta_0,\beta_1,\beta_3) = 864.87096$ and $SS(\beta_3 \mid \beta_0,\beta_1,\beta_2) =$ 485.29656.

6.5 Orthogonal Polynomials

Fundamentals

Suppose, in an experimental design problem, the independent variable(s) is/are of the quantitative type. In this case, the investigator often is interested in the type of mathematical relationship that the response variable has to the independent variable(s). For example, the experimenter may ask the questions: Is the nature of the response linear? or quadratic? or both? or neither? If the response is increasing, is it at an increasing or decreasing or constant rate? Questions such as these, under conditions usually within the control of the investigator, can be answered using **orthogonal polynomials.**

The basic ideas for motivating the orthogonal polynomials technique are contained in the following theorem. It deals with the case where the **X** matrix of a regression model consists of mutually orthogonal column vectors.

Theorem 6.5.1 Consider an intercept linear regression model for n observations

$$\mathbf{Y} = \mathbf{X}\boldsymbol{\beta} + \boldsymbol{\varepsilon} \text{ and its sample form } \mathbf{y} = \mathbf{X}\mathbf{b} + \mathbf{e}.$$

Let us write the matrix $\mathbf{X}$ as a set of column vectors

$$\mathbf{X} = [\mathbf{1} \ \mathbf{x}_1 \ \mathbf{x}_2 \cdots \mathbf{x}_k]$$

where $\mathbf{1}$ represents a column vector with all elements the number one. If the column vectors composing $\mathbf{X}$ are mutually orthogonal, that is, $\mathbf{1}'\mathbf{x}_j = 0$ for all $j = 1, 2, \ldots k$, and $\mathbf{x}_j' \mathbf{x}_\ell = 0$ for all j and ℓ, $j \neq \ell$, then the following results are obtained.

(1) $b_0 = \sum_{i=1}^{n} y_i/n$ and

$$b_j = \frac{\mathbf{x}_j'\mathbf{y}}{\mathbf{x}_j'\mathbf{x}_j} = \frac{\sum_{i=1}^{n} x_{ij}y_i}{\sum_{i=1}^{n} x_{ij}^2} \quad \text{for } j = 1, 2, \ldots, k.$$

(2) The sum of squares for β_j, adjusted for all other coefficients and denoted by $SS(\beta_j)$ is given by

$$SS(\beta_j) = \frac{(\mathbf{x}_j'\mathbf{y})^2}{\mathbf{x}_j'\mathbf{x}_j} = \frac{\sum_{i=1}^{n} (x_{ij}y_i)^2}{\sum_{i=1}^{n} x_{ij}^2} \quad \text{for } j = 1, 2, \ldots k,$$

and $SS(\beta_0) = (\sum_{i=1}^{n} y_i)^2/n = n\bar{y}^2$.

(3) The regression sum of squares, $SSr = \mathbf{b}'\mathbf{X}'\mathbf{y} - n\bar{y}^2$, can then be partitioned as follows:

$$SSr = SS(\beta_1) + SS(\beta_2) + \cdots + SS(\beta_k).$$

Proof. The student can fill in the details of the proof which we now sketch. Because of the orthogonality of the columns of the $\mathbf{X}$ matrix, $\mathbf{X}'\mathbf{X}$ is the diagonal matrix

$$\mathbf{X}'\mathbf{X} = \begin{bmatrix} n & 0 & 0 & \cdots & 0 \\ 0 & \sum_{i=1}^{n} x_{i1}^2 & 0 & & 0 \\ 0 & 0 & \sum_{i=1}^{n} x_{i2}^2 & \cdots & 0 \\ \vdots & \vdots & \vdots & & \vdots \\ 0 & 0 & 0 & \cdots & \sum_{i=1}^{n} x_{ik}^2 \end{bmatrix}$$

Therefore, since $\mathbf{X}'\mathbf{y}$ is the column vector with elements $\sum_{i=1}^{n} y_i$, $\sum_{i=1}^{n} x_{i1}y_i$, $\sum_{i=1}^{n} x_{i2}y_i, \ldots, \sum_{i=1}^{n} x_{ik}y_i$, and $\mathbf{b} = (\mathbf{X}'\mathbf{X})^{-1}\mathbf{X}'\mathbf{y}$, it is easy to see that $b_0 = \sum_{i=1}^{n} y_i/n$ and the general jth element is $b_j = \sum_{i=1}^{n} x_{ij}y_i / \sum_{i=1}^{n} x_{ij}^2$ for $j = 1,2,\ldots,k$. Also, it follows that the first entry in the vector $\mathbf{b}'\mathbf{X}'\mathbf{y}$ is $(\sum_{i=1}^{n} y_i)^2/n$ with the remaining k elements, $(\sum_{i=1}^{n} x_{ij}y_i)^2 / \sum_{i=1}^{n} x_{ij}^2$ for $j = 1, 2, \ldots, k$.

It is important to note that when the $\mathbf{X}$ matrix consists of mutually orthogonal columns, the estimate b_j depends only on the entries in the jth column of $\mathbf{X}$ and the vector $\mathbf{y}$. In the general regression problem, this is not true. Likewise, the sum of squares associated with parameter β_j is a function of the jth column elements together with those in vector $\mathbf{y}$, **independent** of the other column entries. For that reason it is natural to write $SS(\beta_j)$ as the abbreviated symbol for $SS(\beta_j | \beta_0, \beta_1, \ldots, \beta_{j-1}, \beta_{j+1}, \ldots \beta_k)$ and logically permits us to refer to it as the "jth coefficient sum of squares."

To apply the results of Theorem 6.5.1 to the polynomial setting, consider fitting the kth degree polynomial model

$$Y_i = \beta_0 + \beta_1 x_i + \beta_2 x_i^2 + \cdots + \beta_k x_i^k + \varepsilon, \ i = 1, 2, \ldots, n \qquad (6.5.1)$$

to a set of n data points. Let us rewrite model (6.5.1) as

$$Y_i = \alpha_0+\alpha_1 p_1(x_i)+\alpha_2 p_2(x_i)+\cdots+\alpha_k p_k(x_i)+\varepsilon\ ,\ i = 1, 2,\cdots,n, \quad (6.5.2)$$

where $p_j(x_i)$, $j = 1, 2, ..., k$ is a jth degree polynomial in x_i such that

$$\sum_{i=1}^{n} p_j(x_i)p_\ell(x_i) = 0 \text{ for } j \text{ and } \ell,\ j \neq \ell,$$

and
$$\sum_{i=1}^{n} p_j^2(x_i) \neq 0 \text{ for } j = 1, 2,..., k.$$

In matrix notation we can write equation (6.5.2) as

$$\mathbf{Y} = \mathbf{P}\boldsymbol{\alpha} + \boldsymbol{\varepsilon} \qquad (6.5.3)$$

where the matrix $\mathbf{P}$ is a set of orthogonal column vectors

$$\mathbf{P} = [\mathbf{1}\ \ \mathbf{p}_1\ \ \mathbf{p}_2\cdots\mathbf{p}_k].$$

It is obvious that if the $k + 1$ columns of matrix $\mathbf{P}$ are mutually orthogonal, $\sum_{i=1}^{n} p_j(x_i) = 0$ is a necessary restriction on the $p_j(x_i)$. In order to obtain a set of mutually orthogonal vectors $\mathbf{P}$ from the columns of matrix $\mathbf{X}$ corresponding to polynomial model (6.5.1), we can use the Gram–Schmidt Process of Orthogonalization. We now illustrate this through an example which illuminates the essential features of data fitting using the orthogonal polynomial approach.

An Example

Suppose our goal is to fit a third degree polynomial to the $n = 8$ observations of Exercise 3–6,

y	1	3	–2	–4	–3	–5	6	4
x	1	1	2	2	3	3	4	4

We observe that this data set consists of $m = 4$ levels of the independent variable x with each level of x repeated $r = 2$ times. In fitting a third degree model

$$Y = \beta_0 + \beta_1 x + \beta_2 x^2 + \beta_3 x^3 + \varepsilon$$

to these data, we reparametrize to an orthogonal polynomial model

$$Y = \alpha_0 + \alpha_1 p_1(x) + \alpha_2 p_2(x) + \alpha_3 p_3(x) + \varepsilon$$

where $p_j(x)$ is a polynomial of jth degree, $j = 1, 2, 3$. In matrix notation we write the orthogonal polynomial model as

$$\mathbf{Y} = \mathbf{P}\boldsymbol{\alpha} + \varepsilon$$

where $\mathbf{P} = [\mathbf{1}\ \mathbf{p}_1\ \mathbf{p}_2\ \mathbf{p}_3]$ is a set of mutually orthogonal columns. To derive the $\mathbf{p}_j$, we use the Gram–Schmidt Process of Orthogonalization which states that in order to form a set of orthogonal vectors $\{\mathbf{p}_0, \mathbf{p}_1, \mathbf{p}_2, \mathbf{p}_3\}$ from a set of nonzero column vectors $\{\mathbf{x}_0, \mathbf{x}_1, \mathbf{x}_2, \mathbf{x}_3\}$ we let

$$\mathbf{p}_0 = \mathbf{x}_0$$

$$\mathbf{p}_1 = \mathbf{x}_1 - \frac{\mathbf{p}_0'\mathbf{x}_1}{\mathbf{p}_0'\mathbf{p}_0}\mathbf{p}_0$$

$$\mathbf{p}_2 = \mathbf{x}_2 - \frac{\mathbf{p}_0'\mathbf{x}_2}{\mathbf{p}_0'\mathbf{p}_0}\mathbf{p}_0 - \frac{\mathbf{p}_1'\mathbf{x}_2}{\mathbf{p}_0'\mathbf{p}_1}\mathbf{p}_1$$

$$\mathbf{p}_3 = \mathbf{x}_3 - \frac{\mathbf{p}_0'\mathbf{x}_3}{\mathbf{p}_0'\mathbf{p}_0}\mathbf{p}_0 - \frac{\mathbf{p}_1'\mathbf{x}_3}{\mathbf{p}_1'\mathbf{p}_1}\mathbf{p}_1 - \frac{\mathbf{p}_2'\mathbf{x}_3}{\mathbf{p}_2'\mathbf{p}_2}\mathbf{p}_2 .$$

In our problem, the **distinct** levels of x are $x = 1, 2, 3, 4$. Hence the vectors $\{\mathbf{x}_0, \mathbf{x}_1, \mathbf{x}_2, \mathbf{x}_3\}$ are

$$\mathbf{x}_0 = \mathbf{1} = \begin{bmatrix} 1 \\ 1 \\ 1 \\ 1 \end{bmatrix}, \quad \mathbf{x}_1 = \mathbf{x} = \begin{bmatrix} 1 \\ 2 \\ 3 \\ 4 \end{bmatrix}, \quad \mathbf{x}_2 = \mathbf{x}^2 = \begin{bmatrix} 1 \\ 4 \\ 9 \\ 16 \end{bmatrix},$$

and $\mathbf{x}_3 = \mathbf{x}^3 = \begin{bmatrix} 1 \\ 8 \\ 27 \\ 64 \end{bmatrix}$.

Therefore,

$$\mathbf{p}_0 = \begin{bmatrix} 1 \\ 1 \\ 1 \\ 1 \end{bmatrix} \quad \text{or} \quad \mathbf{p}_0(x) = 1 \quad \text{for} \quad x = 1, 2, 3, 4$$

and

$$\mathbf{p}_1 = \mathbf{x}_1 - \frac{5}{2}\mathbf{p}_0 = \begin{bmatrix} -3/2 \\ -1/2 \\ 1/2 \\ 3/2 \end{bmatrix} \quad \text{or} \quad p_1(x) = x - \frac{5}{2} \quad \text{for} \quad x = 1,2,3,4.$$

In practice, it is desirable to avoid fractions which we can do by letting $\mathbf{p}_1^* = 2\mathbf{p}_1$ so that

$$\mathbf{p}_1^* = 2\mathbf{x}_1 - 5\mathbf{x}_0 = \begin{bmatrix} -3 \\ -1 \\ 1 \\ 3 \end{bmatrix} \quad \text{or} \quad p_1^*(x) = 2x - 5 \quad \text{for} \quad x = 1,2,3,4.$$

In the orthogonal polynomials, Table 4 in Appendix B, the factor 2 is denoted by the symbol λ_1. Next we find

$$\mathbf{p}_2 = \mathbf{x}_2 - \frac{15}{2}\mathbf{p}_0 - 5\,\mathbf{p}_1 = \mathbf{x}_2 - \frac{15}{2}\mathbf{x}_0 - 5\,(\mathbf{x}_1 - \frac{5}{2}x_0)$$

$$= \mathbf{x}_2 - 5\mathbf{x}_1 + 5\,\mathbf{x}_0 = \begin{bmatrix} 1 \\ -1 \\ -1 \\ 1 \end{bmatrix} \quad \text{or} \quad p_2(x) = x^2 - 5x + 5$$

for $x = 1, 2, 3, 4$. (Here $\lambda_2 = 1$).

Finally

$$\mathbf{p}_3 = \mathbf{x}_3 - 25\,\mathbf{p}_0 - \frac{104}{5}\mathbf{p}_1 - \frac{15}{2}\mathbf{p}_2$$

$$= \mathbf{x}_3 - 25\mathbf{x}_0 - \frac{104}{5}(\mathbf{x}_1 - \frac{5}{2}\mathbf{x}_0) - \frac{15}{2}(\mathbf{x}_2 - 5\mathbf{x}_1 + 5\mathbf{x}_0)$$

$$= \mathbf{x}_3 - \frac{15}{2}\mathbf{x}_2 + \frac{167}{10}\mathbf{x}_1 - \frac{21}{2}\mathbf{x}_0 = \begin{bmatrix} -3/10 \\ 9/10 \\ -9/10 \\ 3/10 \end{bmatrix}.$$

To obtain integer—valued elements, we let $\lambda_3 = 10/3$ yielding

$$\mathbf{p}_3^* = \frac{10}{3}\,\mathbf{p}_3 = \begin{bmatrix} -1 \\ 3 \\ -3 \\ 1 \end{bmatrix} \quad \text{or } p_3^*(x) = \frac{1}{3}(10x^3 - 75x^2 + 167x - 105)$$

for x = 1, 2, 3, 4.

We are now ready to obtain the coefficients in the least squares estimating function

$$\hat{y} = a_0 + a_1 p_1^*(x) + a_2 p_2(x) + a_3 p_3^*(x).$$

With reference to our earlier general discussion, in this problem the "complete" 8 × 4 matrix $\mathbf{P}^* = [\mathbf{1}\ \ \mathbf{p}_1^*\ \ \mathbf{p}_2\ \ \mathbf{p}_3^*]$ accounting for the two observations at each of the four x levels is

$$\mathbf{P}^* = \begin{bmatrix} 1 & -3 & 1 & -1 \\ 1 & -3 & 1 & -1 \\ 1 & -1 & -1 & 3 \\ 1 & -1 & -1 & 3 \\ 1 & 1 & -1 & -3 \\ 1 & 1 & -1 & -3 \\ 1 & 3 & 1 & 1 \\ 1 & 3 & 1 & 1 \end{bmatrix} \quad \text{and with } \mathbf{y} = \begin{bmatrix} 1 \\ 3 \\ -2 \\ -4 \\ -3 \\ -5 \\ 6 \\ 4 \end{bmatrix}$$

we arrive at

$$\mathbf{a} = (\mathbf{P}^{*\prime}\mathbf{P}^*)^{-1}\mathbf{P}^{*\prime}\mathbf{y} = \begin{bmatrix} 0 \\ 2/5 \\ 7/2 \\ 3/10 \end{bmatrix} = \begin{bmatrix} a_0 \\ a_1 \\ a_2 \\ a_3 \end{bmatrix}.$$

Therefore the estimating regression function in terms of the orthogonal polynomials is

$$\hat{y} = \frac{2}{5}\,p_1^*(x) + \frac{7}{2}\,p_2(x) + \frac{3}{10}\,p_3^*(x).$$

It should be noted that we need not use matrix algebra in order to compute the a_j coefficients. In its stead, the first formula in Theorem 6.5.1,

$$a_j = \frac{\sum_{i=1}^{8} p_j(x_i) y_i}{\sum_{i=1}^{8} p_j^2(x_i)}, \quad j = 1, 2, 3,$$

also yields the required results. To complete this example, we can obtain the desired fitted function in terms of the original independent variable x by substituting the orthogonal polynomial expressions as follows:

$$\hat{y} = \frac{2}{5}(2x - 5) + \frac{7}{2}(x^2 - 5x + 5)$$

$$+ \frac{3}{10}\left[\frac{1}{3}(10x^3 - 75x^2 + 167x - 105)\right]$$

$$= 5 - 4x^2 + x^3.$$

Admittedly, using the Gram—Schmidt Process of Orthogonalization of vectors is not a pleasant task. However, in the case when the levels of the independent variable x are equally spaced <u>and</u> equally replicated, the orthogonal polynomial table, Table 4 in Appendix B can be used. By equally spaced, we mean that the m distinct levels of x can be represented by the sequence

$$x_1, x_1 + d, x_1 + 2d, \ldots, x_1 + (m - 1)d,$$

where x_1 is the first level of x and d is the common difference between successive levels. By equally replicated levels, we mean that the same number of observations are taken at each of the m levels of x. In fact, the first five orthogonal polynomials given in the table are

$$p_0(x_i) = 1$$

$$p_1(x_i) = \lambda_1 \left[\frac{(x_i - \bar{x})}{d}\right]$$

$$p_2(x_i) = \lambda_2 \left[\left(\frac{x_i - \bar{x}}{d}\right)^2 - \frac{m^2 - 1}{12}\right] \tag{6.5.4}$$

$$p_3(x_i) = \lambda_3 \left[\left(\frac{x_i - \bar{x}}{d}\right)^3 - \left(\frac{3m^2 - 7}{20}\right)\left(\frac{x_i - \bar{x}}{d}\right)\right]$$

$$p_4(x_i) = \lambda_4 \left[\left(\frac{x_i - \bar{x}}{d}\right)^4 - \left(\frac{3m^2 - 13}{14}\right)\left(\frac{x_i - \bar{x}}{d}\right)^2 + \frac{3(m^2 - 1)(m^2 - 9)}{560}\right]$$

where $\bar{x} = \sum_{i=1}^{n} x_i/n$ and the λ_j are the constants necessary to create integer valued entries in the p_j vectors. The set of formulas (6.5.4) can be used for the inverse transformation from (6.5.2) back to (6.5.1).

6.6 Response Analysis Using Orthogonal Polynomials

In the analysis of experimental design data, the investigator is seldom interested in the coefficients of the fitted function. Rather, he/she is usually more interested in the **nature** of the response in order to answer questions of the type posed at the beginning of the last section. To obtain the answers, the regression sum of squares is partitioned into polynomial effects sums of squares. The theoretical basis for the partitioning is given in Theorem 6.5.1.

Returning to the kth degree orthogonal polynomial model (6.5.2) for n data points,

$$Y_i = \alpha_0 + \alpha_1 p_1(x_i) + \alpha_2 p_2(x_i) + \cdots + \alpha_k p_k(x_i) + \varepsilon,$$

$i = 1, 2, ..., n,$ we know that the regression sum of squares partitions as

$$SSr = SS(\alpha_1) + SS(\alpha_2) + \cdots + SS(\alpha_k),$$

where the "jth coefficient sum of squares" is given by the formula

$$SS(\alpha_j) = \frac{\left[\sum_{i=1}^{n} p_j(x_i) y_i \right]^2}{\sum_{i=1}^{n} p_j^2(x_i)}, \quad j = 1, 2, ..., k. \tag{6.6.1}$$

In answer to the question, "Is there a jth polynomial effect?", we test the null hypothesis

$$H_0 : \alpha_j = 0 \ .$$

In words, this null hypothesis reads as "no jth polynomial effect." If we apply the principle of conditional error technique, it is easily seen

$$\text{SSe (full model)} - \text{SSe (reduced model)} = SS(\alpha_j) \ . \tag{6.6.2}$$

Therefore, the test statistic value for $H_0 : \alpha_j = 0$ is

$$f = \frac{SS(\alpha_j)}{s^2}, \quad j = 1, 2, ..., k, \tag{6.6.3}$$

where s^2 is the error mean square for the fitted model. The degrees of freedom associated with F–statistic (6.6.3) are 1 and $n - k - 1$.

Example 6.6.1 To continue with the analysis of the data set example introduced in the previous section, we recall the regression analysis of variance relation $SSt = SSr + SSe$. For these $n = 8$ data points

$$SSt = \sum_{i=1}^{n} y_i^2 - \left(\sum_{i=1}^{n} y_i\right)^2 / n = 116.0.$$

and

$$SSr = SS(\alpha_1) + SS(\alpha_2) + SS(\alpha_3)$$

$$= 6.4 + 98.0 + 3.6 = 108.0.$$

Therefore, $SSe = SSt - SSr = 8.0$. The results of these calculations can be conveniently summarized in an "extended" ANOVA table as follows:

Source	df	SS	MS	f
Regression	3	108.0	36.0	18.0
Linear	1	6.4	6.4	3.2
Quadratic	1	98.0	98.0	49.0
Cubic	1	3.6	3.6	1.8
Error	4	8.0	2.0	
Total	7	116.0		

On comparing with the critical values $f_{.05,1,4} = 7.71$ and $f_{.01,1,4} = 21.20$, we conclude that there is a highly significant quadratic trend in the data; the linear and cubic effects, however, are not significant at the 0.05 risk level.

The next question to consider is, "How do we obtain the single degree of freedom polynomial effects when the orthogonal polynomial table cannot be used, i.e., when the independent variable levels are unequally spaced and/or unequally replicated?" The answer is to use a regression program in order to fit successively polynomials of higher and higher degree. To illustrate, we apply this technique to the data set example of Section 6.5 and thereby repeat the analysis done in Example 6.6.1.

On fitting the "full" model

$$Y = \beta_0 + \beta_1 x + \beta_2 x^2 + \beta_3 x^3 + \varepsilon$$

we obtain SSr = 108.0 and SSe = 8.0. Next, in fitting the reduced model

$$Y = \beta_0^* + \beta_1^* x + \varepsilon^*$$

we arrive at

$$SSr^* = SS(\beta_0, \beta_1) = SS(\alpha_1) = 6.4.$$

Then fitting

$$Y = \beta_0^{**} + \beta_1^{**} x + \beta_2^{**} x^2 + \varepsilon^{**},$$

$$SS(\beta_2 | \beta_0, \beta_1) = SSr^{**} - SSr^{*} = SSe^{*} - SSe^{**} = SS(\alpha_2) = 98.0\ .$$

Finally,

$$SS(\beta_3 | \beta_0, \beta_1, \beta_2) = SSr - SSr^{**} = SSe^{**} - SSe = SS(\alpha_3) = 3.6.$$

For those using the SAS GLM program, it is important to note that we only need to fit the full cubic model since the partitioning of the regression sum of squares is given to us through the printout of the Type I sums of squares as follows:

$$SSr = SS(\beta_0, \beta_1) + SS(\beta_2 | \beta_0, \beta_1) + SS(\beta_3 | \beta_0, \beta_1, \beta_2).$$

An Experimental Design Example

To further appreciate the role of a mathematical model in the analysis of the results of a designed experiment, we now turn to an example framed in the experimental design setting. Suppose the following data represent the percent of yield for a chemical process at various levels of pressure, x_1, and temperature, x_2.

Table 6.6.1 Chemical Process Experiment

Pressure, x_1	Temperature, x_2 (^{o}F)		
(atmospheres)	120	150	180
1.0	14.5	20.3	17.8
	15.2	21.2	18.4
2.0	17.3	24.2	20.2
	18.5	25.0	19.6

To analyze these data, let us **postulate** the surface model

$$Y = \beta_0 + \beta_1 x_1 + \beta_2 x_2 + \beta_3 x_2^2 + \beta_4 x_1 x_2 + \beta_5 x_1 x_2^2 + \varepsilon \quad (6.6.4)$$

for which the corresponding orthogonal polynomials model is

$$Y = \alpha_0 + \alpha_1 p_1(x_1) + \alpha_2 p_1(x_2) + \alpha_{22} p_2(x_2) + \alpha_{12} p_1(x_1)\, p_1(x_2)$$
$$+ \alpha_{122} p_1(x_1)\, p_2(x_2) + \varepsilon. \quad (6.6.5)$$

Observe that in this illustration the levels of the independent variables, pressure and temperature, are equally spaced and equally replicated so that use of the orthogonal polynomials table is in order. From this table we obtain the **P** matrix as it relates to the response vector y as follows:

$$
\mathbf{y} = \begin{bmatrix} 14.5 \\ 15.2 \\ 17.3 \\ 18.5 \\ 20.3 \\ 21.2 \\ 24.2 \\ 25.0 \\ 17.8 \\ 18.4 \\ 20.2 \\ 19.6 \end{bmatrix}
\quad
\mathbf{P} = \begin{array}{c} \\ \end{array}
\begin{bmatrix}
 & p_1(x_1) & p_1(x_2) & p_2(x_2) & p_1(x_1)p_1(x_2) & p_1(x_1)p_2(x_2) \\
1 & -1 & -1 & 1 & 1 & -1 \\
1 & -1 & -1 & 1 & 1 & -1 \\
1 & 1 & -1 & 1 & -1 & 1 \\
1 & 1 & -1 & 1 & -1 & 1 \\
1 & -1 & 0 & -2 & 0 & 2 \\
1 & -1 & 0 & -2 & 0 & 2 \\
1 & 1 & 0 & -2 & 0 & -2 \\
1 & 1 & 0 & -2 & 0 & -2 \\
1 & -1 & 1 & 1 & -1 & -1 \\
1 & -1 & 1 & 1 & -1 & -1 \\
1 & 1 & 1 & 1 & 1 & 1 \\
1 & 1 & 1 & 1 & 1 & 1
\end{bmatrix}
$$

Note that the columns of matrix $\mathbf{P}$ are orthogonal and, except for the first column, sum to zero. To examine the underlying structure of the data set, we proceed by calculating the associated analysis of variance table.

The total sum of squares is

$$SSt = \sum_{i=1}^{12} y_i^2 - (\sum_{i=1}^{12} y_i)^2/12 = 4602.60 - (232.2)^2/12 = 109.53000.$$

Using formula (6.6.1), we calculate the dot products of vector $\mathbf{y}$ with the orthogonal polynomial columns of matrix $\mathbf{P}$ and after applying the proper divisors to their squares , we obtain a partitioning of the regression or model sum of squares. Here

$$SSr = SS(\alpha_1) + SS(\alpha_2) + SS(\alpha_{22}) + SS(\alpha_{12}) + SS(\alpha_{122})$$
$$= 25.23000 + 13.78125 + 66.33375 + 0.78125 + 1.35735 = 107.48360.$$

By subtraction, we compute the error sum of squares as

$$SSe - SSt - SSr = 2.04640$$

with $11 - 5 = 6$ degrees of freedom. A summary of these results appear in the extended ANOVA table below where P_L is the symbol for "Pressure, linear", T_L stands for "Temperature, linear", T_Q means "Temperature, quadratic" and P_LT_L and P_LT_Q are sources of variation due to "interaction", a concept introduced and discussed in Chapter 13.

Source	df	SS	MS	f
Model	5	107.48360	21.49672	63.03
P_L	1	25.23000	25.23000	73.97
T_L	1	13.78125	13.78125	40.41
T_Q	1	66.33375	66.33375	194.49
$P_L T_L$	1	0.78125	0.78125	2.29
$P_L T_Q$	1	1.35735	1.35735	3.98
Error	6	2.04640	0.34107	
Total	11	109.53000		

The large value of $f = 63.03$ indicates that the assumed model (6.6.4) does indeed explain most of the variation in the response to pressure and temperature. (Compare with $f_{.01,5,6} = 8.75$.) To determine the mathematical nature of the effect that pressure and temperature have on the response, percent yield, we compare with critical values $f_{.05,1,6} = 5.99$ and $f_{.01,1,6} = 13.7$. This leads us to conclude a highly significant linear effect of pressure on the response; that is, the percent of yield increases with an increase in pressure. Additionally, we conclude that over the experimental domain, the effect of temperature on yield is both linear and quadratic; that is, the percent of yield increases with increasing temperature but at a decreasing rate. Neither of the product ("interaction") terms in the model is significant at the 0.05 level of risk.

In closing, it is appropriate to point out that this example illustrates that "fitting" a mathematical model is central to the analysis of experimental design data, something that students sometimes lose sight of when routinely doing an analysis based upon information contained in an analysis of variance table. This example also gives the reader a glimpse of response surface analysis (Chapter 18) where the goal is to find the combination of levels of the independent variables that provide for an optimal response.

Chapter 6 Exercises

Application Exercises

6–1 Refer to Table 4.4.2 showing the SAS analysis of the energy usage survey data using the regression model

$$Y = \beta_0 + \beta_1 x_1 + \beta_2 x_2 + \beta_3 x_3 + \varepsilon ,$$

where Y is the electric cost response, x_1 is the average outside temperature, x_2 is the thermostat setting, and x_3 is the square footage. Using this printout, give the numerical values of the following quantities.

(a) $SS(\beta_0,\beta_1,\beta_2,\beta_3)$

(b) $SS(\beta_0,\beta_1)$

(c) $SS(\beta_2 | \beta_0,\beta_1)$

(d) $f = b_2^2/s^2 c_{22}$

(e) $t = (b_2 - 3)/\sqrt{s^2 c_{22}}$

(f) The proportion of the variation in electric cost accounted for by the three independent variables.

6–2 In a one–year study to determine if there is a relationship between mortality and air pollution, biweekly suspended particulate readings and sulfate readings were taken in eight metropolitan areas having like population characteristics (e.g., similar in population density, percentage of nonwhites, percentage of age 65 and over, etc.). In the data below, y represents the mortality rate in deaths per 10,000; x_1 represents the mean of the 26 suspended particulate readings in $\mu g/m^3$; and x_2 represents the minimum of the 26 sulfate readings in $\mu g/m^3$.

y	87	98	72	92	106	95	80	90
x_1	95	165	78	132	148	110	101	120
x_2	4.2	5.2	2.1	4.6	7.5	5.8	2.3	3.6

(a) Fit the model $Y = \beta_0 + \beta_1 x_1 + \beta_2 x_2 + \varepsilon$ obtaining the sample regression function $\hat{y} = b_0 + b_1 x_1 + b_2 x_2$.

(b) Test the hypotheses (i) $H_0 : \beta_1 = 0$, (ii) $H_0 : \beta_2 = 0$ and (iii) $H_0 : \beta_2 = 6$ at the 0.01 significance level.

(c) Compute 99% confidence intervals for β_1 and β_2.

(d) Find a 95% confidence interval for the mean mortality rate when $x_1 = 100$ and $x_2 = 5.0$.

(e) Find the value of r^2. Interpret.

6–3 Using a regression program, fit the example data in Section 6.5 successively with first, second, and third degree polynomials and show that $SS(\beta_0,\beta_1) = SS(\alpha_1)$, $SS(\beta_2 | \beta_0,\beta_1) = SS(\alpha_2)$, and $SS(\beta_3 | \beta_0,\beta_1,\beta_2) = SS(\alpha_3)$.
Note: When fitting the third degree polynomial model $Y = \beta_0 + \beta_1 x + \beta_2 x^2 + \beta_3 x^3 + \varepsilon$ using SAS's GLM, $SS(\beta_0,\beta_1)$, $SS(\beta_2 | \beta_0,\beta_1)$ and $SS(\beta_3 | \beta_0,\beta_1,\beta_2)$ are "automatically" printed out under the heading TYPE I SS.

6–4 Refer to the data set in Section 6.5. Plot these data points. Does the conclusion that there exists a quadratic relationship between x and y seem reasonable? Explain why the ANOVA does not reveal a significant linear relationship.

6–5 Refer to the numerical example in Section 6.5. Use the estimating orthogonal polynomials regression function

$$\hat{y} = \frac{2}{5} p_1^*(x) + \frac{7}{2} p_2(x) + \frac{3}{10} p_3^*(x)$$

to obtain a 95% confidence interval for $E(Y)$ when $x = 2$.

6–6 In lieu of the equally spaced levels of the independent variable x in Section 6.5, consider the data set

y	1	3	−2	−4	−3	−5	6	4
x	0	0	2	2	3	3	4	4

where $x = 0$ has replaced $x = 1$ in the original data set.

(a) For the revised data set, use the Gram–Schmidt Process to derive the orthogonal polynomials for fitting a third degree polynomial and obtain the estimating orthogonal polynomials function.
(b) Obtain a 95% confidence interval for E(Y) when $x = 2$.
(c) Calculate the associated analysis of variance table, including the extended analysis for polynomial effects.
(d) Use a regression program on x and y in order to obtain the extended ANOVA table asked for in (c).

6–7 It is well known that when one observation is taken at each level of x, a polynomial of degree $k-1$ can be found that will fit k points exactly. Given the three (x,y) points: (0,1), (2,–4), (4,–5), find the coefficients in the exact equation

$$y = b_0 + b_1 x + b_2 x^2$$

by first deriving the orthogonal polynomials function

$$y = a_0 + a_1 p_1(x) + a_2 p_2(x).$$

6–8 Refer to the Chemical Process Experiment in Section 6.6.
(a) Calculate the estimates of the coefficients in the orthogonal polynomials model (6.6.5.)
(b) Use the transformation equations (6.5.4) in order to obtain the estimating function for (6.6.4), i.e., compute the coefficients in the function

$$\hat{y} = b_0 + b_1 x_1 + b_2 x_2 + b_3 x_2^2 + b_4 x_1 x_2 + b_5 x_1 x_2^2 .$$

(c) The statement was made that model (6.6.4) explains most of the variation in response to pressure and temperature. What proportion of the variation in percent yield is accounted for by the model?

6–9 The data below were obtained in an experimental design setting. In order to determine the effect of a growth stimulant, 16 feedlot beef cattle were given hypodermic injections with 8 of the animals receiving a dosage of 5 units and the other 8 receiving 10 unit doses. Another 8 cattle in the feedlot were given no injections, i.e., a dosage of 0 units. The weight gains in pounds were as follows:

Dose, x	0	5	10
Gain in weight, y	100	115	125
	110	121	140
	92	110	153
	122	130	142
	118	142	130
	98	108	162
	130	112	157
	110	120	160

Using a regression program, fit the following models to these data.

(i) $Y = \beta_0 + \beta_1 x + \beta_2 x^2 + \varepsilon.$

(ii) $Y = \alpha_0 + \alpha_1 x_1 + \alpha_2 x_2 + \varepsilon$

where $x_1 = \begin{cases} 1, \text{if } x = 0 \\ 0, \text{if not} \end{cases}$, $x_2 = \begin{cases} 1, \text{if } x = 5 \\ 0, \text{if not} \end{cases}$.

Note: This example illustrates the identification of three categories using two "dummy" variables.

(iii) $Y = \gamma_0 + \gamma_1 x_1 + \gamma_2 x_2 + \varepsilon$

where $x_1 = \begin{cases} -1, \text{if } x = 0 \\ 0, \text{if } x = 5 \\ 1, \text{if } x = 10 \end{cases}$, $x_2 = \begin{cases} 1, \text{if } x = 0 \\ -2, \text{if } x = 5 \\ 1, \text{if } x = 10 \end{cases}$.

(a) Compare the regression analysis of variance printouts associated with each model.

(b) For each of the three fitted models, estimate the mean weight gain when the level of dosage is 10 units.

(c) Observe that model (iii) utilizes orthogonal polynomials coding for x_1 and x_2. Extend the ANOVA by partitioning the regression sum of squares into linear and quadratic sums of squares. What is the nature of the mathematical relationship between gain in weight and level of growth stimulant dosage?

6–10 Refer to the data set in Exercise 2–6. Use orthogonal polynomials for fitting a fourth degree curve and compute the associated analysis of variance table. Based on the ANOVA, determine the lowest degree polynomial which seemingly "adequately" describes the underlying mathematical relationship between x and y.

6–11 Calculate the average percent yield at each of the three temperature levels in Table 6.6.1. Based on these three means, why did the ANOVA declare both a linear and quadratic effect on yield within the temperature domain from 120° to 180° F?

Theoretical Exercises

6–12 Consider the linear model

$$Y_i = \beta_0 + \varepsilon_i \ , \quad i = 1, 2, ..., n.$$

Show that the least squares estimate for

β_0 is $b_0 = \sum_{i=1}^{n} y_i/n = \bar{y}$.

6–13 In simple linear regression, it can be shown that $SSe = S_{yy} - b_1 S_{xy}$ where $b_1 = S_{xy}/S_{xx}$ and $S_{xy} = \sum_{i=1}^{n} x_i y_i - n\bar{x}\bar{y}$, $S_{xx} = \sum_{i=1}^{n} x_i^2 - n\bar{x}^2$, $S_{yy} = \sum_{i=1}^{n} y_i^2 - n\bar{y}^2 \equiv SSt$.

(a) Show that in this case formula (6.3.5) can be expressed as

$$r^2 = \frac{S_{xy}^2}{S_{xx} S_{yy}} .$$

(b) Here $r = S_{xy}/\sqrt{S_{xx} S_{yy}}$ is called the **sample correlation coefficient.** Deduce the sample correlation coefficient property: $-1 \leq r \leq 1$.

6–14 Consider two sets of data, both of which are analyzed using simple linear regression models. Let us denote the first model by the expression

$$Y_i = \alpha_0 + \alpha_1 x_i + \varepsilon_i \ , \quad i = 1, 2, ..., n_1$$

and the second by

$$Y_j = \beta_0 + \beta_1 x_j + \varepsilon_j \ , \ j = 1, 2, ..., n_2.$$

Suppose we combine the two sets of data into a single set and fit the resulting combined set with the model

$$Y_k = \gamma_0 + \gamma_1 z_{k1} + \gamma_2 z_{k2} + \gamma_3 z_{k1} z_{k2} + \varepsilon_k \ ,$$
$$k = 1, 2, ..., n_1 + n_2.$$

where

$$z_{k1} = \begin{cases} 1, & \text{if observation is in second data set} \\ 0, & \text{if not,} \end{cases}$$

and

$$z_{k2} = \begin{cases} x_j \ , & \text{if observation is in second data set} \\ x_i \ , & \text{if not.} \end{cases}$$

In terms of the parameters of the two simple regression models, produce an argument that shows that

$$\gamma_0 = \alpha_0, \quad \gamma_1 = \beta_0 - \alpha_0, \quad \gamma_2 = \alpha_2 \ \text{ and } \ \gamma_3 = \beta_1 - \alpha_1.$$

6–15 In bioassay problems, it is usually assumed that the response of a biological system to a chemical stimulus is linearly related to the logarithm of the dose. Two preparations of digitalis were given to groups of heart patients in different doses. The data below represent the decrease in heart rate (y) after a certain fixed period of time. The independent variable x found in the table is x = log (dose times 10).

	Preparation A			
Dose, mg.	0.5	0.8	1.0	1.3
x	0.699	0.903	1.000	1.114
Response y	15	25	40	41
	23	30	28	43
	19	35		36

	Preparation B			
Dose, mg.	0.5	0.8	1.0	1.3
x	0.699	0.903	1.000	1.114
Response y	26	29	40	56
		40	44	42
		36		46
		39		

(a) For preparations A and B, fit separately simple linear models

$$Y_i = \alpha_0 + \alpha_1 x_i + \varepsilon_i \ , \quad i = 1, 2, ..., 11$$

and

$$Y_j = \beta_0 + \beta_1 x_j + \varepsilon_j \ , \quad j = 1, 2, ..., 10,$$

and plot the resulting sample response lines on the same set of coordinate axes.

(b) Using the information developed in Exercise 6–14, determine whether there is a statistical difference in the rate of effectiveness within each preparation, i.e., are the two theoretical response lines parallel? (Make the usual normality assumptions concerning ε_i and ε_j.)

(c) Do the two preparations have the same potency, i.e., are the two theoretical response lines coincident? Perform the proper statistical test suggested in Exercise 6–14.

6–16 Show that the vectors $\mathbf{p}_0, \mathbf{p}_1, \mathbf{p}_2, \mathbf{p}_3$ as given by the Gram–Schmidt formulas are mutually orthogonal.

6–17 Refer to the numerical example in Section 6.5 where the fitted function is $\hat{y} = 5 - 4x^2 + x^3$. Since the coefficient of the first degree term is zero, any statistical test would fail to reject the null hypothesis $H_0 : \beta_1 = 0$. In view of this observation, why, in the associated ANOVA, is the linear sum of squares, $SS(\alpha_1)$, not equal to zero?

6–18 Arbitrarily plot sets of three points each which serve as prototypes for illustrating the following relationships between a dependent variable y and an equally spaced independent variable x.

(a) Linear only.
(b) Quadratic only.
(c) Both linear and quadratic.
(d) Neither linear nor quadratic.

6–19 As a follow–up to Exercise 6–9(c), use the x coding for model (iii) in order to obtain vector $\mathbf{X}'\mathbf{y}$. Denote the solution vector to the normal equations by $\mathbf{c}$. Compute $\mathbf{c}'\mathbf{X}'\mathbf{y}$, leaving the result in partitioned form in order to show that $\mathbf{c}'\mathbf{X}'\mathbf{y} = n\bar{y}^2 + SS(\gamma_1) + SS(\gamma_2)$.

6–20 Consider the simple linear regression model

$$Y_i = \beta_0 + \beta_1 x_i + \varepsilon_i \;,\; i = 1, 2, \ldots, n.$$

Let b_0 and b_1 be the least squares estimates of β_0 and β_1, SSe the error sum of squares, r^2 the coefficient of determination, and f the value of the F statistic for testing $H_0 : \beta_1 = 0$.

(a) Suppose we make a linear transformation on Y. That is, let $W_i = aY_i + b$ for some constants a and b with $a \neq 0$. Now consider the simple linear regression model with dependent variable W and independent variable x. Determine how the following quantities for this regression model relate to the corresponding quantities with dependent variable Y:

- (i) the least squares estimates,
- (ii) the error sum of squares,
- (iii) the coefficient of determination,
- (iv) the value of the F statistic for testing the slope is zero.

(b) Repeat part (a) using a linear transformation on independent variable x, i.e., let $v = cx + d$, $c \neq 0$, and then consider the regression model with dependent variable Y and independent variable v.

CHAPTER 7

THE COMPLETELY RANDOMIZED DESIGN

7.1 Experimental Design Nomenclature

The material presented in the previous chapters has provided us with the mathematical foundations needed for the development of the theory associated with experimental designs. In the remainder of this text our objective is to utilize the results of these earlier chapters to determine the proper analysis in order to solve experimental design problems. In general, the method of analysis is dictated by the type of design used in collecting the data. In this and the chapters to follow, the reader will be introduced to the various basic designs used in practice and the analysis of data obtained from these designed experiments.

A Historical Note

In a sense, our story began in 1919 when Ronald A. Fisher, a 19–year–old teacher trained in mathematics and physics, accepted the position of statistician at Rothamsted Experimental Station in England. There he was confronted with the task of interpreting agricultural data obtained in small samples. At the time, many of the tools needed to "make sense" out of the data did not exist and so he set about to invent many of the present–day techniques for conducting and analyzing experiments. By 1925 Fisher had formulated the principles of experimental design and introduced the analysis of variance procedure, the very cornerstone of the subject matter in this text. The general terminology associated with experimental design, to be presented next, dates back to Fisher's formulation of ideas and concepts during the early 1920s.

Basic Vocabulary

A **factor** is an experimental variable of interest that potentially affects the response variable. In a single–factor experiment, different levels or values of the factor are called **treatments**. When an experiment contains two or more factors, a treatment is a combination of the levels of the factors. The goal of most experiments is to measure and compare the "effects" of two or more treatments on the response variable. The object to which a treatment is applied is called an **experimental unit**. To illustrate these three terms, we consider three examples.

In a field experiment, an investigator is interested in the amount of fertilizer needed for optimizing the yield of a certain crop. Five amounts of fertilizer (in pounds per acre) are applied to plots of ground. In this experiment, fertilizer is the factor, the various fertilizer amounts are the treatments, and the plots of ground are the experimental units.

Suppose an experiment is designed to compare the wearing qualities of four brands of automobile tires. Here brands is the variable of interest (the factor), and the four brands are the levels constituting the treatments. The experimental units are the wheels upon which the tires are mounted.

Three different teaching methods are available for teaching individuals (on a one–on–one basis) how to read. A school district wants to evaluate the effectiveness of these three methods. In this study, teaching method is the factor, the three methods are the three treatments, and the students receiving the instruction are the experimental units. If, in addition to teaching method, a potential difference in response by males and females is suspected, sex can be considered to be the second factor. In this two–factor case, the six combinations of methods and sex are the treatments.

Most experiments involve several observations on each treatment. When a treatment is applied to more than one experimental unit, we say the treatments are **replicated**. The method used for randomly assigning the treatments to the experimental units is called the **randomization procedure**. We illustrate this randomization concept with an example.

Example 7.1.1 Let us consider an experiment involving three varieties of wheat (the treatments). Twelve fields for planting (the experimental units) will be required if the treatments are to be **replicated** four times. We illustrate two randomization procedures. First suppose the treatments are applied at random to the fields without restrictions. A possible result of unrestricted randomization is given in Table 7.1.1 where the varieties are labeled as A, B, C and each "square" represents a field. (A simple way of doing the randomization is to draw successively well–mixed twelve slips of paper marked A, B, or C, in equal number, "from a hat.")

Table 7.1.1 Randomization with No Restrictions for Twelve Fields

C	B	C	A
B	A	C	B
C	A	B	A

Next, suppose that each column of fields in Table 7.1.1 are at four experimental stations located at various places in several states. Because of the potential for differences in weather and soil, it is logical to insist that each variety is grown at each station. This imposes a restriction on the randomization, namely, that each variety must occur once and only once in each column. A possible result of such a randomization scheme is given in Table 7.1.2.

Table 7.1.2 Randomization with Restrictions for Four Stations (Columns) — Three Fields Per Station

B	A	B	A
A	B	C	C
C	C	A	B

In Chapter 10 we will learn that the columns in Table 7.1.2 are called **blocks**, a characteristic of a **randomized block design**. On the other hand, the randomization configuration in Table 7.1.1 is "typical" for a **completely randomized design**. The word "design" refers to the various aspects of an experiment including its basic structure such as:

(a) the number and type of factors (which determines the treatments),

(b) the randomization procedure used to determine the assignment of the treatments to the experimental units, and

(c) the number of experimental units assigned a particular treatment (sample sizes).

As we proceed, we will come to understand that it is the randomization procedure that is the primary distinguishing feature in how we classify designs. Also note that in general, in a designed experiment, contrary to a non–designed experiment, the investigator has control over determining the number and types of factors, the specific treatments (levels of factors) to appear in the experiment, and the method of assignment of treatments to the experimental units (the randomization procedure).

In this chapter we will consider the simplest of experimental designs, one with only one factor and with no restriction on the randomization procedure. For this design we will consider an appropriate model with assumptions, estimation ideas, relevant tests and confidence intervals, computer usage, violation of assumptions, appropriateness of the design, and several other notions.

7.2 Completely Randomized Design

In this section we introduce the single—factor completely randomized design. For the sake of brevity, we will frequently refer to this design as a CRD.

Definition 7.2.1. A single—factor **completely randomized design** is an experimental design involving one factor with p levels (treatments). The treatments are assigned to the experimental units with no restrictions on the assignment except that each treatment will be used a specified number of times.

To achieve a completely randomized design, as defined above, we need a single factor with p levels and n experimental units. We then randomly select $n_1, n_2, \ldots, n_p$ experimental units to be used with the first, second,...,pth treatments, respectively. Here $n = \sum_{i=1}^{p} n_i$. The following example is an illustration of a CRD.

Example 7.2.1 A study was designed to determine the effect of copper concentrations in water on the lifetime of aquatic animals. Daphnia magna, small aquatic animals, were selected for the study since they have relatively short lifetimes. Fifteen daphnia, each to be kept in separate containers, were available for the study. Three copper levels for the water in the containers were chosen: no copper, 20 $\mu g/\ell$, and 40 $\mu g/\ell$. The treatments (copper levels) were randomly assigned to the experimental units (daphnia in containers) with no restriction except that each treatment was used on five daphnia. The design, consequently, was a CRD with p = 3. The daphnia were raised in the three copper concentrations and their lifetimes (response variable) were recorded. Unfortunately, one of the no copper specimen was contaminated with copper and lost from the experiment. As a consequence, only four observations were available for the first treatment. The results are given in Table 7.2.1 where we see the resulting n_i values are $n_1 = 4$, $n_2 = 5$, $n_3 = 5$, and n = 14.

Table 7.2.1 Daphnia Lifetimes in Days
Copper Concentration

None	$20\mu g/\ell$	$40\mu g/\ell$
60	58	40
90	74	58
74	50	25
82	65	30
	68	42

If we inspect the data structure in the previous example, we note that we have independent samples from three groups. Hence the resulting data structure for a CRD is essentially the same as that assumed in the one–way classification. For this reason the natural model to use for the CRD is the same as the one used in one–way classification problems. Thus, for the CRD with p treatments and n_i experimental units for the ith treatment, the model is as follows:

$$Y_{ij} = \mu + \tau_i + \varepsilon_{ij} \; , \quad i = 1,2,\ldots,p; \; j = 1,2,\ldots,n_i \qquad (7.2.1)$$

where

- Y_{ij} is the response to the jth experimental unit of the ith treatment,
- μ is the overall average response to all treatments,
- τ_i is the effect of the ith treatment,
- ε_{ij} is the random error associated with the jth experimental unit of the ith treatment. We assume that $E(\varepsilon_{ij}) = 0$ and $Var(\varepsilon_{ij}) = \sigma^2$ for every i and j. Furthermore, we assume the ε_{ij} are uncorrelated.

With this model, for the observation Y_{ij}, we have

$$E(Y_{ij}) = \mu + \tau_i \text{ and } Var(Y_{ij}) = \sigma^2.$$

That is, observations within a treatment have the same mean, but the mean for treatments, $\mu + \tau_i$, for $i = 1,2,\ldots,p$, can be different. All observations (in random form) have the same variance, namely, σ^2. One of our primary objectives in the analysis of a CRD will be to compare the means of the p treatments. For example, the hypothesis H_0: $\tau_1 = \tau_2 = \cdots = \tau_p$, which states that the p treatment means are equal, will generally be of major interest to us.

We recognize the model given in (7.2.1) is the same model first given in Section 2.6 for the one–way classification. In Example 2.6.1, this model is written in matrix notation as a linear model for a specified example. In general, model (7.2.1) can be written in matrix notation by letting

$$Y = \begin{bmatrix} Y_{11} \\ \vdots \\ Y_{1n_1} \\ Y_{21} \\ \vdots \\ Y_{2n_2} \\ \vdots \\ Y_{p1} \\ \vdots \\ Y_{pn} \end{bmatrix}, \quad X = \begin{bmatrix} 1 & 1 & 0 & \cdots & 0 \\ \vdots & \vdots & \vdots & & \vdots \\ 1 & 1 & 0 & \cdots & 0 \\ 1 & 0 & 1 & \cdots & 0 \\ \vdots & \vdots & \vdots & & \vdots \\ 1 & 0 & 1 & \cdots & 0 \\ \vdots & \vdots & \vdots & & \vdots \\ 1 & 0 & 0 & \cdots & 1 \\ \vdots & \vdots & \vdots & & \vdots \\ 1 & 0 & 0 & \cdots & 1 \end{bmatrix} \begin{matrix} \left.\begin{matrix} \\ \\ \\ \end{matrix}\right\} n_1 \text{ rows} \\ \left.\begin{matrix} \\ \\ \\ \end{matrix}\right\} n_2 \text{ rows}, \\ \\ \left.\begin{matrix} \\ \\ \\ \end{matrix}\right\} n_p \text{ rows} \end{matrix} \quad \beta = \begin{bmatrix} \mu \\ \tau_1 \\ \tau_2 \\ \vdots \\ \tau_p \end{bmatrix}, \quad \varepsilon = \begin{bmatrix} \varepsilon_{11} \\ \vdots \\ \varepsilon_{1n_1} \\ \varepsilon_{21} \\ \vdots \\ \varepsilon_{2n_2} \\ \vdots \\ \varepsilon_{p1} \\ \vdots \\ \varepsilon_{pn_p} \end{bmatrix}. \qquad (7.2.2)$$

In this representation, **Y** and ε are n × 1 random vectors, $\boldsymbol{\beta}$ is a (p + 1) × 1 vector of unknown constants, and **X** is a n × (p + 1) matrix of specified constants, 0 or 1. Our CRD model (7.2.1) is now seen to be a linear model $\mathbf{Y} = \mathbf{X}\boldsymbol{\beta} + \varepsilon$, with the assumptions $E(\varepsilon) = \mathbf{0}$ and $\text{Cov}(\varepsilon) = \sigma^2\mathbf{I}$.

In the matrix **X**, called the **design matrix**, the sum of the last p columns equals the first column; therefore, **X** is not full rank. It is easy to see that there are p linearly independent columns in the **X** matrix. Hence the matrix has rank p.

In the remainder of this chapter we will consider various inference procedures and other relevant information on the CRD. Since the CRD and one–way classification have the same model and assumptions, all results obtained in earlier chapters concerning one–way classification also apply to CRD problems.

7.3 Least Squares Results

In this section we detail estimation properties for the completely randomized design. To find least squares estimators, we first find a solution to the normal equations. For a CRD, the normal equations, $\mathbf{X}'\mathbf{X}\mathbf{b} = \mathbf{X}'\mathbf{y}$ are given by

$$\begin{bmatrix} n & n_1 & n_2 & \cdots & n_p \\ n_1 & n_1 & 0 & \cdots & 0 \\ n_2 & 0 & n_2 & \cdots & 0 \\ \cdot & \cdot & \cdot & & \cdot \\ \cdot & \cdot & \cdot & & \cdot \\ \cdot & \cdot & \cdot & & \cdot \\ n_p & 0 & 0 & \cdots & n_p \end{bmatrix} \begin{bmatrix} m \\ t_1 \\ t_2 \\ \cdot \\ \cdot \\ \cdot \\ t_p \end{bmatrix} = \begin{bmatrix} y_{..} \\ y_{1.} \\ y_{2.} \\ \cdot \\ \cdot \\ \cdot \\ y_{p.} \end{bmatrix}$$

where $y_{..} = \sum_{i=1}^{p} \sum_{j=1}^{n_i} y_{ij}$ and $y_{i.} = \sum_{j=1}^{n_i} y_{ij}$.

These normal equations yield the p + 1 equations

$$\begin{array}{lllll} n\,m + n_1t_1 & + n_2t_2 & + \cdots & + n_pt_p & = y_{..} \\ n_1m + n_1t_1 & & & & = y_{1.} \\ n_2m & + n_2t_2 & & & = y_{2.} \\ \cdot & \cdot & & \cdot & \\ n_pm & & & + n_pt_p & = y_{p.} \end{array}$$

We will solve these equations by the side condition approach of Section 3.4. Since the $\mathbf{X}'\mathbf{X}$ matrix has p + 1 columns, but only has rank p, we will need to add (p + 1) − p = 1 side condition to obtain a unique solution. It is easy to see that if we impose the condition $\sum_{i=1}^{p} n_it_i = 0$ on the system we obtain the solution set

$$\begin{cases} m = y_{..}/n = \bar{y}_{..} \\ t_i = \dfrac{y_{i.}}{n_i} - m = \bar{y}_{i.} - \bar{y}_{..},\ i = 1,2,\ldots,p. \end{cases} \tag{7.3.1}$$

Note that our side condition, $\sum_{i=1}^{p} n_it_i = 0$, is of the form $\mathbf{H}\mathbf{b} = 0$, where $\mathbf{H}$ is the $1 \times (p+1)$ vector $(0,n_1,n_2,\cdots,n_p)$.

Next we find the estimable functions for the CRD. Using

Corollary 4.3.1, we know that $\mathbf{X}\boldsymbol{\beta}$ constitutes a set of estimable functions and that all other estimable functions can be expressed as a linear combination of the entries in $\mathbf{X}\boldsymbol{\beta}$. Using the $\mathbf{X}$ and $\boldsymbol{\beta}$ for a CRD given in (7.2.2), we find that the distinct entries in $\mathbf{X}\boldsymbol{\beta}$ are $\mu + \tau_1, \mu + \tau_2, \ldots, \mu + \tau_p$. Hence, $\mu + \tau_i$ is estimable for i = 1,2,...,p. Obviously, $\mu, \tau_1, \ldots, \tau_p$ are not individually estimable since they cannot be expressed as a linear combination of the $\mu + \tau_i$'s. Even though none of the parameters in the CRD model is estimable, we need not be discouraged. Important functions like the treatment means, $\mu + \tau_i$, and the difference between treatment means, $\mu + \tau_1 - (\mu + \tau_2) = \tau_1 - \tau_2$, are estimable.

The next definition presents an important class of estimable functions.

Definition 7.3.1 The function $\sum_{i=1}^{p} c_i \tau_i$ is called a **treatment contrast** if $\sum_{i=1}^{p} c_i = 0$.

There are many contrasts. For example, when p = 4 the functions $\tau_1 - \tau_2$, $\tau_1 + \tau_2 - 2\tau_3$ and $(\tau_1 + \tau_2)/2 - (\tau_3 + \tau_4)/2$ all are treatment contrasts since the sum of the coefficients is zero for each. Functions like $\tau_1 + \tau_2$ and $\tau_1 + \tau_2 - \tau_3$ are not contrasts. To see that all treatment contrasts are estimable, we note that they can be expressed as a linear combination of the estimable $\mu + \tau_i$'s, that is,

$$\sum_{i=1}^{p} c_i \tau_i = \sum_{i=1}^{p} c_i(\mu + \tau_i) \text{ when } \sum_{i=1}^{p} c_i = 0.$$

Recall that for any estimable function $\boldsymbol{\ell}'\boldsymbol{\beta}$ and any solution $\mathbf{b}$ to the normal equations, the least squares estimate $\boldsymbol{\ell}'\mathbf{b}$ is the best linear unbiased estimate (BLUE). Using the solution to the normal equations given in (7.3.1), we find that the BLUE for $\mu + \tau_i$ is $\bar{y}_{..} + (\bar{y}_{i.} - \bar{y}_{..}) = \bar{y}_{i.}$, and the BLUE for $\sum_{i=1}^{p} c_i \tau_i$ is $\sum_{i=1}^{p} c_i(\bar{y}_{i.} - \bar{y}_{..}) = \sum_{i=1}^{p} c_i \bar{y}_{i.}$. These results are summarized as follows.

BLUES for the CRD Model	
Function	BLUE
Treatment Mean $\mu + \tau_i$	$\bar{y}_{i.}$
Treatment Contrast $\sum_{i=1}^{p} c_i \tau_i$	$\sum_{i=1}^{p} c_i \bar{y}_{i.}$
where $\sum_{i=1}^{p} c_i = 0$	

(7.3.2)

Using the solutions to the normal equations (7.3.1) we find that $\hat{y}_{ij} = m + t_i = \bar{y}_{i.}$. Since $\hat{y}_{ij}$ is invariant to the solution set of the normal equations (Section 3.3), we always have

$$\hat{y}_{ij} = \bar{y}_{i.} \tag{7.3.3}$$

for the CRD. Since SSe $= \sum_{i=1}^{p} \sum_{j=1}^{n_i} (y_{ij} - \hat{y}_{ij})^2$ and $s^2 = \text{SSe}/(n - r)$, where r is the rank of $\mathbf{X}$, we have for a CRD

$$\text{SSe} = \sum_{i=1}^{p} \sum_{j=1}^{n_i} (y_{ij} - \bar{y}_{i.})^2$$

and (7.3.4)

$$s^2 = \text{SSe}/(n - p).$$

A computational formula for SSe is

$$\text{SSe} = \sum_{i=1}^{p} \sum_{j=1}^{n_i} y_{ij}^2 - \sum_{i=1}^{p} y_{i.}^2/n_i \, . \tag{7.3.5}$$

Example 7.3.1 Consider the daphnia lifetime data given in Table 7.2.1. The model for this experiment with three treatments is

$$Y_{ij} = \mu + \tau_i + \varepsilon_{ij} \;, \quad i = 1,2,3; \; j = 1,2,\ldots,n_i$$

when $n_1 = 4$, $n_2 = 5$, $n_3 = 5$. For these data, we find the BLUE for the third treatment mean, $\mu + \tau_3$, and the difference between the first and second treatment means, $\tau_1 - \tau_2$. We also find s^2. First we note that

$$\bar{y}_{1\cdot} = 76.5, \ \bar{y}_{2\cdot} = 63.0, \ \bar{y}_{3\cdot} = 39.0 \ \text{and} \ \sum_{i=1}^{3} \sum_{j=1}^{n_i} y_{ij}^2 = 52{,}342.$$

The BLUE for $\mu + \tau_3$ is $\bar{y}_{3\cdot} = 39.0$, and the BLUE for $\tau_1 - \tau_2$ is $\bar{y}_{1\cdot} - \bar{y}_{2\cdot} = 76.5 - 63.0 = 13.5$. Finally,

$$\begin{aligned} \text{SSe} &= \sum_{i=1}^{3} \sum_{j=1}^{n_i} y_{ij}^2 - \sum_{i=1}^{3} y_{i\cdot}^2/n_i \\ &= 52{,}342 - [(306)^2/4 + (315)^2/5 + (195)^2/5] \\ &= 52{,}342 - 50{,}859 = 1483. \end{aligned}$$

Thus, $s^2 = \text{SSe}/(n - p) = 1483/(14 - 3) = 134.8182$.

7.4 Analysis of Variance and F—Test

In this section we obtain the multiparameter test associated with the analysis of variance for a completely randomized design, namely, the hypothesis of no difference in treatment means $H_0 : \tau_1 = \tau_2 = \cdots = \tau_p$. Since τ_i in our model is referred to as the effect of the ith treatment, the above hypothesis is sometimes verbalized as "no difference in treatment effects."

Normal CRD Model

Before any tests of hypotheses are performed or confidence intervals are constructed, we make the assumption of a **normal** CRD model. The normal CRD model is defined as the CRD model (7.2.1) with the added specification that $\boldsymbol{\varepsilon}$ is $MN(\mathbf{0}, \sigma^2\mathbf{I})$. That is, the ε_{ij}'s are independent $N(0, \sigma^2)$ random variables. With this assumption, the Y_{ij} are $N(\mu + \tau_i, \sigma^2)$. Therefore, we essentially have p normal populations with potentially different means but a common variance.

CRD Analysis of Variance

To derive the test statistic for $H_0 : \tau_1 = \tau_2 = \cdots = \tau_p$, we use the principle of conditional error described in Section 5.4. The null hypothesis H_0 can be represented in the form of the general linear hypothesis (5.4.2)

$$\underset{\mathbf{L}}{\begin{bmatrix} 0 & -1 & 1 & 0 & 0 & \cdots & 0 \\ 0 & -1 & 0 & 1 & 0 & \cdots & 0 \\ 0 & -1 & 0 & 0 & 1 & \cdots & 0 \\ \cdot & \cdot & \cdot & \cdot & \cdot & & \cdot \\ \cdot & \cdot & \cdot & \cdot & \cdot & & \cdot \\ \cdot & \cdot & \cdot & \cdot & \cdot & & \cdot \\ 0 & -1 & 0 & 0 & 0 & \cdots & 1 \end{bmatrix}} \underset{\boldsymbol{\beta}}{\begin{bmatrix} \mu \\ \tau_1 \\ \tau_2 \\ \tau_3 \\ \vdots \\ \tau_p \end{bmatrix}} = \underset{\mathbf{0}}{\begin{bmatrix} 0 \\ 0 \\ 0 \\ \cdot \\ \cdot \\ \cdot \\ 0 \end{bmatrix}},$$

where $\mathbf{L}$ is $(p - 1) \times (p + 1)$. We note that $\mathbf{L}$ is of rank $p - 1$ and

$$\mathbf{L}\boldsymbol{\beta} = \begin{bmatrix} \tau_2 - \tau_1 \\ \tau_3 - \tau_1 \\ \cdot \\ \cdot \\ \cdot \\ \tau_p - \tau_1 \end{bmatrix}$$

constitutes a set of $p - 1$ linearly estimable functions. When we impose the hypothesis $\mathbf{L}\boldsymbol{\beta} = \mathbf{0}$ on the model (7.2.1), we find that the $\mu + \tau_i$ are all equal so we can set $\mu + \tau_i \equiv \mu^*$ for $i = 1,2,\ldots,p$. The "reduced" model becomes

$$Y_{ij} = \mu^* + \varepsilon_{ij}^* \ , \quad i = 1,2,\ldots,p \ ; \ j = 1,2,\ldots,n_i \ .$$

In exercise 5—29 we noted that the least squares estimate for μ^* is $\bar{y}_{..}$ so that $\hat{y}_{ij} = \bar{y}_{..}$ for the reduced model. As a result, the error sum of squares for the reduced model is

$$\text{SSe}^* = \sum_{i=1}^{p} \sum_{j=1}^{n_i} (y_{ij} - \bar{y}_{..})^2.$$

Now from (7.3.4) and using some algebra we obtain

$$\begin{aligned} \text{SSe}^* - \text{SSe} &= \sum_{i=1}^{p} \sum_{j=1}^{n_i} (y_{ij} - \bar{y}_{..})^2 - \sum_{i=1}^{p} \sum_{j=1}^{n_i} (y_{ij} - \bar{y}_{i.})^2 \\ &= \sum_{i=1}^{p} n_i (\bar{y}_{i.} - \bar{y}_{..})^2 \equiv \text{SStr}. \end{aligned} \tag{7.4.1}$$

We call SStr the **treatment sum of squares**. The quantity

$$\text{SStr}/(p-1) \equiv \text{MStr} \tag{7.4.2}$$

is called **treatment mean square**. Using Theorem 5.4.2, we obtain the test statistic

$$\begin{aligned} f &= \frac{(\text{SSe}^* - \text{SSe})/q}{s^2} = \frac{\sum_{i=1}^{p} n_i (\bar{y}_{i.} - \bar{y}_{..})^2/(p-1)}{\text{SSe}/(n-p)} \\ &= \frac{\text{MStr}}{\text{MSe}}. \end{aligned} \tag{7.4.3}$$

Accordingly, we reject H_0 if $f = \text{MStr}/\text{MSe} \geq f_{\alpha,p-1,n-p}$.

It is convenient to display the sums of squares, degrees of freedom, mean squares and f in a table called the analysis of variance (ANOVA for short) table. Before doing this, we need to introduce another sum of squares that is always part of the ANOVA table. On adding SSe and SStr we obtain, after some algebraic simplification,

$$\begin{aligned} \text{SSe} + \text{SStr} &= \sum_{i=1}^{p} \sum_{j=1}^{n_i} (y_{ij} - \bar{y}_{i.})^2 + \sum_{i=1}^{p} n_i (\bar{y}_{i.} - \bar{y}_{..})^2 \\ &= \sum_{i=1}^{p} \sum_{j=1}^{n_i} (y_{ij} - \bar{y}_{..})^2 \equiv \text{SSt}. \end{aligned} \tag{7.4.4}$$

The quantity SSt is called the **total sum of squares**.

We are now ready to present the ANOVA table for the CRD and the one–way classification problem. (Fisher used the analysis of variance table for the first time in a 1923 study on crop variation. He explained the table as "merely a convenient way of arranging the arithmetic.") In the table we will use the following conventional abbreviations: "df" for degrees of freedom, "SS" for sum of squares, and "MS" for mean square. The term "degrees of freedom" originates from the chi–square distributions associated with the three sums of squares in the table. We will expand on this later in this section.

Table 7.4.1 ANOVA Table for a CRD

Source	df	SS	MS	f
Treatments	$p - 1$	SStr	$MStr = SStr/(p - 1)$	$MStr/s^2$
Error	$n - p$	SSe	$s^2 = MSe = SSe/(n - p)$	
Total	$n - 1$	SSt		

The formulas for SStr, SSe and SSt are given by (7.4.1), (7.3.4) and (7.4.4), respectively.

The sums of squares SStr and SSe each have an intuitive interpretation for the CRD. The quantity $SStr = \sum_{i=1}^{p} n_i(\bar{y}_{i.} - \bar{y}_{..})^2$ is a measure of the difference in the magnitudes of the sample means $\bar{y}_{1.}, \bar{y}_{2.}, \ldots, \bar{y}_{p.}$. If all p sample means are equal, then $SStr = 0$. However, if the p sample means are vastly different, then SStr is large. Intuitively, we would want to reject H_0: $\tau_1 = \tau_2 = \cdots = \tau_p$ when SStr is large. This coincides with large f values. Since SStr compares $\bar{y}_{1.}, \bar{y}_{2.}, \ldots, \bar{y}_{p.}$, which is a comparison of values between different treatments, it is common to label SStr as **between** sum of squares.

The error of sum of squares $SSe = \sum_{i=1}^{p} \sum_{j=1}^{n_i} (y_{ij} - \bar{y}_{i.})^2$ is a measure of within treatment variability. To see why, we let $s_i^2 = \sum_{j=1}^{n_i} (y_{ij} - \bar{y}_{i.})^2/(n_i - 1)$ be the usual sample variance for the observations in treatment i. Then we see that SSe can be expressed as a weighted sum of the within treatment measures of variability, s_i^2.

That is,

$$SSe = \sum_{i=1}^{p} (n_i - 1)s_i^2 . \qquad (7.4.5)$$

Thus SSe is small only if observations within a treatment are similar in value. Sometimes SSe is called the **within** sum of squares.

Computationally, it is generally easier to calculate the sums of squares using the following computing formulas which the student is asked to derive in the exercises:

$$SSt = \sum_{i=1}^{p} \sum_{j=1}^{n_i} y_{ij}^2 - \left(\sum_{i=1}^{p} \sum_{j=1}^{n_i} y_{ij}\right)^2/n$$

$$SStr = \sum_{i=1}^{p} n_i \bar{y}_{i\cdot}^2 - \left(\sum_{i=1}^{p} \sum_{j=1}^{n_i} y_{ij}\right)^2/n \qquad (7.4.6)$$

$$SSe = \sum_{i=1}^{p} \sum_{j=1}^{n_i} y_{ij}^2 - \sum_{i=1}^{p} y_{i\cdot}^2/n_i = SSt\text{--}SStr.$$

An Example

Example 7.4.1 Consider the daphnia data found in Table 7.2.1. Recall these data come from a CRD with $p = 3$, $n_1 = 4$, $n_2 = 5$ and $n_3 = 5$. Our goal is to test $H_0 : \tau_1 = \tau_2 = \tau_3$. We first obtain the ANOVA table for these data. Some preliminary calculations are

$$\sum_{i=1}^{3} \sum_{j=1}^{n_i} y_{ij} = 816, \quad \sum_{i=1}^{3} \sum_{j=1}^{n_i} y_{ij}^2 = 52{,}342,$$

$$\bar{y}_{1\cdot} = 76.5, \quad \bar{y}_{2\cdot} = 63.0, \quad \bar{y}_{3\cdot} = 39.0.$$

The sums of squares are

$$SSt = 52{,}342 - (816)^2/14 = 4780.8571$$

$$SStr = 4(76.5)^2 + 5(63.0)^2 + 5(39.0)^2 - (816)^2/14$$

$$= 50{,}859 - 47{,}561.1429 = 3297.8571$$

$$SSe = SSt - SStr = 4780.8571 - 3297.8571 = 1483.$$

This yields the following analysis of variance table.

Table 7.4.2 ANOVA Table for Daphnia Data

Source	df	SS	MS	f
Treatments	2	3297.8571	1648.9286	12.23
Error	11	1483.0000	134.8182	
Total	13	4780.8571		

Our 0.05 level test for $H_0: \tau_1 = \tau_2 = \tau_3$ rejects H_0 when the calculated f exceeds $f_{.05,2,11} = 3.98$. Since $f = 12.23$, we reject H_0 and conclude there is a significant difference in treatment effects. To state it another way, we conclude that at least two treatment means are significantly different.

Note that in the previous example the f—test does not identify **which** treatments differ in their effect. This question will be addressed in the next chapter. If we fail to reject H_0 in a problem, the usual statement made is "the treatment means are not significantly different." This, of course, does not mean the **population** treatment means are equal; rather, this implies the **sample** treatment means are not sufficiently different so as to detect a difference in the $\mu + \tau_i$ with the given experiment. That is, we have insufficient evidence to declare a difference in the actual treatment means. We must not forget that the f—test concerns a hypothesis about means as the treatment variances are assumed to be equal.

SAS Computer Printout

Example 7.4.1 presents an analysis of the daphnia data using computing formulas and possibly a hand calculator. We now analyze this same set using SAS and observe that the analysis of variance entries are printed out for us. SAS is not unique in this respect; most standard statistical packages provide the user with the ANOVA "automatically."

Table 7.4.3 SAS Commands and Printout for Daphnia Data

SAS Commands

```
OPTIONS LS = 80;
DATA DAPH;
INPUT COPPER LIFETIME @@;
CARDS;
1  60  1  90  1  74  1  82
2  58  2  74  2  50  2  65  2  68
3  40  3  58  3  25  3  30  3  42
PROC GLM;
CLASS COPPER;                  } Requests the ANOVA Table and
MODEL LIFETIME = COPPER;       } F-test.
MEAN COPPER; } Requests the treatment means
ESTIMATE 'COMPARISON' COPPER 1 -.5 -.5; } Requests information
                                        } on the contrast
```

$\tau_1 - .5\tau_2 - 5_3$.

SAS Printout

```
                    General Linear Models Procedure
                       Class Level Information

                      Class     Levels    Values

                      COPPER         3    1 2 3


              Number of observations in data set = 14

                    General Linear Models Procedure

Dependent Variable: LIFETIME
                                     Sum of           Mean
Source                   DF         Squares         Square   F Value      Pr > F

Model                     2    3297.8571429   1648.9285714     12.23      0.0016

Error                    11    1483.0000000    134.8181818

Corrected Total          13    4780.8571429

               R-Square            C.V.       Root MSE           LIFETIME Mean

               0.689805        19.92104      11.611123               58.285714


Source                   DF       Type I SS    Mean Square   F Value      Pr > F

COPPER                    2    3297.8571429   1648.9285714     12.23      0.0016

Source                   DF     Type III SS    Mean Square   F Value      Pr > F

COPPER                    2    3297.8571429   1648.9285714     12.23      0.0016
```

```
                    General Linear Models Procedure

Dependent Variable: LIFETIME

                                           T for H0:     Pr > |T|    Std Error of
Parameter                     Estimate    Parameter=0                  Estimate

COMPARISON                  25.5000000          3.71       0.0034      6.86923312


              Level of          -----------LIFETIME----------
              COPPER      N          Mean                SD

              1           4      76.5000000        12.7932274
              2           5      63.0000000         9.2736185
              3           5      39.0000000        12.7279221
```

Notice one can perform the F—test by using the information on the printout. There are two ways to do this. The first way is to compare the f—value on the printout to the tabled F—value. That is, since f = 12.23 on the printout exceeds the $f_{.05,2,11} = 3.98$ value, we reject H_0. The second way is to use the "PR > F" value on the printout. The "PR > F" value is the **observed** significance level called the probability value or more commonly the p—value. This observed significance level is the smallest α risk level at which we can reject H_0. Therefore, we reject H_0 if the p—value is α or less and we fail to reject H_0 if the p—value is larger than α. Since the p—value is 0.0016 for this problem, we would reject H_0 at the 0.05 level. Of course we would reject H_0 for any **fixed** α level where $\alpha \geq 0.0016$.

Sums of Squares Distributions

We close this section with further comments concerning the column headed by "degrees of freedom" in the analysis of variance table. The entries in this column are not arbitrary but rather coincide with the degrees of freedom for associated chi—square random variables. From Theorem 5.1.4 we already know that the random variable SSE/σ^2 has a chi—square distribution with $n - p$ degrees of freedom. Hence the error degrees of freedom is set at $n - p$. In Example 5.1.1(c) we found that when H_0 is true, SST/σ^2 has a chi—square distribution with $n - 1$ degrees of freedom; hence, the total degrees of freedom is $n - 1$. Also, when H_0 is true, $SSTR/\sigma^2 = (SSE^* - SSE)/\sigma^2$ is a chi—square random variable with $p - 1$ degrees of freedom. The student is asked to show this in the exercises.

7.5 Confidence Intervals and Tests

In this section we outline the procedure for finding confidence intervals and tests for estimable functions in the completely randomized design setting. Of particular interest are inferences for the ith population mean $\mu + \tau_i$, for a difference between two means, for example $(\mu + \tau_i) - (\mu + \tau_k) = \tau_i - \tau_k$, and for a contrast $\sum_{i=1}^{p} c_i \tau_i$, $\sum_{i=1}^{p} c_i = 0$. The difference between two means is, of course, a special case of a contrast.

In Section 5.3 we found that a confidence interval for an estimable function $\boldsymbol{\ell}'\boldsymbol{\beta}$ is of the form

$$\boldsymbol{\ell}'\mathbf{b} \pm t_{\alpha/2,n-p} \mathrm{Se}(\boldsymbol{\ell}'\hat{\boldsymbol{\beta}}),$$

where $\boldsymbol{\ell}'\mathbf{b}$ is the least squares estimate and $\mathrm{Se}(\boldsymbol{\ell}'\hat{\boldsymbol{\beta}})$ denotes the standard error of $\boldsymbol{\ell}'\hat{\boldsymbol{\beta}}$. Also recall from (5.4.10) that the test statistic for testing the null hypothesis $H_0 : \boldsymbol{\ell}'\boldsymbol{\beta} = \gamma$ is

$$t = (\boldsymbol{\ell}'\mathbf{b} - \gamma)/\mathrm{Se}(\boldsymbol{\ell}'\hat{\boldsymbol{\beta}}).$$

From this, it is evident that to find confidence intervals and perform t–tests we need the least squares estimates and the standard errors for the estimable functions. The least squares estimates were found in Section 7.3 and are: $\bar{y}_{i\cdot}$ for $\mu + \tau_i$, $\bar{y}_{i\cdot} - \bar{y}_{k\cdot}$ for $\tau_i - \tau_k$, and $\sum_{i=1}^{p} c_i \bar{y}_{i\cdot}$ for the contrast $\sum_{i=1}^{p} c_i \tau_i$. To find the standard errors we use Rule 5.3.1 by first finding the variance of the least squares estimators, then replacing σ^2 by s^2 and taking the square root. Now, from elementary statistics we know that

$$\mathrm{Var}(\bar{Y}_{i\cdot}) = \sigma^2/n_i .$$

Hence, we know immediately

$$\mathrm{Se}(\bar{Y}_{i\cdot}) = \sqrt{s^2/n_i} .$$

Next, since $\bar{Y}_{i\cdot}$ and $\bar{Y}_{k\cdot}$ are independent,

$$\mathrm{Var}(\bar{Y}_{i\cdot} - \bar{Y}_{k\cdot}) = \mathrm{Var}(\bar{Y}_{i\cdot}) + \mathrm{Var}(\bar{Y}_{k\cdot}) = \sigma^2/n_i + \sigma^2/n_k.$$

Therefore, $$\mathrm{Se}(\bar{Y}_{i\cdot} - \bar{Y}_{k\cdot}) = \sqrt{s^2\left[\frac{1}{n_i} + \frac{1}{n_k}\right]}.$$

Finally,

$$\mathrm{Var}\left(\sum_{i=1}^{p} c_i\bar{Y}_{i\cdot}\right) = \sum_{i=1}^{p} c_i^2\,\mathrm{Var}(\bar{Y}_{i\cdot}) = \sigma^2 \sum_{i=1}^{p} c_i^2/n_i\,.$$

As a result, the standard error for $\sum_{i=1}^{p} c_i\bar{Y}_{i\cdot}$ is

$$\mathrm{Se}\left(\sum_{i=1}^{p} c_i\bar{Y}_{i\cdot}\right) = \sqrt{s^2 \sum_{i=1}^{p} c_i^2/n_i}\,.$$

We are now in a position to "write down" the inference formulas for estimable functions associated with a completely randomized design for a single factor experiment. First the three types of confidence intervals:

100(1 − α)% Confidence Interval Formulas for a CRD

Function	Interval	
Mean, $\mu + \tau_i$:	$\bar{y}_{i\cdot} \pm t_{\alpha/2,n-p}\sqrt{s^2/n_i}$	
Difference, $\tau_i - \tau_k$:	$(\bar{y}_{i\cdot} - \bar{y}_{k\cdot}) \pm t_{\alpha/2,n-p}\sqrt{s^2(1/n_i+1/n_k)}$	(7.5.1)
Contrast, $\sum_{i=1}^{p} c_i\tau_i$, $\sum_{i=1}^{p} c_i = 0$:	$\sum_{i=1}^{p} c_i\bar{y}_{i\cdot} \pm t_{\alpha/2,n-p}\sqrt{s^2 \sum_{i=1}^{p} c_i^2/n_i}$	

Next we present the essentials for the three classes of tests. In each case, the rejection regions are based on a t distribution with $n - p$ degrees of freedom.

Test Statistics Associated with a CRD

Null Hypothesis	Test Statistic	
$H_0: \mu + \tau_i = \gamma$	$t = (\bar{y}_{i\cdot} - \gamma)/\sqrt{s^2/n_i}$	
$H_0: \tau_i - \tau_k = \gamma$	$t = (\bar{y}_{i\cdot} - \bar{y}_{k\cdot} - \gamma)/\sqrt{s^2(1/n_i + 1/n_k)}$	(7.5.2)
$H_0: \sum_{i=1}^{p} c_i\tau_i = \gamma,\ \sum_{i=1}^{p} c_i = 0$	$t = (\sum_{i=1}^{p} c_i\bar{y}_{i\cdot} - \gamma)/\sqrt{s^2 \sum_{i=1}^{p} c_i^2/n_i}$	

Some Examples

We now illustrate the use of some of the inference formulas given above.

Example 7.5.1 Refer to the daphnia data given in Table 7.2.1 and analyzed in Example 7.4.1. We find a confidence interval for $\mu_1 = \mu + \tau_1$, the population mean lifetime for the no copper group. In addition, we will test the hypothesis $H_0: \tau_1 = \tau_2$ versus $H_1: \tau_1 \neq \tau_2$. First, the confidence interval for $\mu + \tau_1$, given by (7.5.1), is

$$\bar{y}_{1\cdot} \pm t_{\alpha/2,n-p}\sqrt{s^2/n_1}\ .$$

From Example 7.4.1 we have $\bar{y}_{1\cdot} = 76.5$, $s^2 = 134.8182$, $n_1 = 4$ and $n - p = 14 - 3 = 11$. From the t—table, we have $t_{.025,11} = 2.201$. It follows that a 95% confidence interval for the mean lifetime of the no copper group $(\mu + \tau_1)$ is

$$76.5 \pm 2.201\sqrt{134.8182/4} \quad \text{or} \quad (63.72,\ 89.28).$$

To perform the test for $H_0: \tau_1 = \tau_2$ versus $H_1: \tau_1 \neq \tau_2$, we recognize that $\tau_1 = \tau_2$ is equivalent to $\mu + \tau_1 = \mu + \tau_2$ or $\tau_1 - \tau_2 = 0$. That is, the null hypothesis states that the first and second treatment means are equal. Using the second test in (7.5.2), we find that the value of the test statistic for this problem is

$$t = (\bar{y}_{1.} - \bar{y}_{2.}) / \sqrt{s^2(1/n_1 + 1/n_2)}$$
$$= (76.5 - 63.0)/\sqrt{134.8182(1/4 + 1/5)} = 1.73.$$

For this "two–tailed" test, the decision rule is reject H_0 when $|t| \geq t_{\alpha/2,n-p}$. If we let $\alpha = 0.05$, the critical value is $t_{.025,11} = 2.201$. Since the calculated $t = 1.73 < 2.201$, we have insufficient evidence to reject H_0. Based on information obtained in this experiment the average lifetimes for the no copper group and the 20 $\mu g/\ell$ copper group are not significantly different.

Example 7.5.2 The SAS package can be used to perform a t–test for a contrast and to provide the point estimate and standard error needed for a confidence interval. As an illustration, we consider testing the hypothesis H_0: $\tau_1 - \frac{1}{2}(\tau_2 + \tau_3) = 0$ versus H_1: $\tau_1 - \frac{1}{2}(\tau_2 + \tau_3) \neq 0$. The ESTIMATE statement can be used in SAS to obtain the t–test. Table 7.4.3 contains the SAS commands and printout. The line on the printout labeled COMPARISON contains the information for this contrast. From this printout we find the test statistic for H_0: $\tau_1 - \frac{1}{2}(\tau_2 + \tau_3) = 0$ has the value $t = 3.71$ and p–value of $p = 0.0034$. Since $p < 0.05$ we reject H_0 at the $\alpha = 0.05$ risk level. If we desired a 95% confidence interval for the contrast $\tau_1 - \frac{1}{2}(\tau_2 + \tau_3)$, we use formula (5.3.7). namely,

$$\text{point estimate} \pm t_{\alpha/2,n-p} \text{ (standard error)}.$$

Using $t_{.025,11} = 2.201$ and the values on the printout, the 95% confidence interval is

$$25.5 \pm 2.201(6.8692) \text{ or } (10.38,\ 40.62).$$

At a confidence level of 95% we can state that the mean lifetime for daphnia in the no copper group (treatment one) is between 10.38 and 40.62 days higher than the average of the mean lifetimes for the 20 $\mu g/\ell$ and 40 $\mu g/\ell$ groups.

7.6 Reparametrization of a Completely Randomized Design

In some situations it is advantageous to transform a less than full rank experimental design model into a full rank linear model. The most obvious advantage of this is that one can analyze experimental design data using a regression program, a program invariably found at any computing facility where statistical software packages are used. We pass from one model representation to another via a reparametrization procedure. In doing this we find a new full rank model, called the **reparametrized model**, with new parameters which are related to the parameters of the original experimental design model, referred to an the **traditional model**. Using the correspondence between the new and original parameters we are able to perform all estimable and testable procedures available to traditional design analysis.

We now proceed to transform the traditional CRD model having functional form (7.2.1),

$$Y_{ij} = \mu + \tau_i + \varepsilon_{ij} \quad , \quad i = 1,2,\ldots,p \quad ; \quad j = 1,2,\ldots,n_i \quad ,$$

to a model of full rank. Observe that model (7.2.1), which has a design matrix $\mathbf{X}$ of rank p, contains $p + 1$ parameters. Basically, the traditional model is less than full rank because the number of parameters is greater than its rank. Therefore, a full rank CRD model is one which has rank p and p parameters.

To form a reparametrized model, we utilize indicator (dummy) variables, that is, variables which have values of zero and one. For the completely randomized design with a single factor of p levels (p treatments) we will use as the **reparametrized model**,

$$Y = \alpha_0 + \alpha_1 x_1 + \cdots + \alpha_{p-1} x_{p-1} + \varepsilon, \tag{7.6.1}$$

where $$x_i = \begin{cases} 1, & \text{if treatment } i \text{ is applied, } i = 1,2,\ldots,p-1. \\ 0, & \text{if not} \end{cases}$$

In matrix form we write $\mathbf{Y} = \boldsymbol{\mathcal{X}} \boldsymbol{\alpha} + \varepsilon$, where $\boldsymbol{\alpha}' = (\alpha_0, \alpha_1, \ldots, \alpha_{p-1})$. Here we have denoted the p—column "**X** matrix" of the linear model (7.6.1.) by the symbol $\boldsymbol{\mathcal{X}}$ in order to distinguish from the $(p + 1)$—column $\mathbf{X}$ matrix of the traditional CRD model (7.2.1). In fact, the columns of the $\boldsymbol{\mathcal{X}}$ matrix coincide with the first p columns of matrix $\mathbf{X}$ displayed in (7.2.2). It is easy to see that (7.6.1) is a full rank linear model so that the reparametrized model is, in fact, a **multiple regression model**.

In order to carry out inference procedures concerning the parameters of the traditional model, we will need to know the relationships between its parameters and those of the reparametrized (or regression) model. To establish these relationships, we identify the treatment means according to each model as follows:

Table 7.6.1 Treatment Means for Two CRD Models

Treatment	Traditional Model	Reparametrized Model
1	$\mu + \tau_1$	$\alpha_0 + \alpha_1$
2	$\mu + \tau_2$	$\alpha_0 + \alpha_2$
3	$\mu + \tau_3$	$\alpha_0 + \alpha_3$
.	.	.
.	.	.
.	.	.
p	$\mu + \tau_p$	α_0

Equating the expressions for the respective treatment means yields the following correspondence between the parameters of the two models

$$\boldsymbol{\alpha} = \begin{bmatrix} \alpha_0 \\ \alpha_1 \\ \alpha_2 \\ \cdot \\ \cdot \\ \cdot \\ \alpha_{p-1} \end{bmatrix} = \begin{bmatrix} \mu + \tau_p \\ \tau_1 - \tau_p \\ \tau_2 - \tau_p \\ \cdot \\ \cdot \\ \cdot \\ \tau_{p-1} - \tau_p \end{bmatrix} \tag{7.6.2}$$

With (7.6.2) in mind we are able to illustrate how the reparametrized model can be used to obtain the usual tests and confidence intervals for a CRD. We begin with the analysis of variance f–test. From (7.6.2), it is obvious that the hypotheses

$$H_0 : \tau_1 = \tau_2 = \cdots = \tau_p$$

and

$$H_0^* : \alpha_1 = \alpha_2 = \cdots = \alpha_{p-1} = 0$$

are equivalent. The reader will recognize H_0^* to be the null hypothesis associated with the analysis of variance for a multiple regression model having functional form (7.6.1). In Chapter 6 we learned that we reject H_0^* (or equivalently H_0) if

$$f = \frac{SSr/(p - 1)}{SSe/(n - p)} = \frac{MSr}{MSe} \geq f_{\alpha,p-1,n-p}, \tag{7.6.3}$$

where SSr is the regression sum of squares and MSr is the regression mean square. Thus to test for "no difference in treatment effects" in a CRD, we need only test the null hypothesis that "all parameters except for α_0 are simultaneously zero" in the reparametrized regression model.

There is a natural relationship between the sums of squares for the reparametrized and traditional models. Since both involve the same observation, it follows that the total sum of squares, SSt, is the same for both models. In the exercises it is shown that the reparametrized model is in effect the same as the traditional model for a CRD with the side condition $\tau_p = 0$ imposed on the model. As a result, both models have the same error sums of squares, SSe. In addition,

$$SSr(\text{reparametrized model}) = SStr\ (\text{traditional model}).$$

As a consequence, f in (7.6.3) is of the form

$$f = \frac{SStr/(p - 1)}{SSe/(n - p)} = \frac{MStr}{MSe},$$

which is the same f used to test $H_0 : \tau_1 = \tau_2 = \cdots = \tau_p$ in the traditional model.

Individual Tests and Confidence Intervals Illustrated

The tests and confidence intervals for contrasts and other estimable functions presented in the previous section can also be done using the reparametrized model. The procedure is quite simple. For any estimable function involving $\mu, \tau_1, \tau_2, \cdots, \tau_p$, we first express the function as a linear combination of the elements in vector α. Since $\mathbf{Y} = \boldsymbol{X}\boldsymbol{\alpha} + \boldsymbol{\varepsilon}$ is a linear model with full rank, we can use full rank linear model results to obtain a test or confidence interval for the desired linear combination $\boldsymbol{\ell}'\boldsymbol{\alpha}$. This provides a test or confidence interval for the corresponding traditional model estimable function.

As an illustration, let $p = 3$ and consider the contrast $\psi = \tau_1 -$

$2\tau_2 + \tau_3$. In this setting $\boldsymbol{\alpha}' = (\alpha_0, \alpha_1, \alpha_2)$ where $\alpha_0 = \mu + \tau_3$, $\alpha_1 = \tau_1 - \tau_3$, and $\alpha_2 = \tau_2 - \tau_3$. Then $\psi = \alpha_1 - 2\alpha_2$, or $\psi = \boldsymbol{\ell}' \boldsymbol{\alpha}$ where $\boldsymbol{\ell}' = (0, 1, -2)$. Using the result (6.1.4 or 5.3.6), we see that a confidence interval for ψ is

$$\boldsymbol{\ell}' \mathbf{a} \pm t_{\alpha/2,n-p} \sqrt{s^2 \; \boldsymbol{\ell}'(\boldsymbol{X}'\boldsymbol{X})^{-1}\boldsymbol{\ell}}$$

where $\mathbf{a} = (\boldsymbol{X}'\boldsymbol{X})^{-1}\boldsymbol{X}'\mathbf{y}$ is the least squares estimate for $\boldsymbol{\alpha}$.

Example 7.6.1 For a numerical illustration of this reparametrization technique, we consider the daphnia data given in Table 7.2.1. Since there are three treatment groups for these data, we will need two indicator variables for the reparametrized model. The indicator variables are

$$x_1 = \begin{cases} 1 & \text{, if treatment 1} \\ 0 & \text{, if not} \end{cases} , \qquad x_2 = \begin{cases} 1 & \text{, if treatment 2} \\ 0 & \text{, if not} \end{cases} ,$$

and the model is

$$Y_i = \alpha_0 + \alpha_1 x_{i1} + \alpha_2 x_{i2} + \varepsilon_i , \quad i = 1,2,\ldots,14.$$

For inference purposes we assume that the ε_i's are independent $N(0, \sigma^2)$ random variables.

To analyze the daphnia data using the reparametrized model we need to obtain the ANOVA table, the least squares estimates and the $(\boldsymbol{X}'\boldsymbol{X})^{-1}$ matrix. These calculations can be done by hand; however, the reparametrization technique is intended to be used when only a computer regression program is available. Accordingly, we demonstrate using the SAS regression program. The SAS commands for this analysis and corresponding printout are given in Table 7.6.2.

With the reparametrized model we note that $\alpha_0 = \mu + \tau_3$, $\alpha_1 = \tau_1 - \tau_3$, and $\alpha_2 = \tau_2 - \tau_3$. The test for $H_0 : \tau_1 = \tau_2 = \tau_3$ is equivalent to the regression ANOVA hypothesis $H_0 : \alpha_1 = \alpha_2 = 0$. From the printout, we find that the f—value and p—value for this test are $f = 12.23$ and $p = 0.0016$. This is, of course, the same as the values obtained for this test in Example 7.4.1. Note also that the regression sum of squares is 3297.8571, which coincides with the treatment sum of squares obtained earlier. Similarly, the error and total sums of squares agree with the traditional model analysis.

Table 7.6.2

SAS Regression Commands and Printout for Reparametrized Model

SAS COMMANDS

```
OPTIONS LS = 80;
DATA DAPH;
INPUT COPPER LIFETIME @@;
X1 = 0; X2 = 0;
IF COPPER = 1 THEN X1 = 1;
IF COPPER = 2 THEN X2 = 1;
CARDS;
1 60 1 90 1 74 1 82
2 58 2 74 2 50 2 65 2 68
3 40 3 58 3 25 3 30 3 42
PROC PRINT;
PROC GLM;
MODEL LIFETIME = X1 X2/I;
ESTIMATE 'TRT2 MEAN' INTERCEPT 1 X1 0 X2 1;
```

SAS Printout

OBS	COPPER	LIFETIME	X1	X2
1	1	60	1	0
2	1	90	1	0
3	1	74	1	0
4	1	82	1	0
5	2	58	0	1
6	2	74	0	1
7	2	50	0	1
8	2	65	0	1
9	2	68	0	1
10	3	40	0	0
11	3	58	0	0
12	3	25	0	0
13	3	30	0	0
14	3	42	0	0

General Linear Models Procedure

X'X Inverse Matrix

	INTERCEPT	X1	X2	LIFETIME
INTERCEPT	0.2	-0.2	-0.2	39
X1	-0.2	0.45	0.2	37.5
X2	-0.2	0.2	0.4	24
LIFETIME	39	37.5	24	1483

General Linear Models Procedure

Dependent Variable: LIFETIME

Source	DF	Sum of Squares	Mean Square	F Value	Pr > F
Model	2	3297.8571429	1648.9285714	12.23	0.0016
Error	11	1483.0000000	134.8181818		
Corrected Total	13	4780.8571429			

R-Square	C.V.	Root MSE	LIFETIME Mean
0.689805	19.92104	11.611123	58.285714

Source	DF	Type I SS	Mean Square	F Value	Pr > F
X1	1	1857.8571429	1857.8571429	13.78	0.0034
X2	1	1440.0000000	1440.0000000	10.68	0.0075

Source	DF	Type III SS	Mean Square	F Value	Pr > F
X1	1	3125.0000000	3125.0000000	23.18	0.0005
X2	1	1440.0000000	1440.0000000	10.68	0.0075

Parameter	Estimate	T for H0: Parameter=0	Pr > |T|	Std Error of Estimate
TRT2 MEAN	63.0000000	12.13	0.0001	5.19265215

Parameter	Estimate	T for H0: Parameter=0	Pr > |T|	Std Error of Estimate
INTERCEPT	39.00000000	7.51	0.0001	5.19265215
X1	37.50000000	4.81	0.0005	7.78897823
X2	24.00000000	3.27	0.0075	7.34351910

We can easily find a confidence interval for an estimable function involving $\mu_1, \tau_1, \tau_2, \tau_3$. For example, let us find a 95% confidence interval for the second population mean $\mu + \tau_2$ (i.e., the mean lifetime for daphnia in the 20 $\mu g/\ell$ copper group). Since $\mu + \tau_2 = \alpha_0 + \alpha_2$, the point estimate for $\mu + \tau_2$ is $a_0 + a_2 = 39.0 + 24.0 = 63.0$, where a_0 and a_2 denote the least squares estimates for α_0 and α_2, respectively. Note that α_0, α_1 and α_2 are estimable since the reparametrized model is full rank and, consequently, all parameters are estimable. The 95% confidence interval for $\mu + \tau_2 = \alpha_0 + \alpha_2$ is

$$(a_0 + a_2) \pm t_{.025,11} Se(\hat{\alpha}_0 + \hat{\alpha}_2)$$

where $t_{.025,11} = 2.201$. Now $\alpha_0 + \alpha_2 = \boldsymbol{\ell}' \boldsymbol{\alpha}$ where $\boldsymbol{\ell}' = (1,0,1)$ and $\boldsymbol{\alpha}' = (\alpha_0, \alpha_1, \alpha_2)$ so from (5.3.5)

$$\text{Se}(\hat{\alpha}_0 + \hat{\alpha}_2) = \sqrt{s^2 \boldsymbol{\ell}'(\boldsymbol{X}'\boldsymbol{X})^{-1}\boldsymbol{\ell}}\ .$$

The quantities s^2 and $(\boldsymbol{X}'\boldsymbol{X})^{-1}$ can be obtained from the printout where we find $s^2 = 134.8182$ and

$$\boldsymbol{\ell}'(\boldsymbol{X}'\boldsymbol{X})^{-1}\boldsymbol{\ell} = [1 \quad 0 \quad 1] \begin{bmatrix} 0.20 & -0.20 & -0.20 \\ -0.20 & 0.45 & 0.20 \\ -0.20 & 0.20 & 0.40 \end{bmatrix} \begin{bmatrix} 1 \\ 0 \\ 1 \end{bmatrix} = 0.2.$$

Hence, $\text{Se}(\hat{\alpha}_0 + \hat{\alpha}_2) = \sqrt{134.8182(0.2)} = 5.1927$, and the 95% confidence interval for $\mu + \tau_2$ is

$$63.0 \pm 2.201(5.1927) \text{ or } (51.57,\ 74.43).$$

Therefore we are 95% confident that the mean lifetime for the 20 μg/ℓ copper group is between 51.57 and 74.43 days.

We could use SAS to assist us in forming a confidence interval for $\mu + \tau_2 = \alpha_0 + \alpha_2$. As illustrated in Table 7.6.2, the ESTIMATE command yields both the least squares estimate and the standard error for $\hat{\alpha}_0 + \hat{\alpha}_2$.

Our interest in this section has been to illustrate how one can use a regression model to analyze data from a completely randomized design. To reiterate, an obvious advantage of this technique is that the statistical software needed is merely a readily available regression program rather than a specialized ANOVA experimental design package. We do not mean to imply that it is easier to analyze a CRD in this manner. In fact, if a computer is not available, it is generally easier to use traditional formulas in making the "hand calculations." Nevertheless, in some cases it may be a convenient way to utilize available computer software to analyze the data. Further, some of the principles to be studied later can be more clearly elucidated via the reparametrized approach.

7.7 Expected Mean Squares, Restricted Model

In the analysis of complex experiments, in particular those discussed in Chapters 16 through 18, it will be imperative for us to know the expected values of the mean square entries in the ANOVA table. In order to find the expected value of a mean square, referred to as an

expected mean square, we need to look upon a mean square as a random variable, the realization of which for a particular sample appears as a numerical entry in the analysis of variance table. Although we will not utilize this concept a great deal at present, it will provide us with additional insight concerning the f—test procedure.

Finding expected mean squares is not always a simple matter; for a CRD experiment, however, it is rather straightforward. We know from Theorem 5.1.6 that the expected mean square error is given by

$$\mathrm{E(MSE)} = \sigma^2.$$

To find the other expected mean squares, we utilize a rule which we now describe. First recall that a sum of squares can be expressed as a quadratic form. Also recall from Theorem 5.1.5, that for the linear model $\mathbf{Y} = \mathbf{X}\boldsymbol{\beta} + \boldsymbol{\varepsilon}$, where $\boldsymbol{\varepsilon}$ is $\mathrm{MN}(\mathbf{0}, \sigma^2\mathbf{I})$, the expected value of the quadratic form $\mathbf{Y}'\mathbf{A}\mathbf{Y}$ is given by

$$\mathrm{E}(\mathbf{Y}'\mathbf{A}\mathbf{Y}) = \sigma^2\mathrm{trace}(\mathbf{A}) + \{\mathrm{E}(\mathbf{Y})\}'\mathbf{A}\{\mathrm{E}(\mathbf{Y})\}.$$

Now the sum of squares that is of interest to us is one that can be written as SSe* — SSe using the principle of conditional error. It can be shown (see exercises) that when SSE* — SSE corresponds to a sum of squares in the ANOVA table, it can be written as a quadratic form in $\mathbf{Y}$ with an idempotent matrix. Recall that the trace of an idempotent matrix equals the rank of the matrix which equals the degrees of freedom associated with the sum of squares. Putting all these facts together yields the following statements: For a sum of squares $\mathbf{Y}'\mathbf{A}\mathbf{Y}$ with degrees of freedom q, we have

$$\mathrm{E}(\mathbf{Y}'\mathbf{A}\mathbf{Y}) = \sigma^2\mathrm{q} + \{\mathrm{E}(\mathbf{Y})\}'\mathbf{A}\{\mathrm{E}(\mathbf{Y})\}.$$

If we let $\mathrm{W} = \{\mathrm{E}(\mathbf{Y})\}'\mathbf{A}\{\mathrm{E}(\mathbf{Y})\}$, then

$$\mathrm{E}(\mathbf{Y}'\mathbf{A}\mathbf{Y}) = \sigma^2\mathrm{q} + \mathrm{W}.$$

Note that W is the expression for the sum of squares when the Y's are replaced by their expected values. Finally, for the mean square $\mathbf{Y}'\mathbf{A}\mathbf{Y}/\mathrm{q}$, we have its expected value given by

$$\mathrm{E}(\mathbf{Y}'\mathbf{A}\mathbf{Y}/\mathrm{q}) = \sigma^2 + \frac{1}{\mathrm{q}}\mathrm{W}. \qquad (7.7.1)$$

We summarize our findings in the following theorem.

Theorem 7.7.1 Consider the linear model $\mathbf{Y} = \mathbf{X}\boldsymbol{\beta} + \boldsymbol{\varepsilon}$ where $\boldsymbol{\varepsilon}$ is $\mathrm{MN}(\mathbf{0}, \sigma^2\mathbf{I})$. Let SS be the sum of squares in the ANOVA table with q degrees of freedom. The corresponding expected mean square, denoted by E(MS), is given by

$$E(MS) = \sigma^2 + \frac{1}{q}W,$$

where W is the expression for SS when the Y's are replaced by their expected values.

Expected Mean Squares in CRD Analysis of Variance

For an illustration, we turn to the completely randomized design setting. Let us find the associated expected mean squares. Of course,

$$E(MSE) = \sigma^2. \tag{7.7.2}$$

To find the expected mean square for treatments, denoted by E(MSTR), we utilize Theorem 7.7.1. From (7.4.1),

$$SSTR = \sum_{i=1}^{p} n_i(\bar{Y}_{i\cdot} - \bar{Y}_{\cdot\cdot})^2$$

so

$$W = \sum_{i=1}^{p} n_i[E(\bar{Y}_{i\cdot}) - E(\bar{Y}_{\cdot\cdot})]^2.$$

It is easy to show that $E(\bar{Y}_{i\cdot}) = \mu + \tau_i$ and

$$E(\bar{Y}_{\cdot\cdot}) = \mu + \bar{\tau}, \text{ where } \bar{\tau} = \sum_{i=1}^{p} n_i\tau_i/n.$$

Then $W = \sum_{i=1}^{p} n_i[(\mu + \tau_i) - (\mu + \bar{\tau})]^2 = \sum_{i=1}^{p} n_i(\tau_i - \bar{\tau})^2.$

This gives us

$$E(MSTR) = \sigma^2 + \frac{1}{p-1}\sum_{i=1}^{p} n_i(\tau_i - \bar{\tau})^2. \tag{7.7.3}$$

At times it is advantageous to extend the ANOVA table to

include an expected mean square column. For the purpose of illustration, we do that now in Table 7.1.1.

Table 7.7.1 ANOVA for CRD with E(MS)

Source	df	MS	E(MS)
Treatments	$p - 1$	MStr	$\sigma^2 + \sum_{i=1}^{p} n_i(\tau_i - \bar{\tau})^2/(p - 1)$
Error	$n - p$	MSe	σ^2
Total	$n - 1$		

Note that when $H_0 : \tau_1 = \tau_2 = \cdots = \tau_p$ is true, $\tau_i - \bar{\tau} = 0$ is a consequence and then $\sum_{i=1}^{p} n_i(\tau_i - \bar{\tau})^2 = 0$. Thus, when H_0 is true,

$$E(MSTR) = \sigma^2 = E(MSE).$$

In this case the numerator and denominator of the F random variable used in the test for H_0 have the same expected value. On the other hand, when H_0 is false, $\sum_{i=1}^{p} n_i(\tau_i - \bar{\tau})^2 > 0$ resulting in $E(MSTR) > E(MSE)$. This provided us with an intuitive reason for rejecting H_0 when $f = MStr/MSe$ is large.

What we have just demonstrated in a CRD setting is true in general for experimental design problems. That is, whenever an F random variable is used to test a null hypothesis H_0, under H_0 the ratio of the expected values of the numerator and denominator of F is one; this ratio is greater than one when H_0 is not true.

The Restricted Model

In some textbooks the model given for the completely randomized design includes certain restrictions on the parameters of the model. The most common restriction used is

$$\sum_{i=1}^{p} n_i \tau_i = 0. \tag{7.7.4}$$

When this restriction is included as part of the model we refer to the model as the **restricted model**. That is, the restricted model is

$$\begin{cases} Y_{ij} = \mu + \tau_i + \varepsilon_{ij}, \; i = 1,2,\ldots,p; \; j = 1,2,\ldots,n_i . \\ \sum_{i=1}^{p} n_i \tau_i = 0 \end{cases} \tag{7.7.5}$$

In keeping with this terminology, we call the traditional model, namely the one we have been using up to this point, the **unrestricted** CRD model.

Use of the restricted model does not alter the analysis we have described for the unrestricted model in the previous sections of this chapter. For example, the ANOVA table, the f statistic, the confidence interval formulas, and so forth, remain unchanged. This follows since all these quantities are invariant to the side conditions used to solve the system of normal equations.

If we refer back to Section 7.3, we see that the side condition used to obtain a **unique** solution to the normal equations **is** restriction (7.7.4). As a result, when that restriction becomes part of the model, all parameters are **individually** estimable. From (7.3.1) we see that for the restricted CRD model, the BLUE for μ is $\bar{y}_{..}$ and the BLUE for τ_i is $\bar{y}_{i.} - \bar{y}_{..}$. This is a distinct difference from the unrestricted CRD model where $\mu + \tau_i$ is estimable but μ and τ_i separately are not estimable.

There are two other differences between the restricted and unrestricted CRD models that should be noted. The ANOVA null hypothesis $H_0 : \tau_1 = \tau_2 = \cdots = \tau_p$ associated with the unrestricted model "reduces" to $H_0 : \tau_i = 0$ for all i, for the restricted model. In words, for the unrestricted model case, H_0 is "no difference in treatment effects"; for the restricted model, H_0 is "no treatment effects." Also, making (7.7.4) part of the model changes the expected mean squares for treatments. It is easy to see that for the model (7.7.5), the formula for E(MSTR), given in (7.7.3), becomes

$$E(MSTR) = \sigma^2 + \frac{1}{p - 1} \sum_{p=1}^{p} n_i \tau_i^2 .$$

Although the restricted model is used in some textbooks, essentially to avoid the complication of estimability, we will continue to use the unrestricted model in this text for the time being. We recognize, of course, that for the CRD discussed in this chapter it makes little difference in the final analysis which model is utilized.

7.8 Design Considerations

In carrying out experiments, there are two important phases which involve statistics, namely, choosing the design for the experiment, and analyzing the data from the experiment. Thus far in this chapter we have concentrated on the analysis of data for a CRD. It is equally important, however, to discuss how one decides when a CRD is appropriate for an experiment. In this section we will discuss some positive and negative features of the CRD. Being aware of these features will help us decide when to choose the CRD over another experiment.

One feature of a design is the structure and distribution of the data obtained from the design. We recall that the data structure obtained for a CRD and a one–way classification are identical. For a CRD and one–way we have $p \geq 2$ samples of data and these samples are independent. To do the complete analysis we have presented thus far, we must assume that the data are normally distributed and each random observation has the same variance. In the next section we discuss the checking of these assumptions and some of the consequences when they are violated.

Another feature of a design is the randomization process. In the CRD, the experimental units are randomly assigned to the treatment groups, or vice versa. That is, there is essentially no restriction on the randomization except that each treatment is used a specified number of times. Why do we worry about this randomization process? There are two important reasons. First, by randomly assigning the experimental units, we guard against the chance for biases the experimenter may use in the assignment. For example, if the experimenter has a preferred treatment group he/she may subconsciously assign the "best" experimental units to this treatment group, which will bias the results. Second, in a CRD we only consider one factor, namely the treatment group factor. It is rare indeed when a response variable only depends on one factor. That is, there usually are other factors affecting the response variable even though they are not measured. By randomizing, we hope that these other factors are "evened out."

To better understand this "averaging" out of the other factors, let us consider the following example. Consider four drug remedies used with colds. To compare the effectiveness of these drugs, 400 people (experimental units) are available for a study. One hundred people are assigned to each drug group (treatment) and some measurement is made. Now sex of the person is another possible factor. If this is the case, by randomly assigning a large number of people, we should expect that the proportion of men or women in each group to be about the same. As a consequence, even though sex may affect the response variable, the treatment groups are still comparable since the sex factor is balanced out for the four groups.

To see that this balancing out of other factors does not always work, let us consider another example. Consider a field with eight plots used as experimental units. Also consider two treatments, which are two varieties of corn labeled A and B. To complete a CRD we randomly assign the treatments to the plots so that each treatment occurs four times. The results are given as follows:

A	A	B	B
A	A	B	B
good soil		bad soil	

While the result does not appear to be random, it certainly is a possibility. Now suppose there is another factor, say soil quality. Further, suppose that soil quality is good in the four plots to the left and poor in the four plots to the right. Here it is obvious that the soil quality factor does not even out among the two treatment groups. The problem with this design is that if we obtain higher values (corn yield) for treatment A we do not know whether it is because variety A is better or because of the better soil, or both.

This previous example illustrates some important concepts. One cannot expect other factors to even out when there are only a few experimental units in a CRD. Further, other factors not taken into account can lead to wrong conclusions. In many cases it is much better to incorporate these other factors in the design if physically possible. However, there are times when the incorporation of an erroneous factor into a design can reduce the efficiency of the experiment.

In many CRD experiments, one of the treatment groups serves as a **control group.** A control group is essentially a treatment group that serves as a basis of comparison for other treatments. Many times the control group actually gets no treatment at all. As an example, consider the cold remedy discussed earlier in this section. Suppose that we have three drugs used as cold remedies. We may use a control group as a group of subjects who get no drug at all. Then we can compare the other three groups which did get a drug to the control group. This will enable us to determine if the drug actually is effective.

In summary, the CRD is a very useful design. Its major advantages are its ease of application and analysis, and the fact that the sample sizes need not be the same for each treatment. The major disadvantage of a CRD is the risk of excluding potentially important factors from consideration in the experiment. If suspected other factors are present, it is often better to incorporate them into the design to reduce $s^2 = \text{MSe}$, which improves the power of the tests and the precision of the confidence intervals. More will be said about this in Chapter 10. It should also be noted that a CRD may not be possible for some experiments. For example, if observations arise in pairs (e.g., tire wear measured on a bicycle), then a CRD is not applicable.

7.9 Checking Assumptions

The tests and confidence intervals presented in this chapter have been obtained in the setting of the normal linear model, that is $Y_{ij} = \mu + \tau_i + \varepsilon_{ij}$, where $\boldsymbol{\varepsilon}$ is $MN(\mathbf{0}, \sigma^2\mathbf{I})$. It is important to realize that our formulas are dependent on this model and the assumptions made with it. It is, therefore, desirable to determine whether the model and assumptions are appropriate for a particular set of data.

In this section we consider the general problem of checking the model assumptions. We will present techniques that are useful for detecting violations of the assumptions. In addition, we will discuss the problems that can arise when the assumptions are not met. Recall that the assumptions for the normal linear model are primarily used for obtaining tests and confidence intervals; therefore, the effects of any assumption violations will be investigated in terms of the potential effect on these inference procedures. At the end of this section we discuss possible remedies when these assumptions are violated.

There are three particular assumptions that have been made with the normal linear model for the completely randomized design. They are as follows:

1. Equal variance,
2. Normal observations, and
3. Independent observations.

These assumptions apply to both the ε_{ij}'s and the Y_{ij}'s. In addition we will consider another potential problem, namely

4. Bad data and outliers.

We will address each of these topics in turn.

Equal Variances Assumption

In the normal linear model for the CRD we have assumed that $\mathrm{Var}(Y_{ij}) = \sigma^2$, that is, all the observations have the same variance. This is referred to as the homogeneity of variance assumption. In this section we present procedures useful for checking this assumption. The crucial problem is to determine whether the variances for the p treatment groups are the same. If we let $\mathrm{Var}(Y_{ij}) = \sigma_i^2$, for $i = 1,\ldots,p$, then we want to determine whether the hypothesis

$$H_0 : \sigma_1^2 = \sigma_2^2 = \cdots = \sigma_p^2 \tag{7.9.1}$$

is reasonable or not.

There are several procedures available for testing (7.9.1). We will present one such procedure here and two others are presented in the exercises. Hartley (1940 and 1950) proposed a rather simple procedure for testing this hypothesis. It assumes normal distributions and equal sample sizes. We let $n_1 = n_2 \ldots = n_p = r$. To compute the test statistic, we first calculate $s_1^2, s_2^2,\ldots,s_p^2$, the sample variances for the p treatment groups. We let s_{max}^2 and s_{min}^2 be the largest and smallest variances, respectively. The Hartley test statistic is denoted by H, where

$$H = s_{max}^2 / s_{min}^2 . \tag{7.9.2}$$

We reject H_0 using an α—level test if

$$H > h_{\alpha,p,r} .$$

Values of $h_{\alpha,p,r}$ are given in Table 3 of Appendix B. These $h_{d,p,r}$ values represent the upper α cutoff point of the distribution of H when H_0 is true.

There are two drawbacks to Hartley's test. The procedure is only directly applicable to equal sample size problems. If the sample sizes are not the same, then one can use the largest sample size as the r value for the test. This generally results in a reasonable test, especially if the sample sizes are nearly the same; however, the test tends to be somewhat liberal, i.e., the true type one error level is in excess of α. A more serious problem concerns the normality assumption. Hartley's test requires normally distributed observations, and it is known to be quite sensitive to this assumption. That is, if the observations are not normally distributed then the true type one error level for the test can be off by a large amount. Because of this the test should be used with some caution, particularly when the normality assumption is in question.

Example 7.9.1. We refer again to the daphnia lifetime data presented in Table 7.2.1. Our goal is to test $H_0 : \sigma_1^2 = \sigma_2^2 = \sigma_3^2$ for these data. Routine calculations are used to find the following sample variances:

$$s_1^2 = 163.667 \qquad s_2^2 = 86.0 \qquad s_3^2 = 162.0.$$

The value of $H = s_{max}^2 / s_{min}^2 = 163.667 / 86.0 = 1.9$. From Table 3 the .05 cutoff point for H is $h_{.05,3,5} = 15.5$. Note that since the sample sizes are unequal, but nearly the same, we used $r = 5$. Since $H = 1.9 < 15.5$ we fail to reject H_0 and we declare that the sample variances are not significantly different.

There are several other tests available for testing the homogeneity of variance assumption. We presented the Hartley test because it is quite simple to use. Another test that is sometimes used is one proposed by Cochran (1941). This test assumes equal sample sizes and normal data. The test statistic used in Cochran's test is $s_{max}^2 / \sum_{i=1}^{p} s_i^2$. Tables for the distribution of this test statistic are available in Pearson and Hartley (1966). Bartlett (1937) proposed a test for the equality of variances that can be used when the sample sizes are unequal. This procedure is described in the exercises. Unfortunately both Cochran's test and Bartlett's test are adversely affected by non—normality. A test described in Scheffe (1959) is less sensitive to non—normality; however, it is more complicated to use and we will not describe it here.

The effect of unequal variances on the analysis has been investigated by several authors. Cochran (1947) and Box (1953, 1954a, 1954b) have investigated this problem, and Scheffe (1959) has a detailed discussion in his text. Basically these authors have found that when the sample sizes are the same or are nearly the same, the effect of unequal variances on the f—test is minimal. That is, for equal or nearly equal sample sizes the f—test is robust to unequal variances. If the sample sizes are quite different, then large differences in the variances can alter the level of the f—test, although the effect is usually not very great. The effect of unequal variances on a confidence interval for a contrast can be more serious than the effect on the f—test; however, it is also not very serious unless the variances are vastly different. In summary, minor violations in the equal variance assumption can be ignored, especially when the sample sizes are nearly the same.

Normality Assumption

We now turn our attention to checking the normality assumption. For the normal linear model we have assumed that the Y_{ij}'s are $N(\mu + \tau_i, \sigma^2)$. Our first task is to determine whether the normality assumption is reasonable or not. Note that when the normality assumption holds, the observations within a treatment group are a random sample from a normal population. Therefore, to check the normality assumption we need only determine whether it is reasonable to assume that the observations from a treatment group is a random sample from a normal population.

When all sample sizes are large we can check the normality assumption for each sample separately. For each sample this can be accomplished in one of two ways. First, one can obtain graphical evidence for normality. For example, a histogram of the data can be obtained and one can visually inspect whether the graph resembles a normal distribution. One can also obtain a normal probability plot for the data and use it to check the normality assumption. A normal probability plot will be discussed in more detail in an example presented later in this section. Second, one can also obtain a formal test of the hypothesis that the observations come from a normal population. Two tests that are often used are a Kolmogorov type test and the Shapiro—Wilks test. Both tests are described in detail in the text by Conover (1980) and we will not describe them formally here. These tests, as well as the histogram, and normal probability plot can be obtained via SAS by using the PROC UNIVARIATE command. This will be illustrated in an example presented later in this section.

When the sample sizes are small it is unreasonable to check the normality assumption separately for each sample. This is particularly true for the graphical procedures because there is insufficient data to obtain a reasonable graph. In such cases we pool all n observations together and consider the residuals, namely, $e_{ij} = y_{ij} - \bar{y}_{i.}$. If the normality assumption holds, then the e_{ij}'s will mimic the distribution of normal variates. We can then check whether it is reasonable to assume that the distribution of these n residuals comes from a population that is normal or not. One word of caution is in order. These n residuals (in random variable form) are not independent since it is obvious that the residuals within a treatment group are correlated. Also, the variance of the residuals can vary from treatment to treatment if the sample sizes are different. Therefore, we cannot think of the residuals as values from a random sample from a single population. In spite of this fact we can still obtain a crude check for normality using these residuals. As before, we can obtain a visual check of normality using a histogram or a normal probability plot of **all** n residuals. A formal test can be obtained; however, the actual error level and p—value for this test will only be approximate since the observations are not really a single random sample. The next example illustrates this process.

Example 7.9.2. To illustrate a check of the normality assumption we refer to the daphnia lifetime data presented in Table 7.2.1. Since the sample sizes are all relatively small, we will check normality by using all of the residuals together. We will use SAS to obtain a test for normality and to obtain a histogram and a normal probability plot.

These data were previously analyzed using SAS and the commands used are given in Table 7.4.3. The additional commands needed to check the normality assumption are as follows:

```
PROC GLM;
CLASS COPPER;
MODEL LIFETIME = COPPER;
OUTPUT OUT = NEW PREDICTED = YHAT RESIDUAL = RESID;
PROC UNIVARIATE PLOT NORMAL; VAR RESID;
```

The OUTPUT statement creates two variables. The predicted y values are named YHAT, and the residuals are named RESID in our case. The UNIVARIATE statement does many things. Of interest to us are the following: the PLOT statement produces a stem—and—leaf graph (histogram) for the residuals and a normal probability plot; the NORMAL command produces a test for normality. The Shapiro—Wilks test is given when the sample size is under 51, while the Kolmogorov type test is given for other cases. The output produced by these commands for the daphnia data is given in Table 7.9.1.

Table 7.9.1 Output from PROC UNIVARIATE for Daphnia Data

```
                    Moments

N                   14  Sum Wgts           14
Mean                 0  Sum                 0
Std Dev       10.68068  Variance     114.0769
Skewness      0.049718  Kurtosis     -0.75515
USS               1483  CSS              1483
CV                   .  Std Mean     2.854532
T:Mean=0             0  Pr>|T|         1.0000
Num ^= 0            14  Num > 0             8
M(Sign)              1  Pr>=|M|        0.7905
Sgn Rank           1.5  Pr>=|S|        0.9515
W:Normal       0.97258  Pr<W           0.8691


               Quantiles(Def=5)

100% Max            19       99%           19
 75% Q3            5.5       95%           19
 50% Med           1.5       90%         13.5
 25% Q1             -9       10%          -14
  0% Min         -16.5        5%        -16.5
                              1%        -16.5

Range             35.5
Q3-Q1             14.5
Mode             -16.5
```

```
                  Extremes

 Lowest    Obs      Highest     Obs
 -16.5(      1)         5(        9)
   -14(     12)       5.5(        4)
   -13(      7)        11(        6)
    -9(     13)      13.5(        2)
    -5(      5)        19(       11)


Stem Leaf                     #
   1 9                        1
   1 14                       2
   0 56                       2
   0 123                      3
  -0 2                        1
  -0 95                       2
  -1 43                       2
  -1 6                        1
     ----+----+----+----+
Multiply Stem.Leaf by 10**+1
```

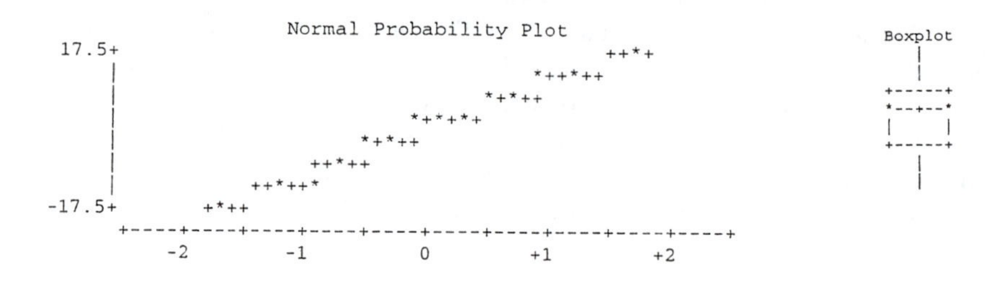

Referring to the printout we first consider the test for normality. The Shapiro—Wilks test has been used to test the null hypothesis H_0 : The data is normal. The p—value on the printout is .842. Recall that this is only an approximate test since the residuals are not really a random sample; therefore, this p—value should be considered as approximate. However, since the p—value is large we can consider this as evidence supporting the normality assumption.

Next we consider the stem—and—leaf graph and the normal probability plot. Since the sample size is rather small it is unreasonable to expect the stem—and—leaf graph to perfectly resemble a normal curve. The graph obtained for these data does not show any irregularities and that is the best that we can expect to get with a sample of this small size. According to the SAS USER'S GUIDE: BASICS the normal probability plot is a quantile–quantile plot of the data. The empirical quantiles are plotted against the quantiles of a standard normal distribution. Asterisks (*) mark the data values. The vertical coordinate is the data value and the horizontal coordinate is

$$\Phi^{-1}(r_i - 3/8)/\ (n + 1/4))$$

where r_i is the rank of the data value, Φ^{-1} is the inverse of the standard normal distribution function, and n is the number of non–missing data values. The plus signs (+) provide a reference straight line that is drawn using the sample mean and standard deviation. If the data are from a normal distribution, they should tend to fall along the reference line.

By inspecting this plot on the printout there does not seem to be any negative evidence to normality.

In summary, the normality assumption seems reasonable for these data. Note that we are not saying that the data set is normal; however, we have not found any evidence to the contrary, and therefore, we are satisfied with the normality assumption for these data.

The normality assumption is necessary for tests and confidence intervals, but not for the estimation results. Pearson (1931) has shown that the F–test is robust to the normality assumption. That is, even if the data are not normal, the F–test generally has reasonable error levels and power. The central limit theorem essentially makes it possible to assume normality (at least approximately) for our tests and confidence intervals when the sample sizes are large. A potential problem occurs for small sample sizes. When the sample sizes are small, the error level for any test will only be approximate when the data are not normal, and there may be a better test in terms of power. Scheffé (1959) has a more detailed discussion on this topic. In summary, for large sample size problems it is reasonable to use the normality assumption. If the sample sizes are small, then the assumption should be checked. If a violation of the assumption is found, then the effect on tests is not all that great; however, an alternative procedure could be considered. See later in this section for details.

Independence Assumption

The assumption we now consider is one in which we assume that the ε_{ij}'s are independent. Of course, if the ε_{ij}'s are independent, then the Y_{ij}'s are also independent. Unfortunately it is difficult to check this assumption using the realized data. In general, it is best to design the experiment in such a way that independence will logically follow because of the manner in which the observations are collected. One problem that can occur in some experiments is a time dependence. If observations are taken in a time process, then it might be reasonable to consider plotting the residuals with respect to time to determine if a time correlation exists. If this residual plot shows a random scatter, then there is no evidence for a time correlation.

Unfortunately the effect of the independence assumption can be quite severe on the f—tests and on confidence intervals. Scheffé (1959) and Box (1954b) have a good discussion on this topic. In general, when the independence assumption is violated, the level and power of the tests can be adversely affected.

Outliers – Detection and Effects

An outlier is an observation that is an extreme value. It differs markedly from the other observations within a treatment group. Outliers can result from bad data collection, an error in experimentation, or it can be just an unusual value that is actually correct.

It is fairly simple to detect potential outliers though it is often much more difficult to determine whether the value is an erroneous value or simply an unusual value. There are formal tests for determining outliers; however, in this text we will use a simple intuitive technique that involves the residuals. In general, for normal data almost all of the observations should lie within three standard deviations of the mean. This suggests that within a treatment group almost all of the observations should lie within the interval $\mu + \tau_i - 3\sigma, \quad \mu + \tau_i + 3\sigma$, which is approximated by $\bar{y}_{i\cdot} - 3s, \ \bar{y}_{i\cdot} + 3s$. This implies that the residuals, $e_{ij} = y_{ij} - \bar{y}_{i\cdot}$, should almost always lie in the interval (—3s, 3s). Basically then, any observation whose residual lies outside the interval (—3s, 3s) is a potential outlier. For reasons that we will explain shortly, the values for $\bar{y}_{i\cdot}$ and s should be based on an analysis with the potential outliers removed.

Though this procedure for identifying potential outliers is not very rigorous, it is simple to use and to explain to others. Further, it requires the user to do two things. First, one must identify potential outliers, and second, one must finally decide on an observation—by—observation basis whether an observation is a "real" outlier. Often one can recheck the original data and find that outliers are just a copy or other kind of error and can be corrected.

Outliers can have a great effect on an analysis affecting both estimation and tests. In general an outlier in treatment i tends to pull the $\bar{y}_{i\cdot}$ value towards it. Outliers tend to increase the SSe value which in turn increases the s^2 and s values. It is for this reason that we suggest removing the suspected outliers from the data set before calculating the $\bar{y}_{i\cdot}$ and s values that are used to check for outliers.

Recall that s^2 is used in most tests and confidence intervals. Since outliers tend to increase s^2, this causes tests to lose power and it causes confidence intervals to be too wide. Both of these effects are extremely bad consequences; therefore, one should check very carefully for outliers in a CRD. We illustrate these effects in the following example.

Example 7.9.3 To see the effect that an outlier can have on an analysis, we consider adding an outlier to the daphnia data of Table 7.2.1. In particular, we will add a daphnia lifetime of 15 to the second treatment group. Our goal is to illustrate two things. First, is this really an obvious outlier, and second, what effect does it have.

Since the lifetime of 15 is a suspected outlier we run the analysis with this data point removed. This corresponds to the analysis given in Table 7.4.3 from which we find $\bar{y}_{2\cdot} = 63$ and $s = 11.61$. Almost all of the residuals should fall in the interval (—3s,3s) which is (—34.83, 34.83). The residual for the observation 15 is $y_{26} - \bar{y}_{2\cdot} = 15 - 63 = -48$. Since this residual falls outside the usual interval we identify this observation as an outlier.

To determine the effect of this outlier, we compare the analysis obtained without the outlier (lifetime of 15) and with the outlier in the data set. The results of this are given in Table 7.9.2.

Table 7.9.2 Comparison of analysis with and without outlier

Item	Value in analysis without outlier	Value in analysis with outlier
$\bar{y}_{2\cdot}$	63.0	55.0
s	134.818	283.583
f	12.23	5.51
p-value for f	.0016	.02

It is clear from the previous table that the single outlier has drastically affected this analysis. In fact it is easy to imagine that the presence of a single outlier can determine whether the f—test is significant or not. Because of this, one should be sure to identify any erroneous data points and to remove them from the data set before the analysis is completed.

Remedies for Violations of Assumptions

When there is evidence that one or more assumption is violated for a set of data then it may be advantageous to take steps to alleviate the effect of such violations. We will describe some techniques that can be used for this purpose.

A common technique that is used when there is evidence of unequal variances is to perform a **transformation**. In a transformation the dependent variable, Y, is modified to be some function of Y, say g(Y). The analysis is then completed with g(Y) as the dependent variable.

Transformations that are often helpful are as follows:

- log transformation: ln (Y) or ln(Y+c) where c is a constant chosen so that Y+c is always positive.
- square root transformation : $\sqrt{Y}$ or $\sqrt{Y+c}$.
- Box—Cox power transformation: $(Y^{\lambda} - 1)/\lambda$, for λ a non—zero constant selected by the user.

More details on the Box power transformation and techniques for selecting λ can be found in Box and Cox (1964).

It is not unusual for the variance or standard deviation of Y_{ij} to be proportional to $E(Y_i)$. For example, in model building sometimes it is reasonable to assume that as $E(Y_i)$ increases, $Var(Y_i) = \sigma_i^2$ increases as well. Under such circumstances the log or square root transformation will often alleviate the problem of unequal variances. An example of this situation will be given later in this section.

Transformations are also used to help alleviate the problem of non—normality. The three previously mentioned transformations are frequently used for this purpose. The log transformation is particularly useful for right skewed data. The square root transformation is sometimes used when the dependent variable has a Poisson distribution.

Another transformation that is sometimes used is arcsin $\sqrt{Y}$, which is helpful when Y is a proportion or percentage.

An example utilizing a transformation follows:

Example 7.9.4. A study was completed to determine the effectiveness of two pesticides, labeled A and B, on controlling insects in soybean fields. The study field, planted in soybeans, was subdivided into 30 fairly homogeneous plots of the same size. Ten randomly selected plots were left untreated for insects (control group), 10 other randomly selected plots were treated with pesticide A, and the remaining 10 plots were treated with pesticide B. At a specified time later in the growing season the soybeans were cut, the insects on the plants were collected and the biomass (weight) of all the insects was recorded. The results are given in the following table.

Table 7.9.3 Biomass of Insects

	Pesticide	
Control/None	A	B
84.9	10.1	22.1
82.0	20.2	11.6
94.9	16.8	31.7
103.2	15.5	33.9
72.2	19.5	20.8
107.6	18.1	17.4
58.8	12.8	22.0
64.0	27.8	17.0
71.6	18.7	24.2
60.0	16.8	28.9

Recognizing that these data have been obtained from a completely randomized design with $p = 3$ treatments, we can utilize SAS for the analysis. Table 7.9.4 contains the SAS printout for these data. An estimate for the contrast $\tau_3 - \tau_2$ is included to compare the effectiveness of pesticides A and B.

Table 7.9.4 SAS Commands and Printout for Pesticide Data

SAS Commands

```
OPTIONS LS = 80;
DATA BIOMASS;
INPUT PEST BIOMASS @@;
CARDS;

1 84.9 182.0 1 94.9 1 103.2 1 72.2 1 107.6 1 58.8 1 64.0 17.6
1 60.0
2 10.1 2 20.2 2 16.8 2 15.5 2 19.5 2 18.1 2 12.8 2 27.8 2
18.7 2 16.8
3 22.1 3 11.6 3 31.7 3 33.9 3 20.8 3 17.4 3 22.0 3 17.0 3
24.2 3 28.9
PROC GLM;
CLASS PEST;
MODEL BIOMASS = PEST;
ESTIMATE 'COMPARE 3 VS 2' PEST 0 -1 1;
OUTPUT OUT = NEW PREDICTED = YHAT RESIDUAL = RESID;
PROC UNIVARIATE PLOT NORMAL; VAR RESID;
PROC MEANS; VAR BIOMASS; BY PEST;
```

SAS Output

```
                    General Linear Models Procedure

Dependent Variable: BIOMASS
                                   Sum of             Mean
Source                DF          Squares           Square    F Value      Pr > F

Model                  2     23842.982000     11921.491000      94.05      0.0001

Error                 27      3422.341000       126.753370

Corrected Total       29     27265.323000

                 R-Square            C.V.         Root MSE          BIOMASS Mean

                 0.874480        28.02708        11.258480             40.170000


Source                DF        Type I SS      Mean Square    F Value      Pr > F

PEST                   2     23842.982000     11921.491000      94.05      0.0001

Source                DF      Type III SS      Mean Square    F Value      Pr > F

PEST                   2     23842.982000     11921.491000      94.05      0.0001

                                        T for H0:    Pr > |T|   Std Error of
Parameter                Estimate     Parameter=0                  Estimate

COMPARE 3 VS 2         5.33000000            1.06      0.2992     5.03494529
```

```
                   Moments

N                 30  Sum Wgts          30
Mean               0  Sum                0
Std Dev     10.86332  Variance    118.0118
Skewness    0.454656  Kurtosis    0.974379
USS         3422.341  CSS         3422.341
CV                 .  Std Mean    1.983362
T:Mean=0           0  Pr>|T|        1.0000
Num ^= 0          30  Num > 0           14
M(Sign)           -1  Pr>=|M|       0.8555
Sgn Rank        -9.5  Pr>=|S|       0.8489
W:Normal    0.960767  Pr<W          0.3672

              Quantiles(Def=5)

 100% Max      27.68        99%      27.68
  75% Q3        4.98        95%      23.28
  50% Med      -0.83        90%      12.96
  25% Q1       -5.96        10%     -13.64
   0% Min     -21.12         5%     -19.92
                             1%     -21.12

 Range          48.8
 Q3-Q1         10.94
 Mode          -0.83
```

```
                   Extremes

Lowest     Obs      Highest     Obs
-21.12(      7)     10.17(       18)
-19.92(     10)     10.94(       24)
-15.92(      8)     14.98(        3)
-11.36(     22)     23.28(        4)
 -8.32(      9)     27.68(        6)

Stem Leaf                      #
   2 8                         1
   2 3                         1
   1 5                         1
   1 01                        2
   0 569                       3
   0 011223                    6
  -0 221111                    6
  -0 888665                    6
  -1 1                         1
  -1 6                         1
  -2 10                        2
     ----+----+----+----+
Multiply Stem.Leaf by 10**+1
```

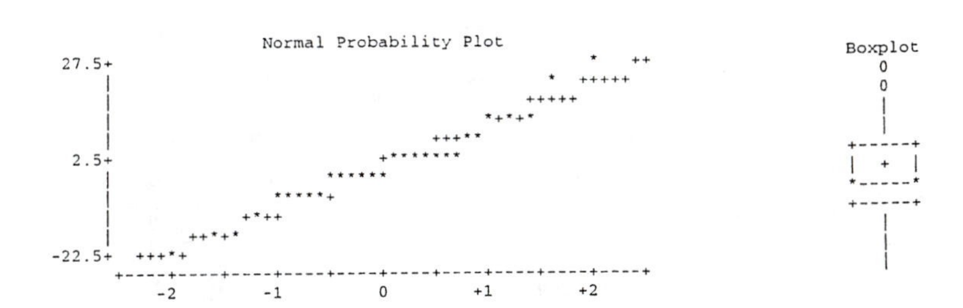

Analysis Variable : BIOMASS

------------------------------------ PEST=1 ------------------------------------

N	Mean	Std Dev	Minimum	Maximum
10	79.9200000	17.5967042	58.8000000	107.6000000

------------------------------------ PEST=2 ------------------------------------

N	Mean	Std Dev	Minimum	Maximum
10	17.6300000	4.7239461	10.1000000	27.8000000

------------------------------------ PEST=3 ------------------------------------

N	Mean	Std Dev	Minimum	Maximum
10	22.9600000	6.9498521	11.6000000	33.9000000

From the printout we observe that the f–test for $H_0 : \tau_1 = \tau_2 = \tau_3$ rejects at the .05 level (p–value = 0.0001 < 0.05); however, the t–test for testing that the contrast $\Psi = \tau_3 - \tau_2$ equals zero yields a p–value of 0.2992. Thus, at the 0.05 level we cannot declare that pesticides A and B have significantly different effectivenesses. Inspecting the output for the PROC UNIVARIATE we do not find any serious concerns about the normality assumption or about possible outliers. Checking for homogeneity of variances we note that the sample standard deviations for the three treatment groups are $s_1 = 17.5967$ $s_2 = 4.7239$ $s_3 = 6.9499$, which are quite different. Calculation of the Hartley test statistic yields

$$H = s^2_{max} \big/ s^2_{min} = (17.5967)^2 \big/ (4.7239)^2 = 13.9.$$

Comparing this value with the tabled 0.05 value for the Hartley test, $h_{.05,3,10} = 5.34$, we reject the hypothesis $H_0 : \sigma_1^2 = \sigma_2^2 = \sigma_3^2$. We, therefore, have compelling evidence that the equal variance assumption has been violated. It may be advisable to transform the data and complete the analysis again.

We will try the log transformation on these data. Table 7.9.5 contains an analysis for these data with ln(Y) as the dependent variable.

Table 7.9.5
SAS Commands and Printout for Pesticide Data – Log Transformation

SAS Commands

```
OPTIONS LS = 80;
DATA BIOMASS;
INPUT PEST BIOMASS @@;
LBIO = LOG(BIOMASS);
CARDS;
1 84.9 1 82.0 1 94.9 1 103.2 1 72.2 1 107.6 1 58.8 1 64.0 1 71.6 1 60.0
2 10.1 2 20.2 2 16.8 2 15.5 2 19.5 2 18.1 2 12.8 2 27.8 2 18.7 2 16.8
3 22.1 3 11.6 3 31.7 3 33.9 3 20.8 3 17.4 3 22.0 3 17.0 3 24.2 3 28.9
PROC GLM;
CLASS PEST;
MODEL LBIO = PEST;
ESTIMATE 'COMPARE 3 VS 2' PEST 0 -1 1;
OUTPUT OUT = NEW PREDICTED = YHAT RESIDUAL = RESID;
PROC UNIVARIATE PLOT NORMAL; VAR RESID;
PROC MEANS; VAR LBIO; BY PEST;
```

SAS Output

```
                    General Linear Models Procedure

Dependent Variable: LBIO
                                     Sum of              Mean
Source                      DF      Squares            Square    F Value      Pr > F

Model                        2   13.31425209       6.65712604      88.41      0.0001

Error                       27  . 2.03298610       0.07529578

Corrected Total             29   15.34723819

                  R-Square          C.V.          Root MSE              LBIO Mean

                  0.867534      8.003340         0.2744008              3.4285782

Source                      DF     Type I SS    Mean Square    F Value      Pr > F

PEST                         2   13.31425209     6.65712604      88.41      0.0001

Source                      DF   Type III SS    Mean Square    F Value      Pr > F

PEST                         2   13.31425209     6.65712604      88.41      0.0001

                                        T for H0:      Pr > |T|    Std Error of
Parameter                 Estimate     Parameter=0                  Estimate

COMPARE 3 VS 2          0.25210407            2.05      0.0497      0.12271575
```

```
               Moments                                      Extremes

N                  30  Sum Wgts          30          Lowest     Obs     Highest     Obs
Mean                0  Sum                0        -0.63821(     22) 0.277268(      4)
Std Dev       0.26477  Variance    0.070103        -0.52458(     11)  0.31902(      6)
Skewness     -0.33986  Kurtosis    0.151236        -0.28767(     17) 0.367098(     23)
USS          2.032986  CSS         2.032986        -0.28526(      7) 0.434196(     24)
CV                  .  Std Mean     0.04834        -0.26506(     10) 0.487921(     18)
T:Mean=0            0  Pr>|T|        1.0000
Num ^= 0           30  Num > 0           16
M(Sign)             1  Pr>=|M|       0.8555
Sgn Rank          9.5  Pr>=|S|       0.8489           Stem Leaf                     #
W:Normal     0.978652  Pr<W          0.8141              4 39                       2
                                                         3 27                       2
                                                         2 78                       2
          Quantiles(Def=5)                               1 0379                     4
                                                         0 015689                   6
 100% Max  0.487921         99%  0.487921               -0 98522                    5
  75% Q3   0.168568         95%  0.434196               -1 0                        1
  50% Med  0.004091         90%  0.343059               -2 997630                   6
  25% Q1   -0.20052         10%  -0.28646               -3
   0% Min  -0.63821          5%  -0.52458               -4
                             1%  -0.63821               -5 2                        1
 Range     1.126135                                     -6 4                        1
 Q3-Q1     0.369086                                        ----+----+----+----+
 Mode      -0.01574                                   Multiply Stem.Leaf by 10**-1
```

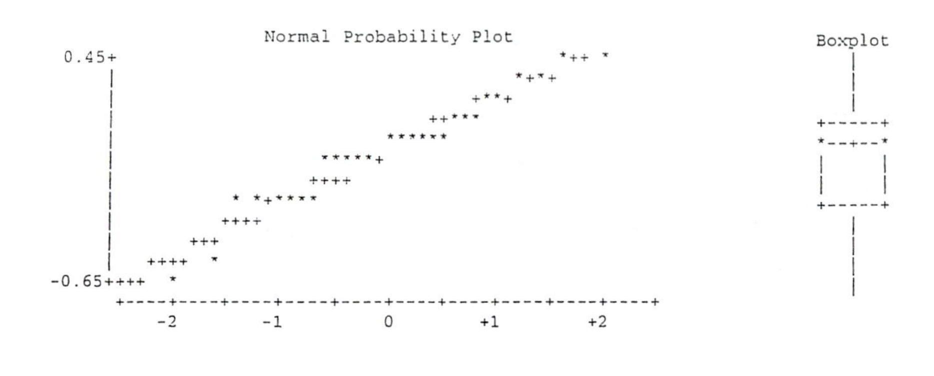


```
Analysis Variable : LBIO
```

------------------------------------ PEST=1 ------------------------------------

N	Mean	Std Dev	Minimum	Maximum
10	4.3594011	0.2188518	4.0741419	4.6784206

------------------------------------ PEST=2 ------------------------------------

N	Mean	Std Dev	Minimum	Maximum
10	2.8371147	0.2715726	2.3125354	3.3250360

------------------------------------ PEST=3 ------------------------------------

N	Mean	Std Dev	Minimum	Maximum
10	3.0892188	0.3228615	2.4510051	3.5234150

For the transformed data, notice that the sample standard deviations are $s_1 = 0.2189$, $s_2 = 0.2716$, $s_3 = 0.3229$, which are relatively more uniform than for the original data. One can easily show that the Hartley test fails to reject $H_0 : \sigma_1^2 = \sigma_2^2 = \sigma_3^2$ for these data at the .05 level. Also note that the output for the PROC UNIVARIATE presents no problems for the normality assumption or for outliers. As a consequence, the assumptions for a CRD seem to have been met for analyzing the transformed data whereas the homogeneity of variance assumption was suspect for the original data. Observing the printout, the f—test for $H_0 : \tau_1 = \tau_2 = \tau_3$ rejects at the 0.05 level (P—value =

0.0001 < 0.05). Checking the t—test for $H_0 : \tau_3 - \tau_2 = 0$ we find a p—value of 0.0497; therefore, at the 0.05 level we can declare that treatments 2 and 3 are significantly different. Recall that we were unable to make such a claim with the analysis of the original data primarily due to the fact that the large variance for the control group caused a loss of power for testing the contrast.

This example illustrates how a transformation can lead to better results. In this case the assumptions are reasonably met after a transformation, and our analysis with the transformed data had increased power for testing a contrast.

Another approach to overcoming the problem of non—normality is to use a procedure that does not require the normality assumption. One such procedure is the **Kruskal—Wallis test** which is a **nonparametric test**, a test that is valid for data from any continuous distribution. The test is introduced in the exercises and we shall not discuss it further at this time.

When the independence assumption is violated, serious problems can arise in the analysis of the data. Unfortunately, there is no simple solution to the problem. Later in this text we will discuss **random effects** models that allow for dependence between certain observations. Another approach that is sometimes used when observations are dependent is **time series analysis**; however, we will not discuss this approach in this text.

In summary, it is important to check whether it is reasonable to assume that the assumptions for a CRD hold. If evidence suggests that the assumptions do not hold, then it is advisable to attempt to remedy the situation by a transformation or some other technique.

7.10 Summary Example — A Balanced CRD Illustration

Many of the procedures presented in this and subsequent chapters are most advantageously used when all treatment sample sizes are the same. This situation is sometimes referred to as the "balanced" case of an experimental design. As we proceed through this text, we will find that, in general, the computations required in analyzing balanced experimental design data are easily carried out. In particular, when the data set is not large, a basic pocket calculator, preferably one that handily accumulates sums of squares, is all that is needed. (In this sense, pocket calculators are the present—day equivalent to the now passe mechanical desk calculators used by R.A. Fisher and his co—workers during their pioneering data analysis development days.) On the other hand, in many of the "unbalanced" design situations (technically, nonorthogonal design cases) to be studied in later chapters, the practitioner welcomes the availability of computers and their accompanying statistical packages. We have already seen that in the multiple regression setting, a computer package is an almost absolute necessity and in Chapter 6 we observed that in order to "get around" horrendous computations, orthogonal polynomials were used in the days before computers for the purpose of response curve and surface data analysis.

As a review of the fundamental principles of CRD analysis and to close this introductory chapter on the subject, we now consider an exercise given by Sincich in his text **Statistics by Example** (Dellen, 1982). In order to present a balanced CRD example, the original data base has been modified so as to end up with an equal replication of treatments.

Sincich reports on an experiment conducted at the University of Melbourne where men and women of various ages were divided into four hair color categories, namely, light blond, dark blond, light brunette and dark brunette. The purpose of the study was to investigate whether hair color and the amount of pain produced by mishaps are related. To determine if there is a relationship, each person in the study was assigned a pain threshold score based upon a pain sensitivity test performance where a high score indicates high tolerance for pain. The modified data set considered here follows:

Table 7.10.1 Pain Threshold Scores

Hair Color

	Light blond	Dark blond	Light Brunette	Dark Brunette
	62	63	42	32
	60	57	50	39
	71	52	41	51
	55	41	37	30
	48	43	43	35
Totals	296	256	213	187

To analyze these data in such a way so as to answer the question concerning the possibility of a relationship between color of hair and pain threshold, we postulate the model

$$Y_{ij} = \mu + \tau_i + \varepsilon_{ij}, \quad i = 1, 2, 3, 4;\ j = 1, 2, 3, 4, 5,$$

where treatments i = 1, 2, 3, 4 represent hair colors light blond, dark blond, light brunette, and dark brunette in that order and j = 1, 2, 3, 4, 5 represent the five individuals (experimental units) chosen for the study within each of the treatments. In order to gain further insight into our problem, a geometric display of the data points is given for us in Figure

7.10.1. From the scatter of the points it is easily seen that the response to hair color dark blond (treatment 2) is most variable and the response to hair color light brunette (treatment 3) is least variable. The reader is invited to check the homogeneity of variance assumption using Hartley's test.

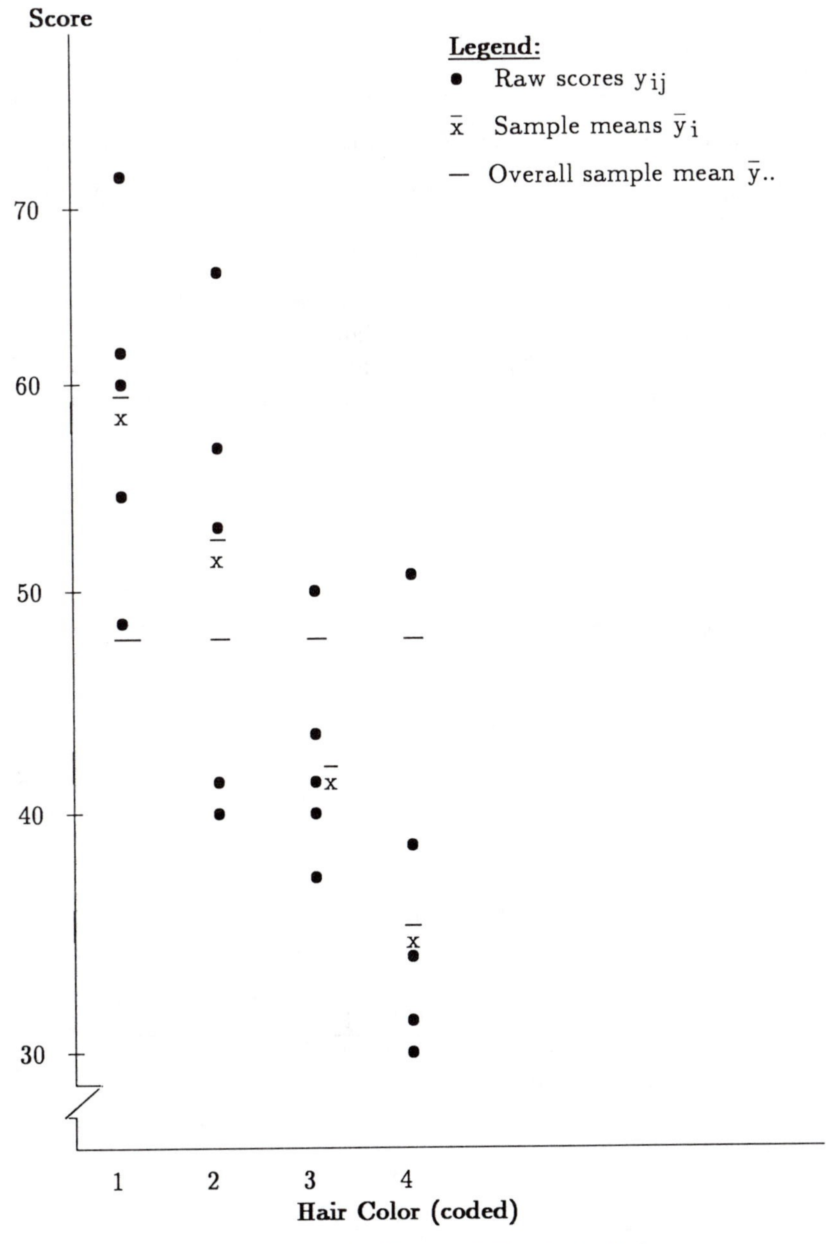

Figure 7.10.1 Pain Threshold by Hair Color

Letting the symbol $\mu_i = \mu + \tau_i$ denote the theoretical population mean score for treatment i, we continue our analysis by posing the null hypothesis $H_0:\mu_1 = \mu_2 = \mu_3 = \mu_4$ (equivalently, $H_0:\tau_1 = \tau_2 = \tau_3 = \tau_4$). From these data, our least squares estimates for $\mu_1, \mu_2, \mu_3, \mu_4$ are

$$\bar{y}_{1\cdot} = 59.2, \quad \bar{y}_{2\cdot} = 51.2, \quad \bar{y}_{3\cdot} = 42.6, \quad \bar{y}_{4\cdot} = 37.4,$$

respectively. These values are also plotted in Figure 7.10.1 in comparison to $\bar{y}_{\cdot\cdot} = 47.6$, our sample estimate of μ, the common value for the μ_i's under H_0. If H_0 is true, intuitively we would expect each of the $\bar{y}_{i\cdot}$ values to cluster around the value for $\bar{y}_{\cdot\cdot}$. Visually, this does not appear to be the case and we now proceed to verify this observation analytically through the analysis of variance.

In using the ANOVA "paper and pencil" computing formulas (7.4.6), we note that in the case of equal sample sizes, the second formula reduces to

$$SStr = \sum_{i=1}^{p} y_{i\cdot}^2/r - (\sum_{i=1}^{p} \sum_{j=1}^{r} y_{ij})^2/n$$

where

$$n_1 = n_2 = \cdots = n_p = r \text{ and } n_1 + n_2 + \cdots + n_p = n.$$

Therefore,

$$SSt = \sum_{i=1}^{4} \sum_{j=1}^{5} y_{ij}^2 - (\sum_{i=1}^{4} \sum_{j=1}^{5} y_{ij})^2/20 = 47{,}700 - (952)^2/20 = 2384.8$$

$$SStr = \sum_{i=1}^{4} y_{i\cdot}^2/5 - (\sum_{i=1}^{4} \sum_{j=1}^{5} y_{ij})^2/20$$

$$= \frac{(296)^2 + (256)^2 + (213)^2 + (187)^2}{5} - \frac{(952)^2}{20} = 1382.8$$

$$SSe = SSt - SStr = 2384.8 - 1382.8 = 1002.0.$$

In summary,

Table 7.10.2 ANOVA for Pain Sensitivity Experiment

Source	df	SS	MS	f
Treatments	3	1382.8	460.933	7.36
Error	16	1002.0	62.625	
Total	19	2384.8		

With experimental value f = 7.36 compared to the one percent critical value $f_{.01,3,16} = 5.29$, we reject H_0 and conclude that, on the basis of the study results, there is a highly significant difference between at least two treatment means. As a consequence, these data provide us with statistical evidence that hair color and pain threshold are related.

There are, of course, other questions the investigators set out to answer through experimentation. One obvious question is: "On average, do blonds and brunettes have different thresholds for pain?" Two other natural questions are: "Is there a difference in pain thresholds between light and dark blonds? between light and dark brunettes?" In the next chapter we discuss how these three "planned" comparisons can be addressed "simultaneously" using a series of f—tests. For now, we will use a t—test to answer the first of the three questions above and then in the next chapter find out that the t and F approaches are statistically equivalent.

The null hypothesis pertinent to the first question is

H_0: Average pain thresholds for blonds and brunettes are the same.

To express H_0 in terms of our model's components, we observe that if light and dark blond haired individuals occur in equal proportions in our population, then the mean pain threshold for blonds is $(\mu + \tau_1 + \mu + \tau_2)/2$. Likewise, if light and dark brunettes occur in equal proportions, then the mean for brunettes is $(\mu + \tau_3 + \mu + \tau_4)/2$. The null hypothesis can now be stated as

$$H_0: (\mu + \tau_1 + \mu + \tau_2)/2 = (\mu + \tau_3 + \mu + \tau_4)/2$$

which algebraically reduces to

$$H_0: \tau_1 + \tau_2 = \tau_3 + \tau_4 \quad \text{or} \quad H_0: \tau_1 + \tau_2 - \tau_3 - \tau_4 = 0.$$

Turning to the third test statistic in set (7.5.2), we calculate

$$\sum_{i=1}^{4} c_i \bar{y}_{i\cdot} = 59.2 + 51.2 - 42.6 - 37.4 = 30.4$$

and $s^2 \sum_{i=1}^{4} c_i^2/n_i = 62.625\,(\frac{1}{5} + \frac{1}{5} + \frac{1}{5} + \frac{1}{5}) = 50.1.$

Hence the value of our t—test statistic is

$$t = \sum_{i=1}^{4} c_i \bar{y}_{i\cdot} \Big/ \sqrt{s^2 \sum_{i=1}^{4} c_i^2/5} = 30.4 \,/\, \sqrt{50.1} = 4.295.$$

If our alternative hypothesis is $H_1\colon \tau_1 + \tau_2 \neq \tau_3 + \tau_4$, our 0.01 risk level critical value is $t_{.005,16} = 2.921$. At this and all higher levels of risk we reject H_0 and conclude that blonds and brunettes do not have the same average threshold for pain.

If our interest is in the magnitude of the difference between the mean for blonds and the mean for brunettes, a confidence interval for the contrast $(\tau_1 + \tau_2)/2 - (\tau_3 + \tau_4)/2$ is in order. Using (7.5.1) we calculate $\sum_{i=1}^{4} c_i \bar{y}_i = (59.2 + 51.2)/2 - (42.6 + 37.4)/2 = 15.2$ and

$$s^2 \sum_{i=1}^{4} c_i^2 /n_i = 62.625\,(\frac{1}{4} + \frac{1}{4} + \frac{1}{4} + \frac{1}{4})\,/5 = 12.525.$$

The above calculations, yield a 99% confidence interval of $15.2 \pm 2.921\sqrt{12.525}$ or (4.86, 25.54) for the difference between the means (blonds — brunettes). The reader should be aware that the construction of this 99% confidence interval can also be used for testing the previous null hypothesis with its two—sided alternative. Since the number 0 is not included in the interval, we would reject $H_0 : (\tau_1 + \tau_2) - (\tau_3 + \tau_4) = 0$ at the 0.01 level of risk.

Having concluded this illustration, we are now ready to explore other inference procedures that are available to us beyond the analysis of variance f—test. Some of these methods are for comparisons planned before "data gathering"; others can be used after "the data are in."

Chapter 7 Exercises

Application Exercises

7–1 A company is considering three different covers for boxes of a brand of cereal. Box type A has a picture of a sports hero eating the cereal, type B has a picture of a child eating the cereal, and type C has a picture of a bowl of the cereal. The company wants to determine which cereal box type provides for the most sales. Eighteen test markets were selected by the company and each box type was randomly assigned to six markets. The number of boxes of this cereal sold per 10,000 population in a specified period is recorded for each test market. The data are as follows:

Type A	52.4	47.8	52.4	51.3	50.0	52.1
Type B	50.1	45.2	46.0	46.5	47.4	46.2
Type C	49.2	48.3	49.0	47.2	48.6	48.2

(a) Explain why this is a CRD.
(b) Find the BLUE for $\mu_A = \mu + \tau_A$.
(c) Find and interpret a 95% confidence interval for $\mu_A = \mu + \tau_A$.
(d) Find the ANOVA table for these data.
(e) Using a 0.05 level, test the hypothesis of no difference in the means for the three treatments.
(f) Using an analysis of residuals and Hartley's test, determine if it is reasonable to assume that the model assumptions hold for this problem.

7–2 Refer to the data in Problem 6–9 in which 24 feedlot beef cattle were randomly assigned to three groups, eight per group. Cattle in group three were given an injection of 10 units of a growth hormone, cattle in group two received an injection of 5 units, and group one received no injection. All 24 cattle were fed identically at the same location. The weight gains, in pounds, after a specified period were as follows:

Group 1	100	110	92	122	118	98	130	110
Group 2	115	121	110	130	142	108	112	120
Group 3	125	140	153	142	130	162	157	160

(a) Explain why this is a completely randomized design.
(b) Find the BLUE for each $\mu + \tau_i$, i = 1, 2, 3.

(c) Find the BLUE for $\tau_2 - \tau_1$.

(d) Find the ANOVA table for these data.

(e) Using a 0.05 risk level, perform the F–test for $H_0 : \tau_1 = \tau_2 = \tau_3$. Interpret the outcome in words.

(f) Find and interpret a 95% confidence interval for $\tau_3 - \tau_1$.

(g) Using an analysis of the residuals and Hartley's test, determine whether it is reasonable to assume that the model assumptions hold for this problem.

7–3 A study involved three groups of women. Group one consisted of 12 anorexic patients, group two consisted of 12 bulemic patients, while group three was a control group of 12 women. Each woman was given Beck's (1961) test for depression. In this test a score of 0 – 9 indicates a nondepressed condition, a score of 10 – 15 indicates a mildly depressed condition, a score of 16 – 23 indicates a moderately depressed condition, and a score of 24 or more indicates a severely depressed condition. The results are given as follows. (The problem is based on a study by S. Myers (master's thesis – Miami University, Ohio, 1986).)

Score on Beck's Depression Test

Group 1	15	29	14	28	26	16	22	34	19	9	20	9
Group 2	28	35	26	18	18	23	24	29	23	5	18	17
Group 3	6	3	7	5	11	4	5	7	6	1	2	6

(a) Find the ANOVA table for these data.

(b) Perform the f–test using a 0.05 level. Report the outcome of this test in words.

(c) Find a 95% confidence interval for the mean of group two.

(d) Find a 95% confidence interval for the difference between the means of groups one and three. What does this interval suggest?

(e) Using a 0.05 level, perform Hartley's test for the equality of variances.

(f) Using the log transformation (i.e., use $\ln Y$ as the dependent variable) repeat parts (a) through (e).

(g) Which dependent variable, Y or $\ln Y$, is preferred for

(i) the f–test,

(ii) the confidence interval for part (c)?

Justify your answers.

7–4 Refer to the data in Problem 3–20 in which students in three IQ groups were given a reading test. An additional student in the low IQ group reports a reading score of 75. This yields the following data:

Reading Scores

Low IQ	39	47	26	36	39	50	75
Medium IQ	49	38	50	49	40	58	
High IQ	61	58	60	57	60	63	

(a) Find the ANOVA table for these data.
(b) Find the ANOVA table for the data without the 75 observation in the low IQ group.
(c) Compare the f–values and the MSe–values for the two problems.
(d) Is there evidence that the 75 observation is an outlier?

7–5 Davis (1974) studied the performance of hard–of–hearing school children on a task involving the knowledge of 50 basic concepts considered necessary for academic development in children age 6 to 8. The Boehm test of basic concepts was administered to a group of children from each age group and the results are given in the table. Higher scores indicate a better knowledge of the 50 concepts. (This data is mentioned in Daniels (1990).)

Boehm test score

Age 6	17	20	24	34	34	38						
Age 7	23	25	27	34	38	47						
Age 8	22	23	26	32	34	34	36	38	38	42	48	50

(a) Find the ANOVA table for these data.
(b) Using a 0.05 level, test whether differences exist between the means for the 3 groups.
(c) Find and interpret a 95% confidence interval for the difference between the mean scores for 8– and 6– year–olds.
(d) Is there any evidence for outliers in these data?
(e) Is there statistical evidence that the mean increase in score from age 7 to 8 is larger than the mean increase from age 6 to 7? **Hint:** Consider a test for the contrast $(\tau_3 - \tau_2) - (\tau_2 - \tau_1) = \tau_3 - 2\tau_2 + \tau_1$.

7–6 Pickles are processed by soaking them in a chemical solution for a period of time. A company has used a standard chemical solution, label it S1; however, the solution is not biodegradable and its disposal is expensive. The company wished to find an alternative solution that could be used to process the pickles which is biodegradable. Three biodegradable solutions, label them S2, S3 and S4, were selected. A total of forty different processing units of the same size were available for the study. Each of the four chemical solutions were assigned randomly to ten processing units. The forty units were all used at the same time. During the soaking process some pickles tend to bloat and must be discarded. One goal of the experiment was to compare the average number of bloater pickles obtained with the four solutions. The data listed in the table consists of the number of bloater pickles in each processing unit.

Number of bloater pickles

Chemical	S1	2	6	7	10	5	6	3	1	4	5
Solution	S2	16	17	8	20	16	14	13	19	21	9
	S3	4	3	10	2	11	4	7	8	11	2
	S4	21	24	13	19	21	22	14	19	20	21

(a) Find the ANOVA table for these data.
(b) Using a 0.05 level, report the outcome of the f–test for the equality of the four treatment means.
(c) Find and interpret a 95% confidence interval for the contrast $\tau_3 - \tau_1$.
(d) Which solution(s) is best statistically?
(e) Should we consider a transformation on these data? If so, which one is theoretically appropriate?

7–7 Consider the pickle processing problem (Exercise 7–6) and the corresponding data.

(a) Write down the reparametrized model for these data.
(b) Using the reparametrized model and a regression computer program find the ANOVA table.
(c) Using a 0.05 level, report the outcome of the f–test for the equality of the four treatment means.
(d) Using the reparametrized model find a 95% confidence interval for $\tau_3 - \tau_1$.

7–8 Refer to Exercise 7–5. Answer parts (a), (c), and (e) using a reparametrized model and a regression program.

7–9 Consider the pain threshold scores given in Table 7.10.1.

(a) Using a 0.05 level perform and interpret Hartley's test for the equality of the variances. (See visual display of data points in Figure 7.10.1.)
(b) Is there any evidence of outliers in these data?
(c) Use an analysis of the residuals to determine if it is reasonable to assume that the model assumptions hold for this problem.

7–10 A sample of water taken from a certain river was "split" into three subsamples and then the subsamples were sent to laboratories 1, 2, and 3, respectively. The laboratories were asked to determine the amount of iron in the subsamples received by them. Each of the laboratories made 5 **determinations** with the following results expressed in micrograms per liter.

	Laboratory		
	1	2	3
Means	125.7	132.3	128.4
Standard deviations	7.4	6.8	8.1

(a) On the basis of these data, is there reason to believe that the three laboratories obtain different results on average? Hint: In order to compute the ANOVA, use formula (7.4.5).
(b) Was a completely randomized design used in obtaining these data? Why or why not?

7–11 Consider an experiment involving 12 persons in which 6 are men and 6 are women. Label these persons $M_1, M_2, \ldots, M_6; W_1, W_2, \ldots, W_6$. These 12 individuals are to serve as the experimental units in a CRD with 3 treatments. Using random numbers (or well–mixed 12 slips of paper labeled as suggested above) find 4 different possible randomizations of the 12 people. Do these randomizations balance out the affect of sex in all 4 randomization outcomes?

7–12 Consider a CRD with $p = 2$ treatments. The two–sample t–test for testing $H_0 : \mu_1 = \mu_2$ is given by

$$t = (\bar{y}_{1\cdot} - \bar{y}_{2\cdot}) \Big/ \sqrt{s_p^2 \left(\frac{1}{n_1} + \frac{1}{n_2}\right)},$$

where $s_p^2 = \dfrac{(n_1 - 1)\, s_1^2 + (n_2 - 1)\, s_2^2}{n_1 + n_2 - 2}$ and

s_1^2 and s_2^2 are the sample variances. This t–statistic has $n_1 + n_2 - 2$ degrees of freedom. Consider the following two samples taken from Exercise 3 – 5:

Sample 1	8	11	12	9
Sample 2	8	6	8	10

(a) Calculate the t–statistic, noting that here it has 6 degrees of freedom.

(b) Calculate the f in the ANOVA table. What are its degrees of freedom?

(c) Observing that $t^2 = f$, what does this suggest?

7–13 (Bartlett's Test) Another test that can be used to test $H_0 : \sigma_1^2 = \cdots = \sigma_p^2$ is Bartlett's test. This test is applicable for equal and unequal sample size cases and is based on the test statistic $b = c^{-1} [(\Sigma(n_i - 1)) \ln MSe - \Sigma(n_i - 1) \ln s_i^2]$; where s_i^2 is the sample variance for treatment i, and

$$c = 1 + \frac{1}{3(p-1)} \left[\left[\sum_{i=1}^{p} \frac{1}{n_i - 1} \right] - \frac{1}{\sum_{i=1}^{p} (n_i - 1)} \right].$$ The test

is based on rejecting H_0 when b exceeds the upper α cutoff point from a chi–square distribution with $p - 1$ degrees of freedom. This yields an approximate α–level test when the n_i

are reasonably large. Use the Bartlett test for the equality of the variances for the data in Exercise 7–5.

Theoretical Exercises

7–14 Consider a CRD with $p = 3$, $n_1 = 4$, $n_2 = 2$, $n_3 = 2$ and the model $Y_{ij} = \mu + \tau_i + \varepsilon_{ij}$, $i = 1, 2, 3$; $j = 1, \cdots, n_i$.

(a) Display the vectors $\mathbf{Y}$, $\boldsymbol{\beta}$, $\boldsymbol{\varepsilon}$ and the matrix $\mathbf{X}$ for the matrix representation of the model (i.e., $\mathbf{Y} = \mathbf{X}\boldsymbol{\beta} + \boldsymbol{\varepsilon}$).

(b) If the treatment means, denoted by μ_i, are $\mu_1 = 10$, $\mu_2 = 14$ and $\mu_3 = 8$, then find possible values for μ, τ_1, τ_2 and τ_3 for the model. Are these values unique? If not, find another set of possible values.

(c) For the treatment means in part (b) find the values for μ, τ_1, τ_2, and τ_3 for the restricted model (i.e., when $\sum_{i=1}^{3} n_i\tau_i = 0$). Are these values unique?

(d) Calculate E(MSTR) using the values given in part (b).

7–15 For a CRD with p treatments and treatment means $\mu_1, \cdots, \mu_p$ the **cell means model** is $Y_{ij} = \mu_i + \varepsilon_{ij}$, $i = 1, \cdots, p$; $j = 1, \cdots, n_i$.

a) Find SSe for this model.

b) Find the form of the f–test for testing $H_0 : \mu_1 = \cdots = \mu_p$ using the assumption that the ε_{ij} are independent $N(0, \sigma^2)$ random variables.

c) Compare the f–test statistic of part (b) to the usual f–test statistic for a CRD given in Equation (7.4.3).

7–16 Verify that the computational formula for SSe given in equation (7.3.5) is algebraically equivalent to the formula for SSe given in Equation (7.3.4).

7–17 For a CRD with model $Y_{ij} = \mu + \tau_i + \varepsilon_{ij}$, $i = 1, \cdots, p$; $j = 1, \cdots, n_i$ determine which of the following functions are estimable:

(i) $\mu + \tau_1 + \tau_2$ (ii) $\mu + (\tau_1 + \tau_2)/2$

(iii) $\mu - \tau_1$ (iv) $\tau_1 + 2\tau_2 - 3\tau_3$.

7–18 Refer to the normal equations for a CRD given in Section 7.3. Impose the side condition $\tau_p = 0$ on the model.

(a) Find the solution to the normal equations satisfying the given side condition.

(b) Using the solution in part (a) find

i) $\hat{y}_{ij}$ ii) SSe iii) $\widehat{\tau_1 - \tau_3}$

(c) Compare the values obtained in part (b) to the values obtained using the side condition $\sum_{i=1}^{p} n_i \tau_i = 0$ that was utilized in Section 7.3.

7–19 Verify the formulas given in the following equations:

(a) The formula for SStr given in Equation (7.4.1).

(b) The formula for SSt given in Equation (7.4.4).

(c) The computational formulas for SSt and SStr given in Equations (7.4.6).

7–20 Consider the normal CRD model and let $\chi^2_{\alpha/2,\, n-p}$ represent the upper $\alpha/2$ cutoff point for a chi–square random variable with $n - p$ degrees of freedom. Use the result of Theorem 5.1.4 to show that the interval

$$\left((n-p)s^2/\chi^2_{\alpha/2,\, n-p} \,,\; (n-p)s^2/\chi^2_{1-\alpha/2,\, n-p} \right)$$

is a $100(1-\alpha)\%$ confidence interval for σ^2.

7–21 Refer to Exercise 7–12 in which the two–sample t–test was compared to the f–test for a CRD with $p = 2$.

(a) Show algebraically that $t^2 = f$.

(b) Explain why the t–test and f–test are equivalent for testing $H_0 : \mu_1 = \mu_2$ versus $H_1 : \mu_1 \neq \mu_2$.

(c) Are the two tests equivalent for testing $H_0 : \mu_1 = \mu_2$ versus the one–sided hypothesis $H_1 : \mu_1 > \mu_2$? Explain why.

7–22 Explain why the reparametrized model for a CRD is in effect the same model as the traditional model for a CRD with the side condition $\tau_p = 0$ imposed.

7–23 Consider the **restricted** model for a CRD given in Equation (7.7.5).

(a) Find the least squares estimators for μ and τ_i .

(b) Show that $\bar{Y}_{..}$ is an unbiased estimator for μ.

(c) Show that $\bar{Y}_{i.} - \bar{Y}_{..}$ is an unbiased estimator for τ_i .

(d) Show that $\mathrm{Var}(\bar{Y}_{i.} - \bar{Y}_{..}) = (\frac{1}{n_i} - \frac{1}{n})\, \sigma^2$.

(e) Based on parts (c) and (d), write down the expression for a $100(1-\alpha)\%$ confidence interval for the i–th treatment effect, τ_i .

(f) Refer to Exercise 7–1 and assume the use of a restricted CRD model for the analysis. Use part (e) to compute a 95% confidence interval for τ_A and compare its width with the width of the confidence interval for $\mu_A = \mu + \tau_A$. Explain intuitively why one would expect the confidence interval for τ_A to be shorter than the confidence interval for μ_A .

7–24 In a CRD problem an alternative to the f–test for testing $H_0 : \tau_1 = \cdots = \tau_p$ is the **Kruskal–Wallis test**. The Kruskal–Wallis test is a **nonparametric test**, which is valid for data from any continuous distribution. Such a test can be useful when the normality assumption is inappropriate. In the Kruskal–Wallis test all n observations are ranked from smallest (rank 1) to largest (rank n). If we let $\bar{R}_i$ represent the average of the ranks for the observations in treatment i, then the Kruskal–Wallis test statistic, KW, is given by

$$KW = \frac{12}{n(n+1)} \sum_{i=1}^{p} n_i (\bar{R}_i - \frac{n+1}{2})^2.$$

An approximate α–level test can be based on rejecting H_0 if $KW \geq \chi^2_{\alpha,p-1}$, where $\chi^2_{\alpha,p-1}$ is the upper α cutoff point for a chi–square random variable with p–1 degrees of freedom. (More details can be found concerning the Kruskal–Wallis test in the text by Hollander and Wolfe (1973).)

(a) Refer to the data in Exercise 7–1. Use the Kruskal–Wallis test to perform a 0.05 level test of $H_0 : \tau_1 = \tau_2 = \tau_3$.

(b) Add the "outlier" sixth observation 60.2 to the Type B data set. Perform the Kruskal–Wallis test at the 0.05 level on the resulting data set, which now includes the outlier.

(c) Perform the F–test on the modified data of part (b) using a 0.05 level. Compare the resulting f statistic with the one obtained in Exercise 7–1(e). (Note that the outlier observation affects the f–test more than the Kruskal–Wallis test.)

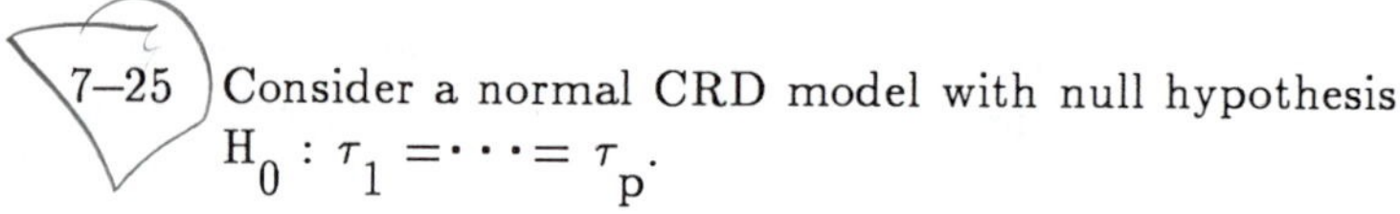

7–25 Consider a normal CRD model with null hypothesis $H_0 : \tau_1 = \cdots = \tau_p$.

(a) Show that if H_0 is true, then $SSTR/\sigma^2$ has a chi–square distribution with $p - 1$ degrees of freedom.

(b) Show that if H_0 is true, then SST/σ^2 has a chi–square distribution with $n - 1$ degrees of freedom.

7–26 Consider a CRD with $p = 4$ treatments and $n = 20$ observations. Our interest is in finding a confidence interval for the contrast $\tau_1 + \tau_2 - \tau_3 - \tau_4$.

(a) Consider the balanced case where $n_1 = n_2 = n_3 = n_4 = 5$. Find the theoretical variance for the estimator of this contrast.

(b) Consider the unbalanced case where $n_1 = 6$, $n_2 = 4$, $n_3 = 6$ and $n_4 = 4$. Find the theoretical variance for the estimator of this same contrast.

(c) Comparing the variances in parts (a) and (b) what implication and conjecture can you draw concerning the best allocation of sample sizes?

(d) Prove your conjecture in part (c).

7–27 (Power of f–test) The **power** of a test is the probability that H_0 is rejected. For the f–test in a normal CRD model the power can be calculated for a given situation in which the sample sizes are fixed and the population means, μ_i, and variance σ^2 are specified. The power of the f–test is based on the **non–central F distribution**, which depends on a parameter called the **non–centrality** parameter. In a normal CRD model the non–centrality parameter is a function of a quantity $\emptyset$ given by

$$\emptyset^2 = \frac{1}{p\sigma^2} \sum_{i=1}^{p} n_i (\tau_i - \bar{\tau})^2 = \frac{1}{p\sigma^2} \sum_{i=1}^{p} n_i (\mu_i - \bar{\mu})^2,$$

where $\bar{\mu} = \sum_{i=1}^{p} n_i \mu_i / n.$

Pearson and Hartley (1951) have provided tables that can be used to find the power of the f–test when $\alpha = 0.05$ or $\alpha = 0.01$, and the quantities p, n and $\emptyset$ are known. In these Pearson–Hartley tables, given as Table 7 in the Appendix, we can find the power on the vertical axis, values of $\emptyset$ are on the horizontal axis, on the left are the $\alpha = 0.05$ "power curves" and on the right are the $\alpha = 0.01$ power curves. Note that in the table, ν_1 is the numerator degrees of freedom $(p - 1)$, and this determines which page in the table one should use. The value ν_2 represents the degrees of freedom denominator $(n - p)$, and determines which particular power curve to use to determine the power.

(a) Suppose that we have a normal CRD model with $p = 3$, $n_1 = n_2 = n_3 = 11$. If $\mu_1 = 8$, $\mu_2 = 9$, $\mu_3 = 13$ and $\sigma^2 = 8$, then show that $\emptyset = 2.53$. Using the Pearson–Hartley power table, observe that the power of the f–test is around .97 for an $\alpha = 0.05$ level test. (The power implies that for this particular situation we have a .97 probability that the data will lead us to reject $H_0 : \tau_1 = \tau_2 = \tau_3$.)

(b) Change the sample sizes in part (a) to $n_1 = n_2 = n_3 = 3$, but leave all other quantities unchanged. Find the power of the f–test. How does the sample size affect the power?

(c) Change the value of σ^2 in part (a) to $\sigma^2 = 16$, but leave all other quantities unchanged. Find the power of the f–test. How does the variance, σ^2, affect the power?

7–28 (Sample Size) One criterion that is sometimes used to select sample sizes in a CRD is to specify a desired power level for a specific configuration of treatment means, and to find the sample size needed to attain the desired power. Recall from Exercise 7–27 that the power can be found by calculating $\emptyset^2$ and then using the Pearson–Hartley tables.

(a) Consider a CRD with $p = 3$, $\mu_1 = 10$, $\mu_2 = 12$, $\mu_3 = 14$ and $\sigma^2 = 5$. Assume that r is the common sample size. Find the power for the $\alpha = .05$ test when $r = 5, 6, 7, 8, 9$. What sample size is needed to insure that the power is at least .90?

(b) Consider a CRD with $p \geq 2$ treatments and equal sample sizes r. Suppose that the difference between at least two treatment means is D or greater. Show that $\emptyset^2 \geq rD^2/(2p\sigma^2)$. (Note that if $\frac{D}{\sigma}$ is specified to be Δ, then $\emptyset^2 \geq r\Delta^2/2p$.)

(c) Consider a CRD with $p = 4$, and suppose that we want to have power of at least 90% when $\Delta = 1.5$. What sample size is needed to attain this power level for a $\alpha = .05$ level test?

7–29 (Sample Size) Another criterion that can be used to select the sample size in a CRD is to find the sample size required to obtain a desired length for a confidence interval. Assume that r is the common sample size. Show that sample size required to obtain a length of L for a confidence interval for the difference between two means, $\tau_i - \tau_k$, is given by

$$r = 8\,\sigma^2\, t^2_{\frac{\alpha}{2},\, n-p} / L^2.$$

CHAPTER 8

PLANNED COMPARISONS

8.1 Introduction

In the previous chapter on the completely randomized design, we were introduced to the analysis of variance F—test used to test the null hypothesis $H_0 : \tau_1 = \tau_2 = \cdots = \tau_p$. Rejection of this hypothesis suggests that there are differences in treatment means but it does not tell us which means are possibly different. When this is of concern to the investigator, as it frequently is, further analysis beyond the F—test is required. In this and the next chapter, we will focus on additional inferential techniques applied to **contrasts** of interest. These auxiliary tests and confidence intervals are generally classified as **planned comparisons** (this chapter) or **multiple comparisons** (next chapter).

By **planned comparisons** we mean tests or confidence intervals that the researcher decided to perform prior to the running of the experiment. For this reason some authors label planned comparisons as **a priori comparisons.** Typically, planned comparisons are those comparisons which are suggested by the original goals of the experiment. For example, in an experiment with a CRD involving three treatments, the researchers may be particularly interested in the mean of treatment one as compared to the means of treatments, two and three. In such a problem, two planned comparisons would be finding confidence intervals for $\tau_1 - \tau_2$ and $\tau_1 - \tau_3$.

In some problems certain comparisons become interesting or logical to consider after an initial analysis of the data has been performed. For example, after calculating the F—test for data arising from a CRD conducted experiment, we may want to determine which of the treatment means are significantly different. This can be accomplished by inspecting all **pairwise contrasts**, namely, $\tau_i - \tau_k$ for all $1 \leq i < k \leq p$. If the number of treatments is $p = 4$, this involves the $\begin{bmatrix} 4 \\ 2 \end{bmatrix} = 6$ contrasts $\tau_1 - \tau_2$, $\tau_1 - \tau_3$, $\tau_1 - \tau_4$, $\tau_2 - \tau_3$, $\tau_2 - \tau_4$, $\tau_3 - \tau_4$. Since we decide to study these contrasts after performing the F—test, these comparisons are called **a posteriori comparisons.** The various procedures used in making inferences of this type are commonly called **multiple comparisons** and are the subject of the next chapter.

8.2 Method of Orthogonal Treatment Contrasts

In Section 7.5 we learned how to perform tests and confidence intervals for a single contrast. (See Definition 7.3.1 for the definition of a treatment contrast.) Of course, one can use these procedures to investigate more than one contrast as a planned comparison. However, when studying more than one contrast, redundant information may be obtained if the contrasts in a set are not chosen carefully. For example, the contrasts $\psi_1 = \tau_1 - \tau_2$, $\psi_2 = \tau_1 - \tau_3$, $\psi_3 = \tau_2 - \tau_3$ and $\psi_4 = \frac{1}{2}(\tau_1 + \tau_2) - \tau_3$ contain redundant information since $\psi_3 = \psi_2 - \psi_1$ and $\psi_4 = \frac{1}{2}(\psi_2 + \psi_3)$. Furthermore, the estimators for ψ_1 and ψ_2, namely $\bar{Y}_{1\cdot} - \bar{Y}_{2\cdot}$ and $\bar{Y}_{1\cdot} - \bar{Y}_{3\cdot}$, respectively, both involve $\bar{Y}_{1\cdot}$ and therefore are correlated. We now introduce a procedure for investigating a set of contrasts which contain no redundancies and whose estimators are uncorrelated. This procedure is based on the concept of **orthogonal contrasts.**

Definition 8.2.1 Two treatment contrasts $\psi_1 = \sum_{i=1}^{p} c_{1i}\tau_i$ and $\psi_2 = \sum_{i=1}^{p} c_{2i}\tau_i$ are said to be **orthogonal contrasts** if $\sum_{i=1}^{p} (c_{1i}c_{2i})/n_i = 0$. In the case of equal sample sizes, i.e., when $n_1 = n_2 = \cdots = n_p = r$, this condition reduces to $\sum_{i=1}^{p} c_{1i}c_{2i} = 0$.

Definition 8.2.2 A set of treatment contrasts $\psi_1, \psi_2, \ldots, \psi_q$ constitute a **mutually orthogonal set** if all pairs in the set are orthogonal.

In an experiment consisting of p treatments, one can find sets of $p - 1$ mutually orthogonal contrasts. To illustrate, in an equal sample size CRD experiment with $p = 4$, the set $\psi_1 = \tau_3 - \tau_1$, $\psi_2 = \tau_4 - \tau_2$ and $\psi_3 = \frac{1}{2}(\tau_1 + \tau_3) - \frac{1}{2}(\tau_2 + \tau_4)$ are mutually orthogonal contrasts. There are other sets of mutually orthogonal contrasts in this case. For example, $\psi_4 = \frac{1}{2}(\tau_2 + \tau_4) - \frac{1}{2}(\tau_1 + \tau_3)$, $\psi_5 = \frac{1}{2}(\tau_1 + \tau_2) - \frac{1}{2}(\tau_3 + \tau_4)$, $\psi_6 = \frac{1}{2}(\tau_1 + \tau_4) - \frac{1}{2}(\tau_2 + \tau_3)$ also constitute a set of mutually orthogonal contrasts when $p = 4$ and $n_1 = n_2 = n_3 = n_4$.

To further illustrate the conditions for contrasts to be mutually orthogonal, let us focus on ψ_1, ψ_2, and ψ_3 above in unequal sample size settings. If $n_1 = 4$, $n_2 = 5$, $n_3 = 4$, and $n_4 = 5$, the three contrasts again constitute a mutually orthogonal set. However, if $n_1 = 4$, $n_2 = 4$, $n_3 = 5$, and $n_4 = 5$, then ψ_1, ψ_2 and ψ_3 are not **mutually** orthogonal. The student should verify these conclusions.

Why are orthogonal contrasts of special interest? We next show that for a set of mutually orthogonal contrasts the normal linear model estimators are statistically independent. This essentially tells us that we are getting uncorrelated (non—redundant) information regarding these contrasts.

Theorem 8.2.1 Consider a set of $q \leq p - 1$ mutually orthogonal contrasts $\psi_1, \psi_2, \ldots, \psi_q$. For the normal CRD model the least squares estimators $\hat{\psi}_1, \hat{\psi}_2, \ldots, \hat{\psi}_q$ are statistically independent.

Proof. Let $\psi_j = \sum_{i=1}^{p} c_{ji} \tau_i$ for $j = 1, 2, \ldots, q$ be a set of mutually orthogonal contrasts. Then for all $j \neq k$ we have $\sum_{i=1}^{p} c_{ji} c_{ki}/n_i = 0$.

The least squares estimators are $\hat{\psi}_j = \sum_{i=1}^{p} c_{ji} \bar{Y}_{i.}$ for $j = 1, 2, \ldots, q$.

Since $\bar{Y}_{1.}, \bar{Y}_{2.}, \ldots, \bar{Y}_{p.}$ are distributed as multivariate normal, it follows that $\hat{\psi}_1, \hat{\psi}_2, \ldots, \hat{\psi}_q$ are also distributed as multivariate normal. From Section 4.5, to establish independence for $\hat{\psi}_1, \hat{\psi}_2, \ldots, \hat{\psi}_q$, we need only to show zero covariance. Now for $j \neq k$,

$$\begin{aligned}
\text{Cov}(\hat{\psi}_j, \hat{\psi}_k) &= \text{Cov}\left(\sum_{i=1}^{p} c_{ji} \bar{Y}_{i\cdot}, \sum_{i=1}^{p} c_{ki} \bar{Y}_{i\cdot}\right) \\
&= \sum_{i=1}^{p} c_{ji} c_{ki} \text{Var}(\bar{Y}_{i\cdot}) \\
&= \sum_{i=1}^{p} c_{ji} c_{ki} \sigma^2/n_i = 0.
\end{aligned}$$

Contrast Sum of Squares

In Section 7.6 we found that the regression sum of squares for the reparametrized CRD model is the same as the treatment sum of squares for the traditional p—treatment CRD model. This fact, together with the results of Theorem 6.5.1, yields an interesting relationship between the treatment sum of squares and the sums of squares associated with a set of $p-1$ mutually orthogonal contrasts. First we define what is meant by a **contrast sum of squares.**

Definition 8.2.3 Consider a treatment contrast $\psi = \sum_{i=1}^{p} c_i \tau_i$ which means $\sum_{i=1}^{p} c_i = 0$. The sum of squares for this contrast is defined by the formula

$$SS(\psi) = (\sum_{i=1}^{p} c_i \bar{y}_{i.})^2/(\sum_{i=1}^{p} c_i^2/n_i). \qquad (8.2.1)$$

The rationale for this definition comes from the F—test for testing $H_0 : \psi = 0$. This can be seen when we refer to the t—tests in the box identified by (7.5.2). If we square the test statistic for $H_0 : \sum_{i=1}^{p} c_i\tau_i = 0$, we obtain

$$f = \frac{(\sum_{i=1}^{p} c_i \bar{y}_{i.})^2}{s^2 \sum_{i=1}^{p} c_i^2/n_i} = \frac{(\sum_{i=1}^{p} c_i \bar{y}_{i.})^2/(\sum_{i=1}^{p} c_i^2/n_i)}{SSe/(n-p)} \qquad (8.2.2)$$

which is the value of an F random variable with 1 and $n-p$ degrees of freedom. It is therefore natural to think of the numerator of the above statistic as the sum of squares for contrast ψ with one degree of freedom.

Returning to the technique of using regression for the analysis of experimental CRD data, if we use a coding scheme for the $p-1$ independent variables corresponding to the c_i coefficients for $p-1$ mutually orthogonal contrasts, the resulting $\boldsymbol{X}$ matrix consists of mutually orthogonal columns putting us in the setting of Theorem 6.5.1. As a consequence, result (3) of this theorem tells us that

$$SStr = SS(\psi_1) + SS(\psi_2) + \cdots + SS(\psi_{p-1}) , \tag{8.2.3}$$

i.e., the treatment sum of squares for p treatments is partitioned into the sums of squares for $p - 1$ mutually orthogonal contrasts. The student will be asked to illustrate this property in the exercises for this chapter. This partitioning of the treatment sum of squares into $p - 1$ statistically independent single degree of freedom sums of squares can also be carried out using the CONTRAST command of the SAS GLM program and will be demonstrated in the forthcoming example. Although this partitioning of the treatment sum of squares has been discussed relative to the CRD model, it is important to note that it is also applicable to more complex experimental designs to be studied in later chapters. Also note that the number of contrasts needed for this partitioning is the same as the degrees of freedom for SStr.

An Example

A company that manufactures word processing equipment has developed four methods for training people to use their equipment. The methods are:

Method 1:	Reading training manuals only
Method 2:	Reading training manuals and watching a videotape
Method 3:	Reading training manuals and personalized instruction
Method 4:	Reading training manuals, watching a videotape and personalized instruction.

The company decided to perform a small study to compare the average effectiveness of these methods. Twenty—four office workers of comparable backgrounds were selected for the study. In a completely randomized design, the four training methods (treatments) were randomly assigned to the office workers (experimental units) so that six workers were assigned to each method. After the workers received their training they were each asked to perform ten different tasks on the same word processing system. Each worker had their overall effectiveness rated by the same supervisor. The ratings were obtained by scoring each task on a scale from 1 = poor to 4 = excellent. The overall rating was the total of the ten scores with the results given in the following table.

Table 8.2.1 Rating for Workers Method

1	2	3	4
18	17	38	37
30	22	24	36
24	29	38	33
22	26	32	30
23	18	37	36
25	33	33	28

Of particular interest in the study is the effect of personalized instruction on the rating. With this in mind, two obvious contrasts of interest are:

$$\psi_1 = \tau_3 - \tau_1 \quad \text{and} \quad \psi_2 = \tau_4 - \tau_2,$$

which are useful in measuring the effect of personalized instruction. These are orthogonal contrasts. If we consider a third contrast

$$\psi_3 = \frac{1}{2}(\tau_1 + \tau_3) - \frac{1}{2}(\tau_2 + \tau_4),$$

then ψ_1, ψ_2, ψ_3 form a set of three mutually orthogonal contrasts. We will demonstrate that $SS(\psi_1) + SS(\psi_2) + SS(\psi_3) = SStr$. From the SAS printout for these data given in Table 8.2.2 to follow, we see that SStr = 552.125.

From Table 8.2.1, we calculate $\bar{y}_{1\cdot} = 23.667$, $\bar{y}_{2\cdot} = 24.167$, $\bar{y}_{3\cdot} = 33.667$, and $\bar{y}_{4\cdot} = 33.333$ and note that $n_1 = n_2 = n_3 = n_4 = 6$. We also note that the coefficients for the three contrasts are $c_{11} = 1$, $c_{12} = 0$, $c_{13} = 1$, $c_{14} = 0$, $c_{21} = 0$, $c_{22} = -1$, $c_{23} = 0$, $c_{24} = 1$, $c_{31} = 0.5$, $c_{32} = -0.5$, $c_{33} = 0.5$, and $c_{34} = -0.5$. Using formula (8.2.1) we obtain $SS(\psi_1) = 300.0000$, $SS(\psi_2) = 252.08333$, and $SS(\psi_3) = 0.04167$. From these figures we see that

$$SS(\psi_1) + SS(\psi_2) + SS(\psi_3) = 552.125 = SStr.$$

Orthogonal Contrasts Using SAS

In Table 8.2.2 we demonstrate the use of the SAS GLM program in analyzing the data of Table 8.2.1 where, in particular, we seek the importance of personalized instruction in the training of office workers. The F VALUE associated with a contrast is an evaluation of formula

(8.2.2) and has 1 and 20 degrees of freedom.

Table 8.2.2
SAS Commands and Printout for Word Processing Training Data

SAS Commands

```
 OPTIONS LS = 80;
 DATA WORDP;
 INPUT METHOD RATING @@ ;
 CARDS;
1  18  1  30  1  24  1  22  1  23  1  25
2  17  2  22  2  29  2  26  2  18  2  33
3  38  3  24  3  38  3  32  3  37  3  33
4  37  4  36  4  33  4  30  4  36  4  28
PROC GLM;
CLASS METHOD;
MODEL RATING = METHOD;
CONTRAST '3 VS 1' METHOD -1 0 1 0 ;
CONTRAST '4 VS 2' METHOD 0 -1 0 1 ;
CONTRAST '1,3 VS 2,4' METHOD .5 -.5 .5 -.5 ;
```

SAS Printout

```
                     General Linear Models Procedure

Dependent Variable: RATING
                                   Sum of            Mean
Source                  DF        Squares          Square   F Value     Pr > F

Model                    3    552.12500000    184.04166667     7.53     0.0015

Error                   20    488.83333333     24.44166667

Corrected Total         23   1040.95833333

              R-Square           C.V.         Root MSE             RATING Mean

              0.530401       17.22096        4.9438514               28.708333


Source                  DF      Type I SS    Mean Square   F Value     Pr > F

METHOD                   3   552.12500000   184.04166667      7.53     0.0015

Source                  DF    Type III SS    Mean Square   F Value     Pr > F

METHOD                   3   552.12500000   184.04166667      7.53     0.0015

Contrast                DF    Contrast SS    Mean Square   F Value     Pr > F

3 VS 1                   1   300.00000000   300.00000000     12.27     0.0022
4 VS 2                   1   252.08333333   252.08333333     10.31     0.0044
1,3 VS 2,4               1     0.04166667     0.04166667      0.00     0.9675
```

To summarize our findings concerning the Word Processing Training Data, we can display the results in an extended analysis of variance table, which includes the auxiliary analyses using the mutually orthogonal contrasts approach. This is given in Table 8.2.3.

Table 8.2.3 ANOVA for Word Processing Training Data

Source	df	SS	MS	f
Methods	3	552.1250	184.0417	7.53
3 vs. 1	1	300.0000	300.0000	12.27
4 vs. 2	1	252.0833	252.0833	10.31
1, 3 vs. 2, 4	1	0.0417	0.0417	< 0.01
Error	20	488.8333	24.4417	
Total	23	1040.9583		

Comparing $f = 7.53$ with the critical value $f_{.01,3,20} = 4.94$, we conclude highly significant differences exist in the effectiveness of some of the four training methods. These differences can be attributed to the increase in effectiveness when the method includes personalized instruction versus the omission of personalized instruction, i.e., Method 3 vs. Method 1 and Method 4 versus Method 2. This is obvious when we compare $f_{.01,1,20} = 8.10$ with the f—value for these contrasts, 12.27 and 10.31, respectively. On the other hand, the difference in effectiveness in the average of the two methods excluding and including watching videotapes (1, 3, versus 2, 4) is insignificant. If we desire to **estimate** the average rating increase which results with the inclusion of personalized instruction, we only need to calculate $\bar{y}_{3\cdot} - \bar{y}_{1\cdot} = 33.667 - 23.667 = 10.000$ and $\bar{y}_{4\cdot} - \bar{y}_{2\cdot} = 33.333 - 24.167 = 9.167$.

In this section we have shown how one can make use of a method for partitioning the treatment sum of squares having $p - 1$ degrees of freedom into $p - 1$ meaningful single degree of freedom sums of squares which can then be used to test supplementary hypotheses of particular interest. The reader may recall that this same method was used in Chapter 6 in order to partition the regression sum of squares into single degree of freedom polynomial sums of squares for the purpose of response analysis. In the next section we will illustrate the use of this idea in the experimental design setting.

8.3 Nature of Response for Quantitative Factors

When a treatment factor in an experiment is of the quantitative type, the experimenter sometimes finds it useful to fit a response curve in order to determine the mathematical nature of the response to levels of the factor. In the case of equally spaced and equally replicated levels, the method of orthogonal polynomials can conveniently be used in order to partition the treatment sum of squares into single degree of freedom sums of squares in the manner described in the last section. This allows one to examine the data for indications of a statistical trend, the nature of which can provide useful information to the investigator. For example, it may be helpful to know whether or not the trend is linear or quadratic or cubic or a combination of these. Beyond this assessment, if desired, the experimenter can produce an equation which relates the response to treatment levels for the purpose of prediction or seeking out an optimal treatment level.

The method of orthogonal polynomials is dealt with in Chapter 6. There we learned the mathematical basis for the development of the table of orthogonal polynomial coefficients, Table 4 of Appendix B, and how it can be used through illustrations and exercises. We also found out "why the method works." With that background in orthogonal polynomial regression to call upon, in this section we will illustrate its usefulness from the viewpoint of experimental design analysis. First we review the formula needed for the partitioning of the treatment sum of squares in the analysis of variance and relate it to the ideas discussed in the previous section.

Recall from Section 6.6, our problem is to fit a kth degree orthogonal polynomial model of the form

$$Y_i = \alpha_0 + \alpha_1 p_1(x_i) + \alpha_2 p_2(x_i) + \cdots + \alpha_k p_k(x_i) + \varepsilon_i \tag{8.3.1}$$

to a set of n data points where $p_j(x_i)$ is a polynomial of degree $j = 1, 2, \ldots, k$. In order to use Table 4, the levels of the independent variable must be equally spaced and the number of observations at each of the levels must be the same. In applying the orthogonal polynomial technique to an experiment consisting of p treatments, we will let the index i in (8.3.1) correspond to the ith treatment so that $i = 1, 2, \ldots, p$, and the number of replications for each treatment is r such that $rp = n$. Also, in this case the number k in (8.3.1) necessarily is $p - 1$. In the previous section the contrast sum of squares formula, (8.2.1), is given in terms of the ith sample mean, $\bar{y}_{i\cdot}$. Here, because of the equal replication of treatment levels, it is convenient to express the contrast sums of squares in terms of treatment totals, $y_{i\cdot}$, for $i = 1, 2,...,p$. Therefore, in using orthogonal polynomials applied to experimental design data analysis, the contrast sum of squares (6.6.1) takes on the form

$$SS(\alpha_j) = \frac{\left[\sum_{i=1}^{p} p_j(x_i) y_{i.}\right]^2}{r \sum_{i=1}^{p} p_j^2(x_i)}, \; j = 1, 2,...,p - 1, \qquad (8.3.2)$$

where $p_j(x_i)$ is the set of p orthogonal polynomial coefficients associated with the ith treatment and jth degree term as found in Table 4. In Chapter 6 we referred to $SS(\alpha_j)$ as the "jth coefficient sum of squares" used to test the null hypothesis of "no jth polynomial effect." For the purpose of calculating the **contrast sum of squares**, $SS(\alpha_j)$, in steps, we will call the expression $\sum_{i=1}^{p} p_j(x_i) y_{i\cdot}$ the **contrast total** and the expression $r \sum_{i=1}^{p} p_j^2(x_i)$ the **contrast divisor**. As noted previously, in this orthogonal setting,

$$SStr = SS(\alpha_1) + SS(\alpha_2) + \cdots + SS(\alpha_{p-1})$$

and

$$f = \frac{SS(\alpha_j)}{MSe}, \quad j = 1, 2,...,p - 1$$

is a value of an F random variable with 1 and $n - p$ degrees of freedom. We now demonstrate the application of orthogonal polynomials in analyzing data obtained in a designed experiment through a series of examples.

Example 8.3.1 Crop fertilizer is made up of three main nutrients: nitrogen, phosphorus and potassium. Of the three, nitrogen is the key ingredient for crop yield. In a study on the cost effectiveness of nitrogen on the yield of corn, various amounts of fertilizer were applied to contiguous acreage following a growing season of alfalfa. See the data in Table 8.3.1. A cursory examination of these data reveals that yield definitely increases with the rate of nitrogen applied but it is suspected that a point of diminishing returns is reached when cost versus sale price is taken into account, i.e., at what point does an additional level of nitrogen provide no economic benefit?

Table 8.3.1 Yields of Corn in Bushels per Acre
Level of Nitrogen in Pounds per Acre

None	40	80	120
116	140	162	160
121	145	151	171
126	136	158	175
112	148	164	168
114	152	167	168

It is a simple matter to calculate the total sum of squares and the treatment sum of squares (with 3 degrees of freedom) as

$$\text{SSt} = 8044.2 \quad \text{and} \quad \text{SStr} = 7480.2 \; .$$

In an effort to answer the research question above, we fit a polynomial of degree 3 to these data. Because the levels of nitrogen are equally spaced and equally replicated, we proceed using the method of orthogonal polynomials. Our first task is to partition SStr into polynomial effects sums of squares. To do this, one can organize the computing of formula (8.3.2) in steps as shown in Table 8.3.2. The orthogonal contrast coefficients come from Appendix Table 4. (**Note:** See Section 6.5 for the step—by—step calculation of these coefficients.)

Table 8.3.2 Calculation of Polynomial Effects Sums of Squares

Nitrogen level	**Treatment totals**	**Orthogonal Contrast Coefficients, $p_j\ (x_i)$**		
		Linear	**Quadratic**	**Cubic**
None	589	–3	1	–1
40	721	–1	–1	3
80	802	1	–1	–3
120	842	3	1	1
Contrast totals		840	–92	10
Contrast divisors		100	20	100
Contrast sums of squares		7056.0	423.2	1.0

We put this information in an extended analysis of variance table as follows:

Table 8.3.3 ANOVA for Corn Yield Data

Source	df	SS	MS	f
Nitrogen levels	3	7480.2	2493.4	70.73
Linear	1	7056.0	7056.0	200.17
Quadratic	1	423.2	423.2	12.01
Cubic	1	1.0	1.0	0.03
Error	16	564.0	35.25	
Total	19	8044.2		

On comparing the calculated values for f with the critical value $f_{.01,1,16} = 8.53$, we see that both linear and quadratic trends are highly significant. This means that although the corn yield increases with an increase in nitrogen level, it does so at a decreasing rate. In the exercises the student will be asked to respond to the economic question posed in this example as indicated by the data given in Table 8.3.1.

Examples Using SAS

Rather than calculating the contrast sums of squares in the previous example with paper, pencil, and hand calculator, the user of SAS finds it a simple matter to duplicate the analysis presented in Tables 8.3.2 and 8.3.3. We now illustrate.

Example 8.3.2 In this example we reproduce the analysis of variance as given in Table 8.3.3 using the SAS GLM program procedure. The demonstrated commands yield the ANOVA lines after rearranging the printout information.

Table 8.3.4 The ANOVA for Corn Yield Data Using SAS GLM

SAS COMMANDS

```
OPTIONS LS = 80;
DATA CYIELD;
INPUT NITROGEN YIELD @@;
CARDS;
1  116   1  121   1  126   1  112
2  140   2  145   2  136   2  148
3  162   3  151   3  158   3  164
4  160   4  171   4  175   4  168
PROC GLM;
CLASS NITROGEN;
MODEL YIELD = NITROGEN
CONTRAST 'LINEAR' NITROGEN -3 -1 1 3;
CONTRAST 'QUADRATIC' NITROGEN 1 -1 -1 1;
CONTRAST 'CUBIC' NITROGEN -1 3 -3 1;
```

SAS Printout

```
                General Linear Models Procedure
                   Class Level Information

                  Class     Levels     Values

                  NITROGEN       4     1 2 3 4

             Number of observations in data set = 20


                General Linear Models Procedure

Dependent Variable: YIELD
                                    Sum of            Mean
Source                   DF        Squares          Square   F Value     Pr > F

Model                     3   7480.2000000    2493.4000000     70.73     0.0001

Error                    16    564.0000000      35.2500000

Corrected Total          19   8044.2000000

                   R-Square           C.V.        Root MSE            YIELD Mean

                   0.929887       4.019750       5.9371710             147.70000

Source                   DF      Type I SS     Mean Square   F Value     Pr > F

NITROGEN                  3   7480.2000000    2493.4000000     70.73     0.0001

Source                   DF    Type III SS     Mean Square   F Value     Pr > F

NITROGEN                  3   7480.2000000    2493.4000000     70.73     0.0001

Contrast                 DF    Contrast SS     Mean Square   F Value     Pr > F

LINEAR                    1   7056.0000000    7056.0000000    200.17     0.0001
QUADRATIC                 1    423.2000000     423.2000000     12.01     0.0032
CUBIC                     1      1.0000000       1.0000000      0.03     0.8684
```

With reference to Table 8.3.3, the above printout provides us with the ANOVA line for nitrogen levels under the row heading MODEL and the printout ERROR corresponds to the ANOVA error line. For the partitioning into linear, quadratic, and cubic effects sums of squares we look to the contrasts so named in the printout. The orthogonal polynomial coefficients programmed in the SAS commands are those given in Table 4 and used in Table 8.3.2.

The reader should note that the discussion and the illustrations up to now feature the use of a table of orthogonal polynomial coefficients. However, we are reminded that such a table is applicable only when the set of quantitative data being analyzed is equally spaced and equally replicated. If this is not the case, a **direct** polynomial regression analysis is "the easy way to go." The use of an ordinary regression program to facilitate the analysis can be found in Section 6.6. There we find an illustrative outline for the procedure of fitting successively polynomials of degree one through degree $p - 1$. It was also noted that if the SAS GLM program is available, the user need only fit a polynomial of degree $p - 1$ to the p treatment levels. This we demonstrate in the next example.

Example 8.3.3 Although data in Table 8.3.1 are equally spaced and equally replicated, we use regression here to duplicate the analyses displayed in the previous tables of this section. In doing so, the reader should appreciate regression analysis to be an "all purpose" procedure whether or not the conditions for use of an orthogonal polynomials table are met. Since our data consist of $p = 4$ treatment levels, we will fit a third degree polynomial to these data using the SAS GLM program.

Table 8.3.5

The ANOVA for Corn Yield Data Using Direct Polynomial Regression

SAS COMMANDS

```
OPTIONS LS = 80;
DATA CYIELD;
INPUT NITROGEN YIELD @@;
XSQ = NITROGEN ** 2;
XCB = NITROGEN ** 3;
CARDS;
0    116  0    121  0    126  0    112  0    114
40   140  40   145  40   136  40   148  40   152
80   162  80   151  80   158  80   164  80   167
120  160  120  171  120  175  120  168  120  168
PROC GLM;
MODEL YIELD = NITROGEN XSQ XCB;
```

SAS Printout

General Linear Models Procedure

Dependent Variable: YIELD

Source	DF	Sum of Squares	Mean Square	F Value	Pr > F
Model	3	7480.2000000	2493.4000000	70.73	0.0001
Error	16	564.0000000	35.2500000		
Corrected Total	19	8044.2000000			

R-Square	C.V.	Root MSE	YIELD Mean
0.929887	4.019750	5.9371710	147.70000

Source	DF	Type I SS	Mean Square	F Value	Pr > F
NITROGEN	1	7056.0000000	7056.0000000	200.17	0.0001
XSQ	1	423.2000000	423.2000000	12.01	0.0032
XCB	1	1.0000000	1.0000000	0.03	0.8684

Source	DF	Type III SS	Mean Square	F Value	Pr > F
NITROGEN	1	351.40566038	351.40566038	9.97	0.0061
XSQ	1	16.17826087	16.17826087	0.46	0.5078
XCB	1	1.00000000	1.00000000	0.03	0.8684

Parameter	Estimate	T for H0: Parameter=0	Pr > |T|	Std Error of Estimate
INTERCEPT	117.8000000	44.37	0.0001	2.65518361
NITROGEN	0.8041667	3.16	0.0061	0.25469549
XSQ	-0.0038125	-0.68	0.5078	0.00562760
XCB	0.0000052	0.17	0.8684	0.00003092

The partitioning of the treatment sum of squares into polynomial effects sums of squares appears in the above printout under the heading TYPE I SS. As in Example 8.3.2, the printout information can be rearranged to produce the ANOVA format labeled Table 8.3.3.

8.4 Error Levels and Bonferroni Procedure

Overall Error Rates and Confidence Levels

Suppose that an experimenter wishes to make more than one comparison of treatment effects using t—tests. These t—tests are each valid at the α risk level only if the decision to make the particular tests is made prior to the running of the experiment. Though each of these tests may have an α significance level individually, the probability of committing at least one type I error, i.e., falsely rejecting the null hypothesis on any of the tests collectively, may be greater than α.

As an illustration, suppose we perform four **independent** tests, each at level α and we let A_i be the event that the ith null hypothesis is accepted (not rejected), $i = 1, 2, 3, 4$. When each null hypothesis is true, the probability of making **at least one type I error** in the set of four tests is

$$\begin{aligned} &1 - P\text{ (no type I error in any of the four tests)} \\ &= 1 - P(A_1 \cap A_2 \cap A_3 \cap A_4) \\ &= 1 - P(A_1)P(A_2)P(A_3)P(A_4) \\ &= 1 - (1 - \alpha)^4. \end{aligned}$$

In general, if c **independent** tests are performed each at level α and each null hypothesis is true, then

$$P\text{ (at least one type I error in the c tests)} = 1 - (1 - \alpha)^c.$$

To gain an appreciation of what this overall level for a family of c tests implies, Table 8.4.1 displays an evaluation of this probability for various values of c when $\alpha = 0.05$.

Table 8.4.1 Overall Type I Error Levels, $\alpha = 0.05$

c	4	6	10	20
P(at least one type I error)	0.19	0.26	0.40	0.64

Although the tests in a designed experiment that are of interest to investigators are usually not independent, the expression $1 - (1 - \alpha)^c$ can be used to approximate the error levels associated with a family of c dependent tests. Since these overall type I error levels can be quite large (much larger than α), it behooves us to consider procedures that will control them to acceptable levels.

First it is necessary for us to distinguish between three different kinds of error levels, namely, the **experimentwise** error level, the **per comparison** error level, and the **overall** type I error level. Consider a family of tests for c distinct null hypotheses. The experimentwise error level is calculated under the **assumption that each null hypothesis is true**. With this assumption, the **experimentwise error level** is the probability of rejecting at least one null hypothesis. The per comparison error level is based on a test for a specific null hypothesis. The **per comparison error level** for the test of the i th hypothesis is the probability of committing a type I error (falsely rejecting the null hypothesis) with this particular test. The **overall type I error level** is the probability of falsely rejecting at least one null hypothesis among the set of c hypotheses. That is, the overall type I error level is the probability of making at least one type I error in the collection of c tests.

To further illustrate these three error levels consider a family of tests for c null hypotheses. As before, let A_i represent the event that hypothesis i is accepted, and A_i' represent the event that the hypothesis is rejected. The experimentwise error level is

$$P\text{ (at least one test falsely rejects)} = 1 - P(A_1 \cap \cdots \cap A_c),$$

calculated under the assumption that each null hypothesis is true. The per comparison error level for the ith comparison is simply $P(A_i')$, calculated under the assumption that the ith null hypothesis is true. Note that while the experimentwise error level is a single number, the per comparison error level might vary from comparison to comparison. Finally, the overall type I error level is P (at least one test falsely rejects) and is calculated in general situations where some null hypotheses are true and others are false. For example, if null hypotheses 1, 2, 3, are true and hypotheses 4 through c are false, then the overall type I error level is P (at least one test falsely rejects) $= 1 - P(A_1 \cap A_2 \cap A_3)$. Note that the overall type I error level and the experimentwise error level are the same for problems in which all c null hypotheses are true.

If our inference objective is interval estimation instead of statistical tests, we can think about a family of c confidence intervals in the same manner. This time let A_i represent the event that the ith confidence interval is correct, i.e., the true parameter value lies inside the confidence interval for $i = 1, 2, \ldots, c$. Then the expression $P(A_1 \cap A_2 \cap \cdots \cap A_c)$ is called the **overall confidence level** or **simultaneous confidence level** and represents the probability that all c confidence intervals contain their true respective parameter values. If the c intervals are independent, each with an individual confidence level of $P(A_i) = 1 - \alpha$, then the simultaneous confidence level will be

$$P\text{ (all } c \text{ confidence intervals are correct)} = (1 - \alpha)^c.$$

We sometimes refer to the number $(1 - \alpha)^c$ as the **family or simultaneous confidence coefficient**.

The previous discussion concerning overall error rates and overall confidence levels was given in terms of independent tests and intervals.

We now turn our attention to a procedure for controlling these overall levels in the more common case when the families of tests and intervals are **not** independent. More procedures of this kind are presented in the next chapter.

Bonferroni Procedure

The Bonferroni procedure for constructing a set of simultaneous confidence intervals is based on the Bonferroni Inequality stated next as a theorem.

Theorem 8.4.1 Let $A_1, A_2, \ldots, A_c$ represent c events. Then

$$P(A_1 \cap A_2 \cap \cdots \cap A_c) \geq 1 - \sum_{i=1}^{c} P(A_i')$$

Proof. From de Morgan's law we know that

$$(A_1 \cap A_2 \cap \cdots \cap A_c)' = A_1' \cup A_2' \cup \cdots \cup A_c'$$

Also, $$P(A_1' \cup A_2' \cup \cdots \cup A_c') \leq \sum_{i=1}^{c} P(A_i').$$ Hence

$$\begin{aligned} P(A_1 \cap A_2 \cap \cdots \cap A_c) &= 1 - P[(A_1 \cap A_2 \cap \cdots \cap A_c)'] \\ &= 1 - P(A_1' \cup A_2' \cup \cdots \cup A_c') \\ &\geq 1 - \sum_{i=1}^{c} P(A_i'). \end{aligned}$$

Utilizing the Bonferroni Inequality we can obtain a set of c confidence intervals with associated simultaneous confidence coefficient of $1 - \alpha$ according to the following argument. As before, let A_i in Theorem 8.4.1 be the event that the ith confidence interval contains the true parameter value, such that $P(A_i) = 1 - \alpha/c$ and $P(A_i') = \alpha/c$ for $i = 1, 2, \ldots, c$. But by the Bonferroni Inequality,

$$P(A_1 \cap A_2 \cap \cdots \cap A_c) \geq 1 - \sum_{i=1}^{c} P(A_i') = 1 - \sum_{i=1}^{c} \alpha/c = 1 - \alpha. \quad (8.4.1)$$

We note that the simultaneous confidence coefficient $1 - \alpha$ in this instance is a lower bound. It follows from (8.4.1) that we can control the overall confidence level for a set of c simultaneous intervals to at least 100 $(1-\alpha)$% by using the $100(1-\alpha/c)$% level in the construction of each of the c individual intervals. Applying this result to equation (5.3.6), we obtain the set of Bonferroni simultaneous confidence intervals which we summarize below for future reference.

<u>Bonferonni Simultaneous Confidence Intervals</u>

Let $\psi_1, \psi_2, \ldots, \psi_c$ be c estimable functions for the normal linear model with estimates $\underline{\ell}_1' \underline{b}$, $\underline{\ell}_2' \underline{b}, \ldots, \underline{\ell}_c' \underline{b}$.

The Bonferroni simultaneous confidence intervals with overall confidence level of at least $100(1-\alpha)$% are

$$\underline{\ell}_i' \underline{b} \pm t_{\alpha/2c,n-r} \, Se(\underline{\ell}_i' \hat{\underline{\beta}}) \quad , \quad i = 1, 2, \ldots, c. \tag{8.4.2}$$

Application to the CRD

To discover how Bonferroni simultaneous confidence intervals (SCI) can be used in the analysis of the results of a designed experiment, consider the CRD with model

$$Y_{ij} = \mu + \tau_i + \varepsilon_{ij} \ , \quad i = 1, 2, \ldots, p \ ; \ j = 1, 2, \ldots, n_i \ ,$$

where ε is $MN(\mathbf{0}, \sigma^2\mathbf{I})$. Suppose we are interested in finding SCI for **all** possible pairwise differences between treatment effects, namely $\tau_i - \tau_k$ for all $1 \leq i < k \leq p$. For these $\binom{p}{2}$ pairwise differences, the point estimates are $\bar{y}_{i\cdot} - \bar{y}_{k\cdot}$, and the standard errors are $s\sqrt{\frac{1}{n_i} + \frac{1}{n_k}}$. Therefore, by (8.4.2) the Bonferroni SCI for pairwise treatment differences in a CRD experiment are

$$(\bar{y}_{i\cdot} - \bar{y}_{k\cdot}) \pm t_{\alpha/2c,n-p} \, s\sqrt{\frac{1}{n_i} + \frac{1}{n_k}} \ , \ 1 \leq i < k \leq p \ , \tag{8.4.3}$$

where c is the number of comparisons.

We can use the Bonferroni SCI for testing purposes. For example, in a CRD experiment suppose that we want to carry out tests for all pairwise differences in treatment means (effects), that is, we want to test the $c = \binom{p}{2}$ hypotheses of the form $H_0 : \tau_i = \tau_k$, where $1 \leq i < k \leq p$. By using the Bonferroni SCI (8.4.3), we can reject $H_0 : \tau_i = \tau_k$ if zero is not in the SCI for $\tau_i - \tau_k$, and fail to reject H_0 if zero is in the interval. This is equivalent to rejecting H_0 if

$$|\bar{y}_{i.} - \bar{y}_{k.}| \geq t_{\alpha/2c,n-p}\, s\sqrt{\frac{1}{n_i} + \frac{1}{n_k}}, \quad c = \binom{p}{2}. \tag{8.4.4}$$

We will refer to this procedure as the **Bonferroni multiple comparison procedure** for the CRD. In the exercises it is shown that the overall type I error level for this procedure is less than or equal to α.

Values for $t_{\alpha/2c}$ where $c = \binom{p}{2}$ are not found in standard t–tables familiar to the reader. For this reason we have provided Table 5 in Appendix B which contains many of these values.

Example 8.4.1 Refer to the daphnia data given in Table 7.2.1 and analyzed in Example 7.4.1. Some summary information for these data from a CRD is $p = 3$, $n_1 = 4$, $n_2 = n_3 = 5$, $n = 14$, $\bar{y}_{1\cdot} = 76.5$, $\bar{y}_{2\cdot} = 63.0$, $\bar{y}_{3\cdot} = 39.0$, $s = 11.611$. Let us find the Bonferroni SCI for the pairwise treatment differences with overall confidence level of at least 95%. With 11 degrees of freedom for error and $c = \binom{3}{2} = 3$, interpolation in Table 5 yields $t_{.05/6,11} = 2.83$. Therefore, in order to calculate the three SCI from (8.4.3), we use

$$(\bar{y}_{i\cdot} - \bar{y}_{k\cdot}) \pm 2.83\,(11.611)\sqrt{\frac{1}{n_i} + \frac{1}{n_k}}$$

with the results as follows:

Treatment Differences	Simultaneous Confidence Intervals		
$\tau_1 - \tau_2$	13.5 ± 22.04	or	(–8.54, 35.54)
$\tau_1 - \tau_3$	37.5 ± 22.04	or	(15.46, 59.54)
$\tau_2 - \tau_3$	24.0 ± 20.78	or	(3.22, 44.78)

If we use these intervals to determine which treatment effects are different, we readily see that treatments 1 and 3 as well as treatments 2 and 3 are significantly different. Alternately, we reach the same conclusions using the **critical difference** approach (8.4.4) since $13.5 < 22.04$, $37.5 > 22.04$ and $24.0 > 20.78$.

Traditionally, a table has been used in reporting significant differences between treatment effects. This practice calls for listing the sample treatment means in ascending (or descending) numerical order and the means that are **not** significantly different are then "joined" by an underlining segment. For the daphnia data, the table would appear as follows:

$$\begin{array}{lccc} \text{Treatments} & 3 & 2 & 1 \\ \text{Means} & 39.0 & \multicolumn{2}{c}{\underline{\;63.0 \qquad\quad 76.5\;}} \end{array}$$

The reader perhaps has observed that the widths of the Bonferroni simultaneous intervals in Example 8.4.1 are large. For example, the width of the Bonferroni SCI for $\tau_1 - \tau_2$ is $2(22.04) = 44.08$ units. This compares with a width of 34.28 units for the **individual** 95% confidence interval using (7.5.1). (See Exercise 8—7.) Since the SCI uses the tabled t—value $t_{\alpha/2c}$, $c = \begin{bmatrix} p \\ 2 \end{bmatrix}$, and the individual CI uses $t_{\alpha/2}$, it is clear that the SCI will always be wider than that for the individual CI. This is essentially the price that is being paid to control the overall type I error level.

As noted, the value of $t_{\alpha/2c}$ is a contributing factor to the width of a confidence interval and that its value increases as c, the number of comparisons, increases. Therefore, if one attempts to find SCI for too many comparisons, the interval widths in the set may be too wide to be of any practical use. For example, in a CRD with $p = 10$ treatments, there are a total of 45 pairwise treatment differences to consider and the number $t_{\alpha/90}$ could be so large as to produce rather meaningless intervals. The moral is that the user needs to be careful when selecting the number of comparisons to be considered in an SCI procedure.

In some cases we may be interested in only several of the $\begin{bmatrix} p \\ 2 \end{bmatrix}$ pairwise comparisons of the treatment means. For example, in the daphnia data of Example 8.4.1 we might only be interested in comparing treatment 1 versus 2, and treatment 1 versus 3. This would suggest testing $H_0 : \tau_1 = \tau_2$ and $H_0 : \tau_1 = \tau_3$. In this specific case we would use $c = 2 < \binom{p}{2}$ to find our $t_{\alpha/2c}$ value from the table for use in the simultaneous confidence intervals (SCI). Since $t_{\alpha/2c}$ decreases as c decreases, the length of the SCI will be shorter when a smaller c is used. As a consequence, it is best for a user to plan in advance which treatment comparisons are of interest and to find SCI (or multiple tests) only for these interesting comparisons. The following example features planned comparisons using computer output to implement the Bonferroni procedure.

A Bonferroni Example Using SAS

For this example refer to the experiment discussed earlier in Exercise 7—6 where a company examines the results obtained from processing pickles in four chemical solutions. The first solution, S1, is currently being used by the company but it is not biodegradable while alternative solutions S2, S3, and S4 possess this desirable property. The data set, as presented in Exercise 7—6 and repeated as part of the SAS COMMAND section of Table 8.4.2 to follow, consists of the number of bloater pickles in 40 distinct processing units, 10 for each solution. We recognize this experimental set—up as that for a completely randomized design with four treatments. (Here the experimental units are the processing units and the four solutions are the treatments.) To analyze these data, the appropriate model therefore is

$$Y_{ij} = \mu + \tau_i + \varepsilon_{ij}, \quad i = 1, 2, 3, 4; \quad j = 1, 2, \ldots, 10,$$

where $\boldsymbol{\varepsilon}$ is $MN(\mathbf{0}, \sigma^2\mathbf{I})$ and treatment i corresponds to solution i.

A particular objective of this experiment is to make a comparison of the standard solution, S1, with the alternative solutions S2, S3, and S4, i.e., planned comparisons $\tau_2 - \tau_1$, $\tau_3 - \tau_1$, and $\tau_4 - \tau_1$ are of interest. We will use the Bonferroni multiple comparison procedure for testing these three planned comparisons with an overall error level of 0.05 or less. Specifically, using (8.4.4) with $c = 3$ rather than $\binom{4}{2} = 6$, we will reject $H_0 : \tau_k = \tau_1$ if

$$|\bar{y}_{k.} - \bar{y}_{1.}| \geq t_{.05/6,36}\, s\sqrt{\frac{1}{n_k} + \frac{1}{n_1}}$$

or

$$\frac{|\bar{y}_{k.} - \bar{y}_{1.}|}{s\sqrt{\frac{1}{n_k} + \frac{1}{n_1}}} \geq t_{.05/6,36} = 2.516$$

for $k = 2, 3, 4$. The left side of the second inequality above is recognized to be the absolute value of the usual t—statistic given in Equation (7.5.2) for testing $H_0 : \tau_k = \tau_1$ versus $H_1 : \tau_k \neq \tau_1$.

The SAS printout given in Table 8.4.2 shows an analysis of the bloater pickle data and, in particular, contains the t—tests for the three planned comparisons. Since the t—values for testing $H_0 : \tau_2 = \tau_1$ and $H_0 : \tau_4 = \tau_1$ exceed 2.516, we conclude that chemical solutions two and four have significantly different mean number of bloater pickles than the standard solution. Note that the mean for S3 is not significantly different than the mean for S1.

An alternate way to carry out the three Bonferroni tests is to take advantage of the p—values shown on the printout. The Bonferroni multiple comparisons procedure for testing c two—sided tests is equivalent to rejecting H_0 if the p—value is α/c or less. (See Exercise 8—14.) In the pickle problem just discussed, we would reject any of the null hypotheses for which the p—value $\leq 0.05/3 = 0.0167$. One can easily determine that we obtain the same results for the set of three tests using this p—value criteria.

In this chapter we have considered post analysis of variance f—test procedures for carrying out treatment comparisons planned in advance of the experimental results. In the next chapter, post f—test multiple comparison procedures are studied where the selected set of comparisons may not have been planned beforehand but rather have been dictated by the results of the experiment.

Table 8.4.2 SAS Commands and Printout for Pickle Data

<u>SAS COMMANDS</u>

```
OPTIONS LS = 80;
DATA PICKLE;
INPUT SOLUTION NBLOATS @@ ;
CARDS;
1 2    1 6   1 7    1 10   1 5    1 6    1 3    1 1    1 4    1 5
2 16   2 7   2 8    2 20   2 16   2 14   2 13   2 19   2 21   2 9
3 4    3 3   3 10   3 2    3 11   3 4    3 7    3 8    3 11   3 2
4 21   4 4   4 13   4 19   4 21   4 22   4 14   4 19   4 20   4 21
PROC GLM;
CLASS SOLUTION;
MODEL NBLOATS = SOLUTION;
ESTIMATE 'TRT 2 VS TRT 1' SOLUTION -1  1  0  0 ;
ESTIMATE 'TRT 3 VS TRT 1' SOLUTION -1  0  1  0 ;
ESTIMATE 'TRT 4 VS TRT 1' SOLUTION -1  0  0  1 ;
```

SAS Printout

```
                         General Linear Models Procedure

Dependent Variable: NBLOATS
                                     Sum of               Mean
Source                     DF       Squares             Square    F Value      Pr > F

Model                       3    1484.9000000        494.9666667     38.82      0.0001

Error                      36     459.0000000         12.7500000

Corrected Total            39    1943.9000000

                     R-Square            C.V.           Root MSE          NBLOATS Mean

                     0.763877        31.18528          3.5707142             11.450000

Source                     DF       Type I SS        Mean Square    F Value      Pr > F

SOLUTION                    3    1484.9000000        494.9666667      38.82      0.0001

Source                     DF     Type III SS        Mean Square    F Value      Pr > F

SOLUTION                    3    1484.9000000        494.9666667      38.82      0.0001

                                         T for H0:     Pr > |T|     Std Error of
Parameter                   Estimate    Parameter=0                  Estimate

TRT 2 VS TRT 1            10.4000000           6.51      0.0001       1.59687194
TRT 3 VS TRT 1             1.3000000           0.81      0.4209       1.59687194
TRT 4 VS TRT 1            14.5000000           9.08      0.0001       1.59687194
```

Chapter 8 Exercises

Application Exercises

8–1 Refer to the pain threshold data given in Table 7.10.1 and the analysis of variance summary in Table 7.10.2. In a study such as this, the "natural" goal is to compare the thresholds for the two types of blonds, the two types of brunettes, and blonds versus brunettes. Make these three planned comparisons by partitioning the treatment sum of squares into single degree of freedom sums of squares. Observe that the calculated value of the F–statistic for comparing blonds with brunettes is the square of the t–value obtained in Section 7.10 for testing the null hypothesis that the average pain thresholds for blonds and brunettes are the same.

8–2 Individuals who engage in sporatic, strenuous physical activity often experience muscle soreness for 48 hours or more after such activity, known as delayed onset muscle soreness (DMOS). In an experiment to determine the effect of massage on the signs of DMOS as measured by deficits in maximal voluntary contraction (MVC), 10 subjects were tested for baseline MVC. Then each subject completed a bout of high intensity exercises followed by a controlled massage treatment. Twenty–four hours later the same subjects were retested for MVC after which massage again was given. At 48 hours after the high intensity exercises the subjects were tested for MVC again. The three MVC readings as measured in ft.–lbs. for each subject recorded at 0, 24, and 48 hours are given in the following table.

0	24	48
35	32	32
30	21	24
36	24	20
24	22	22
30	25	26
26	15	15
26	13	13
36	27	25
28	20	23
31	19	23

Calculate the analysis of variance table for these data, partitioning the treatment sum of squares into linear and quadratic effects. Interpret the results.

8–3 Consider the data in Example 8.3.1. Use the information in Table 8.3.2 together with Section 6.5 to calculate the estimating function $\hat{y} = a_0 + a_1 p_1(x) + a_2 p_2(x) + a_3 p_3(x)$ and then express this function in the form $\hat{y} = b_0 + b_1 x + b_2 x^2 + b_3 x^3$ where x is the level of nitrogen used, x = 0, 40, 80, 120. This final cubic equation should agree with that shown in the computer printout for Example 8.3.3.

8–4 Consider the objective of the corn growing experiment discussed in Example 8.3.1. Assume that the cost of fertilizer is $0.25 per pound and that the sale price of corn will be $2.80 per bushel.

Based upon the estimating function $\hat{y} = a_0 + a_1p_1(x) + a_2p_2(x) + a_3p_3(x)$ found in Exercise 8–3, seek out the optimum level of nitrogen to the nearest 10 pounds per acre after which the cost of additional fertilizer exceeds the additional income generated through the use of additional nitrogen. (Note: Extrapolate beyond the nitrogen levels used in the experiment, if necessary.)

8–5 Refer to the data given in Table 7.2.1 and the summary analysis of variance as displayed in Table 7.4.2. Extend the ANOVA table by partitioning the treatment sum of squares into linear and quadratic effects.

8–6 At a large university an experiment was completed to compare various methods of teaching pre–calculus. The following five teaching methods were used for the study.

Method 1: Lecture method of instruction, large class
Method 2: Lecture method of instruction, large class with smaller recitation sections once a week
Method 3: Lecture method of instruction, small class
Method 4: Half lecture, half group work, small class
Method 5: All group work for instruction, small class

Thirty sections of pre–calculus were available for the study and each teaching method was randomly assigned to six sections. For each section a student survey yielded a single numerical value indicating their satisfaction with the course (higher scores indicate more satisfaction). The means for the six sections used for each teaching method are as follows:

Method	1	2	3	4	5
Mean	6	8	9	12	10

Assuming a CRD design the total sum of squares is 165. Four comparisons of the methods that are of interest are:

i) 1, 2 vs. 3, 4, 5
ii) 1 vs. 2
iii) 3, 4 vs. 5
iv) 3 vs. 4

a) Show that these four comparisons involve orthogonal contrasts.
b) Perform an f–test at the .05 level for each comparison.

8–7 Refer to Example 8.4.1 comparing the effects of copper on the lifetimes for daphnia. In this example, Bonferroni simultaneous confidence intervals were found for $\tau_1 - \tau_2$, $\tau_1 - \tau_3$, and $\tau_2 - \tau_3$. The length of the $\tau_1 - \tau_2$ interval is 44.08. Find an **individual** 95% confidence interval (using 7.5.1) for $\tau_1 - \tau_2$ and compare the width of this interval to the length of the Bonferroni SCI.

8–8 Refer to the data in Exercise 7–1. Find the 95% Bonferroni SCI for $\tau_1 - \tau_2$, $\tau_1 - \tau_3$ and $\tau_2 - \tau_3$. Deduce which treatments are significantly different.

8–9 Refer to the data in Exercise 7–3. Use the Bonferroni multiple comparison procedure (8.4.4) at a .05 level to determine which treatment means are significantly different.

8–10 In a treatment versus control experiment one treatment is identified as the control group, and a major purpose of the experiment is to compare the other treatments to the control. Refer to the pesticide data given in Example 7.9.3 and let designated treatments 1, 2, 3 denote the control, group A and B, respectively. Use the log of the biomass as the response variable.

a) The "treatment versus control" comparisons are: $\tau_2 - \tau_1$ and $\tau_3 - \tau_1$. Find the Bonferroni SCI for these comparisons using a 95% confidence level.
b) Find the Bonferroni SCI for all three pairwise treatment comparisons using a 95% confidence level.
c) Compare the length of the SCI in parts a and b. Discuss why "treatment versus control" comparisons might be preferable in some cases to comparing **all** pairwise comparisons.

Theoretical Exercises

8–11 Refer to the illustrations following Definition 8.2.2. Verify that ψ_1, ψ_2, ψ_3 are also mutually orthogonal when $n_1 = 4$, $n_2 = 5$, $n_3 = 4$, and $n_4 = 5$ but not when $n_1 = 4$, $n_2 = 4$, $n_3 = 5$, and $n_4 = 5$.

8–12 Let $n_1 = n_2 = n_3 = 4$ in an equally replicated CRD experiment with $p = 3$ treatments. Consider the contrasts.

$$\psi_1 = \tau_1 - \tau_2$$
$$\psi_2 = \tau_1 - \tau_3$$
$$\psi_3 = \tau_1 - 2\tau_2 + \tau_3.$$

(a) Denoting $y_{i\cdot}$ to be the total for treatment i, show that

$SS(\psi_1) = (y_{1\cdot} - y_{2\cdot})^2/8$, $SS(\psi_2) = (y_{1\cdot} - y_{3\cdot})^2/8$ and

$SS(\psi_3) = (y_{1\cdot} - 2y_{2\cdot} + y_{3\cdot})^2/24$.

(b) Next show that $SS(\psi_1) + SS(\psi_3) \neq SStr$ whereas $SS(\psi_2) + SS(\psi_3) = SStr$. Why was this result expected?

8–13 Verify that formula (6.6.1) takes the form of (8.3.2) in the experimental design setting with equal replication per treatment, i.e., when $r\,p = n$.

8–14 Explain why the Bonferroni multiple comparison procedure for testing "c" two–sided tests of the form $H_0 : \tau_1 = \tau_k$ is equivalent to rejecting H_0 if the t–test for H_0 has p–value of α/c or less.

8–15 Refer to the Bonferroni multiple comparison procedure for a CRD as presented in (8.4.4). Show that this procedure has an overall type I error level of α or less.

CHAPTER 9

MULTIPLE COMPARISONS

9.1 Introduction

In the setting of a CRD, if we fail to reject $H_0 : \tau_1 = \cdots = \tau_p$ with the f—test, we conclude that the means are "not significantly different" and this **usually** ends our interest in any further analysis of the data. On the other hand, rejection of H_0 provides evidence that some difference between the means exist but the f—test does not indicate **which** of the means are different. To determine this, we need to utilize an additional analysis and this will usually involve a **multiple comparison** procedure.

Our goal in this chapter is to present several multiple comparison procedures. The primary objective of these procedures will be to indicate which treatment means are different with some specified error level. Generally these procedures will involve comparing all possible pairs of treatment means, and since many comparisons will potentially be made, the name "multiple" comparison seems appropriate. These procedures are sometimes referred to as **a posteriori** or "data snooping" type techniques as they are particularly appropriate for unplanned comparisons dictated by the results of the f—test.

Several competing multiple comparison procedures have been developed, based on different properties. Some procedures provide both simultaneous confidence intervals and tests, while others provide only tests. Some procedures apply directly to equal sample size problems, while others apply to both equal and unequal sample size problems. Some procedures proceed in a stepwise manner, while others are completed in one step. However, the major reason for having several competing techniques pertains to different philosophies on how we should control the three different error levels (experimentwise, per comparison, overall type I) that were introduced in Section 8.4.

In the remainder of this chapter we will present the most commonly used multiple comparison procedures in the setting of a normal CRD model with the knowledge that they can also be used in analyzing data from other models to be studied later. Our focus will be on pairwise comparison of means. The properties of these procedures will be presented, and at the end of the chapter, we will present a comparison of the procedures. In the first few sections we will present procedures that provide both simultaneous confidence intervals and tests.

9.2 Bonferroni and Fisher's Least Significant Difference Procedures

In this section we consider two procedures that are based on the t—distribution: the Bonferroni multiple comparison procedure and Fisher's "protected" least significant difference (LSD) procedure. The Bonferroni procedure, as presented in Section 8.4, controls the overall type I error level, while the Fisher's protected LSD controls the per comparison error level.

For the normal CRD model the Bonferroni procedure can yield both tests and simultaneous confidence intervals for all $c = \binom{p}{2}$ pairwise comparisons of treatment means. The formulas are given in Equations 8.4.3 and 8.4.4. For completeness, these formulas are repeated below.

Bonferroni Procedures for Normal CRD Model

100(1—α)% Simultaneous Confidence Intervals for all

pairs $\tau_i - \tau_k$:

$$(\bar{y}_{i.} - \bar{y}_{k.}) \pm t_{\alpha/2c}\, s\sqrt{\frac{1}{n_i} + \frac{1}{n_k}}, \quad 1 \le i < k \le p.$$

Multiple Comparison tests for $H_0 : \tau_i = \tau_k$:

(9.2.1)

Decide $\tau_i \neq \tau_k$ if $|\bar{y}_{i.} - \bar{y}_{k.}| \geq t_{\alpha/2c}\, s\sqrt{\frac{1}{n_i} + \frac{1}{n_k}}$, $1 \le i < k \le p$.

Note: $c = \binom{p}{2}$. Values for $t_{\alpha/2c}$ can be found in Table 5 of Appendix B with n—p degrees of freedom.

It was demonstrated in Section 8.4 that the simultaneous confidence intervals have overall confidence level of 100(1—α)% for all pairwise comparisons. As a family of $c = \binom{p}{2}$ tests, the Bonferroni multiple comparison has overall type I error level of α or less, as was claimed in Exercise 8—15. It follows that the experimentwise error level is also α or less.

Three notes are in order at this time. First, notice that the Bonferroni procedure is applicable for both equal and unequal sample size problems. Second, as was illustrated in Section 8.4, the Bonferroni procedure can be used for more than just pairwise comparisons. For example, when $p = 3$ we could use the Bonferroni procedure to find tests for the three pairwise comparisons $\tau_1 - \tau_2$, $\tau_1 - \tau_3$, $\tau_2 - \tau_3$ and for the contrast $\frac{1}{2}(\tau_1 + \tau_2) - \tau_3$. In this case we would use $c = 4$. Finally, the Bonferroni procedure can be implemented using SAS. In Section 9.6 we will illustrate how SAS can be used to perform the Bonferroni and all of the other procedures we present in this chapter.

The Fisher's protected LSD procedure is based on the F–test and a series of t–tests, each at level α. In this procedure pairwise comparison of means are made **only** if the F–test rejects $H_0 : \tau_1 = \cdots = \tau_p$ at the α–level. If H_0 is rejected, then each pair of treatment means is compared via an α–level t–test. This is outlined as follows:

Fisher's Protected LSD Multiple Comparison

Step 1: Test $H_0 : \tau_1 = \cdots = \tau_p$ with an α–level F–test. If H_0 is not rejected, then declare that the means are not significantly different and stop. If H_0 is rejected, then proceed to Step 2.

Step 2: For $1 \leq i < k \leq p$, test $H_0 : \tau_i = \tau_k$ with an α–level t–test. That is, decide that $\tau_i \neq \tau_k$ if

$$|\bar{y}_{i\cdot} - \bar{y}_{k\cdot}| \geq t_{\alpha/2} s \sqrt{\frac{1}{n_i} + \frac{1}{n_k}} \qquad (9.2.2)$$

for $1 \leq i < k \leq p$, where the t has $n - p$ degrees of freedom.

Some properties of the Fisher's protected LSD are given. First, the LSD refers to the quantity $t_{\alpha/2}\ s\sqrt{\frac{1}{n_i} + \frac{1}{n_k}}$ which is the smallest (least) difference required between treatment means to be declared significantly different. Second, Fisher's test is called "protected" in that we have screened (protected) the procedure with the f—test in that we only perform the multiple t—tests of the Fisher procedure if the f—test rejects. Third, since this procedure is based on comparing two treatments with an α—level t—test, the per comparison error level is $\leq \alpha$, the inequality is because of the protection of the f—test. Also, because of the f—test protection, the experimentwise error level is α or less. The overall type I error level is unknown and can exceed α by fairly large margins; however, the protection of the f—test does tend to reduce this overall type I error level somewhat. Fourth, the procedure is applicable for both equal and unequal sample size problems. Finally, it should be clear that the Fisher's protected LSD can potentially find more differences between the means than the Bonferroni procedure. This follows since the cutoff point used by Fisher's is $t_{\alpha/2}$, which is clearly smaller than the cutoff point, $t_{\alpha/2c}$, that is used by the Bonferroni. Therefore, we say that the Bonferroni is more "conservative" than the Fisher LSD.

We close this section with an example which illustrates the use of both the Bonferroni and Fisher's protected LSD. Naturally, an experimenter would generally only use one of these competing procedures depending on whether control of the overall type I error level (Bonferroni) or control of the per comparison error level (Fisher) is desired.

Example 9.2.1 Refer to the pain threshold data given in Table 7.10.1 in which we investigated the relationship between people's hair color and their tolerance to pain. A summary of the results of this analysis of a CRD with $p = 4$ and each $n_i = 5$ is as follows:

Treatment	1	2	3	4
Hair Color	Light Blond	Dark Blond	Light Brunette	Dark Brunette
$\bar{y}_{i\cdot}$	59.2	51.2	42.6	37.4

Recall that the f—test rejected $H_0 : \tau_1 = \cdots = \tau_4$ at the $\alpha = 0.05$ level and that $s^2 = 62.625$ with 16 degrees of freedom.

We complete both the Bonferroni and Fisher's protected multiple comparison procedures at the 0.05 level. With $\alpha = 0.05$ and $c = \binom{4}{2} = 6$, the appropriate cutoff points for a t are $t_{\alpha/2c} = 3.018$ (from Table 5 using interpolation on the 15 and 20 degrees of freedom values) and $t_{\alpha/2} = 2.120$ (from Table 2). Using Equation (9.2.1) the Bonferroni multiple comparison procedure will decide that $\tau_i \neq \tau_k$ if $|\bar{y}_{i.} - \bar{y}_{k.}| \geq 3.018 \sqrt{62.625} \sqrt{\frac{1}{5} + \frac{1}{5}} = 15.1$. This results in the following "separation of means" displayed in tabular form.

Treatment	4	3	2	1
Mean	37.4	42.6	51.2	59.2

For the Fisher's protected LSD we first note that the F—test rejects H_0 ; therefore, we proceed with the comparison of means. Using Equation (9.2.2) we decide that $\tau_i \neq \tau_k$ if $|\bar{y}_{i.} - \bar{y}_{k.}| \geq$ $2.10 \sqrt{62.625} \sqrt{\frac{1}{5} + \frac{1}{5}} = 10.6$. This results in the following separation of means:

Treatment	4	3	2	1
Mean	37.4	42.6	51.2	59.2

Note that the Fisher's protected "LSD" value (10.6) is substantially smaller than the corresponding "LSD" value for the Bonferroni (15.1). This accounts for Fisher's procedure declaring more differences between the means than the Bonferroni procedure in this particular example. Of course, the Fisher's procedure also has a potentially higher overall type I error level than the Bonferroni.

9.3 Tukey Multiple Comparison Procedure

Another commonly used multiple comparison procedure is one proposed by John Tukey. This procedure is particularly effective when we want to find simultaneous confidence intervals for **all** possible differences between means. It will generally produce shorter confidence intervals in such a case than the Bonferroni procedure.

The Tukey procedure is based on the **Studentized range** distribution which we now describe.

Definition 9.3.1 Suppose that $Y_1,\ldots,Y_p$ are independent normal random variables with mean μ and variance σ^2. Let $\hat{\sigma}^2$ be an estimator of σ^2 that is independent with the Y_i's and such that $\nu\,\hat{\sigma}^2/\sigma^2$ has a chi–square distribution with ν degrees of freedom. If $\max Y_i$ and $\min Y_i$ represent the largest of the Y_i's and the smallest of the Y_i's, respectively, then the random variable

$$Q = (\max Y_i - \min Y_i)/\hat{\sigma}$$

has a **Studentized range** distribution with parameters p and ν.

We will let $q_{\alpha,p,\nu}$ represent the upper α cutoff point for a Studentized range random variable with parameters p and ν. Values for $q_{\alpha,p,\nu}$ are given in Table 6 of Appendix B.

This Studentized range distribution can be used in a normal CRD experiment to find simultaneous confidence intervals for all pairwise differences, $\tau_i - \tau_k$ for $1 \leq i < k \leq p$, when the sample sizes are equal. We let $n_1 = \cdots = n_p = r$ and note that the random variables

$$\sqrt{r}(\bar{Y}_{1\cdot} - \mu_1),\ \sqrt{r}(\bar{Y}_{2\cdot} - \mu_2),\ldots,\sqrt{r}(\bar{Y}_{p\cdot} - \mu_p)$$

are independent normal random variables with a mean of 0 and variance σ^2. We also recall that $(n - p)\,\hat{\sigma}^2/\sigma^2$ has a chi–square distribution with $n - p$ degrees of freedom (Theorem 5.1.4). Since $\bar{Y}_{i\cdot}$ is the least squares estimator for the estimable function $\mu + \tau_i$, it follows from Theorem 5.2.3 that $(n - p)\hat{\sigma}^2/\sigma^2$ is independent with the random variables $\sqrt{r}(\bar{Y}_{i.} - \mu_i)$, $i = 1,\cdots,p$. Using Definition 9.3.1 we observe that the random variable

$$[\max \sqrt{r}(\bar{Y}_{i.} - \mu_i) - \min\sqrt{r}(\bar{Y}_{i.} - \mu_i)] \,/\, \hat{\sigma}$$

has a Studentized range distribution with parameters p and $n - p$.

To find simultaneous confidence intervals for $\tau_i - \tau_k$ for $1 \leq i < k \leq p$ we proceed as follows. First observe that

$$P\,\{[\max\sqrt{r}(\bar{Y}_{i.} - \mu_i) - \min\sqrt{r}(\bar{Y}_{i.} - \mu_i)]/\hat{\sigma} \leq q_{\alpha,p,n-p}\} = 1 - \alpha.$$

Next, note that $[\max\sqrt{r}(\bar{Y}_{i.} - \mu_i) - \min\sqrt{r}(\bar{Y}_{i.} - \mu_i)]/\hat{\sigma} \leq q_{\alpha,p,n-p}$ is equivalent to

$$\{|\bar{Y}_{i.} - \mu_i) - (\bar{Y}_{k.} - \mu_k)| \leq q_{\alpha,p,n-p}\ \hat{\sigma}/\sqrt{r} \text{ for all } 1 \leq i < k \leq p\},$$

which in turn is equivalent to

$$-\,q_{\alpha,p,n-p}\hat{\sigma}/\sqrt{r} \leq (\bar{Y}_{i\cdot} - \bar{Y}_{k\cdot}) - (\mu_i - \mu_k) \leq q_{\alpha,p,n-p}\hat{\sigma}/\sqrt{r}$$

$$\text{for all } 1 \leq i < k \leq p.$$

Hence, on reordering the terms,

$$(\bar{Y}_{i\cdot} - \bar{Y}_{k\cdot}) - q_{\alpha,p,n-p}\hat{\sigma}/\sqrt{r} \leq \mu_i - \mu_k \leq (\bar{Y}_{i\cdot} - \bar{Y}_{k\cdot}) + q_{\alpha,p,n-p}\hat{\sigma}/\sqrt{r}$$

$$\text{for all } 1 \leq i < k \leq p.$$

It follows that

$$P[(\bar{Y}_{i\cdot} - \bar{Y}_{k\cdot}) - q_{\alpha,p,n-p}\hat{\sigma}/\sqrt{r} \leq (\mu_i - \mu_k) \leq (\bar{Y}_{i\cdot} - \bar{Y}_{k\cdot}) + q_{\alpha,p,n-p}\hat{\sigma}/\sqrt{r}$$

$$\text{for all } 1 \leq i < k \leq p] = 1 - \alpha.$$

Noting that the sample value for $\hat{\sigma}$ is s, and that $\mu_i - \mu_k = \tau_i - \tau_k$, the previous equation yields the $100(1-\alpha)\%$ simultaneous confidence intervals for $\begin{bmatrix} p \\ 2 \end{bmatrix}$ pairwise differences in means $\tau_i - \tau_k$ to be

$$(\bar{y}_{i\cdot} - \bar{y}_{k\cdot}) \pm q_{\alpha,p,n-p}\ s/\sqrt{r} \quad \text{for } 1 \leq i < k \leq p. \tag{9.3.1}$$

If we are interested in testing each $H_0 : \tau_i = \tau_k$ for $1 \leq i < k \leq p$ we can use the Tukey simultaneous confidence intervals in (9.3.1). We decide that $\tau_i \neq \tau_k$ if 0 is not in the interval for $\tau_i - \tau_k$. This is equivalent to deciding that $\tau_i \neq \tau_k$ if $|\bar{y}_{i.} - \bar{y}_{k.}| > q_{\alpha,p,n-p} s/\sqrt{r}$. We summarize this as follows:

Tukey Procedures for Normal CRD Model

$$(n_1 = \cdots = n_p = r)$$

$100(1 - \alpha)\%$ Simultaneous Confidence Intervals for all pairs $\tau_i - \tau_k$:

$$(\bar{y}_{i\cdot} - \bar{y}_{k\cdot}) \pm q_{\alpha,p,n-p}\, s/\sqrt{r},\ 1 \leq i < k \leq p \qquad (9.3.2)$$

Multiple Comparison tests for $H_0 : \tau_i = \tau_k$:

Decide $\tau_i \neq \tau_k$ if

$$|\bar{y}_{i\cdot} - \bar{y}_{k\cdot}| > q_{\alpha,p,n-p} s/\sqrt{r},\ 1 \leq i < k \leq p.$$

Since the Tukey multiple comparison is based on $100(1-\alpha)\%$ simultaneous confidence intervals, it follows that the overall type one error level is controlled to be α or less. The experimentwise error level is α. The per comparison error levels are less than α, and these levels can be considerably smaller than α.

The following example illustrates the Tukey simultaneous confidence intervals and multiple comparison.

Example 9.3.1 As in Example 9.2.1, refer to the pain threshold data given in Table 7.10.1 in which we investigated the relationship between people's hair color and their tolerance to pain. Again, recall that in this example $p = 4$, $n - p = 16$, $s^2 = 62.625$, and $n_1 = n_2 = n_3 = n_4 = r = 5$. To find the 95% Tukey simultaneous confidence intervals we first find $q_{.05,4,16}\, s/\sqrt{r} = (4.05)\sqrt{62.625}/\sqrt{5} = 14.3$. The simultaneous confidence intervals are $(\bar{y}_{i.} - \bar{y}_{k.}) \pm 14.3$ for $1 \leq i < k \leq p$ which are:

Comparison	Interval	Comparison	Interval
$\tau_1 - \tau_2$	8 ± 14.3	$\tau_2 - \tau_3$	8.6 ± 14.3
$\tau_1 - \tau_3$	16.6 ± 14.3	$\tau_2 - \tau_4$	13.8 ± 14.3
$\tau_1 - \tau_4$	21.8 ± 14.3	$\tau_3 - \tau_4$	5.2 ± 14.3

The result of the Tukey multiple comparison at an 0.05 level in tabular form is:

$$\begin{array}{lcccc}
\text{Treatment} & 4 & 3 & 2 & 1 \\
\text{Mean} & 37.4 & 42.6 & 51.2 & 59.2 \\ \cline{2-4}
 & & & & \\ \cline{4-5}
\end{array}$$

Notice that the result of the Tukey multiple comparison matches the result of the Bonferroni procedure. Both control the overall type I error level at 0.05. Notice, however, that the Tukey simultaneous confidence intervals, $(\tau_i - \tau_k) \pm 14.3$ are shorter intervals than the corresponding Bonferroni simultaneous confidence intervals $(\tau_i - \tau_k) \pm 15.1$. This is the strong feature of the Tukey procedure, namely, shorter intervals when **all** pairwise differences are investigated.

When the sample sizes are unequal we can adjust the Tukey procedure by using the **Tukey—Kramer** method. In this method we replace the quantity $s/\sqrt{r}$ in Equation (9.3.2) by the quantity $s\sqrt{\frac{1}{2}\left[\frac{1}{n_i} + \frac{1}{n_k}\right]}$. Our simultaneous confidence intervals for $\tau_i - \tau_k$ become

$$(\bar{y}_{i\cdot} - \bar{y}_{k\cdot}) \pm q_{\alpha,p,n-p}\, s\sqrt{\frac{1}{2}\left[\frac{1}{n_i} + \frac{1}{n_k}\right]},\ 1 \leq i < k \leq p. \tag{9.3.3}$$

In our multiple comparison procedure we decide that $\tau_i \neq \tau_k$ if

$$|\bar{y}_{i\cdot} - \bar{y}_{k\cdot}| > q_{\alpha,p,n-p}\, s\sqrt{\frac{1}{2}\left[\frac{1}{n_i} + \frac{1}{n_k}\right]},\ 1 \leq i < k \leq p. \tag{9.3.4}$$

Hayter (1984) has shown that this Tukey—Kramer adjustment is conservative. That is, the overall confidence level is greater than $100(1-\alpha)\%$ and the overall type I error level is less than α.

There is an adjustment that can be made to the Tukey procedure to enable us to find simultaneous confidence intervals for general contrasts. We will not consider this here since the Bonferroni and the Scheffe procedures (next section) give better results for general contrasts.

9.4 Scheffé Multiple Comparison Procedure

Henry Scheffé proposed a general purpose multiple comparison procedure that can be used to find simultaneous confidence intervals for all possible contrasts. The procedure is based upon the F distribution, and can be used for both equal and unequal sample size cases.

For the Scheffé procedure consider **all possible** contrasts ψ_k, where $\psi_k = \sum_{i=1}^{p} c_{ki}\tau_i$ and $\sum_{i=1}^{p} c_{ki} = 0$. The Scheffé simultaneous confidence intervals are given below in (9.4.1). Note that we can also use these simultaneous confidence intervals to test $H_0 : \psi_k = 0$ by rejecting H_0 if the interval for ψ_k does not contain zero.

Scheffé Procedures for Normal CRD Model

The Scheffé simultaneous confidence intervals for all possible contrasts $\sum_{i=1}^{p} c_{ki}\tau_i$, $\sum_{i=1}^{p} c_{ki} = 0$ are:

$$\sum_{i=1}^{p} c_{ki}\bar{y}_{i\cdot} \pm \sqrt{(p-1)f_{\alpha,p-1,n-p}}\sqrt{s^2 \sum_{i=1}^{p} \frac{c_{ki}^2}{n_i}} . \tag{9.4.1}$$

For multiple comparisons we reject $H_0 : \psi_k = 0$ if the simultaneous confidence interval does not contain zero, that is, if

$$\left| \sum_{i=1}^{p} c_{ki}\bar{y}_{i\cdot} \right| \geq \sqrt{(p-1)f_{\alpha,p-1,n-p}}\sqrt{s^2 \sum_{i=1}^{p} \frac{c_{ki}^2}{n_i}} .$$

Note: The overall confidence level is at least $100(1-\alpha)\%$.

Several comments are in order. First, observe that the overall confidence level is at least $100(1-\alpha)\%$. (A verification of this fact is available in the exercises.) Notice that this implies that the multiple comparison test has overall type I error of α or less. Second, the form of the Scheffé interval is somewhat familiar. For a contrast $\psi_k = \sum_{i=1}^{p} c_{ki}\tau_i$, the quantity $\sum_{i=1}^{p} c_{ki}\bar{y}_{i\cdot}$ is the least squares estimate for ψ_k, and the quantity $\sqrt{s^2 \sum_{i=1}^{p} \frac{c_{ki}^2}{n_i}}$ is the standard error for this least squares estimate. As a consequence, the Scheffé simultaneous confidence interval is actually of the form: estimate $\pm$ standard error, here $\sqrt{(p-1)f_{\alpha,p-1,n-p}}$. Third, Scheffé (1959) points out an interesting relationship between the F–test for $H_0 : \tau_1 = \cdots = \tau_p$ and his multiple comparison procedure. He shows that the F–test rejects at level α if and only if there is at least one contrast significantly different from zero using the Scheffé multiple comparison procedure (9.4.1). (The contrast that is significant need not be a "pairwise" contrast).

To compare this Scheffé procedure with the others mentioned in this chapter let us consider an example.

Example 9.4.1 To facilitate comparison with the other procedures let us refer again to the pain threshold data give in Table 7.10.1. Recall that we investigated the relationship between people's hair color and their tolerance to pain. For these data, $p = 4$, $n - p = 16$, $s^2 = 62.625$, $n_i = 5$, $\bar{y}_{i\cdot} = 59.2$, $\bar{y}_{2\cdot} = 51.2$, $y_{3\cdot} = 42.6$ and $\bar{y}_{4\cdot} = 37.4$. Using $\alpha = 0.05$, here the Scheffé simultaneous confidence intervals for all possible contrasts are:

$$\sum_{i=1}^{4} c_{ki}\bar{y}_{i.} \pm \sqrt{(4-1)f_{.05,3,16}} \sqrt{62.625 \sum_{i=1}^{4} \frac{c_{ki}^2}{5}} .$$

In particular, for pairwise contrasts this becomes

$$(\bar{y}_{i\cdot} - \bar{y}_{k\cdot}) \pm \sqrt{3(3.24)} \sqrt{62.625 \left[\frac{1}{5} + \frac{1}{5}\right]}$$

$$\text{or } (\bar{y}_{i\cdot} - \bar{y}_{k\cdot}) \pm 15.6.$$

Therefore the resulting confidence intervals are:

Comparison	Interval	Comparison	Interval
$\tau_1 - \tau_2$	8 ± 15.6	$\tau_2 - \tau_3$	8.6 ± 15.6
$\tau_1 - \tau_3$	16.6 ± 15.6	$\tau_2 - \tau_4$	13.8 ± 15.6
$\tau_1 - \tau_4$	21.8 ± 15.6	$\tau_3 - \tau_4$	5.2 ± 15.6

The results of the Scheffé multiple comparison on the pairwise contrasts for these data are:

Treatment	4	3	2	1
Mean	37.4	42.6	51.2	59.2

Upon comparing these results with the Tukey procedure in Example 9.3.1, and the Bonferroni and LSD procedures in Example 9.2.1, we notice that the Scheffé procedure gives wider confidence intervals for pairwise differences than the other two. This means that it might find fewer pairwise differences. This example illustrates the fact that the Scheffé procedure is more conservative than the LSD, Bonferroni and Tukey for pairwise comparisons. The beauty of the Scheffe procedure is its ability to find simultaneous confidence intervals for **all possible** contrasts (not just the pairwise ones). More will be stated later in this chapter about the comparison of the Scheffé procedure to the other procedures.

9.5 Stepwise Multiple Comparison Procedures

In the multiple comparison procedures presented thus far the **same** critical difference is used in making **all** comparisons. For example, in the Tukey procedure the critical difference $q_{\alpha,p,n-p} s/\sqrt{r}$ is used for all pairwise comparisons. In this section we present procedures in which the critical difference for a comparison depends on how "far apart" the pair of sample means are when they are ranked in order of magnitude. These procedures are called stepwise procedures. We first describe how to perform a stepwise procedure, and then discuss two such procedures: Newman–Keuls and Duncan. Finally, we briefly mention a newer procedure proposed by Einot and Gabriel.

To describe the stepwise procedure we assume a normal CRD model in which the number of replications is r for each of the p treatments. Let $\bar{y}_{1\cdot},\cdots,\bar{y}_{p\cdot}$ be the treatment means **ordered** according to magnitude from smallest to largest. (Note that we might need to **renumber** the treatments to achieve this order.) The stepwise procedures are based upon comparing treatments of **range** k with a critical difference w_k. The ordered treatments are said to be of range k if they are k "steps apart." For example, after renumbering according to order, $\bar{y}_{3\cdot}$ and $\bar{y}_{1\cdot}$ are of range 3 as are $\bar{y}_{5\cdot}$ and $\bar{y}_{3\cdot}$, while $\bar{y}_{3\cdot}$ and $\bar{y}_{2}$ are of range 2.

Stepwise procedures involve the following steps:

Step 1: (Range of p) We compare ordered treatments p and 1 and reject $H_0 : \tau_p = \tau_1$ if $\bar{y}_{p\cdot} - \bar{y}_{1\cdot} \geq w_p$. If we reject H_0, then we declare that τ_p and τ_1 are significantly different and we proceed to the next step. If we fail to reject H_0, then we declare that τ_p and τ_1, and all of the ordered treatments between them, not significantly different. The procedure stops in this latter case.

Step 2: (Range of p − 1) In this step we compare ordered treatments of range p − 1, that is, ordered treatment pair p and 2, as well as ordered pair p − 1 and 1. We reject $H_0 : \tau_{p-1} = \tau_1$ if $\bar{y}_{p-1\cdot} - \bar{y}_{1\cdot} \geq w_{p-1}$, and we reject $H_0 : \tau_p = \tau_2$ if $\bar{y}_{p\cdot} - \bar{y}_{2\cdot} \geq w_{p-1}$. If we reject one or both H_0 we proceed to the next step. If we fail to reject an H_0, then the two treatments are declared not significantly different. Further, all ordered treatments between these two are also declared not significantly different. It is important to realize that treatments declared not significantly different are not compared in subsequent steps.

Subsequent Steps: We proceed in a similar manner for steps of range p − 2, p − 3,...,2. The procedure ends at the step comparing ranges of 2, or at an earlier step if no significant differences are found.

Different stepwise procedures are found by selecting the critical difference w_k according to different criteria. We describe the criteria proposed by Newman (1939) and Keuls (1952), and by Duncan (1955).

Newman–Keuls Procedure

Newman–Keuls Procedure for Normal CRD Model

Proceed in steps. For a range of k, say $\bar{y}_{i\cdot}$ and $\bar{y}_{\ell\cdot}$, reject $H_0: \tau_i = \tau_\ell$ if $|\bar{y}_{\ell\cdot} - \bar{y}_{i\cdot}| \geq w_k$, where

$$w_k = q_{\alpha,k,n-p} \frac{s}{\sqrt{r}} . \tag{9.5.1}$$

For the Newman–Keuls procedure notice that the w_k value is based upon the studentized range statistic "q," where the parameter k matches the range, and the parameter $n - p$ equals the degrees of freedom for error in a CRD. Also note that this procedure uses the same critical difference as the Tukey procedure (9.3.2) for a range of p, but uses a different (smaller) value for all other steps. Because of this fact, the Newman–Keuls procedure can potentially find more differences between treatments than the Tukey procedure.

The Newman–Keuls procedure controls the experimentwise error level (when all treatments are equal) and the per comparison error level to α, but the overall type I error level can exceed α. Hochberg and Tamkane (1987) show that the upper bound on the type I error is $1 - (1-\alpha)^{[\![p/2]\!]}$ or approximately $p\alpha/2$, where $[\![\ ]\!]$ is the greatest integer function. For example, when $p = 6$ and $\alpha = 0.05$ this upper bound is 0.143, well above 0.05. Therefore, the price paid by the Newman–Keuls in finding potentially more differences than the Tukey is a higher type I error level.

Duncan Procedure

Duncan Procedure for Normal CRD Model

Proceed in steps. For a range of k, say $\bar{y}_{i\cdot}$ and $\bar{y}_{\ell\cdot}$, reject $H_0 : \tau_i = \tau_\ell$ if $|\bar{y}_{\ell\cdot} - \bar{y}_{i\cdot}| \geq w_k$, where

$$w_k = d_{\alpha,k,n-p} \frac{s}{\sqrt{r}} . \tag{9.5.2}$$

Values of $d_{\alpha,k,n-p}$ are found in Table 8 of Appendix B.

Notice that the Duncan procedure requires the use of a new table to obtain the critical differences. To compare this procedure with the Newman—Keuls and with other procedures we consider the pain threshold data.

Example 9.5.1 For the pain threshold of Table 7.10.1 we recall that $p = 4$, $n - p = 16$, $s^2 = 62.625$, $n_i = r = 5$, and $\bar{y}_{4\cdot} = 37.4$, $\bar{y}_{3\cdot} = 42.6$, $\bar{y}_{2\cdot} = 51.2$, $\bar{y}_{1\cdot} = 59.2$. To perform the Duncan procedure we first order the treatment means, which means reversing the labels for the treatments. Next we find the critical differences for each range using $\alpha = 0.05$. The values are given in the following table.

Critical Difference for the Duncan Procedure

Range k	4	3	2
$d_{.05,k,16}$	3.235	3.144	2.998
$w_k = d_{.05,k,16} \dfrac{s}{\sqrt{r}}$	11.4	11.1	10.6

The results for the steps of the Duncan procedure follow.

Step 1: (Range of 4) We reject $H_0 : \tau_4 = \tau_1$ since $|\bar{y}_{4\cdot} - \bar{y}_{1\cdot}| = 59.2 - 37.4 = 21.8 \geq 11.4$.

Step 2: (Range of 3) We reject $H_0 : \tau_3 = \tau_1$ since $|\bar{y}_{3\cdot} - \bar{y}_{1\cdot}| = 51.2 - 37.4 = 13.8 \geq 11.1$. We also reject $H_0 : \tau_4 = \tau_2$ since $|\bar{y}_{4\cdot} - \bar{y}_{2\cdot}| = 59.2 - 42.6 = 16.6 \geq 11.1$.

Step 3: (Range of 2) We fail to reject $H_0 : \tau_4 = \tau_3$ since $|\bar{y}_{4\cdot} - \bar{y}_{3\cdot}| = 59.2 - 51.2 = 8 < 10.6$. Similarly we fail to reject $H_0 : \tau_3 = \tau_2$ and $H_0 : \tau_2 = \tau_1$. The final result of the Duncan procedure is:

Mean	37.4	42.6	51.2	59.2
	———	———		
		———	———	
			———	———

We note that for this example, Duncan's procedure finds more differences than the Tukey and Bonferroni procedures, but the same as Fisher's protected LSD. In the exercises we compare the Duncan and Newman–Keuls for these data. From Exercise 9–4 results we see that the Duncan can find potentially more differences than the Newman–Keuls procedure.

The Duncan procedure is based on a special table which controls the overall type I error level to $1 - (1 - \alpha)^{p-1}$. This corresponds to the error level for $p - 1$ independent tests each performed at level α. For example, when $p = 6$ and $\alpha = 0.05$ this error level is 0.226. The per comparison error level is α or less. In comparison to other procedures, the Duncan has a higher type I error level than the Newman–Keuls, Tukey, and Bonferroni, but a lower level than for Fisher's LSD.

A couple of comments are in order. The Duncan and Newman–Keuls both assume equal sample sizes. If the sample sizes are unequal, then one can use the Kramer adjustment based on replacing $s/\sqrt{r}$ by $s\sqrt{\frac{1}{2}\left[\frac{1}{n_i} + \frac{1}{n_\ell}\right]}$. We also should note that the Duncan and Newman–Keuls procedures do not provide any simultaneous confidence intervals.

Einot and Gabriel (1975) proposed a modification in the stepwise procedure to achieve a type I error level of α. They recommend using the following critical differences w_k:

$$w_k = \begin{cases} q_{\alpha,p,n-p}\; s/\sqrt{r}\;, \text{ range of } p \\ q_{\alpha,p-1,n-p}\; s/\sqrt{r}\,, \text{ range of } p-1 \\ q_{\alpha_k,k,n-p}\; s/\sqrt{r}\;, \text{ range of } k = 2,\cdots,p-2, \\ \text{where } \alpha_k = 1 - (1-\alpha)^{k/p} \;. \end{cases}$$

By using these values we obtain a procedure that controls the overall error level, as does the Tukey and Bonferroni, but potentially finds more differences than these other two procedures.

9.6 Computer Usage for Multiple Comparisons

All of the multiple comparison procedures mentioned thus far in this chapter can be performed using SAS. One only needs to add a line to the SAS program which requests the means for the treatments, and a specific multiple comparison procedure. For example, to obtain a Duncan multiple comparison for the pain threshold data (Table 7.10.1) we would use the following commands:

```
PROC      GLM;
CLASS     HAIR;
MODEL     PAIN = HAIR;
MEANS     HAIR/DUNCAN; .
```

Notice that in these commands "HAIR" is the treatment variable, and "PAIN" is the dependent variable.

To obtain other multiple comparisons we simply replace "DUNCAN" with the appropriate choice. We use "LSD" for the LSD procedure, "SNK" for the Newman–Keuls, "REGWQ" for the Einot–Gabriel procedure, "TUKEY" for the Tukey procedure, "BON" for the Bonferroni, and "SCHEFFE" for the Scheffé procedure. You can also request simultaneous confidence intervals. For example, for Tukey simultaneous confidence intervals use the command:

```
MEANS HAIR/TUKEY CLDIFF; .
```

We illustrate the usage of SAS for multiple comparisons with the pain threshold data. The printout that follows contains the Duncan procedure, the Tukey procedure, and the Tukey simultaneous intervals. Observe that the printout indicates significant treatment differences by "separating the means" using the familiar underlining tabular form technique. The reader can compare these computer results with the ones we obtained earlier in Examples 9.3.1 and 9.5.1. Finally, it should be pointed out that one would usually perform only a single multiple comparison procedure on a set of data. We have used both the Duncan and Tukey for illustration purposes.

Table 9.6.1 SAS Commands and Printout for Pain Threshold Data

SAS Commands

```
OPTIONS LS = 80;
DATA PAIN;
INPUT HAIR PAIN @@;
CARDS
1  62    1  60    1  71    1  55    1  48
2  63    2  57    2  52    2  41    2  43
3  42    3  50    3  41    3  37    3  43
4  32    4  39    4  51    4  30    4  35
PROC GLM;
CLASS HAIR;
MODEL PAIN = HAIR;
MEANS HAIR / DUNCAN;
MEANS HAIR / TUKEY;
MEANS HAIR / TUKEY CLDIFF;
```

SAS Printout

```
                        General Linear Models Procedure

Dependent Variable: PAIN
                                     Sum of            Mean
Source                    DF        Squares          Square    F Value      Pr > F

Model                      3   1382.8000000     460.9333333       7.36      0.0026

Error                     16   1002.0000000      62.6250000

Corrected Total           19   2384.8000000

                    R-Square           C.V.        Root MSE             PAIN Mean

                    0.579839       16.62520       7.9135959             47.600000


Source                    DF     Type I SS     Mean Square    F Value      Pr > F

HAIR                       3  1382.8000000     460.9333333       7.36      0.0026

Source                    DF   Type III SS     Mean Square    F Value      Pr > F

HAIR                       3  1382.8000000     460.9333333       7.36      0.0026
```

```
                General Linear Models Procedure

         Duncan's Multiple Range Test for variable: PAIN

   NOTE: This test controls the type I comparisonwise error rate, not
         the experimentwise error rate

                Alpha= 0.05  df= 16  MSE= 62.625

              Number of Means      2     3     4
              Critical Range   10.61 11.13 11.45

   Means with the same letter are not significantly different.

         Duncan Grouping              Mean      N  HAIR

                       A            59.200      5  1
                       A
                  B    A            51.200      5  2
                  B
                  B    C            42.600      5  3
                       C
                       C            37.400      5  4
```

```
                General Linear Models Procedure

     Tukey's Studentized Range (HSD) Test for variable: PAIN

   NOTE: This test controls the type I experimentwise error rate, but
         generally has a higher type II error rate than REGWQ.

                Alpha= 0.05  df= 16  MSE= 62.625
            Critical Value of Studentized Range= 4.046
             Minimum Significant Difference= 14.319

   Means with the same letter are not significantly different.

         Tukey Grouping               Mean      N  HAIR

                       A            59.200      5  1
                       A
                  B    A            51.200      5  2
                  B
                  B                 42.600      5  3
                  B
                  B                 37.400      5  4
```

```
                      General Linear Models Procedure

          Tukey's Studentized Range (HSD) Test for variable: PAIN

        NOTE: This test controls the type I experimentwise error rate.

             Alpha= 0.05  Confidence= 0.95  df= 16  MSE= 62.625
                  Critical Value of Studentized Range= 4.046
                    Minimum Significant Difference= 14.319

      Comparisons significant at the 0.05 level are indicated by '***'.

                        Simultaneous                 Simultaneous
                            Lower      Difference       Upper
               HAIR       Confidence     Between      Confidence
            Comparison      Limit         Means         Limit

          1    - 2         -6.319         8.000        22.319
          1    - 3          2.281        16.600        30.919    ***
          1    - 4          7.481        21.800        36.119    ***

          2    - 1        -22.319        -8.000         6.319
          2    - 3         -5.719         8.600        22.919
          2    - 4         -0.519        13.800        28.119

          3    - 1        -30.919       -16.600        -2.281    ***
          3    - 2        -22.919        -8.600         5.719
          3    - 4         -9.119         5.200        19.519

          4    - 1        -36.119       -21.800        -7.481    ***
          4    - 2        -28.119       -13.800         0.519
          4    - 3        -19.519        -5.200         9.119
```

When the sample sizes for the treatments are unequal, SAS handles the situation in the following manner. If a procedure gives simultaneous confidence intervals, then SAS automatically provides them for unequal sample size cases. For other procedures (i.e., DUNCAN, SNK, REGWQ) SAS uses the harmonic mean of the sample sizes. The harmonic mean of $n_1,\ldots,n_p$ is given by n^*, where $n^* = p/\sum_{i=1}^{p}\frac{1}{n_i}$.

SAS then replaces $s/\sqrt{r}$ by $s/\sqrt{n^*}$ in the formulas for these multiple comparisons. One can obtain the tabular forms for the output of the multiple comparison for LSD, TUKEY, BON or SCHEFFE by using the "LINES" command. For example, the command MEANS HAIR/ TUKEY LINES; will provide a tabular form for the output of the multiple comparison. SAS uses the harmonic means when "LINES" are requested. For example, for the Tukey, SAS uses the critical difference

$q_{\alpha,p,n-p}\ s/\sqrt{n^*}$.

The following SAS printout contains a Tukey multiple comparison for the daphnia data given in Table 7.2.1. Note that the sample sizes for these data are: $n_1 = 4$, $n_2 = 5$, $n_3 = 5$.

Table 9.6.2 SAS Commands and Printout for Daphnia Data

SAS Commands

```
OPTIONS LS = 80;
DATA DAPHNIA;
INPUT COPPER LIFETIME @@;
CARDS
1   60    1   90    1   74    1   82
2   58    2   74    2   50    2   65    2   68
3   40    3   58    3   25    3   30    3   42
PROC GLM;
CLASS COPPER;
MODEL LIFETIME = COPPER;
MEANS COPPER / TUKEY;
MEANS COPPER / TUKEY LINES;
```

SAS Printout

```
                      General Linear Models Procedure

Dependent Variable: LIFETIME
                                    Sum of            Mean
Source                  DF         Squares          Square   F Value      Pr > F

Model                    2    3297.8571429    1648.9285714     12.23      0.0016

Error                   11    1483.0000000     134.8181818

Corrected Total         13    4780.8571429

                  R-Square            C.V.        Root MSE         LIFETIME Mean

                  0.689805        19.92104       11.611123             58.285714


Source                  DF       Type I SS     Mean Square   F Value      Pr > F

COPPER                   2    3297.8571429    1648.9285714     12.23      0.0016

Source                  DF     Type III SS     Mean Square   F Value      Pr > F

COPPER                   2    3297.8571429    1648.9285714     12.23      0.0016
```

```
              General Linear Models Procedure

   Tukey's Studentized Range (HSD) Test for variable: LIFETIME

   NOTE: This test controls the type I experimentwise error rate.

       Alpha= 0.05  Confidence= 0.95  df= 11  MSE= 134.8182
            Critical Value of Studentized Range= 3.820

Comparisons significant at the 0.05 level are indicated by '***'.


                      Simultaneous               Simultaneous
                          Lower     Difference      Upper
          COPPER       Confidence    Between     Confidence
        Comparison        Limit       Means         Limit

     1     - 2          -7.537       13.500       34.537
     1     - 3          16.463       37.500       58.537    ***

     2     - 1         -34.537      -13.500        7.537
     2     - 3           4.167       24.000       43.833    ***

     3     - 1         -58.537      -37.500      -16.463    ***
     3     - 2         -43.833      -24.000       -4.167    ***
```

```
              General Linear Models Procedure

   Tukey's Studentized Range (HSD) Test for variable: LIFETIME

   NOTE: This test controls the type I experimentwise error rate, but
         generally has a higher type II error rate than REGWQ.

              Alpha= 0.05  df= 11  MSE= 134.8182
           Critical Value of Studentized Range= 3.820
            Minimum Significant Difference= 20.643
              WARNING: Cell sizes are not equal.
            Harmonic Mean of cell sizes= 4.615385


  Means with the same letter are not significantly different.

        Tukey Grouping              Mean      N  COPPER

                     A            76.500      4  1
                     A
                     A            63.000      5  2

                     B            39.000      5  3
```

There are several other multiple comparison procedures available on SAS. One of particular interest is Dunnett's procedure. Dunnett (1955) proposed a multiple comparison procedure for the treatment versus control problem. In such a problem we have a single control group and $p - 1$ other treatment groups. Dunnett's procedure compares the $p - 1$ treatments individually to the control group. Interested readers can learn more about Dunnett's procedure, and several other procedures, in Miller's (1981) book.

9.7 Comparison of Procedures, Recommendations

Competing Power

In the previous sections we have presented a number of multiple comparison procedures. Our purpose in this section is to briefly compare these procedures in terms of power (ability to detect differences) and in terms of controlling overall error level.

Let us first concentrate on comparing pairwise differences between means. For this we will focus on the critical differences calculated for each procedure, that is, the value c for which we decide $\tau_i \neq \tau_k$ if $|\bar{y}_{i\cdot} - \bar{y}_{k\cdot}| \geq c$. The student is asked to do this in Exercise 9—3, using the pain threshold data found in Table 7.10.1 and further analyzed in Sections 2, 3, 4, 5, and 6 of this chapter. If we label the procedure with the lowest c value **liberal** (or most powerful), i.e., more potential differences detected with higher overall error, and the procedure with the highest c value most **conservative**, i.e., less potential differences detected with lower overall error, the student will find in completing Exercise 9—3, that the ranking from liberal to conservative is LSD, Tukey, Bonferroni and Scheffé in that order. It is important to realize that we have only looked at one example to obtain this ordering; however, this order has also been confirmed in a study by Carmer and Swanson (1973) for many different situations. We note that although the LSD has the desired property of being most powerful among these four non—stepwise procedures, it fails to control the overall error rate which reduces its attractiveness to statistical practitioners.

Among the stepwise procedures, in doing Exercise 9—4, again using the pain threshold data, the student will find the Duncan to be most powerful followed by the Newman—Keuls, and then by the Einot—Gabriel procedure. Also note that each procedure has generally lower power than the LSD procedure, but generally more power than the Tukey procedure. It is important to recall, however, that the Einot—Gabriel is the only one of these three stepwise procedures that controls the overall error level.

Recommendations

For pairwise comparisons of means we recommend the Tukey procedure since it is easy to use, it controls the overall error level and admits simultaneous confidence intervals, if desired. The Einot—Gabriel procedure has somewhat better power than the Tukey while controlling the overall error, but it is more complicated to use and does not admit SCI. If more power is desired and at the same time a higher error level than α can be tolerated, and SCI are not sought, then the Duncan procedure should suffice.

For comparisons among contrasts that are not pairwise, the situation is more complex. The LSD, Bonferroni, and Scheffé procedures can routinely be used for more general contrasts and there is a modification of the Tukey procedure that can also be used. We will use the Bonferroni procedure for general contrasts since it is easy to use and it controls the overall error level. The Bonferroni procedure also generally has more power than the Scheffé procedure, especially when there is a small number of contrasts to consider.

Chapter 9 Exercises

Application Exercises

9–1 Refer to the cereal box data in Exercise 7.1. Use the Tukey procedure to make all pairwise comparisons of the means at the 0.05 level.

(a) Report the outcome in tabular form.
(b) Find the 95% simultaneous confidence intervals for all pairwise differences of the means.

9–2 Refer to the feedlot cattle data in Exercise 7.2. Use a 0.05 level to make all pairwise comparisons of the means for the following procedures:

(a) Protected LSD
(b) Duncan
(c) Tukey
(d) Bonferroni
(e) Scheffé
(f) Compare the critical difference used for the various procedures.

9–3 Refer to the pain threshold data found in Table 7.10.1. For all pairwise comparisons of means, use SAS to determine the critical differences for the four non–stepwise multiple comparison procedures introduced in this chapter using a 0.05 level. From the values obtained, order the procedures from "liberal" to "conservative" thereby verifying the conclusions given in Section 9.7.

9–4 Refer to Exercise 9–3 above. Do the same for the three stepwise procedures (Duncan, Newman–Keuls, Einot–Gabriel) discussed in this chapter. Here the critical differences will need to be determined for the three possible ranges, namely, 2, 3, and 4.

9–5 Refer to the word processing training data in Table 8.2.1. In Section 8.2, three contrasts: $\tau_3 - \tau_1$, $\tau_4 - \tau_2$ and $\frac{1}{2}\left(\tau_1 + \tau_3\right) - \frac{1}{2}\left(\tau_2 + \tau_4\right)$ were considered.

(a) Find the 95% simultaneous confidence intervals for these three contrasts using the Bonferroni procedure.
(b) Repeat part (a) using the Scheffé procedure.
(c) Compare the lengths of these intervals for the Bonferonni and Scheffé procedures.

9–6 Refer to the bloater pickle data in Exercise 7.6. Suppose that our interest is in comparing the last three treatments (S2, S3, and S4) with the first treatment (S1). Specifically, our interest is in the three contrasts $\tau_2 - \tau_1$, $\tau_3 - \tau_1$, and $\tau_4 - \tau_1$. (This is essentially the treatment versus control problem.) Determine which treatments (S2, S3, S4) are significantly different from S1 using the Bonferroni procedure.

9–7 Refer to the daphnia lifetime data in Table 7.2.1.

(a) Use the Tukey–Kramer method with a 0.05 level to determine which treatment means are significantly different.
(b) Use the harmonic mean approach described in Section 9.6 to perform the Tukey procedure using a 0.05 level. Report the outcome in tabular form.

Theoretical Exercises

9–8 Explain why the Newman–Keuls procedure controls the experimentwise error level (when $H_0 : \tau_1 = \cdots = \tau_p$ is true) to α.

9–9 Consider a balanced CRD. Show that the critical difference for the range of 2 step for the Newman–Keuls procedure is equivalent to the critical difference for the LSD procedure.

9–10 Consider a normal CRD model with $p = 4$ treatments. Suppose that $\tau_1 < \tau_2 = \tau_3 < \tau_4$. Find the overall type I error level for the LSD procedure for this situation.

9–11 (Scheffé) The purpose of this problem is to derive the Scheffé procedure given in Section 4. The derivation uses the following fact about numbers: for all constants a_i, w_i and $c > 0$, the inequality $\left|\sum_{i=1}^{p} a_i w_i\right| \le c\left(\sum_{i=1}^{p} a_i^2\right)^{1/2}$ holds for all $a_i, \cdots, a_p$ if and only if $\sum_{i=1}^{p} w_i^2 \le c^2$.

(a) Show for the normal CRD model that $\sum_{i=1}^{p} n_i(\bar{Y}_{i\cdot} - \tau_i - \bar{Y}_{\cdot\cdot} + \bar{\tau})^2/[(p-1)\,\hat{\sigma}^2]$ has an F–distribution with $p - 1$ and $n - p$ degrees of freedom, where $\bar{\tau} = \sum_{i=1}^{p} n_i\tau_i/n$. Deduce that

$$P\left[\sum_{i=1}^{p} n_i(\bar{Y}_{i.} - \tau_i - \bar{Y}_{\cdot\cdot} + \bar{\tau})^2 \le (p-1)\,\hat{\sigma}^2 f_{\alpha,p-1,n-p}\right] = 1 - \alpha.$$

(b) Now consider all possible contrasts of the form $\sum_{i=1}^{p} c_{ki}\bar{y}_{i\cdot}$, where $\sum_{i=1}^{p} c_{ki} = 0$. Let $c^2 = (p-1)\hat{\sigma}^2 f_{\alpha,p-1,n-p}$, $w_i = \sqrt{n_i}\left(\bar{y}_i - \tau_i - \bar{y}_{\cdot\cdot} + \bar{\tau}\right)$, and $a_i = c_{ki}/\sqrt{n_i}$ for $i = 1,\cdots,p$. Use the fact about numbers and the probability statement in part (a) to show that

$$P\left[\left|\sum_{i=1}^{p} c_{ki}(\bar{Y}_{i\cdot} - \tau_i)\right| \le \sqrt{\left[(p-1)\,\hat{\sigma}^2 f_{\alpha,p-1,n-p}\right] \sum_{i=1}^{p} c_{ki}^2/n_i} \text{ for all contrasts}\right] \ge 1 - \alpha.$$

(c) Using part (b) derive the Scheffé formulas given in (9.4.1).

CHAPTER 10

RANDOMIZED COMPLETE BLOCK DESIGN

10.1 Blocking

In the last three chapters we studied the analysis of a single—factor completely randomized design experiment. In many experimental situations, however, more than one factor can have an effect on the response variable. In those cases, it is generally advantageous to incorporate the additional factors in the model and the design. In this and the next chapter, we will consider designs involving a treatment factor (the variable of primary interest) and an extraneous factor called a blocking factor. (The term "block" dates back to the time of R.A. Fisher and his use of blocks of land in the conduct of agricultural field experiments.) This second factor is "extraneous" in the sense that it is not the subject of the investigation; yet, its presence should logically be recognized because of its potential in accounting for some of the variation in the experimental results. In Chapter 12 our main concern will be a design which accommodates two blocking factors.

An Illustration

The concept of blocking was first introduced in our discussion concerning Table 7.1.2, which involved three varieties of wheat and fields at different experimental stations. The need for blocking was further alluded to in Section 7.8 where the shortcoming of ignoring an important factor in a completely randomized design was pointed out. To add to our appreciation of the blocking concept, we now turn to several examples.

A manufacturing company has three makes of machines which are used to manufacture metal brackets. Let us call them A, B, and C. These machines have been made by different companies and operate somewhat differently. Each machine requires one skilled worker for its operation. The manufacturing company wants to determine which machine is most efficient in producing the brackets. That is, over a fixed period of time, on which of the machines can a worker produce the greatest number of acceptable brackets? To do this, the company decides to set up an experiment. Three workers are available for the study, and to estimate the possible variation in results per worker, each worker will use a machine three times. In this experiment, the three machines (A,B,C) are the treatments, and there are nine experimental units corresponding to the nine observations.

Suppose, in carrying out the experiment, each machine is randomly assigned three times to the workers without restriction; that is, a completely randomized design is used. It is possible that the resulting randomization plan turns out to be the following.

Table 10.1.1 Completely Random Plan

Worker		
1	2	3
B	A	C
A	A	C
B	B	C

With the above randomization scheme, the number of acceptable metal brackets produced on the machines by the workers during a two—hour period appears in the table below.

Table 10.1.2 Number of Acceptable Brackets

Machine		
A	B	C
72	59	90
81	73	95
69	82	89

At first glance the data seem to indicate that machine C is the most efficient machine. But since all machine C data values were produced by the third worker, how do we know whether the higher values for machine C are due to machine C's relative efficiency or due to worker 3 being a faster and more skilled worker? We do not know. In fact, this has turned out to be a poor experiment because a poor design has been used. The design deficiency here is that an important factor, namely worker, has been omitted from design consideration. To make more valid treatment (machine) comparisons, it is necessary to try to "remove" or "equalize" the worker effect by treating workers as a blocking factor. In this experiment, this can be done by randomly assigning each treatment (machine) once to each block (worker). This guarantees that each machine will be operated by every worker; the randomization within a block dictates the order in which a machine is used by the worker such as illustrated in Table 10.1.3. When the randomization of treatments to experimental units is restricted in this manner, the manufacturing company is using a randomized block design, which on the basis of logic alone, is preferable to the plan of Table 10.1.1.

Table 10.1.3 Randomized Block Plan

Worker (Block)		
1	2	3
B	C	C
A	B	A
C	A	B

An important goal in the use of blocking is to divide the experimental units into groups constituting relatively homogeneous sets. These groups are blocks and are typically based on levels of a blocking factor. In the previous example, the three workers constitute the three blocking levels. Later in this chapter we will follow up on this example with the analysis of a machine–worker numerical data set.

R. A. Fisher recognized the advantages of blocking in field experiments. To illustrate, we return to that setting in the next example.

Other Blocking Examples

An experiment is to be conducted to compare the effect of three fertilizers on the yield of a crop. These fertilizers, identified as A, B, and C, differ in their respective proportions of nitrogen, phosphoric acid and soluble potash. The three treatments (fertilizers A, B, and C) are to be compared on twelve plots (experimental units) grouped into four blocks of three contiguous plots. This is done since it is reasonable to assume that the three contiguous plots are more uniform with respect to fertility gradients than the twelve–plot areas as a whole. Then the treatments are assigned to the plots in a manner requiring each block to contain a complete one–replicate set of treatments, randomly arranged within the block. Figure 10.1.1 illustrates a typical field layout for this experiment.

Block 1	A	C	B	B	C	A	Block 2
Block 3	B	A	C	A	B	C	Block 4

Figure 10.1.1 Randomized Block Field Layout

Before leaving this example, it should be noted that a primary purpose of blocking in this and other block experiments is to reduce the experimental error $s^2 = MSe$ since s^2 is to be the "yardstick" against which to measure the significance of treatment differences. In this field experiment, comparison of fertilizers are made within a block and differences in yields on plots within a block are combined with the differences between treatments obtained in the other blocks. If the variability of observations within a block is less than the variability of observations in all plots taken as a whole, then a smaller experimental error will result. To demonstrate this important feature of a randomized block experiment, we turn to an example of the type encountered by the reader in a first course in statistics.

Example 10.1.1 A person on the Weight—Watchers' diet can expect to lose between 8 to 10 pounds during the first month. Eight college women interested in losing weight before the school's spring break adopted the prescribed program. The weights of these eight individuals immediately before and after one month of dieting are given in Table 10.1.4.

Table 10.1.4 Weights of College Women on Diet Program

Woman	1	2	3	4	5	6	7	8
Weight before, y	142	128	166	178	153	135	158	188
Weight after, x	135	124	154	168	144	132	153	174
Difference, d	7	4	12	10	9	3	5	14

Since the before and after weights are taken on the same subject, these data in elementary texts are referred to as matched—paired or paired—difference data. In the nomenclature of experimental designs for a randomized block experiment the women 1 through 8 serve as blocks, and "weights before" and "weights after" are the "treatments." For this two—treatment, eight—block experiment, the error mean square is given by the familiar formula

$$s_d^2 = \sum_{i=1}^{8} \frac{(d_i - \bar{d})^2}{n - 1} = \frac{108}{7} = 15.43.$$

Now, if we consider the "weight before" and "weight after" data sets as completely independent sets, the analysis is completed using the two—sample t—test which coincides with the analysis for a completely randomized design experiment. In this case, the formula for computing the error mean square is

$$s_p^2 = \frac{(n_y - 1)s_y^2 + (n_x - 1)s_x^2}{n_y + n_x - 2} = \frac{7(434.57) + 7(307.71)}{14} = 371.14.$$

On comparing the numerical values of s_d^2 and s_p^2, we see that the implementation of blocking greatly reduced the error mean square. Decreasing the experimental error by design in this way increases the power of subsequent tests and the precision of subsequent confidence intervals. In the exercises, using the data in Table 10.1.4, the student will be asked to calculate analysis of variance tables and confidence intervals associated with a randomized blocks analysis and a completely randomized analysis.

Both in the machine—worker illustration and in the diet example, repeated measures were taken on individuals serving as blocks. A set—up such as this is sometimes called a **repeated measures** experiment, a type prevalent in the fields of psychological and pharmaceutical testing. Experiments involving the use of individual subjects, however, need not always result in the subjects serving as blocks. For example, consider an experiment in education in which the effectiveness of three teaching methods is to be compared. In order to guard against possible confounding of results due to differences in student abilities, the students in the experiment can be grouped into blocks of size three using some measure such as IQ or pre—test score. In this case, the three teaching methods (the treatments) will be randomly assigned to the three students constituting a group (block) of similar ability. Here students are the homogeneous experimental units within blocks of different ability levels.

From the examples given, we can summarize the characteristics of experiments utilizing the concept of blocking. Randomized block designs expressly give recognition to treatment and block variables. The experimental units are grouped into blocks with levels of the blocking factor providing for uniformity of units within blocks. The randomization procedure provides for a random assignment of the treatments to the experimental units within each block. Generally, the primary interest in such designs is the comparison between treatments and little interest is given to the comparison of blocks since block effects are expected to be different. The purpose of blocking is to logically make the comparability between treatments more valid and to numerically decrease experimental error in order to "sharpen" the results of inference procedures. In this chapter we consider the analysis for a special randomized block design, namely, the randomized **complete** block design. In the next chapter, we focus our attention on **incomplete** block designs including the case of missing data.

10.2 Randomized Complete Block Design

Now that we have gained a familiarity with the concept of blocking and its use in planning an experiment, we next focus on the analytical aspects associated with the most "trouble–free" of the randomized block designs, namely, the randomized **complete** block design. For the purpose of brevity, we refer to this particular design as an RCBD.

Definition 10.2.1 A **randomized complete block design** is an experimental design in which the experimental units are divided into relatively homogeneous sets, called blocks. The number of experimental units per block is equal to the number of treatments. The treatments are randomly assigned to the experimental units within a block such that each treatment occurs exactly once in every block.

The word **complete** appears in the name for this design because every block contains each treatment exactly once. In incomplete blocks, discussed in the next chapter, there is at least one block in which at least one treatment does not appear. Notice that in an RCBD, the number of experimental units in each block, called the **block size**, is the same as the number of treatments in the experiment.

For an example of an RCBD, we return to the metal bracket problem introduced in the last section, but this time let us suppose that four workers instead of three are available for the operation of machines A, B, and C. If each worker can be used in three trials, then the workers serve as blocks of size three corresponding to the three treatments, i.e., the three machines. For the RCBD, each worker will use each machine once and the order in which they use the machines is randomized. Note that the randomized order helps to eliminate any possible order effects such as fatigue. Recall that in this problem the measurement variable is the number of acceptable brackets produced by the worker during two hours of machine operation. The resulting experimental data, now with four workers, are given in the next table.

Table 10.2.1 Number of Acceptable Brackets Produced

		Machine		
		A	B	C
Worker	1	82	59	62
	2	91	75	75
	3	98	74	81
	4	89	69	76

Before getting into the details of analyzing this particular data set, we first turn to a discussion of the analysis of an RCBD in general. We begin with the RCBD model.

Model for an RCBD

To write down a general model for the observations obtained in an RCBD experiment, we let p represent the number of treatments and r represent the number of blocks. Since the blocks and treatments can both potentially affect the response variable, we need parameters in the model corresponding to each. The randomized complete block design model is given by

$$Y_{ij} = \mu + \tau_i + \beta_j + \varepsilon_{ij}, \qquad i = 1, 2, \ldots, p \; ; \; j = 1, 2, \ldots, r \; , \tag{10.2.1}$$

where

- Y_{ij} is the response to the ith treatment in the jth block,
- μ is the overall average response to all treatments,
- τ_i is the effect of the ith treatment,
- β_j is the effect of the jth block, and
- ε_{ij} is the random error associated with the ith treatment in the jth block. We assume that $E(\varepsilon_{ij}) = 0$, $Var(\varepsilon_{ij}) = \sigma^2$ for every i and j, and that the ε_{ij} are uncorrelated.

The last assumption in the above model immediately yields the moments for observations Y_{ij}. They are

$$E(Y_{ij}) = \mu + \tau_i + \beta_j \; ,$$

and

$$Var(Y_{ij}) = \sigma^2 \text{ for every } i \text{ and } j \, .$$

This tells us that all observations have the same variance σ^2. The mean, however, can vary according to the treatment and block levels. To test for no difference in treatment effects, the null hypothesis is

$$H_0 : \tau_1 = \tau_2 = \cdots = \tau_p .$$

This is the hypothesis of primary interest and will be carried out through an analysis of variance. Note that when this hypothesis is true, the observations **within a block** have the same mean; however, observations in different blocks can have different means due to the block level effect, β_j. Though it is usually of much less interest, the analysis of variance will also yield, as a by–product, the value of the test statistic for testing

$$H_0 : \beta_1 = \beta_2 = \cdots = \beta_r .$$

This test for no difference in block effects indicates whether or not the blocking scheme was effective.

Because of the structure of the RCBD model (10.2.1), it is said to be an **additive model**; that is, the response level is related to the sum of the two identified effects, treatments (τ_i) and blocks (β_j). This structure is an assumption which may or may not be appropriate. More will be said about this later. Meanwhile, to better understand the nature of this additivity assumption, we consider the following:

$$\begin{aligned} E(Y_{ij} - Y_{kj}) &= (\mu + \tau_i + \beta_j) - (\mu + \tau_k + \beta_j) \\ &= \tau_i - \tau_k , \quad \text{for all j.} \end{aligned}$$

This tells us that for model (10.2.1), the mean of the difference between observations associated with treatments i and k within a block is the same for all blocks. We will encounter non–additive models later in this text.

Matrix Representation of the RCBD Model

Since model (10.2.1) is a linear model, it can be written in the form $\mathbf{Y} = \mathbf{X}\boldsymbol{\beta} + \boldsymbol{\varepsilon}$. To illustrate, we consider an RCBD experiment with $p = 3$ treatments and $r = 4$ blocks. The metal bracket problem presented in Table 10.2.1 is an example of such a design. Corresponding to the numbers shown in that table, we display the $pr = 12$ observations in the same two–way format as follows:

	Treatment 1	Treatment 2	Treatment 3
Block 1	Y_{11}	Y_{21}	Y_{31}
Block 2	Y_{12}	Y_{22}	Y_{32}
Block 3	Y_{13}	Y_{23}	Y_{33}
Block 4	Y_{14}	Y_{24}	Y_{34}

Model (10.2.1) can then be written in matrix form as

$$\underset{\mathbf{Y}}{\begin{bmatrix} Y_{11} \\ Y_{12} \\ Y_{13} \\ Y_{14} \\ Y_{21} \\ Y_{22} \\ Y_{23} \\ Y_{24} \\ Y_{31} \\ Y_{32} \\ Y_{33} \\ Y_{34} \end{bmatrix}} = \underset{\mathbf{X}}{\begin{bmatrix} 1 & 1 & 0 & 0 & 1 & 0 & 0 & 0 \\ 1 & 1 & 0 & 0 & 0 & 1 & 0 & 0 \\ 1 & 1 & 0 & 0 & 0 & 0 & 1 & 0 \\ 1 & 1 & 0 & 0 & 0 & 0 & 0 & 1 \\ 1 & 0 & 1 & 0 & 1 & 0 & 0 & 0 \\ 1 & 0 & 1 & 0 & 0 & 1 & 0 & 0 \\ 1 & 0 & 1 & 0 & 0 & 0 & 1 & 0 \\ 1 & 0 & 1 & 0 & 0 & 0 & 0 & 1 \\ 1 & 0 & 0 & 1 & 1 & 0 & 0 & 0 \\ 1 & 0 & 0 & 1 & 0 & 1 & 0 & 0 \\ 1 & 0 & 0 & 1 & 0 & 0 & 1 & 0 \\ 1 & 0 & 0 & 1 & 0 & 0 & 0 & 1 \end{bmatrix}} \underset{\boldsymbol{\beta}}{\begin{bmatrix} \mu \\ \tau_1 \\ \tau_2 \\ \tau_3 \\ \beta_1 \\ \beta_2 \\ \beta_3 \\ \beta_4 \end{bmatrix}} + \underset{\boldsymbol{\varepsilon}}{\begin{bmatrix} \varepsilon_{11} \\ \varepsilon_{12} \\ \varepsilon_{13} \\ \varepsilon_{14} \\ \varepsilon_{21} \\ \varepsilon_{22} \\ \varepsilon_{23} \\ \varepsilon_{24} \\ \varepsilon_{31} \\ \varepsilon_{32} \\ \varepsilon_{33} \\ \varepsilon_{34} \end{bmatrix}}$$

Notice that in the design matrix **X**, the sum of the second, third, and fourth columns equals column one. Also, the sum of columns five through eight equals column one. It therefore follows that this eight—column **X** matrix has rank 6. This establishes the fact that the traditional RCBD model is a less than full rank model.

In general, for the matrix representation of the RCBD model, **Y** and $\boldsymbol{\varepsilon}$ are $pr \times 1$ vectors, $\boldsymbol{\beta}$ is a $(p + r + 1) \times 1$ vector and **X** is a $pr \times (p + r + 1)$ matrix. The rank of the **X** matrix will be $p + r - 1$ since the linearly independent columns are: column one, $p - 1$ columns associated with the τ_i's and $r - 1$ columns associated with the β_j's.

In the remainder of this chapter we will continue our discussion of the analysis of data sets obtained in RCBD experiments. Our primary interest will be focused on inferences concerning the treatment effects. The data in Table 10.2.1 will serve as a source for numerical illustrations.

10.3 Least Squares Results

Our goal in this section is to find a solution to the normal equations associated with the analysis of an RCBD experiment. In addition, we will find the least squares estimators for estimable functions and the error sum of squares in this setting. First we introduce some needed notation.

The sample form for the RCBD model (10.2.1) is given by the relation

$$y_{ij} = m + t_i + b_j + e_{ij}, \qquad i = 1, 2,\dots,p\ ; \ \ j = 1, 2,\dots,r . \tag{10.3.1}$$

We denote various functions of the realized observations y_{ij} as follows.

$$\begin{aligned} y_{\cdot\cdot} &= \sum_{i=1}^{p} \sum_{j=1}^{r} y_{ij} \quad \text{and} \quad \bar{y}_{\cdot\cdot} = \frac{y_{\cdot\cdot}}{rp}\ ; \\ y_{i\cdot} &= \sum_{j=1}^{r} y_{ij} \quad \text{and} \quad \bar{y}_{i\cdot} = \frac{y_{i\cdot}}{r}\ . \\ y_{\cdot j} &= \sum_{i=1}^{p} y_{ij} \quad \text{and} \quad \bar{y}_{\cdot j} = \frac{y_{\cdot j}}{p}\ . \end{aligned} \tag{10.3.2}$$

With this notation, we are now able to write normal equations $\mathbf{X}'\mathbf{X}\mathbf{b} = \mathbf{X}'\mathbf{y}$ for a general RCBD model with p treatments and r blocks. First in matrix algebra form and then in "expanded" form they are

$$\overset{\mathbf{X}'\mathbf{X}}{\begin{bmatrix} rp & r & r & \cdots & r & p & p\cdots p \\ r & r & 0 & \cdots & 0 & 1 & 1\cdots 1 \\ r & 0 & r & \cdots & 0 & 1 & 1\cdots 1 \\ \cdot & \cdot & \cdot & & \cdot & \cdot & \cdot\quad\cdot \\ \cdot & \cdot & \cdot & & \cdot & \cdot & \cdot\quad\cdot \\ \cdot & \cdot & \cdot & & \cdot & \cdot & \cdot\quad\cdot \\ r & 0 & 0 & \cdots & r & 1 & 1\cdots 1 \\ p & 1 & 1 & \cdots & 1 & p & 0\cdots 0 \\ p & 1 & 1 & \cdots & 1 & 0 & p\cdots 0 \\ \cdot & \cdot & \cdot & & \cdot & \cdot & \cdot\quad\cdot \\ \cdot & \cdot & \cdot & & \cdot & \cdot & \cdot\quad\cdot \\ \cdot & \cdot & \cdot & & \cdot & \cdot & \cdot\quad\cdot \\ p & 1 & 1 & \cdots & 1 & 0 & 0\cdots p \end{bmatrix}} \overset{\mathbf{b}}{\begin{bmatrix} m \\ t_1 \\ t_2 \\ \cdot \\ \cdot \\ \cdot \\ t_p \\ b_1 \\ b_2 \\ \cdot \\ \cdot \\ \cdot \\ b_r \end{bmatrix}} \overset{\mathbf{X}'\mathbf{y}}{\begin{bmatrix} y_{..} \\ y_{1.} \\ y_{2.} \\ \cdot \\ \cdot \\ \cdot \\ y_{p.} \\ y_{.1} \\ y_{.2} \\ \cdot \\ \cdot \\ \cdot \\ y_{.r} \end{bmatrix}}$$

or

$$\begin{cases} rmp + r(t_1 + t_2 + \cdots + t_p) + p(b_1 + b_2 + \cdots + b_r) = y_{..} \\ rm + rt_1 + b_1 + b_2 + \cdots + b_r = y_{1.} \\ rm + rt_2 + b_1 + b_2 + \cdots + b_r = y_{2.} \\ \quad\cdot \qquad\qquad \cdot \qquad\qquad \cdot \\ rm + rt_p + b_1 + b_2 + \cdots + b_r = y_{p.} \\ pm + t_1 + t_2 + \cdots + t_p + pb_1 = y_{.1} \\ pm + t_1 + t_2 + \cdots + t_p + pb_2 = y_{.2} \\ \quad\cdot \qquad\qquad \cdot \qquad\qquad \cdot \\ pm + t_1 + t_2 + \cdots + t_p + pb_r = y_{.r} \end{cases}$$

Note that, as in the case of the CRD and all other designs to follow, the $\mathbf{X}'\mathbf{y}$ vector is a vector of totals. For example, the entry $y_{..}$ is the total of all rp observations, $y_{1.}$ is the total of the r observations associated with treatment 1, $y_{.1}$ is the total of the p observations in block 1, and so on.

To solve these normal equations, we utilize side conditions. Since the $\mathbf{X}'\mathbf{X}$ matrix has $p + r + 1$ columns and has rank $p + r - 1$, we will need $(p + r + 1) - (p + r - 1) = 2$ linearly independent side conditions to solve the system of equations. After examining the set of equations in expanded form, it appears that the following two side conditions will greatly simplify the algebra needed to find a solution:

$$\sum_{i=1}^{p} t_i = 0 \quad \text{and} \quad \sum_{j=1}^{r} b_j = 0 \; .$$

With these side conditions, we obtain the solution set

$$\begin{aligned} m &= \bar{y}_{\cdot\cdot} \, , \\ t_i &= \bar{y}_{i\cdot} - \bar{y}_{\cdot\cdot} \, , \; i = 1, 2, \ldots, p. \\ b_j &= \bar{y}_{\cdot j} - \bar{y}_{\cdot\cdot} \, , \; j = 1, 2, \ldots, r. \end{aligned} \tag{10.3.3}$$

We use these solutions to find least squares estimates but first we need to identify which functions are estimable.

Recall that the estimable functions are the entries of the $\mathbf{X}\boldsymbol{\beta}$ vector and all linear combinations of these entries. For the RCBD model (10.2.1), we find that $\mathbf{X}\boldsymbol{\beta}$ is

$$\begin{bmatrix} \mu + \tau_1 + \beta_1 \\ \mu + \tau_1 + \beta_2 \\ \vdots \\ \mu + \tau_1 + \beta_r \\ \vdots \\ \mu + \tau_p + \beta_1 \\ \mu + \tau_p + \beta_2 \\ \vdots \\ \mu + \tau_p + \beta_r \end{bmatrix}$$

Therefore, the estimable functions are $\mu + \tau_i + \beta_j$ for all i,j and all linear combinations of these functions. For example, $\tau_i - \tau_k$ and $\beta_j - \beta_\ell$ are estimable since $\tau_i - \tau_k = (\mu + \tau_i + \beta_1) - (\mu + \tau_k + \beta_1)$ and $\beta_j - \beta_\ell = (\mu + \tau_1 + \beta_j) - (\mu + \tau_1 + \beta_\ell)$. Also, all treatment contrasts, which are functions of the form $\sum_{i=1}^{p} c_i \tau_i$ where $\sum_{i=1}^{p} c_i = 0$, are estimable. This follows from the fact that

$$\sum_{i=1}^{p} c_i\tau_i = \sum_{i=1}^{p} c_i(\mu + \tau_i + \beta_j),$$

which is a linear combination of the estimable functions $\mu + \tau_i + \beta_j$, $i = 1,2,\ldots,p$. Similarly, block contrasts, which are functions of the form $\sum_{j=1}^{r} h_j\beta_j$ where $\sum_{j=1}^{r} h_j = 0$, are also estimable.

Many functions are not estimable. For example, μ, τ_i, and β_j separately are not estimable since they cannot be written as linear combinations of the $\mu + \tau_i + \beta_j$ functions. Nor is $\mu + \tau_i$ estimable in the RCBD model although this function is estimable in the CRD model.

To find the least squares estimate for a treatment contrast $\sum_{i=1}^{p} c_i\tau_i$, we simply use the solution to the normal equations given in (10.3.3). That is, the least squares estimate for $\sum_{i=1}^{p} c_i\tau_i$ is

$$\sum_{i=1}^{p} c_i t_i = \sum_{i=1}^{p} c_i(\bar{y}_{i\cdot} - \bar{y}_{\cdot\cdot}) = \sum_{i=1}^{p} c_i\bar{y}_{i\cdot}\ .$$

From the Gauss—Markoff Theorem, we know that this is the best linear unbiased estimate for a treatment contrast. It is not difficult to find the variance and the standard error of the corresponding estimator, $\sum_{i=1}^{p} c_i\bar{Y}_{i\cdot}$. Since the Y_{ij}'s are uncorrelated, we have

$$\mathrm{Var}\left(\sum_{i=1}^{p} c_i\bar{Y}_{i\cdot}\right) = \sum_{i=1}^{p} c_i^2\mathrm{Var}(\bar{Y}_{i\cdot}) = \sum_{i=1}^{p} c_i^2\frac{\sigma^2}{r} = \frac{\sigma^2}{r}\sum_{i=1}^{p} c_i^2 \qquad (10.3.4)$$

Using rule 5.3.1, we find the standard error by replacing σ^2 by s^2 and then taking the square root. We summarize these important results.

BLUE for the RCBD Model

Function		Blue	Standard Error	
Treatment Contrast	$\sum_{i=1}^{p} c_i \tau_i$	$\sum_{i=1}^{p} c_i \bar{y}_{i.}$	$\sqrt{\frac{s^2}{r} \sum_{i=1}^{p} c_i^2}$	(10.3.5)

where $\sum_{i=1}^{p} c_i = 0$.

As an illustration, the best linear unbiased estimate for the difference between the effects of treatments i and k, namely $\tau_i - \tau_k$, is simply the difference between the treatment sample means, $\bar{y}_{i.} - \bar{y}_{k.}$.

Next we find a formula for SSe in this setting. Using the solution set (10.3.3), we get

$$\hat{y}_{ij} = m + t_i + b_j = \bar{y}_{..} + (\bar{y}_{i.} - \bar{y}_{..}) + (\bar{y}_{.j} - \bar{y}_{..})$$

which simplifies to

$$\hat{y}_{ij} = \bar{y}_{i.} + \bar{y}_{.j} - \bar{y}_{..} \quad . \tag{10.3.6}$$

As a consequence,

$$\text{SSe} = \sum_{i=1}^{p} \sum_{j=1}^{r} (y_{ij} - \hat{y}_{ij})^2 \tag{10.3.7}$$

$$= \sum_{i=1}^{p} \sum_{j=1}^{r} (y_{ij} - \bar{y}_{i.} - \bar{y}_{.j} + \bar{y}_{..})^2$$

In the exercises, the reader will show that a computational formula for SSe is

$$\text{SSe} = \sum_{i=1}^{p} \sum_{j=1}^{r} y_{ij}^2 - \sum_{i=1}^{p} \frac{y_{i.}^2}{r} - \sum_{j=1}^{r} \frac{y_{.j}^2}{p} + \frac{y_{..}^2}{rp} \quad . \tag{10.3.8}$$

The degrees of freedom for SSe is the number of observations minus the rank of **X**, which works out to be $pr - (p + r - 1) = (p - 1)(r - 1)$.

As a result, the error mean square in an RCBD problem is given by the formula

$$s^2 = SSe/[(p-1)(r-1)] . \qquad (10.3.9)$$

Example 10.3.1 Let us turn to the metal bracket production problem with the numerical results given in Table 10.2.1. Recall these data were obtained in an RCBD experiment with $p = 3$ treatments (machines) and $r = 4$ blocks (workers). From the table we routinely calculate the column totals as $y_{1\cdot} = 360$, $y_{2\cdot} = 277$, $y_{3\cdot} = 294$, the row totals as $y_{\cdot 1} = 203$, $y_{\cdot 2} = 241$, $y_{\cdot 3} = 253$, $y_{\cdot 4} = 234$ and the grand total as $y_{\cdot\cdot} = 931$. To compare the production efficiency of machine A with machine C, we find that the best linear unbiased estimate for the contrast $\tau_1 - \tau_3$ is

$$\bar{y}_{1\cdot} - \bar{y}_{3\cdot} = 90.0 - 73.5 = 16.5.$$

Since $\tau_1 - \tau_3$ represents the mean difference in two hours of production on machines A and C, we estimate that the mean number of acceptable brackets produced on machine A is 16.5 more than the mean number produced on machine C during a two—hour period. Next we calculate SSe and s^2. Using computational formula (10.3.8), we get

$$SSe = \sum_{i=1}^{3} \sum_{j=1}^{4} y_{ij}^2 - \sum_{i=1}^{3} y_{i\cdot}^2/4 - \sum_{j=1}^{4} y_{\cdot j}^2/3 + y_{\cdot\cdot}^2/12$$

$$= 73{,}679 - 73{,}191.25 - 72{,}685 + 72{,}230.083 = 32.833.$$

Then the error mean square is

$$s^2 = SSe/[(p-1)(r-1)] = 32.833/6 = 5.472.$$

Finally, the associated standard error for the treatment contrast $\tau_1 - \tau_3$ is

$$Se\,(\widehat{\tau_1 - \tau_3}) = \sqrt{\frac{s^2}{r} \sum_{i=1}^{3} c_i^2}$$

$$= \sqrt{\frac{5.472}{4} [1^2 + 0^2 + (-1)^2]} = 1.654.$$

10.4 Analysis of Variance and F—Tests

In this section we use the principle of conditional error to obtain tests for the hypothesis H_0 : "no difference in treatment effects" and H_0 : "no difference in block effects." Recall, we need to assume the **normal** linear model to perform tests and properly interpret confidence intervals. Therefore, for the remainder of this chapter we add the assumption that the ε_{ij} in model (10.2.1) are independent $N(0, \sigma^2)$ random variables.

Derivations of F—statistics

First we consider testing H_0 : "no difference in treatment effects," which can be written as

$$H_0 : \tau_1 = \tau_2 = \cdots = \tau_p. \tag{10.4.1}$$

To use the principle of conditional error, we need to write H_0 in the general linear hypothesis form H_0: $\mathbf{L}\boldsymbol{\beta} = \mathbf{0}$. This can be accomplished by writing $\mathbf{L}\boldsymbol{\beta}$ as

$$\mathbf{L}\boldsymbol{\beta} = \begin{bmatrix} 0 & -1 & 1 & 0 & 0 & \cdots & 0 & 0 & \cdots & 0 \\ 0 & -1 & 0 & 1 & 0 & \cdots & 0 & 0 & \cdots & 0 \\ 0 & -1 & 0 & 0 & 1 & \cdots & 0 & 0 & \cdots & 0 \\ \cdot & \cdot & \cdot & \cdot & \cdot & & \cdot & \cdot & & \cdot \\ \cdot & \cdot & \cdot & \cdot & \cdot & & \cdot & \cdot & & \cdot \\ \cdot & \cdot & \cdot & \cdot & \cdot & & \cdot & \cdot & & \cdot \\ 0 & -1 & 0 & 0 & 0 & \cdots & 1 & 0 & \cdots & 0 \end{bmatrix} \begin{bmatrix} \mu \\ \tau_1 \\ \cdot \\ \cdot \\ \cdot \\ \tau_p \\ \beta_1 \\ \cdot \\ \cdot \\ \cdot \\ \beta_r \end{bmatrix} = \begin{bmatrix} \tau_2 - \tau_1 \\ \tau_3 - \tau_1 \\ \tau_4 - \tau_1 \\ \cdot \\ \cdot \\ \cdot \\ \tau_p - \tau_1 \end{bmatrix}$$

(p columns under the τ columns; r columns under the β columns)

It is easily seen that $\mathbf{L}$ is a $(p - 1) \times (p + r + 1)$ matrix and the vector $\mathbf{L}\boldsymbol{\beta}$ constitutes a set of $p - 1$ linearly independent estimable functions since all treatment contrasts are estimable.

To proceed with the principle of conditional error technique (5.4.16), we recall that the "full" model has functional form

$$Y_{ij} = \mu + \tau_i + \beta_j + \varepsilon_{ij}, \quad i = 1,2,\ldots,p \; ; \; j = 1,2,\ldots,r,$$

with the error sum of squares

$$SSe = \sum_{i=1}^{p} \sum_{j=1}^{r} y_{ij}^2 - \sum_{i=1}^{p} \frac{y_{i\cdot}^2}{r} - \sum_{j=1}^{r} \frac{y_{\cdot j}^2}{p} + \frac{y_{\cdot\cdot}^2}{rp},$$

as given by (10.3.8). For the "reduced" model, with $H_0 : \tau_1 = \tau_2 = \cdots = \tau_p$, we can let $\mu + \tau_i = \mu^*$ for all i. Then the reduced model becomes

$$Y_{ij} = \mu^* + \beta_j + \varepsilon^*_{ij}, \quad i = 1,2,\ldots,p \; ; \; j = 1,2,\ldots,r.$$

The reader will recognize this reduced model as a CRD model (7.2.1) with a blocking factor in lieu of a treatment factor. The error sum of squares for this reduced model is then found by making appropriate modifications to formula (7.3.5) resulting in

$$SSe^* = \sum_{i=1}^{p} \sum_{j=1}^{r} y_{ij}^2 - \sum_{j=1}^{r} y_{\cdot j}^2/p.$$

Hence

$$SSe^* - SSe = \sum_{i=1}^{p} y_{i\cdot}^2/r - y_{\cdot\cdot}^2/rp = \sum_{j=1}^{p} r(\bar{y}_{i\cdot} - \bar{y}_{\cdot\cdot})^2. \qquad (10.4.2)$$

We call (10.4.2) **the treatment sum of squares**, denoted by SStr. Its associated degrees of freedom is $p - 1$ since the $\mathbf{L}$ matrix has rank $p - 1$. The symbol MStr will represent the treatment mean square defined as $MStr = SStr/(p - 1)$. From Theorem 5.4.2, our test statistic for the null hypothesis $H_0 : \tau_1 = \tau_2 = \cdots = \tau_p$ is

$$f = \frac{(SSe^* - SSe)/(p - 1)}{SSe/[(p - 1)(r - 1)]} = \frac{MStr}{MSe}. \qquad (10.4.3)$$

We reject H_0 if $f = MStr/MSe > f_{\alpha,p-1,(p-1)(r-1)}$.

In a similar manner, we can obtain a test for $H_0 : \mathbf{L}_1\boldsymbol{\beta} = \mathbf{0}$, where $\mathbf{L}_1$ is a $(r - 1) \times (p + r + 1)$ matrix and

$$\mathbf{L}_1\boldsymbol{\beta} = \begin{bmatrix} \beta_2 - \beta_1 \\ \beta_3 - \beta_1 \\ \cdot \\ \cdot \\ \cdot \\ \beta_r - \beta_1 \end{bmatrix}$$

constitutes a set of linearly estimable functions. Imposing the hypothesis $\mathbf{L}_1\boldsymbol{\beta} = \mathbf{0}$ on the "full" model, we obtain the "reduced" model

$$Y_{ij} = \mu^{**} + \tau_i + \varepsilon_{ij}^{**}, \quad i = 1,2,\ldots,p \; ; \; j = 1,2,\ldots,r ,$$

where $\mu^{**} = \mu + \beta_j$ for all j. We observe that this reduced model is the CRD model (7.2.1). As a result, the conditional error sum of squares can immediately be written from (7.3.5) as

$$SSe^{**} = \sum_{i=1}^{p} \sum_{j=1}^{r} y_{ij}^2 - \sum_{i=1}^{p} y_{i\cdot}^2/r .$$

Let SSbl = SSe** − SSe, called the **block sum of squares**. Then

$$SSbl = \sum_{j=1}^{r} y_{\cdot j}^2/p - y_{\cdot\cdot}^2/rp = \sum_{j=1}^{r} p(\bar{y}_{\cdot j} - \bar{y}_{\cdot\cdot})^2. \qquad (10.4.4)$$

Hence the statistic used to test $H_0 : \beta_1 = \beta_2 = \cdots = \beta_r$ is

$$f = \frac{SSbl/(r-1)}{SSe/[(p-1)\;(r-1)]} = \frac{MSbl}{MSe} . \qquad (10.4.5)$$

Accordingly, we reject H_0 if $f = MSbl/MSe > f_{\alpha,r-1,(p-1)(r-1)}$.

Analysis of Variance for an RCBD

Before summarizing the above tests in an ANOVA table, there is one more sum of squares needed and that is the total sum of squares introduced earlier in Chapters 6 and 7. As a reminder, its formula is

$$SSt = \sum_{i=1}^{p} \sum_{j=1}^{r} (y_{ij} - \bar{y}_{\cdot\cdot})^2 = \sum_{i=1}^{p} \sum_{j=1}^{r} y_{ij}^2 - y_{\cdot\cdot}^2/rp. \qquad (10.4.6)$$

Table 10.4.1 ANOVA for an RCBD

Source	df	SS	MS	f
Blocks	r–1	SSbl	MSbl=SSbl/(r–1)	MSbl/MSe
Treatments	p–1	SStr	MStr=SStr/(p–1)	MStr/MSe
Error	(r–1)(p–1)	SSe	MSe=SSe/[(r–1)(p–1)]	
Total	rp − 1	SSt		

where the computing formulas for SSbl, SStr, and SSt are given by (10.4.4), (10.4.2), and (10.4.6), respectively, and SSe = SSt − SSbl − SStr.

Several comments concerning Table 10.4.1 are in order. First, it is easy to see from the computational formula for SSe given by (10.3.8) and those noted for SSt, SSbl, and SStr, that

$$SSt = SSbl + SStr + SSe. \tag{10.4.7}$$

As a consequence, SSe can be obtained by subtraction as indicated in the last line of the table. Notice that the degrees of freedom have the same additive relationship. Second, except for $n_i = r$ for all i, the formulas for SStr and SSt are the same as those given for the CRD. Hence these quantities are the same regardless of the design used. Again, using the computing formulas for SSe and SSbl, we see that their sum is

$$SSe + SSbl = \sum_{i=1}^{p} \sum_{j=1}^{r} y_{ij}^2 - \sum_{i=1}^{p} y_{i\cdot}^2/r .$$

The right–hand side of the above equation is the computing formula for the error sum of squares of a CRD (7.3.5). Hence SSe(RCBD) = SSe(CRD) − SSbl, which implies that SSe(RCBD) ≤ SSe(CRD). This confirms a point made in the first section of this chapter and demonstrated in Example 10.1.1, namely, a primary purpose for blocking is to reduce the experimental error. Interestingly, the same relationship does not necessarily hold for the error mean squares associated with the two designs because of the degrees of freedom element. More will be said about this later in the chapter.

Example 10.4.1 Refer to the metal bracket production data given in Table 10.2.1 obtained in an RCBD experiment with machines as treatments and workers as blocks. We will construct the ANOVA table for these data. Previously we had obtained the following summary calculations: $p = 3$, $r = 4$, $y_{..} = 931$, $y_{1.} = 360$, $y_{2.} = 277$, $y_{3.} = 294$, $y_{.1} = 203$, $y_{.2} = 241$, $y_{.3} = 253$, $y_{.4} = 234$ and $\sum_{i=1}^{3} \sum_{j=1}^{4} y_{ij}^2 = 73{,}679$. To three decimal places, the sums of squares are

$$SSt = 73{,}679 - (931)^2/12 = 1{,}448.917$$

$$SStr = [(360)^2 + (227)^2 + (294)^2]/4 - (931)^2/12 = 961.167$$

$$SSbl = [(203)^2 + (241)^2 + (253)^2 + (234)^2]/3 - (931)^2/12 = 454.917$$

$$SSe = SSt - SSb1 - SStr = 1448.917 - 961.167 - 454.917 = 32.833.$$

With these basic calculations, we are able to complete the ANOVA table.

Table 10.4.2 ANOVA Table for Metal Bracket Production

Source	df	SS	MS	f
Workers (blocks)	3	454.917	151.639	27.71
Machines (treatments)	2	961.167	480.583	87.82
Error	6	32.833	5.472	
Total	11	1,448.917		

The 0.05 level test for H_0: $\tau_1 = \tau_2 = \tau_3$ rejects H_0 if $f > f_{.05, 2, 6} = 5.14$. Since $f = 87.82$, we reject H_0 and conclude that there is a significant difference in treatment (machine) effects. Notice, however, we do not know which of the machines differ from one another in efficiency. This problem will be addressed in the next section where multiple comparisons are discussed in a RCBD set—up. The 0.05 level test for H_0: $\beta_1 = \beta_2 = \beta_3 = \beta_4$ rejects H_0 if $f > f_{.05, 3, 6} = 4.76$.

Since f = 27.71, we reject H_0. Our interest in this latter hypothesis is primarily to substantiate our feeling that machine output on average will differ from worker to worker. Rejection of H_0, therefore, is justification for using workers as a blocking factor.

Use of SAS in an RCBD Analysis

The metal bracket data can be analyzed using SAS programs. In Table 10.4.3 we find that SAS commands and the printout in response to the commands through MODEL AMTPROD = WORKER MACHINE. Excerpts of the printout in response to the last three commands appear in the next section.

Table 10.4.3 SAS Command and ANOVA Tests for Metal Bracket Data

SAS Commands

```
OPTIONS LS = 80;
DATA RCBD;
INPUT WORKER MACHINE AMTPROD @@;
CARDS;
1 1 82 1 2 59 1 3 62
2 1 91 2 2 75 2 3 75
3 1 98 3 2 74 3 3 81
4 1 89 4 2 69 4 3 76
PROC GLM; CLASS WORKER MACHINE;
MODEL AMTPROD = WORKER MACHINE;
MEANS MACHINE;
ESTIMATE 'TRT 1 VS TRT 2' MACHINE 1 -1 0;
MEANS MACHINE/TUKEY CLDIFF;
```

SAS Printout

```
General Linear Models Procedure
    Class Level Information

Class      Levels     Values

WORKER          4     1 2 3 4

MACHINE         3     1 2 3

Number of observations in data set = 12
```

General Linear Models Procedure

Dependent Variable: AMTPROD

Source	DF	Sum of Squares	Mean Square	F Value	Pr > F
Model	5	1416.0833333	283.2166667	51.76	0.0001
Error	6	32.8333333	5.4722222		
Corrected Total	11	1448.9166667			

R-Square	C.V.	Root MSE	AMTPROD Mean
0.977339	3.015181	2.3392781	77.583333

Source	DF	Type I SS	Mean Square	F Value	Pr > F
WORKER	3	454.91666667	151.63888889	27.71	0.0007
MACHINE	2	961.16666667	480.58333333	87.82	0.0001

Source	DF	Type III SS	Mean Square	F Value	Pr > F
WORKER	3	454.91666667	151.63888889	27.71	0.0007
MACHINE	2	961.16666667	480.58333333	87.82	0.0001

The first thing we notice is that the SAS printout is not in the same ANOVA format as given in Table 10.4.2. However, the essentials for presenting the ANOVA table in standard form (Table 10.4.1) are there. Recognizing that the machines are treatments and workers are blocks, we find that SStr = 961.167, SSbl = 454.917, SSe = 32.833, and SSt = 1448.917, the latter called the "corrected" total sum of squares. Notice that the degrees of freedom corresponding to these sums of squares are also printed out. The only mean square appearing is the error mean square, MSe = 5.472. However, the F—values together with the p—values are given. For example, in testing H_0 : "No difference in treatment (machine) effects," we see that $f = 87.82$ with $p = 0.0001$. Since the p—value is so small, we have strong evidence for rejecting H_0. Similarly, in testing H_0 : "No difference in block (workers) effects," we have $f = 27.71$ and $p = 0.0007$ which means we would reject H_0 for all fixed α levels greater than 0.0007.

Before closing this subsection on the SAS computer printout, it should be pointed out that in an RCBD problem, SAS Type I and Type III sums of squares are the same. This is not the case in all situations as we shall see in the next chapter. More light will be shed on this subject later in this chapter when the notions of adjusted and unadjusted sums of squares in an experimental design setting are discussed.

Finally, out of curiosity, the reader is probably wondering about the "model" sum of squares, 1416.083. It is the sum SStr + SSbl. The corresponding F—test generally is of little interest to us although it does give us information concerning the appropriateness of the model, a consideration important in model building, particularly in regression analysis. Here we have assumed the structure of the model to be (10.2.1) and in the analysis of variance we present a partitioning of the model sum of squares into block and treatment sums of squares.

10.5 Inference for Treatment Contrasts

In an RCBD, the estimable functions of primary interest are the treatment contrasts. In this section we present the forms of a confidence interval and a t—test for an individual treatment contrast, $\sum_{i=1}^{p} c_i \tau_i$, where $\sum_{i=1}^{p} c_i = 0$. We then conclude this section with an illustration of the Tukey multiple comparison procedure.

Confidence Intervals and Tests

Recall from Section 5.3 that the form of a confidence interval for an estimable function is

$$\boldsymbol{\ell}'\mathbf{b} \pm t_{\alpha/2\,,\,(p-1)\,(r-1)} \mathrm{Se}(\boldsymbol{\ell}'\hat{\boldsymbol{\beta}})\ ,$$

where $\boldsymbol{\ell}'\mathbf{b}$ is the least squares estimate. Using the information from (10.3.5) , the confidence interval for a general treatment contrast is

$$\sum_{i=1}^{p} c_i \bar{y}_{i\cdot} \pm t_{\alpha/2\,,\,(p-1)(r-1)} \sqrt{\frac{s^2}{r} \sum_{i=1}^{p} c_i^2}\ .$$

For the "pairwise" contrast $\tau_i - \tau_k$, the above confidence interval "reduces" to the expression

$$(\bar{y}_{i\cdot} - \bar{y}_{k\cdot}) \pm t_{\alpha/2\,,\,(p-1)(r-1)} \sqrt{2s^2/r}.$$

For future reference, we summarize these facts as follows:

100(1 − α)% Confidence Intervals for a RCBD	
Function	**Interval**
Difference, $\tau_i - \tau_k$:	$(\bar{y}_{i.} - \bar{y}_{k.}) \pm t_{\alpha/2,(p-1)(r-1)}\sqrt{2s^2/r}$ (10.5.1)
Contrast, $\sum_{i=1}^{p} c_i\tau_i$, $\sum_{i=1}^{p} c_i = 0$:	$\sum_{i=1}^{p} c_i\bar{y}_{i.} \pm t_{\alpha/2,(p-1)(r-1)}\sqrt{s^2 \sum_{i=1}^{p} c_i^2/r}$

In section 5.4, the t—test for one estimable function, $\boldsymbol{\ell}'\boldsymbol{\beta}$, was derived. Recall that the form of the test statistic is

$$t = (\boldsymbol{\ell}'\mathbf{b} - \gamma)/\text{Se}\,(\boldsymbol{\ell}'\hat{\boldsymbol{\beta}})$$

for testing the null hypothesis $H_0 : \boldsymbol{\ell}'\boldsymbol{\beta} = \gamma$, where γ is some constant. In the RCBD setting, these formulas work out as follows:

Test Statistics Associated with RCBD	
Null Hypothesis	**Test Statistic**
$H_0: \tau_i - \tau_k = \gamma$	$t = (\bar{y}_{i.} - \bar{y}_{k.} - \gamma)/\sqrt{2s^2/r}$ (10.5.2)
$H_0: \sum_{i=1}^{p} c_i\tau_i = \gamma$, $\sum_{i=1}^{p} c_i = 0$	$t = (\sum_{i=1}^{p} c_i\bar{y}_{i.} - \gamma)/\sqrt{s^2 \sum_{i=1}^{p} c_i^2/r}$
Note: All rejection regions are based on the t—distribution with (p − 1)(r − 1) degrees of freedom.	

Example 10.5.1 Consider again the metal bracket production data given in Table 10.2.1 and the associated ANOVA table displayed in Example 10.4.1. Let us find a 95% confidence interval for $\tau_1 - \tau_2$ from these data. Using (10.5.1) with $p = 3$ and $r = 4$, we find the confidence interval is

$$(\bar{y}_{1.} - \bar{y}_{2.}) \pm t_{.025,6}\sqrt{2s^2/4}.$$

Since $\bar{y}_{1.} = 360/4 = 90$, $\bar{y}_{2.} = 277/4 = 69.25$ and $s^2 = 5.472$, this interval is

$$(90 - 69.25) \pm 2.447/\sqrt{2(5.472)/4}$$

or

$$(16.70,\ 24.80).$$

Now $\tau_1 - \tau_2$ represents the mean difference in two hours of production on machines A and B. Therefore, we are estimating that the average number of brackets produced on machine A during a two—hour period is between 16.7 and 24.8 more than that produced on machine B during the same period. We can also arrive at the above confidence interval using SAS. In the list of commands in Table 10.4.3 we find

ESTIMATE 'TRT 1 VS TRT 2' MACHINE 1 —1 0 ; .

The response to this ESTIMATE command pertinent to this example follows:

ESTIMATE	T FOR H0: PARAMETER = 0	PR > |T|	STD ERROR OF ESTIMATE
20.75000000	12.54	0.0001	1.65411944

Since a $100(1-\alpha)\%$ confidence interval is of the form point estimate ± $t_{\alpha/2,\nu}$ (standard error), where ν is the error degrees of freedom, we see from the printout that a $100(1-\alpha)\%$ confidence interval for $\tau_1 - \tau_2$ is

$$20.75 \pm t_{\alpha/2,6}\,(1.65411944).$$

We need only look up the proper $t_{\alpha/2,6}$ value from a t—table in order to complete the computation.

Multiple Comparison Example

In Chapter 9 we studied various multiple comparison procedures which can be used in order to detect which treatment pairs in a CRD experiment are significantly different. In assuming the normal linear model for the RCBD, we know that $\bar{Y}_{1\cdot}, \bar{Y}_{2\cdot}, \ldots, \bar{Y}_{p\cdot}$ are independent normal random variables with the same variance. It follows that these same procedures can be used in the case of an RCBD experiment. Here we will illustrate the construction of Tukey simultaneous confidence intervals in the present setting. For the RCBD, the Tukey simultaneous confidence intervals for pairwise treatment contrasts $\tau_i - \tau_k$ with overall confidence level of $100(1-\alpha)\%$ are the set

$$(\bar{y}_{i\cdot} - \bar{y}_{k\cdot}) \pm q_{\alpha,p,(p-1)(r-1)} \sqrt{s^2/r} \quad \text{for all } 1 \leq i < k \leq p. \qquad (10.5.3)$$

Example 10.5.2 Let us set out to find the 95% Tukey simultaneous confidence intervals for the pairwise treatment contrasts in the metal bracket production example. From Example 10.4.1, we have $p = 3$,

$r = 4$, $\bar{y}_{1\cdot} = 90.00$, $\bar{y}_{2\cdot} = 69.25$, $\bar{y}_{3\cdot} = 73.50$, and $s^2 = 5.472$. The Studentized range table in Appendix B yields the remaining piece of information needed for formula (10.5.3), namely, $q_{.05,3,6} = 4.34$. Therefore, the 95% simultaneous confidence intervals are $(\bar{y}_{i\cdot} - \bar{y}_{k\cdot}) \pm 4.34\sqrt{5.472/4}$ and displayed as follows:

Treatment Pair	95% Tukey Simultaneous Interval
$\tau_1 - \tau_2$	20.75 ± 5.08 or (15.67, 25.83)
$\tau_1 - \tau_3$	16.50 ± 5.08 or (11.42, 21.58)
$\tau_2 - \tau_3$	-4.25 ± 5.08 or (-9.33, 0.83).

Since zero is only contained in the third interval, we conclude that production using machine A (treatment 1) is significantly different than that using machines B and C (treatments 2 and 3). At the $\alpha = 0.05$ level, we have not found the production difference between machines B and C to be significant.

Treatment means and Tukey simultaneous confidence intervals can also be obtained on SAS in the RCBD setting. For the above example, this is done by using the commands.

MEANS MACHINE;

and

MEANS MACHINE/TUKEY CLDIFF;

as given in Table 10.4.3. The corresponding printouts follow.

```
                General Linear Models Procedure

      Level of              -----------AMTPROD----------
      MACHINE      N          Mean                 SD

      1            4       90.0000000          6.58280589
      2            4       69.2500000          7.32006375
      3            4       73.5000000          8.10349719
```

```
                    General Linear Models Procedure

      Tukey's Studentized Range (HSD) Test for variable: AMTPROD

      NOTE: This test controls the type I experimentwise error rate.

        Alpha= 0.05  Confidence= 0.95  df= 6  MSE= 5.472222
             Critical Value of Studentized Range= 4.339
               Minimum Significant Difference= 5.0751

  Comparisons significant at the 0.05 level are indicated by '***'.

                     Simultaneous                Simultaneous
                         Lower      Difference      Upper
        MACHINE       Confidence     Between     Confidence
       Comparison        Limit         Means        Limit

      1     - 3          11.425       16.500       21.575    ***
      1     - 2          15.675       20.750       25.825    ***

      3     - 1         -21.575      -16.500      -11.425    ***
      3     - 2          -0.825        4.250        9.325

      2     - 1         -25.825      -20.750      -15.675    ***
      2     - 3          -9.325       -4.250        0.825
```

10.6 Reparametrization of an RCBD

In this section we deal with the reparametrization of the traditional model (10.2.1) so that regression techniques can be brought to bear in the analysis of a randomized block experiment.

Introduction

We have already seen that in the case of a moderate–sized RCBD data set, obtaining the analysis of variance and performing auxiliary inference procedures "by hand" with a pocket calculator is not an unreasonable task. In fact, one may legitimately raise the question: "Why even use the computer?" The answer is that there are times when the realization of an experiment does not materialize as planned. For example, an investigator may have designed an experiment involving litters of rats where a litter serves as a block and rats within litters are the experimental units. Let us further suppose, for simplicity and balance purposes, the litters are all the same size so that the experimental design is that of a randomized complete block. Now if one or more of the rats die during the course of the investigation, the resulting data no longer is that for an RCBD and the computing formulas in previous sections no longer hold. In addition, the complexity of a proper analysis increases greatly. It is in a case such as this that the use of regression analysis can be used to considerable advantage, particularly in the absence of specialized computer programs such as SAS GLM. This will be a major theme of the next chapter. Also, this is an opportune time for the reader to first become acquainted with the mechanics of analyzing data using a general regression approach which provides tools for the possible "rescue" of undesigned or poorly designed experiments or of experiments suffering mishaps of one kind or another.

In the process of our discussion, the reader will also be introduced to the concept of orthogonality in a design and reintroduced to the notion of adjusted and unadjusted sums of squares.

An Example

For comparison purposes, we illustrate the procedure for reparametrizing an RCBD model using the metal bracket production set—up. Recall that the form of the traditional model for this experiment consisting of $p = 3$ treatments and $r = 4$ blocks is

$$Y_{ij} = \mu + \tau_i + \beta_j + \varepsilon_{ij}, \; i = 1, 2, 3, \; ; \; j = 1, 2, 3, 4. \tag{10.6.1}$$

In Section 10.2 we noted that this model of eight parameters has rank six. To come up with a full—rank model, that is, a linear model with six parameters, we define the following variables:

$$x_1 = \begin{cases} 1, & \text{if treatment 1 is applied} \\ 0, & \text{if not} \end{cases}$$

$$x_2 = \begin{cases} 1, & \text{if treatment 2 is applied} \\ 0, & \text{if not} \end{cases}$$

$$x_3 = \begin{cases} 1, & \text{if observation is in block 1} \\ 0, & \text{if not} \end{cases}$$

$$x_4 = \begin{cases} 1, & \text{if observation is in block 2} \\ 0, & \text{if not} \end{cases}$$

$$x_5 = \begin{cases} 1, & \text{if observation is in block 3} \\ 0, & \text{if not.} \end{cases}$$

Then the reparametrized model for this problem is

$$Y = \alpha_0 + \alpha_1 x_1 + \alpha_2 x_2 + \alpha_3 x_3 + \alpha_4 x_4 + \alpha_5 x_5 + \varepsilon, \tag{10.6.2}$$

with the distribution assumptions concerning ε as before. In matrix form, we write

$\mathbf{Y} = \mathbf{X}\boldsymbol{\alpha} + \varepsilon$, where the vectors and matrix are given by

$$\underset{\mathbf{Y}}{\begin{bmatrix} Y_{11} \\ Y_{12} \\ Y_{13} \\ Y_{14} \\ Y_{21} \\ Y_{22} \\ Y_{23} \\ Y_{24} \\ Y_{31} \\ Y_{32} \\ Y_{33} \\ Y_{34} \end{bmatrix}} = \underset{\boldsymbol{\mathcal{X}}}{\begin{bmatrix} 1 & 1 & 0 & 1 & 0 & 0 \\ 1 & 1 & 0 & 0 & 1 & 0 \\ 1 & 1 & 0 & 0 & 0 & 1 \\ 1 & 1 & 0 & 0 & 0 & 0 \\ 1 & 0 & 1 & 1 & 0 & 0 \\ 1 & 0 & 1 & 0 & 1 & 0 \\ 1 & 0 & 1 & 0 & 0 & 1 \\ 1 & 0 & 1 & 0 & 0 & 0 \\ 1 & 0 & 0 & 1 & 0 & 0 \\ 1 & 0 & 0 & 0 & 1 & 0 \\ 1 & 0 & 0 & 0 & 0 & 1 \\ 1 & 0 & 0 & 0 & 0 & 0 \end{bmatrix}} \underset{\boldsymbol{\alpha}}{\begin{bmatrix} \alpha_0 \\ \alpha_1 \\ \alpha_2 \\ \alpha_3 \\ \alpha_4 \\ \alpha_5 \end{bmatrix}} + \underset{\boldsymbol{\varepsilon}}{\begin{bmatrix} \varepsilon_{11} \\ \varepsilon_{12} \\ \varepsilon_{13} \\ \varepsilon_{14} \\ \varepsilon_{21} \\ \varepsilon_{22} \\ \varepsilon_{23} \\ \varepsilon_{24} \\ \varepsilon_{31} \\ \varepsilon_{32} \\ \varepsilon_{33} \\ \varepsilon_{34} \end{bmatrix}}$$

When we compare matrix $\boldsymbol{\mathcal{X}}$ of the reparametrized model with matrix $\mathbf{X}$ of the traditional model (Section 10.2), it is easy to see that the matrix $\boldsymbol{\mathcal{X}}$ results from matrix $\mathbf{X}$ after deleting the columns of $\mathbf{X}$ associated with the last treatment (here treatment 3) and the last block (here block 4). That is, the reparametrized model is essentially the traditional model with the side conditions $\tau_3 = 0$ and $\beta_4 = 0$ imposed. This makes $\boldsymbol{\mathcal{X}}$ full rank.

To identify the relationships between the parameters of the traditional model (10.6.1) and the reparametrized model (10.6.2), we equate the $E(Y_{ij})$ for both models. Suppose we choose observations Y_{34}, Y_{14} and Y_{31}. Then

Traditional Model	Reparametrized Model
$E(Y_{34}) = \mu + \tau_3 + \beta_4$	$E(Y_{34}) = \alpha_0$
$E(Y_{14}) = \mu + \tau_1 + \beta_4$	$E(Y_{14}) = \alpha_0 + \alpha_1$
$E(Y_{31}) = \mu + \tau_3 + \beta_1$	$E(Y_{31}) = \alpha_0 + \alpha_3$

On equating these expected values and solving for the α_i, we get $\alpha_0 = \mu + \tau_3 + \beta_4$, $\alpha_1 = \tau_1 - \tau_3$ and $\alpha_3 = \beta_1 - \beta_4$. For reference purposes we now list the correspondence between the parameters of the two models obtained by equating all twelve $E(Y_{ij})$ and then solving simultaneously. For this illustration where $p = 3$ treatments and $r = 4$ blocks, the correspondence follows.

$$\begin{cases} \alpha_0 = \mu + \tau_3 + \beta_4 \\ \alpha_1 = \tau_1 - \tau_3 \\ \alpha_2 = \tau_2 - \tau_3 \\ \alpha_3 = \beta_1 - \beta_4 \\ \alpha_4 = \beta_2 - \beta_4 \\ \alpha_5 = \beta_3 - \beta_4 \end{cases} \tag{10.6.3}$$

The reader should be aware that relationships (10.6.3) are not unique in the sense they depend upon the definitions given to the dummy variable x_i. To find the analysis of variance table and f—tests in this problem using the reparametrized model, we note from (10.6.3) that

$$H_0 : \tau_1 = \tau_2 = \tau_3 \text{ is equivalent to } H_0 : \alpha_1 = \alpha_2 = 0.$$

Also,

$$H_0 : \beta_1 = \beta_2 = \beta_3 = \beta_4 \text{ is equivalent to } H_0 : \alpha_3 = \alpha_4 = \alpha_5 = 0.$$

Hence, to obtain the ANOVA corresponding to the tests of these hypotheses we use the following models:

$$\text{Full Model: } Y = \alpha_0 + \overbrace{\alpha_1 x_1 + \alpha_2 x_2}^{\text{treatments}} + \overbrace{\alpha_3 x_3 + \alpha_4 x_4 + \alpha_5 x_5}^{\text{blocks}} + \varepsilon$$

Reduced Model 1: $Y = \alpha_0^* + \alpha_3^* x_3 + \alpha_4^* x_4 + \alpha_5^* x_5 + \varepsilon^*$ (block variables only)

Reduced Model 2: $Y = \alpha_0^{**} + \alpha_1^{**}x_1 + \alpha_2^{**}x_2 + \varepsilon^{**}$ (treatment variables only).

In fitting the full model, the obtained regression SSe is identical to the RCBD traditional model SSe. The fitting of reduced models 1 and 2 yields SSe* and SSe**, respectively. Then by the principle of conditional error we get SStr = SSe* − SSe and SSbl = SSe** − SSe. Of course, the fitting of all three models produces the same total sum of squares, SSt.

To illustrate these calculations with a numerical example, we next present SAS printout excerpts from the analyses for the full and the two reduced models using the metal bracket production data contained in Table 10.2.1. Of course, any other regression computer program can be utilized here as well. In Table 10.6.1, the SAS commands necessary to yield the printout excerpts found in the remainder of this section are given first. This is followed by the display of the data and the GLM printout for the fitted full regression (reparametrized) model.

Table 10.6.1 SAS Regression Commands and Full Model Printout

<u>SAS Commands</u>

```
OPTIONS LS = 80;
DATA BLK;
INPUT  Y  X1  X2  X3  X4  X5  @@ ;
CARDS;
82  1  0  1  0  0  91  1  0  0  1  0
98  1  0  0  0  1  89  1  0  0  0  0
59  0  1  1  0  0  75  0  1  0  1  0
74  0  1  0  0  1  69  0  1  0  0  0
62  0  0  1  0  0  75  0  0  0  1  0
81  0  0  0  0  1  76  0  0  0  0  0
PROC PRINT;
PROC GLM;
MODEL  Y = X1  X2  X3  X4  X5  /  I ;
ESTIMATE  'TRT  1  VS  TRT  2 '  X1  1  X2  -1  ;
PROC GLM ;  MODEL  Y = X3  X4  X5 ;
PROC GLM ;  MODEL  Y = X1  X2 ;
```

SAS Printout

OBS	Y	X1	X2	X3	X4	X5
1	82	1	0	1	0	0
2	91	1	0	0	1	0
3	98	1	0	0	0	1
4	89	1	0	0	0	0
5	59	0	1	1	0	0
6	75	0	1	0	1	0
7	74	0	1	0	0	1
8	69	0	1	0	0	0
9	62	0	0	1	0	0
10	75	0	0	0	1	0
11	81	0	0	0	0	1
12	76	0	0	0	0	0

General Linear Models Procedure

X'X Inverse Matrix

	INTERCEPT	X1	X2	X3	X4
INTERCEPT	0.5	-0.25	-0.25	-0.333333333	-0.333333333
X1	-0.25	0.5	0.25	-5.55112E-17	-5.55112E-17
X2	-0.25	0.25	0.5	-5.55112E-17	-5.55112E-17
X3	-0.333333333	-5.55112E-17	-5.55112E-17	0.6666666667	0.3333333333
X4	-0.333333333	-5.55112E-17	-5.55112E-17	0.3333333333	0.6666666667
X5	-0.333333333	-5.55112E-17	-5.55112E-17	0.3333333333	0.3333333333
Y	73.916666667	16.5	-4.25	-10.33333333	2.3333333333

	X5	Y
INTERCEPT	-0.333333333	73.916666667
X1	-5.55112E-17	16.5
X2	-5.55112E-17	-4.25
X3	0.3333333333	-10.33333333
X4	0.3333333333	2.3333333333
X5	0.6666666667	6.3333333333
Y	6.3333333333	32.833333333

General Linear Models Procedure

Dependent Variable: Y

Source	DF	Sum of Squares	Mean Square	F Value	Pr > F
Model	5	1416.0833333	283.2166667	51.76	0.0001
Error	6	32.8333333	5.4722222		
Corrected Total	11	1448.9166667			

R-Square	C.V.	Root MSE	Y Mean
0.977339	3.015181	2.3392781	77.583333

Source	DF	Type I SS	Mean Square	F Value	Pr > F
X1	1	925.04166667	925.04166667	169.04	0.0001
X2	1	36.12500000	36.12500000	6.60	0.0424
X3	1	393.36111111	393.36111111	71.88	0.0001
X4	1	1.38888889	1.38888889	0.25	0.6324
X5	1	60.16666667	60.16666667	10.99	0.0161

Source	DF	Type III SS	Mean Square	F Value	Pr > F
X1	1	544.50000000	544.50000000	99.50	0.0001
X2	1	36.12500000	36.12500000	6.60	0.0424
X3	1	160.16666667	160.16666667	29.27	0.0016
X4	1	8.16666667	8.16666667	1.49	0.2677
X5	1	60.16666667	60.16666667	10.99	0.0161

Parameter	Estimate	T for H0: Parameter=0	Pr > |T|	Std Error of Estimate
TRT 1 VS TRT 2	20.7500000	12.54	0.0001	1.65411944

Parameter	Estimate	T for H0: Parameter=0	Pr > |T|	Std Error of Estimate
INTERCEPT	73.91666667	44.69	0.0001	1.65411944
X1	16.50000000	9.98	0.0001	1.65411944
X2	-4.25000000	-2.57	0.0424	1.65411944
X3	-10.33333333	-5.41	0.0016	1.91001260
X4	2.33333333	1.22	0.2677	1.91001260
X5	6.33333333	3.32	0.0161	1.91001260

From the above printout, we see that the least squares equation $\hat{y} = a_0 + a_1x_1 + a_2x_2 + a_3x_3 + a_4x_4 + a_5x_5$ is

$$\hat{y} = 73.917+16.500x_1-4.250x_2-10.333x_3+2.333x_4+6.333x_5 \quad (10.6.4)$$

and we observe that the model accounts for more than 97.7% of the variation in the observations. Also, we see that SSe = 32.833 with 6 degrees of freedom and SSt = 1448.917 agree with Table 10.4.2 entries.

To continue our task of computing the analysis of variance using regression, we fit the reduced model containing block variables only. In this case the estimates of the coefficients of the fitted function $Y = \alpha_0^* + \alpha_3^*x_3 + \alpha_4^*x_4 + \alpha_5^*x_5 + \varepsilon^*$ are immaterial to our analysis but the analysis of variance is important since it yields SSe*. Table 10.6.2 displays this ANOVA printout and we find that SSe* = 994.000 with 8 degrees of freedom.

Table 10.6.2 ANOVA for Fitting $Y = \alpha_0^* + \alpha_3^* x_3 + \alpha_4^* x_4 + \alpha_5^* x_5 + \varepsilon^*$

General Linear Models Procedure

Dependent Variable: Y

Source	DF	Sum of Squares	Mean Square	F Value	Pr > F
Model	3	454.91666667	151.63888889	1.22	0.3635
Error	8	994.00000000	124.25000000		
Corrected Total	11	1448.91666667			

Finally, we fit the model with treatment variables only, $Y = \alpha_0^{**} + \alpha_1^{**} x_1 + \alpha_2^{**} x_2 + \varepsilon^{**}$, to obtain SSe** = 487.750 with 9 degrees of freedom. The SAS printout of the associated ANOVA is given in Table 10.6.3.

Table 10.6.3 ANOVA for fitting $Y = \alpha_0^{**} + \alpha_1^{**} x_1 + \alpha_2^{**} x_2 + \varepsilon^{**}$

General Linear Models Procedure

Dependent Variable: Y

Source	DF	Sum of Squares	Mean Square	F Value	Pr > F
Model	2	961.16666667	480.58333333	8.87	0.0075
Error	9	487.75000000	54.19444444		
Corrected Total	11	1448.91666667			

We are now able to complete the analysis of variance table as previously directed. We find that

$$SStr = SSe^* - SSe = 994.000 - 32.833 = 961.167$$

with $8 - 6 = 2$ degrees of freedom, and

$$SSbl = SSe^{**} - SSe = 487.750 - 32.833 = 454.917$$

which has $9 - 6 = 3$ degrees of freedom. These values coincide with the values found in Table 10.4.2. The remaining entries in the ANOVA table can now be easily calculated.

The full rank reparametrized model (10.6.2) can also be used to find point and interval estimates. To demonstrate this, we turn to some examples previously worked out using the traditional model formulas. In

Example 10.3.1 we obtained the point estimate for $\tau_1 - \tau_3$ and the standard error of its estimator. Now from (10.6.3), we know that $\alpha_1 = \tau_1 - \tau_3$. Therefore the point estimate of $\tau_1 - \tau_3$ is $a_1 = 16.500$. Also, the standard error read from the second entry of the STD ERROR OF ESTIMATE column in the Table 10.6.1 printout is

$$\text{Se}\,(\widehat{\tau_1 - \tau_3}) = \text{Se}(\hat{\alpha}_1) = 1.654.$$

Again, these values agree with those previously obtained.

Next, let us redo Example 10.5.1 where we calculated a 95% confidence interval for

$$\tau_1 - \tau_2 = (\tau_1 - \tau_3) - (\tau_2 - \tau_3) = \alpha_1 - \alpha_2.$$

Therefore, the point estimate of $\tau_1 - \tau_2$ is

$$a_1 - a_2 = 16.500 - (-4.250) = 20.750.$$

To find the standard error, we calculate

$$\text{Se}(\hat{\alpha}_1 - \hat{\alpha}_2) = \sqrt{s^2 \boldsymbol{\ell}' (\boldsymbol{X}' \boldsymbol{X})^{-1} \boldsymbol{\ell}}$$

where $\boldsymbol{\ell}' = (0, 1, -1, 0, 0, 0)$, the $(\boldsymbol{X}' \boldsymbol{X})^{-1}$ matrix is given for the full reparametrized model in Table 10.6.1, and $s^2 = \text{MSe} = 5.472$. It is easy to determine that $\boldsymbol{\ell}' (\boldsymbol{X}' \boldsymbol{X})^{-1} \boldsymbol{\ell} = 0.500$. Using (6.1.4) with $t_{.025,6} = 2.447$, the 95% confidence interval for $\tau_1 - \tau_2 = \alpha_1 - \alpha_2$ is

$$20.750 \pm 2.447\sqrt{(5.472)(0.500)}$$

or

$$(16.70, \quad 24.80).$$

If we use SAS we can obtain the above essentials without doing many of the described hand calculations by using the command

```
ESTIMATE 'TRT 1 VS TRT 2' X1  1  X2  -1 ;  .
```

In response to this, we find the point estimate for $\alpha_1 - \alpha_2$ and the standard error of the estimator printed out on the last line of Table 10.6.1. There we also find $t = 12.54$ which is the value of the t—statistic for testing $H_0 : \alpha_1 - \alpha_2 = 0$ or equivalently $H_0 : \tau_1 - \tau_2 = 0$, which, for a two—tailed test, has a p—value of 0.0001.

Let us pose another problem which the reader can work out numerically as an exercise. Although, without further conditions, we cannot estimate the mean of treatment 1, $\mu + \tau_1$, we can estimate the mean of treatment 1 in a given block, say block 1. This theoretical mean is $E(Y_{11}) = \mu + \tau_1 + \beta_1$. From (10.6.3) we know that

$$\mu + \tau_1 + \beta_1 = (\mu + \tau_3 + \beta_4) + (\tau_1 - \tau_3) + (\beta_1 - \beta_4) = \alpha_0 + \alpha_1 + \alpha_3.$$

As a result, the point estimate of $\mu + \tau_1 + \beta_1$ is $a_0 + a_1 + a_3$ and its standard error is

$$\mathrm{Se}(\hat{\alpha}_0 + \hat{\alpha}_1 + \hat{\alpha}_3) = \sqrt{s^2 \boldsymbol{\ell}' (\boldsymbol{X}'\boldsymbol{X})^{-1} \boldsymbol{\ell}}$$

where $\boldsymbol{\ell}' = (1, 1, 0, 1, 0, 0)$.

Orthogonal Properties

In order to more easily perceive the $(\boldsymbol{X}'\boldsymbol{X})^{-1}$ entry pattern for the matrix $(\boldsymbol{X}'\boldsymbol{X})^{-1}$ exhibited in Table 10.6.1, let us rewrite this matrix expressing the elements in fractional form and identifying the rows and columns associated with treatment and block variables.

$$
\begin{array}{c}
\hphantom{(\boldsymbol{X}'\boldsymbol{X})^{-1} = \frac{1}{2}\quad} \overbrace{\hphantom{-\frac{1}{4}\quad -\frac{1}{4}}}^{\text{treatments}} \quad \overbrace{\hphantom{-\frac{1}{3}\quad -\frac{1}{3}\quad -\frac{1}{3}}}^{\text{blocks}} \hphantom{\quad\text{treatments}} \\
(\boldsymbol{X}'\boldsymbol{X})^{-1} = \left[\begin{array}{c|cc|ccc}
\frac{1}{2} & -\frac{1}{4} & -\frac{1}{4} & -\frac{1}{3} & -\frac{1}{3} & -\frac{1}{3} \\ \hline
-\frac{1}{4} & \frac{1}{2} & \frac{1}{4} & 0 & 0 & 0 \\
-\frac{1}{4} & \frac{1}{4} & \frac{1}{2} & 0 & 0 & 0 \\ \hline
-\frac{1}{3} & 0 & 0 & \frac{2}{3} & \frac{1}{3} & \frac{1}{3} \\
-\frac{1}{3} & 0 & 0 & \frac{1}{3} & \frac{2}{3} & \frac{1}{3} \\
-\frac{1}{3} & 0 & 0 & \frac{1}{3} & \frac{1}{3} & \frac{2}{3}
\end{array}\right]
\begin{array}{l}
\vphantom{\frac{1}{2}} \\
\left.\vphantom{\begin{array}{c}\frac{1}{2}\\\frac{1}{4}\end{array}}\right\}\text{treatments} \\
\left.\vphantom{\begin{array}{c}\frac{2}{3}\\\frac{1}{3}\\\frac{1}{3}\end{array}}\right\}\text{blocks}
\end{array}
\end{array}
$$

We see that the entry corresponding to the covariance between a treatment variable and a block variable is zero in all cases. This means that the estimators for the treatment and block parameters are uncorrelated. Because of this property, we say that treatments and blocks are **orthogonal** and that the RCBD is an **orthogonal design**. This leads to the following definition concerning multi–factor experiments.

Definition 10.6.1 Two factors are said to be **orthogonal** when the covariance between an estimator for a parameter associated with factor one and an estimator for a parameter associated with factor two is always zero. If all factors in an experimental design are mutually orthogonal, the design is said to be an **orthogonal design**.

With respect to an RCBD, we now list the ramifications of orthogonality in the design. First, the relationship

$$\text{SSt} = \text{SStr} + \text{SSbl} + \text{SSe}$$

is a result of this orthogonality property. For instance, the above relationship does not hold in an experiment with incomplete blocks. Second, the estimators for the treatment (block) parameters are the same whether one uses the full model or a reduced model. For example, in the metal bracket production illustration, had we included all of the **complete** SAS printouts we would have observed that $a_1 = a_1^{**}$, $a_2 = a_2^{**}$, $a_3 = a_3^*$, $a_4 = a_4^*$ and $a_5 = a_5^*$. Third, in an RCBD, the adjusted and unadjusted sums of squares for a given factor are the same, Again, this is not true in an incomplete block experiment. In order to further explore this third result in the RCBD setting, we employ the SS( ) notation introduced in Section 4 of Chapter 6.

SS( ) for a Randomized Block Design

Recall from Chapter 6 that SS( ) denotes the regression sum of squares when fitting the model indicated by the parameters specified within the parentheses. In the case of a randomized block design, the full model (regression) sum of squares will be denoted by $SS(\mu, \tau, \beta)$. Accordingly, $SS(\mu, \tau)$ is the regression sum of squares for the reduced model with treatment variables only and $SS(\mu, \beta)$ denotes the regression sum of squares for the model containing only the block variables. We say $SS(\mu, \tau)$ is the "treatment sum of squares **unadjusted** for blocks" and $SS(\mu, \beta)$ is the "block sum of squares **unadjusted** for treatments." In addition, we use the notation

$$SS(\tau | \mu, \beta) = SS(\mu, \tau, \beta) - SS(\mu, \beta)$$

which we call the sum of squares for treatment effects **adjusted** for block effects. Notice that $SS(\tau | \mu, \beta)$ is the reduction in sums of squares due to including treatments in the model in addition to blocks and the overall mean as opposed to leaving treatments out. This reduction can be expressed in terms of the error sum of squares where we have previously used the "star notation" in the illustration of this section. That is,

$$SS(\tau \mid \mu, \beta) = SSe^* - SSe,$$

where SSe* is the error sum of squares obtained in fitting the (reduced) model with block variables only and SSe is the full model error sum of squares. Likewise, the sum of squares for block effects **adjusted** for treatment effects is

$$SS(\beta \mid \mu, \tau) = SS(\mu, \tau, \beta) - SS(\mu, \tau) = SSe^{**} - SSe,$$

where SSe** is the error sum of squares obtained in fitting the (reduced) model with treatment variables only. To "nail down" these concepts we now refer to the numerical example presented earlier in this section.

Example 10.6.1 Table 10.6.1 gives us the results of fitting the full RCBD reparametrized model. There we find the model (regression) sum of squares to be

$$SS(\mu, \tau, \beta) = 1416.083.$$

In Table 10.6.2 we note that $SS(\mu, \beta) = 454.917$ and from Table 10.6.3 we get $SS(\mu, \tau) = 961.167$. Rounding to three decimal places, we observe that

$$SS(\tau \mid \mu, \beta) = SS(\mu, \tau, \beta) - SS(\mu, \beta) = 1416.083 - 454.917 = 961.167,$$

$$SS(\beta \mid \mu, \tau) = SS(\mu, \tau, \beta) - SS(\mu, \tau) = 1416.083 - 961.167 = 454.917.$$

Hence, in this case $SS(\tau \mid \mu, \beta) = SS(\mu, \tau)$ and $SS(\beta \mid \mu, \tau) = SS(\mu, \beta)$. That is, the adjusted and unadjusted sums of squares for both factors are the same. For that reason we have referred to $SS(\tau \mid \mu, \beta) = SS(\mu, \tau)$ simply as "the treatment sum of squares" denoted by SStr. In like manner, $SS(\beta \mid \mu, \tau) = SS(\mu, \beta)$ is called "the block sum of squares," denoted by the symbol SSbl.

Because the adjusted and the unadjusted sums of squares are equal in the RCBD case, we really only need to run two regression models to find the ANOVA table. The full reparametrized model yields SSt and SSe. The reduced model containing only the treatment variables gives us $SStr = SS(\mu, \tau)$ and SSe**. Then $SSbl = SS(\beta \mid \mu, \tau) = SSe^{**} - SSe$.

Before leaving this topic, it is appropriate to point out the difference in SAS Type I and Type III sums of squares using SS( ) notation. Let us turn back to Table 10.4.3 where we find the Type I and Type III sums of squares to be identical. Recall that workers constitute the blocking factor and that machines are our treatments. In terms of symbols, the printed sums of squares are

	Type I SS	Type III SS
Worker	$SS(\mu, \beta)$	$SS(\beta \mid \mu, \tau)$
Machine	$SS(\tau \mid \mu, \beta)$	$SS(\tau \mid \mu, \beta)$.

But $SS(\beta \mid \mu, \tau) = SS(\mu, \beta)$ in the case of an RCBD, which explains the reason why the two sets of sums of squares are identical. This will not be true in the next chapter where we deal with incomplete blocks. It is important to note that Type III SS entries here are the sum of squares for a factor effect **adjusted for the other factor(s)**.

Since the ideas presented in this section are so important for a basic understanding of the material in the next and subsequent chapters, we close with a recapitulation of the fundamentals for future reference. Although the presentation here is made relative to a randomized block design, in general, the notions are easily extended to more complex situations.

General Reparametrization Procedures and Results for Randomized Block Design

Step 1. Define p—1 treatment variables. (10.6.5)

$$x_i = \begin{cases} 1, & \text{if treatment } i \text{ applied}, \; i = 1, 2, \ldots, p-1 \\ 0, & \text{if not.} \end{cases}$$

Step 2. Define r-1 block variables.

$$x_{p-1+j} = \begin{cases} 1, & \text{if observation is in block } j, \; j = 1, 2, \ldots, r-1 \\ 0, & \text{if not.} \end{cases}$$

Step 3. Fit the full model

$$Y = \alpha_0 + \overbrace{\alpha_1 x_1 + \cdots + \alpha_{p-1} x_{p-1}}^{\text{treatments}} + \overbrace{\alpha_p x_p + \cdots + \alpha_{p+r-2} x_{p+r-2}}^{\text{blocks}} + \varepsilon$$

and find $SSr = SS(\mu, \tau, \beta)$, SSe and SSt.

Step 4. Fit the reduced model containing block variables only

$$Y = \alpha_0^* + \alpha_p^* x_p + \cdots + \alpha_{p+r-2}^* x_{p+r-2} + \varepsilon^*$$

to obtain $SSr^* = SS(\mu, \beta) = SSbl$ (unadjusted) and SSe^*.
Then SStr (adjusted) $= SS(\tau \mid \mu,\beta) = SS(\mu,\tau,\beta) - SS(\mu,\beta)$.

Step 5. Fit the reduced model containing treatment variables only

$$Y = \alpha_0^{**} + \alpha_1^{**} x_1 + \cdots + \alpha_{p-1}^{**} x_{p-1} + \varepsilon^{**}$$

to obtain SSr** = SS(μ, τ) = SStr (unadjusted) and SSe**. Then SSbl (adjusted) = SS($\beta | \mu,\tau$) = SS(μ,τ,β) − SS(μ,τ).

Step 6. The full reparametrized model (Step 3) can be used for parametric inference utilizing the parametric relationships

$\alpha_0 = \mu + \tau_p + \beta_r$

$\alpha_i = \tau_i - \tau_p$, i = 1 , 2 ,..., p − 1

$\alpha_{p-1+j} = \beta_j - \beta_r$, j = 1 , 2 ,..., r − 1.

It should be noted that in order to complete an analysis using reparametrization, it generally is not necessary to implement all of the steps in outline (10.6.5). The required steps depend upon the goals of the experiment and whether or not the data set is complete (or "balanced") in the design sense.

10.7 Expected Mean Squares, Restricted RCBD Model

In Section 7 of Chapter 7 we introduced the notion of the theoretical expected mean squares associated with the analysis of variance table in the CRD case. There we find Theorem 7.7.1 which provides us with the formula for calculating E(MS), where E(MS) denotes, in general, the expected value of a mean square entry in the ANOVA table summarizing the results of a designed experiment. Before proceeding, the reader should review this theorem and its application to a CRD experiment. According to the theorem, our task is to find the expressions for W as related to the RCBD quantities E(MSTR) and E(MSBL). Let us focus on deriving the expression for E(MSTR); the expression for E(MSBL) is easily obtained in the same manner.

Expected Mean Squares in RCBD Analysis of Variance

From formula (10.4.2) we know that the random variable SSTR is

$$\text{SSTR} = \sum_{i=1}^{p} r\,(\bar{Y}_{i\cdot} - \bar{Y}_{\cdot\cdot})^2$$

so that by Theorem 7.7.1

$$W = \sum_{i=1}^{p} r\,[E(\bar{Y}_{i\cdot}) - E(\bar{Y}_{\cdot\cdot})]^2.$$

In terms of the RCBD model (10.2.1), the expectations work out to be

$$E(\bar{Y}_{i\cdot}) = E(Y_{i\cdot}/r) = E\left(\sum_{j=1}^{r} Y_{ij}/r\right) = \mu + \tau_i + \bar{\beta}$$

and

$$E(\bar{Y}_{\cdot\cdot}) = E(Y_{\cdot\cdot}/rp) = E\left(\sum_{i=1}^{p} \sum_{j=1}^{r} Y_{ij}/rp\right) = \mu + \bar{\tau} + \bar{\beta}$$

where $\bar{\beta} = \sum_{j=1}^{r} \beta_j/r$ and $\bar{\tau} = \sum_{i=1}^{p} \tau_i/p$.

On substituting into the expression for W above and applying Theorem 7.7.1, we get

$$E(MSTR) = E[SSTR/(p-1)] = \sigma^2 + \sum_{i=1}^{p} r(\tau_i - \bar{\tau})^2/(p-1). \qquad (10.7.1)$$

In like manner we obtain

$$E(MSBL) = E[SSBL/(r-1)] = \sigma^2 + \sum_{j=1}^{r} p(\beta_j - \bar{\beta})^2/(r-1). \qquad 10.7.2)$$

The next table includes this information in a theoretical analysis of variance for the RCBD.

Table 10.7.1 Theoretical ANOVA for RCBD with E(MS)

Source	df	MS	E(MS)
Blocks	r–1	MSBL	$\sigma^2 + p\sum_{j=1}^{r}(\beta_j-\bar{\beta})^2/(r-1)$
Treatments	p–1	MSTR	$\sigma^2 + r\sum_{i=1}^{p}(\tau_i-\bar{\tau})^2/(p-1)$
Error	(r–1)(p–1)	MSE	σ^2
Total	rp–1		

The Restricted RCBD Model

Back in Section 10.3 we used the side conditions $\sum_{i=1}^{p} t_i = 0$ and

$\sum_{j=1}^{r} b_j = 0$ to obtain a unique solution to the normal equations of the RCBD model. If we assume that the model **includes** these conditions, we then are working with the **restricted** RCBD model

$$\begin{cases} Y_{ij} = \mu + \tau_i + \beta_i + \varepsilon_{ij}, \quad i = 1,2,\dots,p \; ; \; j = 1,2,\dots,r \\ \sum_{i=1}^{p} \tau_i = 0 = \sum_{j=1}^{r} \beta_j \end{cases} \tag{10.7.3}$$

From (10.3.3), for model (10.7.3), we observe that the parameters μ, τ_i and β, are individually estimable with the BLUE for μ being $\bar{y}_{\bullet\bullet}$, the BLUE for τ_i being $\bar{y}_{i\bullet} - \bar{y}_{\bullet\bullet}$ and the BLUE for β_j being $\bar{y}_{\bullet j} - \bar{y}_{\bullet\bullet}$. Also the ANOVA hypotheses for the unrestricted RCBD model,

$$H_0 : \tau_1 = \tau_2 = \cdots = \tau_p \quad \text{and} \quad H_0 : \beta_1 = \beta_2 = \cdots = \beta_r,$$

become in the restricted case

$$H_0 : \tau_1 = \tau_2 = \cdots = \tau_p = 0 \quad \text{and} \quad H_0 : \beta_1 = \beta_2 = \cdots = \beta_r = 0$$

and therefore (10.7.1) and (10.7.2) reduce to

$$E(MSTR) = \sigma^2 + r \sum_{i=1}^{p} \tau_i^2/(p-1) \quad \text{and} \quad E(MSBL) = \sigma^2 + p \sum_{j=1}^{r} \beta_j^2/(r-1)$$

respectively. We note that use of the restricted RCBD model avoids the complication of estimability, since all parameters are estimable. As to the ANOVA f—tests, the only difference between the analyses in the unrestricted and restricted cases are the null hypotheses, in words, "no difference in effects" as opposed to "no effects."

10.8 Design Considerations

As a design, an RCBD is a two—factor design involving treatments and blocks. In this design the experimental units are grouped into blocks, the size of a block equals the number of treatments, and the treatments are randomly assigned to the experimental units within a block such that each treatment occurs exactly once in every block. Because there is a restriction on the randomization (each treatment occurs once in each block), an RCBD is clearly a different design than a CRD.

In an RCBD the purpose of the random assignment of treatments in a block is to reduce the potential effect of other factors. For example, in the metal bracket production problem in Table 10.2.1, each worker is used three times. By randomizing the treatments (machines A, B, C) to the worker, some workers will use machine A first, some will use B first and some C. Randomization will help to eliminate (even out) any possible order effect. We should recognize, however, that this "evening out" will be more effective when the number of blocks is large.

In an RCBD the purpose of blocking is to potentially reduce the experimental error. To observe how this occurs, let us compare the error for a CRD and an RCBD. Let SSe* and MSe* be the sum of squares and mean square for error, respectively, for a CRD. Let SSe and MSe be the corresponding values for an RCBD. It was pointed out earlier in this chapter that $SSe = SSe^* - SSbl$; therefore, $SSe \leq SSe^*$. In other words, if $SSbl > 0$, then blocking does reduce the experimental error. However, in tests and confidence intervals, MSe is used. Since the degrees of freedom for error in an RCBD is $(r - 1)(p - 1)$ which is smaller than the corresponding value in a CRD, namely $p(r-1)$, the mean square error for an RCBD can be smaller or larger than the corresponding value in a CRD. Therefore, one criterion for comparing an RCBD to a CRD is to compare their mean square error values. The following example illustrates this.

Example 10.8.1 Refer to the metal bracket production data given in Table 10.2.1 and analyzed in Example 10.4.1. The values for SSe, degrees of freedom, MSe, f used in testing for treatments, and the corresponding p—value for this test are given for a CRD analysis and the RCBD analysis in the following table:

Design	SSe	df	MSe	f	p—value
CRD	487.75	9	54.19	8.87	.0075
RCBD	32.83	6	5.47	87.82	.0001

For these data we note that the MSe for the RCBD is much smaller than the corresponding value for a CRD. This leads to more power for the f—test (smaller p—value), and it is clear that confidence intervals for treatment contrasts will also be shorter in an RCBD for these data.

In the previous example it was clearly beneficial to include blocks as part of the design. In some problems the f—test for treatments will be significant for an RCBD, but not for a CRD due to the decrease in MSe by including blocks. On the other hand, if SSbl is very small (i.e., blocks have little effect), then the MSe in an RCBD may be larger than the corresponding value in a CRD. In such a case blocking has not been advantageous. Examples of these ideas are given in the exercises.

To quantify the comparison of an RCBD and CRD a measure of **efficiency** is sometimes calculated. One such measure is given by

$$E = \frac{MSe^*}{MSe}.$$

An efficiency greater than one would suggest that the RCBD is the better design. For the metal bracket production data considered in the example of this section the efficiency is $E = 54.19/5.47 = 9.9$, which clearly indicates that the RCBD is more efficient than the CRD. An interpretation for E is that it is approximately the number of times larger the number of replications for a CRD needs to be in order to have the CRD as efficient as an RCBD.

In summary, when blocking can significantly reduce the experimental error the RCBD is a better choice for a design than a CRD. In such a case the RCBD will have better power for tests and shorter confidence intervals for contrasts.

10.9 Summary Example

In this section we consider an additional example illustrating conclusions drawn from the analysis of RCBD data. We also demonstrate how we can check the model assumptions for an RCBD.

Our example is based on comparing methods for processing pickles. Pickles are processed by soaking them in a chemical solution for a period of time. A pickle processing company has used a standard solution, label it S1; however, the solution is not biodegradable and its disposal is expensive. The company wished to find an alternative solution for processing pickles that is biodegradable. Four biodegradable solutions (label them S2, S3, S4, and S5) were selected, and a study was completed to compare the taste of pickles processed with these five solutions. In this study pickles were processed using each of the five solutions. To compare the taste of these pickles the company used their eight "taste testers" who were individuals employed by the company to serve as expert judges for food products. Each judge (taste tester) was given three pickles processed from each solution, and the judge gave a single rating from 1 to 10 for the quality of the taste, with 10 being best. The order in which the judges rated the pickles from the five solutions was randomized to help eliminate any possible order effect. The results are given in Table 10.9.1.

Table 10.9.1 Taste Results for Pickles

		Solution				
		S1	S2	S3	S4	S5
	1	7	4	8	6	3
	2	6	2	7	5	2
	3	8	5	9	7	5
Judge	4	7	5	8	7	2
	5	6	3	9	6	4
	6	7	4	8	5	3
	7	6	5	8	7	3
	8	8	4	9	6	3

One can readily observe that the design used in this problem is an RCBD, where the treatments are the five solutions, and the blocks are the eight judges. The model to be used is:

$$Y_{ij} = \mu + \tau_i + \beta_j + \varepsilon_{ij} \quad , \quad i = 1, \cdots, 5; \; j = 1, \cdots, 8,$$

where Y represents the rating. As a quick comparison of the five solutions we calculate the treatment means, $\bar{y}_{i\cdot}$, and find $\bar{y}_{1\cdot}$ = 6.875, $\bar{y}_{2\cdot}$ = 4.000, $\bar{y}_{3\cdot}$ = 8.250, $\bar{y}_{4\cdot}$ = 6.125, $\bar{y}_{5\cdot}$ = 3.125. There seems to be evidence of differences in taste. Since our major goal is to compare the taste of the pickles for the solutions, we will perform the f—test comparing the five treatment means. To determine which, if any, means are statistically different we will follow the f—test with a Tukey multiple comparison.

To perform the f—test we first calculate the ANOVA table. Using the equations in Table 10.4.1 we arrive at the following ANOVA summary:

Table 10.9.2 ANOVA Table for Pickle Taste Data

Source	df	SS	MS	f
Treatments	4	140.6500	35.1625	78.45
Blocks	7	15.5750	2.2250	4.96
Error	28	12.5500	.4482	
Total	39	168.775		

We note that we reject $H_0 : \tau_1 = \cdots = \tau_5$, since $f = 78.45$ exceeds the 0.05 critical value $f_{.05,4,28} = 2.71$. We conclude, therefore, that the mean taste rating is significantly different for at least two solutions (treatments). Similarly, the f—test for blocks rejects since $f = 4.96 > f_{.05,7,28} = 2.36$ indicating the blocking to be effective.

To determine which solutions have different population means we perform the Tukey multiple comparison using a 0.05 level. From (10.5.3) we decide that $\tau_i \neq \tau_k$ if

$$|\bar{y}_{i\cdot} - \bar{y}_{k\cdot}| \geq q_{\alpha,p,(p-1)(r-1)} \sqrt{\frac{s^2}{r}} .$$

In our problem $q_{.05,5,28} \sqrt{\frac{s^2}{8}} = 4.1 \sqrt{\frac{.4482}{8}} = .97$, which leads to the following result in tabular form:

Solution	5	2	4	1	3
$\bar{y}_{i\cdot}$	3.125	4.000	6.125	6.875	8.250

(Underlines join 5 and 2, and 4 and 1.)

From this result we can conclude that in terms of taste, solution 3 is rated the best, while solution 4 is not significantly different from the currently used solution 1. Clearly, solutions 2 and 5 are rated lower in taste.

We should point out that these pickles were also judged on other criteria. Exercise 7—6 considered data from this experiment based on the number of pickles that bloat. Solution 5 was removed from that portion of the experiment due to poor taste results.

To carry out the tests used in this section we needed to assume that the ε_{ij}'s are independent $N(0,\sigma^2)$. Analyzing the residuals for these data can aid in checking the assumptions. Residuals can be analyzed using the univariate command on SAS. In this problem we use the following SAS commands:

```
PROC GLM; CLASS JUDGE SOLUTION;
MODEL RATING = JUDGE SOLUTION;
OUTPUT OUT = NEW P = YHAT R = RESID;
PROC UNIVARIATE PLOT NORMAL; VAR RESID;
```

The SAS output for the univariate command is given in Table 10.9.3. The normality assumption seems reasonable as evidenced by the test (W : Normal, p = .2423), by the stem—and—leaf plot and the normal probability plot. There is no evidence of an outlier. (Observe the stem—and—leaf plot). Checking for equal variances is difficult since there are no replications in an RCBD. Finally, the independence assumption is also difficult to check. We might be concerned that since a judge rates all five solutions there may be a correlation among the ratings. Precautions were taken to guard against this correlation in this experiment by allowing for time between the taste tests of the five sets of pickles. In summary, there appears to be no problems with the assumptions.

Table 10.9.3 SAS Printout for PROC UNIVARIATE

```
Variable=RESID

                              Moments

             N                 40  Sum Wgts           40
             Mean               0  Sum                 0
             Std Dev      0.56727  Variance     0.321795
             Skewness    -0.13085  Kurtosis     -0.51755
             USS            12.55  CSS             12.55
             CV                 .  Std Mean     0.089693
             T:Mean=0           0  Pr>|T|         1.0000
             Num ^= 0          40  Num > 0            21
             M(Sign)            1  Pr>=|M|        0.8746
             Sgn Rank           2  Pr>=|S|        0.9790
             W:Normal    0.960373  Pr<W           0.2423

                           Quantiles(Def=5)

              100% Max       0.95       99%       0.95
               75% Q3         0.4       95%      0.875
               50% Med   7.77E-16       90%     0.8125
               25% Q1      -0.375       10%     -0.825
                0% Min      -1.25        5%    -0.9625
                                         1%      -1.25
              Range           2.2
              Q3-Q1         0.775
              Mode         -0.375

                              Extremes

               Lowest    Obs      Highest    Obs
               -1.25(     20)       0.8(      36)
                  -1(     31)     0.825(      23)
              -0.925(     22)     0.875(      17)
               -0.85(     29)     0.875(      32)
                -0.8(     21)      0.95(      25)

         Stem Leaf                      #             Boxplot
            8 02885                     5                |
            6 555                       3                |
            4 002                       3             +-----+
            2 08                        2             |     |
            0 00228555                  8             *--+--*
           -0 82555                     5             |     |
           -2 888255                    6             +-----+
           -4 55                        2                |
           -6 2                         1                |
           -8 250                       3                |
          -10 0                         1                |
          -12 5                         1                |
              ----+----+----+----+
          Multiply Stem.Leaf by 10**-1
```

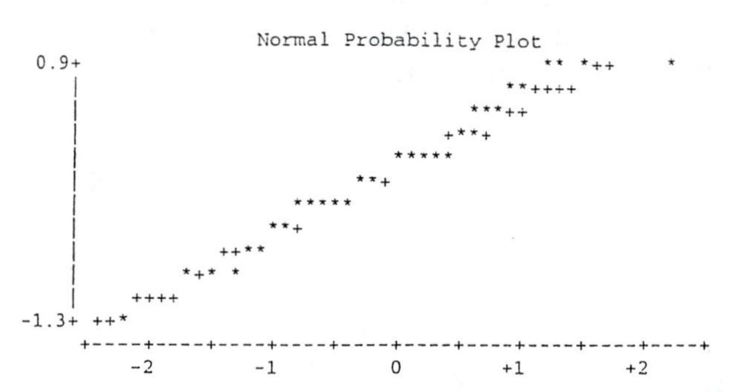

Chapter 10 Exercises

Application Exercises

10–1 A company completed a study to determine how many breaks (none, one, or two) to provide for secretaries during the morning. Six secretaries were available for the study. A secretary worked for three weeks using no break for one week, one break for another week, and two breaks for the remaining week. The order in receiving the treatments (0, 1 or 2 breaks) was randomized for each secretary. Each week the secretary was given an efficiency rating on a six–point scale (higher values are better). The data are as follows:

		Secretary					
		1	2	3	4	5	6
Number of breaks	0	3.6	3.1	4.6	4.9	4.3	3.6
	1	4.1	3.7	5.2	5.8	4.9	4.5
	2	3.2	2.9	4.6	5.0	4.1	3.7

(a) Find the ANOVA Table and perform each f–test using a

0.05 level.

(b) Find a 95% confidence interval for $\tau_2 - \tau_1$ and interpret the results.

(c) Perform a Tukey multiple comparison on the breaks using a 0.05 level.

(d) Analyze the data as a CRD using breaks as the treatments and ignoring secretaries. Perform the f–test for treatments using a 0.05 level. Compare the result to the one obtained in part (a).

(e) Calculate and interpret the efficiency measure of the RCBD compared to a CRD.

10–2 Scheaffer and McClave (1982) considered the following problem. Due to increased energy shortages and costs, utility companies are stressing ways in which home and apartment utility bills can be cut. One utility company reached an agreement with the owner of a new apartment complex to conduct a test of energy saving plans for apartments. The tests were to be conducted before the apartments were rented. Four apartments were chosen that were identical in size, amount of shade, and direction faced. Four plans were to be tested, one on each apartment. The thermostat was set at 75° in each apartment and the monthly utility bill was recorded for each of the three summer months. The results are listed in the table below.

Month	Treatment			
	1	2	3	4
June	$74.44	$68.75	$71.34	$65.47
July	89.96	73.47	83.62	72.33
August	82.00	71.23	79.88	70.87

Treatment 1: No insulation
Treatment 2: Insulation in walls and ceilings
Treatment 3: No insultation; awnings for windows
Treatment 4: Insulation and awnings for windows

(a) Is there evidence that the four treatments have a different effect on the mean monthly utility bills? Use a test with a 0.05 level.

(b) Show that the three contrasts $\tau_1 - \tau_3$, $\tau_2 - \tau_4$ and τ_1

$+ \tau_3 - \tau_2 - \tau_4$ are orthogonal.

(c) Find the sum of squares for each contrast in part (b). Show that these partition the treatment sum of squares.

(d) Perform an f–test using a 0.05 level for each contrast in part (b). Interpret each outcome.

(e) Use the Tukey procedure with a 0.05 level to "separate the means."

10–3 A seed company performs an experiment to compare four varieties of wheat. Five fields are available for the study and each field is subdivided into four plots of equal size. In each field each variety is randomly assigned to a plot, and the yield in bushels is recorded as follows.

		Variety of Wheat			
		1	2	3	4
Field	1	45	47	53	38
	2	37	41	47	32
	3	41	38	50	40
	4	48	46	56	43
	5	32	37	45	29

(a) Find the ANOVA Table and perform each f–test using a 0.05 level.

(b) Perform a Tukey multiple comparison on the varieties using a 0.05 level.

(c) Use the residuals to check the assumptions of the model.

10–4 A high school wants to compare its students' performances on national tests in English, Math, and Science. Each student takes a test in each subject and the percentile score is recorded as follows:

		Student						
		1	2	3	4	5	6	7
	English	78	42	45	68	24	15	96
Subject	Math	65	33	40	57	27	18	79
	Science	79	38	51	62	36	28	88

(a) Use a test at the 0.05 level to determine if the students perform differently in the subject areas.
(b) Use a Fisher's protected LSD with level 0.05 to separate the means for the three subjects.
(c) Use the residuals to check the assumptions of the model.

10–5 The paired–difference experiment encountered in a previous statistics course is in fact a special case of a randomized complete block experiment where $p = 2$ and each "pair" constitute a "block." To appreciate this, consider the Weight–Watchers' diet data considered in Example 10.1.1. Let μ_1 represent the population mean before, and μ_2 the population mean after. Suppose that we desire to test $H_0 : \mu_D = \mu_1 - \mu_2 = 0$ versus $H_1 : \mu_D \neq 0$.

(a) Calculate the $n - 1$ degrees of freedom test statistic

$$t = \frac{\bar{d}}{s_d/\sqrt{n}} .$$

(b) Using an RCBD, calculate the ANOVA and compare $f = MStr/MSe$ with t^2 obtained in (a). Comment.
(c) Incorrectly analyze the data as a CRD experiment and compare the ANOVA with that obtained in (b). Which analysis gives better results?
(d) Using both the RCBD analysis and the CRD analysis, calculate a 95% confidence interval for $\mu_1 - \mu_2$ and observe the difference in lengths of the two intervals. Comment.

10–6 Redo problem 10–1 parts (a) and (b) using a reparametrized approach.

10–7 Refer to the metal bracket data and the reparametrized model of Section 10.6. Use the printout in Table 10.6.1 to help find a 95% confidence interval for $\tau_1 - \tau_3$.

10–8 Refer to the pickle taste data in Table 10.9.1 that has been analyzed in Section 10.9. Using hand calculations verify that SSt = 168.775, SStr = 140.65, SSbl = 15.575 and SSe = 12.55.

10–9 In the pickle processing problem in Section 10.9 four treatments (solutions 2, 3, 4, 5) were compared to a standard (solution 1). In Section 10.9 we analyzed this data by performing the f–test for treatments followed by a Tukey multiple comparison for **all** pairwise differences between means. Another way to approach these data is to compare the four new treatments with the control treatment (solution 1). To this end, a Bonferroni procedure can be used to test the "treatment versus control" comparisons: $\tau_1 - \tau_2$, $\tau_1 - \tau_3$, $\tau_1 - \tau_4$, and $\tau_1 - \tau_5$. Perform these four comparisons using the Bonferroni procedure with $\alpha = 0.05$.

10–10 An experiment was conducted to study the effects of temperature on the life in hours of a component. Four temperature levels and five ovens (serving as blocks) were used in the experiment. The order for the temperatures was randomized for each oven. The following results were recorded.

	Temperature (degrees)			
Oven	200	300	400	500
1	340	324	307	274
2	361	338	312	281
3	346	328	298	276
4	358	332	315	285
5	343	321	294	269

(a) Test the hypothesis that there is no difference in the effects of temperature on the life of the component. Use a 0.05 level.

(b) Determine the nature of the response (i.e., linear, quadratic, cubic) to temperature.

Theoretical Exercises

10–11 For the traditional RCBD model (10.2.1) the cell mean is $\mu_{ij} = \mu + \tau_i + \beta_j$.

(a) Show that the BLUE for μ_{ij} is $\bar{Y}_{i\cdot} + \bar{Y}_{\cdot j} - \bar{Y}_{\cdot\cdot}$.

(b) Show that the variance of this estimator is $\sigma^2(p + r - 1)/(rp)$, and use this to deduce the form of a confidence interval for μ_{ij} .

10–12 Verify that the equations given in (10.3.3) actually form a solution set for the normal equations for an RCBD.

10–13 Find the solution to the normal equations that corresponds to the side condition $\tau_p = 0$, $\beta_r = 0$. Use this solution to find the least squares estimate for the contrast $\sum_{i=1}^{p} c_i \tau_i$, and compare the result to the one given in (10.3.5). (The solution that SAS gives for the normal equations corresponds to these side conditions.)

10–14 Verify the computational formula for SSe given in (10.3.8).

10–15 Verify the formula given for SSbl in (10.4.4).

10–16 Verify algebraically that in an RCBD, SSt = SStr + SSbl + SSe.

10–17 Show algebraically that in the paired–difference experiment described in Exercise 10–5, $t^2 = f$. Deduce from this that the t–test and f–test are equivalent for testing $H_0 : \mu_D = 0$ versus $H_1 : \mu_D \neq 0$. Are they equivalent for a one–sided test? Explain why.

10–18 Consider the restricted RCBD model discussed in Section 10.7.

(a) Explain why all model parameters are estimable.

(b) Find the BLUE for $\mu + \tau_i$.

(c) Find the variance for the estimator in part (b) and use it to find the form of a confidence interval for $\mu + \tau_i$.

(d) Repeat parts (b) and (c) for the parameter μ .

(e) Repeat parts (b) and (c) for the parameter τ_i .

10–19 Derive E(MSBL) for the restricted RCBD model.

10–20 Another method for reparametrizing an RCBD is to use different "coding" variables. For example, consider an RCBD with $p = 3$, $r = 2$ and the following coding variables:

$$x_1 = \begin{cases} 1 & \text{if treatment } 1 \\ 0 & \text{if treatment } 2 \\ -1 & \text{if treatment } 3 \end{cases} \qquad x_2 = \begin{cases} 0 & \text{if treatment } 1 \\ 1 & \text{if treatment } 2 \\ -1 & \text{if treatment } 3 \end{cases}$$

$$x_3 = \begin{cases} 1 & \text{if block } 1 \\ -1 & \text{if block } 2 \end{cases} .$$

The reparametrized model is then $Y = \alpha_0 + \alpha_1 x_1 + \alpha_2 x_2 + \alpha_3 x_3 + \varepsilon$.

(a) Is this reparametrized model full rank?

(b) Find the relationship between the parameters of the traditional RCBD (10.2.1) and this reparametrized model.

(c) Explain how to test for no treatment effect using this reparametrized model.

(d) Explain how to find a confidence interval for $\tau_1 - \tau_3$ using this reparametrized model.

10–21 In an RCBD the blocks and treatments are orthogonal. As a consequence, an estimator for a treatment contrast is uncorrelated with an estimator for a block contrast. Verify that the correlation (or covariance) is zero for the estimators for the two contrasts $\tau_1 - \tau_2$ and $\beta_1 - \beta_2$ when $p = r = 2$.

(Hint: Replace the Y_{ij} values by $\mu + \tau_i + \beta_j + \varepsilon_{ij}$.)

10–22 (Power of f–test) In Exercise 7–27 we calculated the power of the f–test for a CRD using the Pearson–Hartley tables given in the Appendix. In a similar manner we can calculate the power for the f–test for treatments in an RCBD by using $\emptyset^2 = r \sum_{i=1}^{p} (\tau_i - \bar{\tau})^2/(p\sigma^2)$, where $\bar{\tau} = \sum_{i=1}^{p} \tau_i/p$. Note that $\nu_1 = p - 1$ and $\nu_2 = (p - 1)(r - 1)$.

(a) Find the power of the f–test for treatments when $p = 4$, $r = 6$, $\tau_1 = 5$, $\tau_2 = 7$, $\tau_3 = 4$, $\tau_4 = 4$, $\sigma^2 = 2$, and $\alpha = 0.05$.

(b) What is the power for the f–test in part (a) when the number of blocks is increased to $r = 8$?

10–23 (Sample Size) In Exercise 7–28 we calculated the treatment sample size necessary to attain a desired power for the f–test in a CRD. In an RCBD the treatment sample size is the number of blocks, r. Use the procedure described in Exercise 10–22 for finding power to help answer the following.

(a) Consider an RCBD with $p = 3$, $\sigma^2 = 4$, $\tau_1 = 10$, $\tau_2 = 14$, $\tau_3 = 16$. How many blocks are required for the 0.05 level f–test to have power of .9 or higher?

(b) As in Exercise 7–28 part (b), if the difference between two treatment means is D or greater, then $\emptyset^2 \geq rD^2/(2p\sigma^2)$. If $D/\sigma = \Delta$, then $\emptyset^2 \geq r\Delta^2/(2p)$. For an RCBD with $p = 4$, how many blocks are needed to have the 0.05 level f–test have a power of .9 or higher when $\Delta = 2$?

CHAPTER 11

INCOMPLETE BLOCK DESIGNS

11.1 Incomplete Blocks

If in a randomized block design the number of experimental units in a block is less than the number of treatments, then the design is called an **incomplete block design (IBD)**. Obviously in such designs one or more treatment block combinations are missing. This situation can arise in two ways. First, the experiment could have been planned as an RCBD, but during the experiment one or more observations had been lost. For example, in a study of the effectiveness of three drugs (treatments) each subject (blocks) could be required to use each drug. For some reason one or more subjects could have only used two drugs in the experiment, and this would cause one or more missing treatment, block combination(s). A second reason for having incomplete blocks is because of design **constraints**. As an illustration of a design with a constraint, consider an experiment aimed at comparing three brands of bicycle tires. If we think of the brands of tires as the treatments and the bicycles as the blocks, then clearly only two treatments can be used in a block at one time. An incomplete block design in such as experiment could result in the following arrangement of data (x represents a single observation):

Brand of Tire (Treatment)

		A	B	C
	1	x	x	
Bicycle	2		x	x
	3	x		x

There are several difficulties that we encounter with incomplete blocks. We list them here without justification; however, most will be justified or illustrated later in this chapter or in the exercises. Except for the total sum of squares, the formulas for sum of squares given in Chapter 10 are no longer valid. The treatments and blocks are no longer orthogonal and, as a consequence, the adjusted and unadjusted sum of squares can be different. The treatment means, $\bar{y}_{i\cdot}$'s, are no longer comparable in that the least squares estimate for $\tau_i - \tau_k$ might not be $\bar{y}_{i\cdot} - \bar{y}_{k\cdot}$. As a consequence, the formulas in Chapter 10 given for treatment contrasts and multiple comparisons no longer hold.

Since most of the formulas given in Chapter 10 for an RCBD do not hold for incomplete block designs (IBD), it is clear that we need to consider how to analyze data from such designs. We start with the model and its sums of squares.

The IBD Model and Associated Sums of Squares

To write down a model we let Y_{ij} represent the observation for the ith treatment in the jth block and we define a set D such that D = {(i,j): there is an observation in cell (i,j)}. Our model then is

$$Y_{ij} = \mu + \tau_i + \beta_j + \varepsilon_{ij},$$

$$i = 1, 2, \ldots, p; \quad j = 1, 2, \ldots, r \text{ and } (i,j) \in D. \tag{11.1.1}$$

In the next section we deal with the analysis of IBD data but before doing so, let us first review the concepts of adjusted and unadjusted sums of squares together with notation introduced in Section 10.6. Although this notation was used in Section 10.6 in conjunction with a reparametrized model, it is also convenient for discussing traditional model analysis where we will use the phrase "model sum of squares" in place of "regression sum of squares."

Recall that $SS(\mu, \tau, \beta)$ stands for the full model (11.1.1) sum of squares and that $S(\mu, \beta)$ denotes the sum of squares for the reduced model

$$Y_{ij} = \mu^* + \beta_j^* + \varepsilon_{ij}^*, \tag{11.1.2}$$

and is called the sum of squares for blocks **unadjusted** for treatments. Then the treatment sum of squares **adjusted** for blocks is given by

$$SS(\tau \mid \mu,\beta) = SS(\mu,\tau,\beta) - SS(\mu,\beta). \tag{11.1.3}$$

If we denote the error sum of squares associated with these full and reduced models by the symbols SSe and SSe*, respectively, $SS(\tau \mid \mu, \beta)$ can also be expressed in terms of the principle of conditional error sums of squares as follows:

$$SS(\tau \mid \mu, \beta) = SSe^* - SSe. \tag{11.1.4}$$

By symmetry, the symbol $SS(\mu, \tau)$ represents the treatment sum of squares **unadjusted** for blocks and is the sum of squares for the reduced model which contains treatment variables only, ignoring blocks. Hence the relation

$$SS(\beta | \mu, \tau) = SS(\mu, \tau, \beta) - SS(\mu, \tau) \tag{11.1.5}$$

gives us the block sum of squares **adjusted** for treatments. It is through their mathematical expectations that we truly appreciate the difference in the two types of sums of squares. (See Exercise 11—7.)

11.2 Analysis for Incomplete Blocks — Linear Models Approach

To analyze data from an incomplete block design we use the linear models techniques described in prior chapters. Unlike the SStr and SSbl formulas for a RCBD, closed form formulas for the traditional model IBD case, in general, do not exist. Because of this, in our discussion we will assume the availability of a computer program package. To perform statistical inference relative to the IBD model (11.1.1), we shall also assume that the ε_{ij} are independent $N(0,\sigma^2)$ random variables.

It is routine to check for estimability of treatment contrasts. For $(i,j) \in D$, we know that $\mu + \tau_i + \beta_j$ is estimable and that any linear combination of these quantities is also estimable. Using this information, it is easy to check whether pairwise treatment contrasts or other contrasts are estimable. Unless there are many empty cells, the treatment contrasts will be estimable but one should always check before completing the analysis. For example, if each treatment pair occurs at least once together in a block, then estimability of the treatment contrasts is guaranteed. This fact and some examples where estimability is a difficulty are considered further in the exercises. Throughout this chapter we will assume that there is sufficient data to ensure that treatment contrasts are estimable.

To test $H_0 : \tau_1 = \tau_2 = \cdots = \tau_p$, we use the principle of conditional error. Recall that this H_0 is a general linear hypothesis involving $p - 1$ linearly independent estimable functions. The error sum of squares for the full model (11.1.1) is

$$SSe = \sum_{(i,j)} \sum_{\in D} (y_{ij} - \hat{y}_{ij})^2, \tag{11.2.1}$$

where $\hat{y}_{ij}$ is obtained from any solution of the normal equations. SSe has degrees of freedom $n - (\text{rank of } X) = n - p - r + 1$, where n is the number of observations. Using a computer program, we fit the full model (11.1.1) to the data which gives us $SS(\mu, \tau, \beta)$ and SSe. By fitting the reduced model (11.1.2) we obtain $SS(\mu, \beta)$ and SSe*. In reference to (11.1.4) and (11.1.3), the test statistic for H_0 is

$$f = \frac{(SSe^* - SSe)/(p-1)}{SSe/(n-p-r+1)} = \frac{SS(\tau|\mu, \beta)/(p-1)}{MSe} = \frac{MStr}{MSe},$$

where MStr is the treatment mean square adjusted for blocks. In like manner, the test statistic for $H_0 : \beta_1 = \beta_2 = \cdots = \beta_r$ is

$$f = \frac{SS(\beta|\mu, \tau)/(r-1)}{MSe} = \frac{MSbl}{MSe},$$

where $SS(\beta|\mu, \tau)$ is given by formula (11.1.5). We summarize in an analysis of variance table.

Table 11.2.1 ANOVA Table for Incomplete Blocks (Adjusted Sum of Squares)

Source	df	SS	MS	f
Treatments	p–1	$SS(\tau\|\mu,\beta)$	MStr	MStr/MSe
Blocks	r–1	$SS(\beta\|\mu,\tau)$	MSbl	MSbl/MSe
Error	n–p–r+1	SSe	$MSe = s^2$	
Total	n–1			

As before, the hypotheses "all treatment effects equal" and "all block effects equal" are rejected if the calculated ANOVA of values exceed tabulated f_α values with $p - 1,\ n - p - r + 1$, and $r - 1,\ n - p - r + 1$ degrees of freedom, respectively.

The reader may have noticed that the total sum of squares, $SSt = \sum_{(i,j) \varepsilon D} \sum (y_{ij} - \bar{y}..)^2$, was omitted from Table 11.2.1. The reason for this omission is that in the IBD case the sum of squares entries "do not add up"; that is, $SS(\tau|\mu, \beta) + SS(\beta|\mu, \tau) + SSe \neq SSt$. Instead, in the exercises, the reader is asked to verify that

$$SS(\tau|\mu, \beta) + SS(\mu, \beta) + SSe = SSt.$$

To find a confidence interval or test for a treatment contrast

$\sum_{i=1}^{p} c_i \tau_i$ we use the approach outlined in Chapter 5. From (5.3.7) we observe that a confidence interval is of the form

$$\sum_{i=1}^{p} c_i t_i \pm t_{\frac{\alpha}{2}, n-p-r+1} \text{Se}\,(\sum_{i=1}^{p} c_i \hat{\tau}_i), \tag{11.2.2}$$

where $\sum_{i=1}^{p} c_i t_i$ is the least squares estimate for the contrast, and

$\text{Se}\,(\sum_{i=1}^{p} c_i \hat{\tau}_i)$ represents the standard error of this least squares estimate.

Similarly, the form of the t—test for testing $H_0 : \sum_{i=1}^{p} c_i \tau_i = \gamma$ is

$$t = (\sum_{i=1}^{p} c\, t - \gamma)/\text{Se}\,(\sum_{i=1}^{p} c_i \hat{\tau}_i). \tag{11.2.3}$$

To perform multiple comparisons we first note that the $\bar{y}_{i\cdot}$'s are not comparable; therefore, we should not use the Tukey procedure described in (10.5.3). Instead we recommend using the Bonferroni approach described in (8.4.2). To determine which treatments are different we consider the $c = \binom{p}{2}$ pairwise contrasts $\tau_i - \tau_k$. We decide that $\tau_i \neq \tau_k$ if

$$|t_i - t_k| \geq t_{\frac{\alpha}{2c}, n-p-r+1} \text{Se}\,(\widehat{\tau_i - \tau_k}), \tag{11.2.4}$$

where $t_i - t_k$ is the least squares estimate for $\tau_i - \tau_k$. Note that we could also obtain simultaneous confidence intervals of the form

$(t_i - t_k) \pm t_{\frac{\alpha}{2c}} \text{Se}\,(\widehat{\tau_i - \tau_k})$, or we could obtain the Fisher's LSD procedure by replacing $t_{\frac{\alpha}{2c}}$ by $t_{\frac{\alpha}{2}}$ in (11.2.4).

To calculate the tests and confidence intervals presented thus far we can proceed in one of two ways. First, if a linear models computer program such as SAS is available, then we can use SAS GLM to perform the needed calculations. Alternatively, we can reparametrize the model to a full rank regression model and use regression techniques to perform the calculations. We illustrate both of these approaches with an example.

Linear Models (SAS) Approach

Example 11.2.1 Three drugs that induce sleep are being compared in terms of their effectiveness. Six subjects are available for the study. Each subject uses each drug in a random order on different days. The minutes required to fall asleep, Y, is recorded in each case. Unfortunately, due to illness, subject 4 only used two of the drugs. The data obtained appears in the table below.

Table 11.2.2 Data for Sleep Inducing Drugs

		Subject					
		1	2	3	4	5	6
Drug	1	15	20	21	8	17	10
	2	27	24	30	12	25	15
	3	12	17	22		14	8

We note that the design planned in this experiment is an RCBD with drugs serving as the treatments and subjects as the blocks. However, due to the missing cell we ended up with an IBD.

Our goal is to determine if there are differences between the three drugs; therefore, we will perform an f—test for the treatments and follow it with a Bonferroni multiple comparison. Using SAS GLM we obtain the necessary calculations and the computer output is give in Table 11.2.3.

Table 11.2.3 SAS Commands and Output for Sleep Inducing Drug Study

<u>**SAS Commands**</u>

```
OPTIONS LS = 80;
DATA ICB;
INPUT DRUG  SUBJ Y@@;
CARDS;
1  1  15  1  2  20  1  3  21  1  4   8  1  5  17  1
2  1  27  2  2  24  2  3  30  2  4  12  2  5  25  2
3  1  12  3  2  17  3  3  22            3  5  14  3
PROC GLM; CLASS SUBJ DRUG;
MODEL Y = SUBJ DRUG;
ESTIMATE 'DI VS D2' DRUG  1 -1  0;
ESTIMATE 'DI VS D3' DRUG  1  0 -1;
ESTIMATE 'D2 VS D3' DRUG  0  1 -1;
LSMEANS DRUG/PDIFF;
```

SAS Printout

```
                         General Linear Models Procedure

Dependent Variable: Y
                                      Sum of            Mean
Source                  DF           Squares          Square   F Value      Pr > F

Model                    7       667.36862745     95.33837535     22.08      0.0001

Error                    9        38.86666667      4.31851852

Corrected Total         16       706.23529412

                  R-Square              C.V.        Root MSE                Y Mean

                  0.944966          11.89487       2.0781045             17.470588

Source                  DF         Type I SS     Mean Square   F Value      Pr > F

SUBJ                     5      408.23529412     81.64705882     18.91      0.0002
DRUG                     2      259.13333333    129.56666667     30.00      0.0001

Source                  DF       Type III SS     Mean Square   F Value      Pr > F

SUBJ                     5      462.00000000     92.40000000     21.40      0.0001
DRUG                     2      259.13333333    129.56666667     30.00      0.0001

                                          T for H0:     Pr > |T|    Std Error of
Parameter                 Estimate      Parameter=0                   Estimate

D1 VS D2               -7.00000000            -5.83       0.0002      1.19979422
D1 VS D3                2.30000000             1.79       0.1075      1.28663596
D2 VS D3                9.30000000             7.23       0.0001      1.28663596


                         General Linear Models Procedure
                              Least Squares Means

                DRUG              Y    Pr > |T| H0: LSMEAN(i)=LSMEAN(j)
                             LSMEAN    i/j      1         2        3

                1        15.1666667    1    .        0.0002   0.1075
                2        22.1666667    2  0.0002      .       0.0001
                3        12.8666667    3  0.1075   0.0001      .

      NOTE: To ensure overall protection level, only probabilities associated with
            pre-planned comparisons should be used.
```

An advantage in using the SAS GLM program is that we gain information ready for further analysis because of its Type III SS feature which obviates the theoretical need for also fitting the "blocks only" reduced model. Recall from Section 10.6, the Type III SS column provides us with the sum of squares adjusted for the other factor. Hence $SS(\tau|\mu, \beta) = 259.1333$ and $SS(\beta|\mu, \tau) = 462.000$. (Although not needed here to complete an analysis, the sequential Type I SS column reveals that the unadjusted sum of squares for blocks is $SS(\mu, \beta) = 408.2353$ indicating that in this experiment blocks and treatments are not orthogonal.)

The test for treatments, $H_0 : \tau_1 = \tau_2 = \tau_3$, has $f = 30.00$ with $p = 0.0001$ so we reject H_0 and conclude that at least two drugs have significantly different mean effects. To determine which drugs are different we perform the Bonferroni procedure (11.2.4) for the $c = 3$ pairwise contrasts $\tau_1 - \tau_2$, $\tau_1 - \tau_3$, $\tau_2 - \tau_3$. From the Bonferroni table we find $t_{\frac{\alpha}{2c}, n-p-r+1} = t_{\frac{.05}{2(3)}, 9} \approx 2.87$. The point estimates and standard errors are found on the printout and lead to the following:

Contrast	Point Estimate	2.87(standard error)	Significant?
$\tau_1 - \tau_2$	−7.0	3.44	Yes
$\tau_1 - \tau_3$	2.3	3.69	No
$\tau_2 - \tau_3$	9.3	3.69	Yes

As a consequence, we conclude that effectiveness of drugs one and three are not significantly different, while drug two has significantly higher values than the other drugs.

SAS Least Squares Means

In the example we used the ESTIMATE statement on SAS to perform the Bonferroni procedure. An easier method is to use the SAS LSMEANS statement. The LSMEANS statement calculates **least squares means** for factors. In an incomplete block design the least squares mean for treatment i, LSM_i, is given by

$$LSM_i = \frac{1}{r} \sum_{j=1}^{r} (m + t_i + b_j), \tag{11.2.5}$$

where m, t_i, b_j are from any solution to the normal equations. Note that the summation is completed over all cells, even if there are missing observations. Since $LSM_i - LSM_k = t_i - t_k$, the least squares estimate for $\tau_i - \tau_k$, we can use these least squares means to compare the treatments. For the data from Example 11.2.1, Table 11.2.3 contains the least squares means (LSMEAN) and the p—value for testing the equality of every possible pair. As noted before, testing the equality of two least squares means for treatments i and k is equivalent to testing $H_0 : \tau_i - \tau_k = 0$. For example, from the printout we note that the p—value for testing $H_0 : \tau_1 - \tau_3 = 0$ is $p = .1075$. To perform the Bonferroni procedure for "c" contrasts we should reject H_0 if the p—value $\leq \alpha/c$. For the data in Example 11.2.1 the Bonferroni procedure for the pairwise treatment contrasts decides that $\tau_i \neq \tau_k$ if the p—value for the LSMEANS test is $\alpha/c = .05/3 = .0167$ or less. From the printout we can conclude that treatments one and three are not significantly different, but treatment two is significantly different from the other treatments. Note that this agrees with our previous result for the Bonferroni in Example 11.2.1. Finally, note that we could perform a Fisher's LSD procedure using the LSMEANS by deciding that $\tau_i \neq \tau_k$ if the p—value for the LSMEANS test is $\leq \alpha$.

11.3 Analysis for Incomplete Blocks — Reparametrized Approach

An alternative method that can be used to analyze data from an incomplete block design is to reparametrize the model to a full rank regression model and then use regression techniques to carry out the analysis. This approach is particularly useful when one does not have access to computer software that handles **general** linear models. Section 10.6 contains a six—step outline of this procedure for us to follow. The complete example carried through in that section provides the reader with the rationale and demonstrates the details involved. Here we illustrate by reanalyzing the data of Example 11.2.1 using regression.

Alternative Analysis for Example 11.2.1

Recall that the data set in Example 11.2.1 consists of three treatments and six blocks. Let us refer to the summary steps given at the end of Section 10.6. If we define (0,1) x_i variables as suggested in Steps 1 and 2, the full reparametrized model (Step 3) becomes

$$Y = \alpha_0 + \alpha_1 x_1 + \alpha_2 x_2 + \alpha_3 x_3 + \alpha_4 x_4 + \alpha_5 x_5 + \alpha_6 x_6 + \alpha_7 x_7 + \varepsilon$$

where x_1 and x_2 are associated with treatments and x_3 through x_7 are associated with blocks, i.e., the reduced model containing **blocks only** (Step 4) is

$$Y = \alpha_0^* + \alpha_3^* x_3 + \alpha_4^* x_4 + \alpha_5^* x_5 + \alpha_6^* x_6 + \alpha_7^* x_7 + \varepsilon^*$$

and the reduced model containing **treatments only** (Step 5) is

$$Y = \alpha_0^{**} + \alpha_1^{**} x_1 + \alpha_2^{**} x_2 + \varepsilon^{**}.$$

In order to carry out the analysis we need to know the relationship between the parameters of the reparametrized model and the traditional model. From Step 6, as explained in the derivation of (10.6.3), the (0,1) coding results in the following correspondences:

$$\alpha_0 = \mu + \tau_3 + \beta_6,\ \alpha_1 = \tau_1 - \tau_3,\ \alpha_2 = \tau_2 - \tau_3,$$

$$\alpha_3 = \beta_1 - \beta_6,\ \ldots,\ \alpha_7 = \beta_5 - \beta_6.$$

Therefore, testing $H_0 : \tau_1 = \tau_2 = \tau_3$ is equivalent to testing $H_0 : \alpha_1 = \alpha_2 = 0$ and finding a confidence interval for $\tau_1 - \tau_2$ is equivalent to finding a confidence interval for $\alpha_1 - \alpha_2$.

For the data of Example 11.2.1, Table 11.3.1 contains the SAS GLM program and the complete printout for the full model; only partial printouts pertinent to our discussion are given for the two reduced models.

Table 11.3.1 SAS Commands and Printout for Reparametrized Model

SAS Commands

```
OPTIONS LS = 80;
DATA ICB;
INPUT  DRUG  SUBJ  Y  @@;
X1 = 0;  X2 = 0;  X3 = 0;  X4 = 0;  X5 = 0;  X6 = 0;  X7 = 0;
IF DRUG = 1 THEN X1 = 1;  IF DRUG = 2 THEN X2 = 1;
IF SUBJ = 1 THEN X3 = 1;  IF SUBJ = 2 THEN X4 = 1;
IF SUBJ = 3 THEN X5 = 1;  IF SUBJ = 4 THEN X6 = 1;
IF SUBJ = 5 THEN X7 = 1;
CARDS;
1  1  15  1  2  20  1  3  21  1  4   8  1  5  17  1
2  1  27  2  2  24  2  3  30  2  4  12  2  5  25  2
3  1  12  3  2  17  3  3  22              3  5  14  3
PROC  GLM;
MODEL Y = X1  X2  X3  X4  X5  X6  X7  /  I;
PROC  GLM;
MODEL Y = X3  X4  X5  X6  X7;
PROC  GLM;
MODEL Y = X1 X2;
```

SAS Printout

General Linear Models Procedure

X'X Inverse Matrix

	INTERCEPT	X1	X2	X3	X4
INTERCEPT	0.4666666667	-0.2	-0.2	-0.333333333	-0.333333333
X1	-0.2	0.3833333333	0.2166666667	-2.22045E-16	-2.21177E-16
X2	-0.2	0.2166666667	0.3833333333	-2.22045E-16	-2.22045E-16
X3	-0.333333333	-2.22045E-16	-2.22045E-16	0.6666666667	0.3333333333
X4	-0.333333333	-2.21177E-16	-2.22045E-16	0.3333333333	0.6666666667
X5	-0.333333333	-2.19443E-16	-2.2031E-16	0.3333333333	0.3333333333
X6	-0.266666667	-0.1	-0.1	0.3333333333	0.3333333333
X7	-0.333333333	-2.22045E-16	-2.22045E-16	0.3333333333	0.3333333333
Y	7.1333333333	2.3	9.3	7	9.3333333333

	X5	X6	X7	Y
INTERCEPT	-0.333333333	-0.266666667	-0.333333333	7.1333333333
X1	-2.19443E-16	-0.1	-2.22045E-16	2.3
X2	-2.2031E-16	-0.1	-2.22045E-16	9.3
X3	0.3333333333	0.3333333333	0.3333333333	7
X4	0.3333333333	0.3333333333	0.3333333333	9.3333333333
X5	0.6666666667	0.3333333333	0.3333333333	13.333333333
X6	0.3333333333	0.8666666667	0.3333333333	-2.933333333
X7	0.3333333333	0.3333333333	0.6666666667	7.6666666667
Y	13.333333333	-2.933333333	7.6666666667	38.866666667

General Linear Models Procedure

Dependent Variable: Y

Source	DF	Sum of Squares	Mean Square	F Value	Pr > F
Model	7	667.36862745	95.33837535	22.08	0.0001
Error	9	38.86666667	4.31851852		
Corrected Total	16	706.23529412			

R-Square	C.V.	Root MSE	Y Mean
0.944966	11.89487	2.0781045	17.470588

Source	DF	Type I SS	Mean Square	F Value	Pr > F
X1	1	49.22014260	49.22014260	11.40	0.0082
X2	1	156.14848485	156.14848485	36.16	0.0002
X3	1	1.73153153	1.73153153	0.40	0.5423
X4	1	38.49145697	38.49145697	8.91	0.0153
X5	1	268.19923372	268.19923372	62.10	0.0001
X6	1	65.41111111	65.41111111	15.15	0.0037
X7	1	88.16666667	88.16666667	20.42	0.0015

Source	DF	Type III SS	Mean Square	F Value	Pr > F
X1	1	13.80000000	13.80000000	3.20	0.1075
X2	1	225.62608696	225.62608696	52.25	0.0001
X3	1	73.50000000	73.50000000	17.02	0.0026
X4	1	130.66666667	130.66666667	30.26	0.0004
X5	1	266.66666667	266.66666667	61.75	0.0001
X6	1	9.92820513	9.92820513	2.30	0.1638
X7	1	88.16666667	88.16666667	20.42	0.0015

Parameter	Estimate	T for H0: Parameter=0	Pr > |T|	Std Error of Estimate
INTERCEPT	7.13333333	5.02	0.0007	1.41961567
X1	2.30000000	1.79	0.1075	1.28663596
X2	9.30000000	7.23	0.0001	1.28663596
X3	7.00000000	4.13	0.0026	1.69676526
X4	9.33333333	5.50	0.0004	1.69676526
X5	13.33333333	7.86	0.0001	1.69676526
X6	-2.93333333	-1.52	0.1638	1.93461005
X7	7.66666667	4.52	0.0015	1.69676526

Regression ANOVA for Blocks Only Model
General Linear Models Procedure

Dependent Variable: Y

Source	DF	Sum of Squares	Mean Square	F Value	Pr>F
Model	5	408.23529412	81.64705882	3.01	0.0592
Error	11	298.00000000	27.09090909		
Corrected Total	16	706.23529412			

Regression ANOVA for Treatments Only Model
General Linear Models Procedure

Dependent Variable: Y

Source	DF	Sum of Squares	Mean Square	F Value	Pr>F
Model	2	205.36862745	102.68431373	2.87	0.0902
Error	14	500.86666667	35.77619048		
Corrected Total	16	706.23529412			

We now point out information contained in these printouts which provide us with an alternative method for analyzing the data of Example 11.2.1. From the full model output we observe that $SS(\mu, \tau, \beta) =$

667.3686 and SSe = 38.8667. Also, the regression ANOVA for the blocks only model yields SS(μ, β) = 408.2353 and SSe* = 298.0000 while we obtain SS(μ, τ) = 205.3686 and SSe** = 500.8667 from the fitted treatments only model. These quantities allow us to calculate the adjusted sums of squares as

$$SS(\tau | \mu, \beta) = 667.3686 - 408.2353 = 298.0000 - 38.8667 = 259.1333$$

and

$$SS(\beta | \mu, \tau) = 667.3686 - 205.3686 = 500.8667 - 38.8667 = 462.0000.$$

These numbers agree with the adjusted sum of squares displayed in the "Type III SS" column of Table 11.2.3.

To find confidence intervals or tests for treatment contrasts, we rewrite the contrasts in terms of the regression parameters. For example, if we wish to perform a statistical inference procedure concerning $\tau_1 - \tau_3$, we recall that $\alpha_1 = \tau_1 - \tau_3$. From the X1 line of the parameter estimate section of the full model printout in Table 11.3.1, we find that the point estimate for $\alpha_1 = \tau_1 - \tau_3$ is 2.3000 and its standard error is $Se(\hat{\alpha}_1) = 1.2866$. Note that this X1 line is identical to the D1 vs. D3 line of the traditional model analysis presented in Table 11.2.3.

For a more complex example, let us calculate a 95% confidence interval for $\tau_1 - \tau_2 = \alpha_1 - \alpha_2$. From the full model printout in Table 11.3.1, we find the least squares estimate for $\alpha_1 - \alpha_2$ to be $2.3 - 9.3 = -7.0$. The standard error of this estimate is given by $Se(\hat{\alpha}_1 - \hat{\alpha}_2) = \sqrt{s^2 \boldsymbol{\ell}' (\boldsymbol{X}'\boldsymbol{X})^{-1} \boldsymbol{\ell}}$ where $\boldsymbol{\ell}' = (0, 1, -1, 0, 0, 0, 0, 0)$. Using the printout quantities for s^2 and $(\boldsymbol{X}'\boldsymbol{X})^{-1}$, $Se(\hat{\alpha}_1 - \hat{\alpha}_2) = \sqrt{4.3185(0.3333)} = 1.200$ and since $t_{.025,9} = 2.262$, the 95% confidence interval for $\tau_1 - \tau_2$ is $-7.00 \pm 2.262\ (1.200)$ or $(-9.71, 4.29)$. We should observe that if one uses SAS GLM, as we did here, the estimate and its standard error can be programmed for output using the ESTIMATE statement as was demonstrated in the Section 10.6 example. In a similar manner we can find confidence intervals or perform tests for other contrasts or carry out the Bonferroni multiple comparisons procedure.

Finally, it should be pointed out that if treatments in an IBD experiment are of the quantitative type and the investigator is interested in the mathematical nature of the response to these treatments, the reparametrized approach can be used to attack the problem. See Exercise 11—5.

11.4 Balanced Incomplete Block Design

In some cases due to constraints on the design an experimenter will be unable to have every treatment present in each block. In such a case a design that may be appropriate, having hand computable formulas, is a balanced incomplete block (BIB) design.

Definition 11.4.1 A balanced incomplete block design is a design in which

- there are p treatments,
- there are r blocks,
- there are $k < p$ experimental units in each block, k called the block size,
- each treatment occurs in the same number, say t, blocks,
- each pair of treatments occur together in the same block the same number, say λ, of times,
- each block, k different treatments are randomly assigned to the experimental units.

As an example, consider an experiment in which we aim to compare five brands of perfumes. People were to rate the perfumes on a ten—point scale but it was decided that a person should only rate three perfumes at a time. This resulted in the use of a BIB design with the perfumes being the treatments ($p = 5$), judges serving as blocks and each judge used three times (block size $k = 3$). The design with the obtained data are given in Table 11.4.1.

Table 11.4.1 BIB for Perfume Study

	Person (Judge)									
	1	2	3	4	5	6	7	8	9	10
Perfume 1	4	4	3	6	5	8				
2	5	4	4				6	7	5	
3	8			9	8		9	10		6
4		6		8		10	9		8	7
5			5		6	7		8	6	4

The reader should check to see that in this design each treatment occurs

$t = 6$ times and each pair of treatments occur together in $\lambda = 3$ blocks.

There are some relationships that exist involving p, r, k, t, λ and sample size n. For example, n can be expressed as $n = pt$ or $n = kr$. In the exercises it is shown that $\lambda(p-1) = t(k-1)$. We therefore have

$$pt = kr \quad \text{and} \quad \lambda = t(k-1)/(p-1). \tag{11.4.1}$$

Since λ must be an integer, it is clear that only certain combinations of p, r, k, t, and λ are possible. It also can be shown that a BIB design exists only when $r \geq p$. We can use (11.4.1) to help determine the number of blocks that will be needed to have a BIB design. As an illustration, suppose that we have $p = 6$ treatments and blocks of size $k = 4$. From (11.4.1) we have $6t = 4r$ and $\lambda = 3t/5$. It is clear that the choice of $t = 6$ and $r = 9$ will not work since λ is not an integer; however, $t = 10$ and $r = 15$ with $\lambda = 6$ fulfills the structural requirements. Fortunately, tables of BIB designs have been prepared giving the design for various choices of p and k . One good source is Cochran and Cox (1957).

Using the linear model computer approaches of Sections 11.2 and 11.3, the analysis of data from a BIB design is routine. Since a BIB design has special properties, we are able to obtain closed form formulas for sums of squares, tests and confidence intervals. The model for a BIB design is the same as the one considered earlier in this chapter for general incomplete blocks, namely (11.1.1), and with m, t_i and b_j used as the realizations of μ, τ_i and β_j, the normal equations for a BIB design are:

$$\begin{cases} nm + t\sum_{i=1}^{p} t_i + k\sum_{j=1}^{r} b_j = y_{..} \\ tm + tt_i + \sum_{j=1}^{r} n_{ij}\, b_j = y_{i.}\ , \quad i = 1, 2, \ldots, p \\ km + \sum_{i=1}^{p} n_{ij}\, t_i + k\, b_j = y_{.j}\ , \quad j = 1, 2, \ldots, r \end{cases} \tag{11.4.2}$$

$$\text{where } n_{ij} = \begin{cases} 1\ , & \text{if treatment } i \text{ occurs in block } j \\ 0\ , & \text{otherwise.} \end{cases}$$

If we impose the side conditions $\sum_{i=1}^{p} t_i = 0$ and $\sum_{j=1}^{r} b_j = 0$, we then obtain

$$m = \bar{y}_{\cdot\cdot} \quad \text{and} \quad t_i = Q_i/\lambda p \tag{11.4.3}$$

$$\text{where} \quad Q_i = k\, y_{i\cdot} - \sum_{j=1}^{r} n_{ij}\, y_{\cdot j}\,.$$

As a notational note, the symbol Q_i, adopted here, is widely used in statistical literature.

Starting with (11.4.3), we can obtain the computing formulas for the BIB design ANOVA sums of squares as listed below. Their derivations will be left to the exercises. First, another note on notation. In the previous sections of this chapter, the sums of squares were obtained by fitting various models to the data using a computer. In recognition of this, for example, we represented the treatment sum of squares adjusted for blocks by the symbol $SS(\tau|\mu, \beta)$. Here we will obtain these same sum of squares by "plugging" into a formula. To differentiate between these two methods, we use the notation SStr (adj) in place of $SS(\tau|\mu, \beta)$. With this in mind, the hand—calculating formulas for the sums of squares associated with a BIB design are as follows:

$$SSt = \sum_{(i,j)\varepsilon D}\sum (y_{ij} - \bar{y}_{\cdot\cdot})^2 = \sum_{(i,j)\varepsilon D}\sum y_{ij}^2 - y_{\cdot\cdot}^2/n$$

$$SStr\ (adj) = \sum_{i=1}^{p} Q_i^2/\lambda pk \tag{11.4.4}$$

$$SSbl\ (unadj) = \sum_{j=1}^{r} k\,(\bar{y}_{\cdot j} - \bar{y}_{\cdot\cdot})^2 = \sum_{j=1}^{r} y_{\cdot j}^2/k - y_{\cdot\cdot}^2/rk$$

$$SSe = SSt - SStr\ (adj) - SSbl\ (unadj).$$

Using these formulas on the BIB data in Table 11.4.1, the student is asked to verify, in the exercises, that SSt = 111.5000, SStr (adj) = 42.5333, SSbl (unadj) = 64.1667 and SSe = 4.8000 which culminates in an f—value of 35.44 with 4 and 16 degrees of freedom.

For auxiliary analyses beyond the analysis of variance, such as confidence intervals for treatment contrasts $\sum_{i=1}^{p} c_i\tau_i$, $\sum_{i=1}^{p} c_i = 0$, we can use the least squares estimate and its standard error

$$\text{estimate:} \qquad \sum_{i=1}^{p} c_i Q_i / \lambda p$$

$$\text{standard error:} \qquad \sqrt{s^2 \frac{k}{\lambda p} \sum_{i=1}^{p} c_i^2} \, . \qquad (11.4.5)$$

Using (11.4.5), multiple comparisons can be made by the Bonferroni and Fisher's LSD methods to obtain appropriate confidence intervals and t–tests. There is a modification of the Tukey procedure that can be used but we shall not present it here. See Scheffé (1959) for details. The student will be asked to use Fisher's protected LSD procedure in order to compare the perfume data means, Table 11.4.1, in the exercises.

In summary, the analysis of data from a BIB design can be done with a linear models computer program such as SAS, or by reparametrizing the model and using regression techniques, or using the special formulas presented in this section. All three approaches, of course, yield the same results.

Chapter 11 Exercises

Application Exercises

11–1 In an experiment to compare four brands of tires an RCBD was used with brands as treatments and cars as blocks. Each car was driven a specified number of miles and the tire wear recorded. The resulting data are presented in the accompanying table.

		Tire Brand			
		A	B	C	D
Car	1	20	13	24	12
	2	27	19	29	22
	3	18	38	25	15
	4	32	23	34	24
	5	28	23	31	23

(a) Analyze the data as an RCBD and report the outcome of the f–test for brands using a 0.05 level.

(b) Use residuals to check the model assumptions and observe the presence of an outlier in observation y_{23}. What effect does the outlier have?

(c) Delete the outlier and rerun the analysis as an incomplete block design.

(i) Report the outcome of the f–test for brands using $\alpha = 0.05$.

(ii) What effect did removing the outlier have?

(iii) Check the model assumptions.

(iv) Perform a Bonferroni multiple comparison for brands using a 0.05 level. Which brand(s) seem best?

(v) (SAS Required) Use the Tukey command on SAS GLM for these data. Explain why the resulting output is not appropriate.

11–2 Refer to the sleep–inducing drug data in Table 11.2.2 and analyzed in Example 11.2.1 and again in Section 11.3.

(a) Use the LSMEANS on the SAS printout in Table 11.2.3 to perform a Fisher's protected LSD for the three drugs. Use $\alpha = 0.05$.

(b) Using the reparametrized model and the printout in Table 11.3.1, perform the Bonferroni multiple comparison for drugs at the 0.05 level.

11–3 A company is installing a new piece of equipment and has four operators who can potentially use the equipment. To compare the performance of these operators a study was completed over a four–day period. Due to time constraints only three operators can use the equipment on a given day and since day was considered to be a potential factor an IBD was employed. The results are given in the table to follow with the response variable being an efficiency rating with higher values being better.

		Treatment			
		A	B	C	D
	1	65	82	79	–
Day	2	73	–	90	93
	3	–	81	88	95
	4	71	70	–	87

(a) Using a linear models computer program (e.g., SAS), answer the following:
   (i) Find the ANOVA table and report the outcome of the f–tests for operators and for days.
   (ii) Use least squares means to perform the Bonferroni multiple comparison using $\alpha = 0.05$.
   (iii) Perform a t–test comparing the average of operators A, B, C as new employees with the average for experienced operator D. Use the 0.05 level.
   (iv) Use the residuals to check some model assumptions.

(b) Use a reparametrized model in order to obtain the analysis of variance displayed as in Table 11.2.1 and also use this model to calculate a 95% confidence interval to estimate the mean response to operator B on day 2.

(c) Recognizing that the design used was actually a BIB design, use the special formulas presented in Section 11.4 to find the following:
   (i) The ANOVA table. In what respect does this table differ from that calculated in (a) and (b) above? Why?
   (ii) The Bonferroni multiple comparison for operators.

11–4 Refer to the perfume data in Table 11.4.1.

(a) Using formulas (11.4.4), verify the sums of squares results given immediately following these formulas in the text and use these numbers to test $H_0 : \tau_1 = \tau_2 = \tau_3 = \tau_4 = \tau_5$ at the 0.05 level.

(b) Using (11.4.5), calculate a 95% confidence interval for $\tau_3 - \tau_5$. What conclusion can be drawn?

(c) Perform a Fisher's protected LSD for the perfumes using the 0.05 level and present your results in graphical form.

11–5 Refer to Exercise 10–10. Suppose the observations Oven 1, Temperature 200° and Oven 5, Temperature 500° are missing due to mishaps. For the resulting IBD, use a regression program in order to respond to the following parts:

(a) Using reparametrizing (0,1) coding as demonstrated in Section 10.6 and again in Section 11.3, print out the $(\boldsymbol{X}'\boldsymbol{X})^{-1}$ matrix and obtain and write out the ANOVA formatted in Table 11.2.1. In what way does the

$(\mathbf{X}'\mathbf{X})^{-1}$ matrix prove that the two factors, ovens and temperatures, are not orthogonal?

(b) Partition the treatment (temperature) sum of squares into single degree of freedom sum of squares using the following regression model.

$$Y = \alpha_0 + \alpha_1 x_1 + \alpha_2 x_1^2 + \alpha_3 x_1^3 + \alpha_4 x_4 + \alpha_5 x_5 + \alpha_6 x_6 + \alpha_7 x_7 + \varepsilon$$

where

$$x_1 = \begin{cases} 1, & \text{if temp 200} \\ 2, & \text{if temp 300} \\ 3, & \text{if temp 400} \\ 4, & \text{if temp 500} \end{cases} \quad x_4 = \begin{cases} 1, & \text{if oven 1} \\ 0, & \text{if not} \end{cases} \quad x_5 = \begin{cases} 1, & \text{if oven 2} \\ 0, & \text{if not} \end{cases} \quad x_6 = \begin{cases} 1, & \text{if oven 3} \\ 0, & \text{if not} \end{cases} \quad x_7 = \begin{cases} 1, & \text{if oven 4} \\ 0, & \text{if not} \end{cases}.$$

Could we have used this model to obtain the ANOVA as well as your model in (a)?

(c) Suppose the four treatment temperatures had been 200, 300, 400 and 600. What coding for x_1 in part (b) would you use?

Theoretical Exercises

11–6 Verify that for an incomplete block design

$$SSt = SS(\tau \mid \mu,\beta) + SS(\mu,\beta) + SSe .$$

Hint: Use (11.1.4) and recall that SSe* is the error sum of squares for an appropriate reduced model.

11–7 Consider an RCBD with 2 treatments and 4 blocks. From Chapter 10 we know that $E(\bar{Y}_{2\cdot} - \bar{Y}_{1\cdot}) = \tau_2 - \tau_1$. Now suppose Y_{24} is missing. For this IBD situation, answer the following parts:

(a) Find $E(\bar{Y}_{2\cdot} - \bar{Y}_{1\cdot})$.

(b) From the result in part (a), explain why the two treatment means are not "comparable."

(c) Let SSTR (unadj) denote the treatment sum of squares random variable unadjusted for blocks. Write down the "natural" formula for SSTR (unadj) and derive E[SSTR (unadj)].

(d) Calculate E[SSTR (adj)] recalling that SSTR (adj) = SST − SSBL (unadj) − SSE.

(e) By observing the answers to parts (c) and (d) explain why SStr (adj) is sometimes called the "clean" treatment sum of squares.

11–8 If every treatment pair occurs together in at least one block of an incomplete block design, show that in this case all linear treatment contrasts are estimable.

11–9 Verify the relationship between the parameters of the traditional model and the reparametrized model given in Section 11.3.

11–10 Consider the two designs given below and the traditional model for an incomplete block design. Here * represents an observation.

Design I

BLK \ TRT	1	2	3	4
1	*	*		*
2		*	*	
3	*	*		

Design II

BLK \ TRT	1	2	3	4
1	*	*		
2			*	*
3	*	*		

(a) Show that for design I all pairwise treatment contrasts are estimable.

(b) Are all pairwise treatment contrasts estimable for design II?

11–11 For a BIB design verify that $\lambda(p-1) = t(k-1)$.

11–12 Verify that the normal equations for a BIB are given by (11.4.2).

11–13 Find the solution to the normal equations for a BIB that correspond to the side condition $\sum_{i=1}^{p} t_i = 0$ and $\sum_{j=1}^{r} b_j = 0$. As a result verify (11.4.3).

11–14 Verify the least squares estimate and standard error given for a linear contrast in (11.4.5).

11–15 Verify the formula (11.4.4) given for SStr (adj).

11–16 Find a formula for SSbl (adj) in a BIB design. (Hint: Reverse the rolls of blocks and treatments.)

CHAPTER 12

LATIN SQUARE DESIGNS

12.1 Latin Squares

In reviewing the designs considered thus far, the CRD is a one–factor design aimed at comparing $p \geq 2$ treatments, while the RCBD allows us to incorporate another factor into the design through blocking. In this chapter we consider a design that allows for the comparison of $p \geq 2$ treatments that incorporate **two** blocking factors into the experiment.

The design we consider is called a Latin square design and is described in the following definition.

Definition 12.1.1 A **Latin square design** involves $p \geq 2$ treatments and the experimental units are arranged according to two blocking factors which we label as rows and columns. There are p rows and p columns. Each treatment appears exactly once in each row and each column.

As an illustration consider a Latin square design (LSD) with $p = 3$ and treatments labeled as A, B, C. One possible such LSD follows. The letter indicates which treatment is applied to the particular row, column.

		Column		
		1	2	3
	1	A	B	C
Row	2	B	C	A
	3	C	A	B

We note that other Latin squares are possible by assigning the treatments in a different order. Ideally in practice one should choose the Latin square at random. We also note that only some of the row, column, treatment combinations are present in a Latin square. If all such combinations were present, then p^3 observations would be required. A Latin square requires only p^2 observations, which is substantially smaller than p^3. For example, when $p = 4$ a Latin square requires only 16 observations while we would need 64 observations to have all possible row, column, treatment combinations present. The following example illustrates the use of an LSD for an experiment.

Example 12.1.1 In the Chapter 11 exercises we considered the problem of comparing four brands of tires (A, B, C, D). An RCBD was employed using cars as blocks. Since we know that position on a car does affect tire wear (the reason for tire rotation) we should obtain a better result (smaller experimental error) if we incorporate it into the design. To this end, following an example in Hicks (1973), an LSD is employed with two blocking factors: cars and position. The data are given in the table with the response variable representing the tire wear after each car is driven the same specified distance. The treatment (brand of tire) applied is given in parentheses.

Table 12.1.1 Latin Square Design for Tire Wear

		Position			
		L Front	R Front	L Rear	R Rear
Car	1	30 (B)	36 (A)	25 (C)	22 (D)
	2	24 (D)	34 (C)	18 (A)	15 (B)
	3	35 (A)	30 (D)	15 (B)	28 (C)
	4	32 (C)	24 (B)	13 (D)	14 (A)

In this Latin square with $p = 4$ our major goal in the experiment is to compare the mean tire wear for the four brands.

To facilitate the analysis of data from an LSD we introduce a model to explain the variation in the results. We let $Y_{ij(k)}$ represent the response to treatment k applied to the experimental unit in the ith row and jth column. The subscript k is placed in parentheses to indicate that k depends on i and j; that is, only one k value corresponds to a particular i, j combination. Our model is then:

$$Y_{ij(k)} = \mu + \rho_i + \gamma_j + \tau_k + \varepsilon_{ij(k)}$$

$$i = 1,2,\cdots,p;\ j = 1,2,\cdots,p;\ k = 1,2,\cdots,p; \qquad (12.1.1)$$

where

μ is the overall average response to all treatments,

ρ_i is the effect of the ith row,

γ_j is the effect of the jth column,

τ_k is the effect of the kth treatment,

$\varepsilon_{ij(k)}$ is the random error associated with the kth treatment

applied to the ith row and jth column with $E(\varepsilon_{ij(k)}) = 0$ and $Var(\varepsilon_{ij(k)}) = \sigma^2$ for every i and j. Furthermore, it is assumed that the $\varepsilon_{ij(k)}$ are uncorrelated.

We note that this model implies that

$$E[Y_{ij(k)}] = \mu + \rho_i + \gamma_j + \tau_k \quad \text{and} \quad Var[Y_{ij(k)}] = \sigma^2.$$

All observations have the same variance, but the mean can vary according to the row, column, and treatment. Our major interest will be in comparing the p treatments. A test for $H_0 : \tau_1 = \cdots = \tau_p$ should be obtained as well as tests and confidence intervals for treatment contrasts.

12.2 Least Squares Results

Our objective in this section is to find the normal equations for an LSD, a solution to these equations, and to use the solution to find least square estimates. We first note that the model (12.1.1) for an LSD is a linear model and can be written as $\mathbf{Y} = \mathbf{X}\boldsymbol{\beta} + \boldsymbol{\varepsilon}$. To illustrate this fact consider the LSD when $p = 3$ with observations described as follows:

Column

	$y_{11(1)}$	$y_{12(2)}$	$y_{13(3)}$
Row	$y_{21(2)}$	$y_{22(3)}$	$y_{23(1)}$
	$y_{31(3)}$	$y_{32(1)}$	$y_{33(2)}$

The model in matrix form is then

$$
\underset{\mathbf{Y}}{\begin{bmatrix} Y_{11(1)} \\ Y_{12(2)} \\ Y_{13(3)} \\ Y_{21(2)} \\ Y_{22(3)} \\ Y_{23(1)} \\ Y_{31(3)} \\ Y_{32(1)} \\ Y_{33(2)} \end{bmatrix}} = \underset{\mathbf{X}}{\begin{bmatrix} 1 & 1 & 0 & 0 & 1 & 0 & 0 & 1 & 0 & 0 \\ 1 & 1 & 0 & 0 & 0 & 1 & 0 & 0 & 1 & 0 \\ 1 & 1 & 0 & 0 & 0 & 0 & 1 & 0 & 0 & 1 \\ 1 & 0 & 1 & 0 & 1 & 0 & 0 & 0 & 1 & 0 \\ 1 & 0 & 1 & 0 & 0 & 1 & 0 & 0 & 0 & 1 \\ 1 & 0 & 1 & 0 & 0 & 0 & 1 & 1 & 0 & 0 \\ 1 & 0 & 0 & 1 & 1 & 0 & 0 & 0 & 0 & 1 \\ 1 & 0 & 0 & 1 & 0 & 1 & 0 & 1 & 0 & 0 \\ 1 & 0 & 0 & 1 & 0 & 0 & 1 & 0 & 1 & 0 \end{bmatrix}} \underset{\boldsymbol{\beta}}{\begin{bmatrix} \mu \\ \rho_1 \\ \rho_2 \\ \rho_3 \\ \gamma_1 \\ \gamma_2 \\ \gamma_3 \\ \tau_1 \\ \tau_2 \\ \tau_3 \end{bmatrix}} + \underset{\boldsymbol{\varepsilon}}{\begin{bmatrix} \varepsilon_{11(1)} \\ \varepsilon_{12(2)} \\ \varepsilon_{13(3)} \\ \varepsilon_{21(2)} \\ \varepsilon_{22(3)} \\ \varepsilon_{23(1)} \\ \varepsilon_{31(3)} \\ \varepsilon_{32(1)} \\ \varepsilon_{33(2)} \end{bmatrix}}
$$

As expected the $\mathbf{X}$ matrix is not full rank. Note that the sum of columns 2, 3, 4 equals column one. The same is true for columns 4, 5, 6 and columns 7, 8, 9. As a consequence, the rank of $\mathbf{X}$ for this example is $1 + (3-1) + (3-1) + (3-1) = 7$. It is easy to generalize and conclude that the rank of $\mathbf{X}$ for any LSD will be $1 + 3(p-1) = 3p - 2$.

To find the normal equations we need some notation. Our sample form for the model is:

$$
y_{ij(k)} = m + r_i + c_j + t_k + e_{ij(k)}, \tag{12.2.1}
$$

$$
\text{for} \quad i = 1,2,\ldots,p; \;\; j = 1,2,\ldots,p; \;\; k = 1,2,\ldots,p.
$$

As before we let $y_{i\cdot\cdot}$, $y_{\cdot j\cdot}$ and $y_{\cdot\cdot k}$ represent the sum of the observations in the ith row, jth column, and kth treatment, respectively. Similarly, $y_{\cdot\cdot\cdot}$ is the sum of all p^2 observations. To denote the corresponding average we use bars over the symbol. For example, $\bar{y}_{\cdot\cdot k}$ represents the average of the p observations in treatment k. With this notation it can be shown the set of normal equations for the LSD in general is

$$p^2 m + p\sum_{i=1}^{p} r_i + p\sum_{j=1}^{p} c_j + p\sum_{k=1}^{p} t_k = y_{\cdot\cdot\cdot}$$

$$pm + pr_i + \sum_{j=1}^{p} c_j + \sum_{k=1}^{p} t_k = y_{i\cdot\cdot}, \quad i = 1,2,\cdots,p$$

$$pm + \sum_{i=1}^{p} r_i + pc_j + \sum_{k=1}^{p} t_k = y_{\cdot j\cdot}, \quad j = 1,2,\cdots,p \tag{12.2.2}$$

$$pm + \sum_{i=1}^{p} r_i + \sum_{j=1}^{p} c_j + pt_k = y_{\cdot\cdot k}, \quad k = 1,2,\cdots,p.$$

In the exercises the student will be asked to verify (12.2.2) for the case when $p = 3$.

Since the **X** matrix is rank deficient by three, we need three side conditions to obtain a unique solution. If we impose the conditions

$\sum_{i=1}^{p} r_i = \sum_{j=1}^{p} c_j = \sum_{k=1}^{p} t_k = 0$ on the system we obtain the solution set:

$$m = \bar{y}_{\cdot\cdot\cdot}$$

$$r_i = \bar{y}_{i\cdot\cdot} - \bar{y}_{\cdot\cdot\cdot}, \quad i = 1,\cdots,p$$

$$c_j = \bar{y}_{\cdot j\cdot} - \bar{y}_{\cdot\cdot\cdot}, \quad j = 1,\cdots,p \tag{12.2.3}$$

$$t_k = \bar{y}_{\cdot\cdot k} - \bar{y}_{\cdot\cdot\cdot}, \quad k = 1,\cdots,p .$$

We can now find a formula for SSe. In general SSe =

$\sum_{i=1}^{p} \sum_{j=1}^{p} (y_{ij(k)} - \hat{y}_{ij(k)})^2$, and since $\hat{y}_{ij(k)} = m + r_i + c_j + t_k = \bar{y}_{i\cdot\cdot} + \bar{y}_{\cdot j\cdot} + \bar{y}_{\cdot\cdot k} - 2\bar{y}_{\cdot\cdot\cdot}$, it follows that

$$SSe = \sum_{i=1}^{p} \sum_{j=1}^{p} (y_{ij(k)} - \bar{y}_{i\cdot\cdot} - \bar{y}_{\cdot j\cdot} - \bar{y}_{\cdot\cdot k} + 2\bar{y}_{\cdot\cdot\cdot})^2 \qquad (12.2.4)$$

The degrees of freedom for SSe is $n - \text{rank } \mathbf{X} = p^2 - (3p-2) = (p-1)(p-2)$. The error mean square for an LSD is then

$$s^2 = SSe/[(p-1)(p-2)]. \qquad (12.2.5)$$

To find least squares estimates for quantities we first need to identify what is estimable. We know that $\mu + \rho_i + \gamma_j + \tau_k$ is estimable for every (i, j, k) for which an observation is present. In the exercises the student is asked to use this information to prove that all treatment, row and column contrasts are estimable in a Latin square design.

We can now find the BLUE for a treatment contrast, $\sum_{k=1}^{p} c_k \tau_k$, $\sum_{k=1}^{p} c_k = 0$. Using the solution (12.2.3) the BLUE for the contrast is $\sum_{k=1}^{p} c_k (\bar{y}_{\cdot\cdot k} - \bar{y}_{\cdot\cdot\cdot}) = \sum_{k=1}^{p} c_k \bar{y}_{\cdot\cdot k}$. To find the standard error we first find the variance. Since the observations are independent we obtain

$$\text{Var}\left(\sum_{k=1}^{p} c_k \bar{Y}_{\cdot\cdot k}\right) = \sum_{k=1}^{p} c_k^2 \,\text{Var}(\bar{Y}_{\cdot\cdot k}) = \sum_{k=1}^{p} c_k^2 \sigma^2/p.$$

Using rule 5.3.1 we replace σ^2 by s^2 and take the square root to obtain the standard error. Next we summarize these results and their application will be illustrated in the next section.

BLUE for Latin Square

Function	BLUE	Standard Error
Treatment Contrast		(12.2.6)
$\sum_{k=1}^{p} c_k \tau_k$, $\sum_{k=1}^{p} c_k = 0$	$\sum_{k=1}^{p} c_k \bar{y}_{\cdot\cdot k}$	$\sqrt{\frac{s^2}{p} \sum_{k=1}^{p} c_k^2}$

12.3 Inference for an LSD

To make statistical inferences for data from an LSD we need to assume the **normal LSD model**, namely the model (12.1.1) with the added assumption that the $\varepsilon_{ij(k)}$'s are independent $N(0,\sigma^2)$ random variables. Our goal in this section is to find the form of tests and confidence intervals for an LSD.

We first turn our attention to the f—tests and ANOVA table for an LSD. To test H_0 : "No differences in treatment effects" we use the principle of conditional error. The hypothesis $H_0 : \tau_1 = \cdots = \tau_p$, can be written in the form of a general linear hypothesis $H_0 : \mathbf{L\beta} = \mathbf{0}$. Here $\mathbf{L}$ is a $(p-1) \times (3p + 1)$ matrix such that $\mathbf{L\beta}$ contains as its $p-1$ rows the linearly independent estimable functions $\tau_2 - \tau_1$, $\tau_3 - \tau_1, \cdots, \tau_p - \tau_1$. Imposing the hypothesis $\mathbf{L\beta} = \mathbf{0}$ on the full model (12.1.1) we obtain the reduced model:

$$Y_{ij(k)} = \mu^* + \rho_i + \gamma_j + \varepsilon^*_{ij(k)}; \quad i = 1,\cdots,p; \quad j = 1,\cdots,p.$$

Since there is only one k—value for each i,j pair, we observe that this reduced model is an RCBD model. From (10.3.7) and using the notation for an LSD we arrive at

$$SSe^* = \sum_{i=1}^{p} \sum_{j=1}^{p} (y_{ij(k)} - \bar{y}_{i\cdot\cdot} - \bar{y}_{\cdot j\cdot} + \bar{y}_{\cdot\cdot\cdot})^2.$$

The **treatment sum of squares**, SStr, is then obtained as $SStr = SSe^{*} - SSe$, where SSe is given by (12.2.4). After some algebra we arrive at

$$SStr = p \sum_{k=1}^{p} (\bar{y}_{..k} - \bar{y}_{...})^2. \tag{12.3.1}$$

Hence the statistic used to test $H_0 : \tau_1 = \cdots \tau_p$ is

$$f = \frac{SStr/(p-1)}{s^2} = \frac{MStr}{MSe}, \tag{12.3.2}$$

and we reject H_0 if $f \geq f_{\alpha,p-1,(p-1)(p-2)}$. Note that this SStr is actually an adjusted sum of squares; however, as we shall discover later on, an LSD is orthogonal. As a consequence the adjusted and unadjusted sum of squares are identical.

In a similar manner the hypothesis $H_0 : \rho_1 = \cdots \rho_p$ can be tested using the statistic

$$f = \frac{SSro/(p-1)}{s^2} = \frac{MSro}{MSe}, \tag{12.3.3}$$

where $SSro = p \sum_{i=1}^{p} (\bar{y}_{i..} - \bar{y}_{...})^2$ is called the **row sum of squares**. Also the hypotheses $H_0 : \gamma_1 = \cdots = \gamma_p$ can be tested using the statistic

$$f = \frac{SSco/(p-1)}{s^2} = \frac{MSco}{MSe}, \tag{12.3.4}$$

where $SSco = p \sum_{j=1}^{p} (\bar{y}_{.j.} - \bar{y}_{...})^2$ is called the **column sum of squares**. These tests are summarized in the following ANOVA table.

Table 12.3.1 ANOVA Table for LSD

Source	df	SS	MS	f
Rows	p−1	SSro	MSro = SSro/(p−1)	MSro/MSe
Columns	p−1	SSco	MSco = SSco/(p−1)	MSco/MSe
Treatments	p−1	SStr	MStr = SStr/(p−1)	MStr/MSe
Error	(p−1)(p−2)	SSe	MSe = SSe/(p−1)(p−2)	
Total	p^2-1	SSt		

Formulas for these sums of squares are given in (12.3.1), (12.3.3), (12.3.4), and (12.2.4). Computational formulas for these quantities are as follows:

$$\text{SSro} = \sum_{i=1}^{p} \frac{y_{i..}^2}{p} - \frac{y_{...}^2}{p^2}, \quad \text{SSco} = \sum_{j=1}^{p} \frac{y_{.j.}^2}{p} - \frac{y_{...}^2}{p^2}$$

$$\text{SStr} = \sum_{k=1}^{p} \frac{y_{..k}^2}{p} - \frac{y_{...}^2}{p^2}, \quad \text{SSt} = \sum_{i=1}^{p} \sum_{j=1}^{p} y_{ij(k)}^2 - \frac{y_{...}^2}{p^2}, \tag{12.3.5}$$

$$\text{SSe} = \text{SSt} - \text{SSro} - \text{SSco} - \text{SStr} .$$

These formulas are justified in the exercises.

We now turn our attention to tests and confidence intervals for treatment contrasts. From Section 5.3 we know that for the normal LSD model the confidence interval for an estimable function is given by: (point estimate) ± $t_{\frac{\alpha}{2},(p-1)(p-2)}$ (standard error). Using (12.2.6) it is easy to find a confidence interval for any treatment contrast and, in particular, for a pairwise treatment difference. In a similar way, t—tests can be obtained. These results are summarized as follows:

Tests and Confidence Intervals for Contrasts, LSD

Function	Confidence Interval
$\tau_k - \tau_t$	$(\bar{y}_{..k} - \bar{y}_{..t}) \pm t_{\frac{\alpha}{2},(p-1)(p-2)} \sqrt{\frac{2s^2}{p}}$
Contrast, $\sum_{k=1}^{p} c_k \tau_k$, $\sum_{k=1}^{p} c_k = 0$	$\sum_{k=1}^{p} c_k \bar{y}_{..k} \pm t_{\frac{\alpha}{2},(p-1)(p-2)} \sqrt{\frac{s^2}{p} \sum_{k=1}^{p} c_k^2}$

(12.3.6)

Null Hypothesis	Test Statistic
$H_0 : \tau_k - \tau_t = \gamma$	$t = (\bar{y}_{..k} - \bar{y}_{..t} - \gamma) \Big/ \sqrt{\frac{2s^2}{p}}$
$H_0 : \sum_{k=1}^{p} c_k \tau_k = \gamma$, $\sum_{k=1}^{p} c_k = 0$	$t = (\sum_{k=1}^{p} c_k \bar{y}_{..k} - \gamma) \Big/ \sqrt{\frac{s^2}{p} \sum_{k=1}^{p} c_k^2}$

Note: All rejection regions are based upon a t–distribution with (p–1) (p–2) degrees of freedom.

For multiple comparisons of the treatments we consider the Tukey procedure. Since $\bar{Y}_{..1}, \cdots, \bar{Y}_{..p}$ are independent normal random variables with variance σ^2/p, it follows from Section 9.3 that we can use the Tukey procedure. It is summarized in (12.3.7).

We close this section with elements of a numerical example. Hand–calculating details omitted here can be filled in by the reader.

Tukey Procedure for Latin Square

$100\,(1-\alpha)\%$ Simultaneous Confidence Intervals

for all pairs $\tau_k - \tau_t$:

$$(\bar{y}_{\cdot\cdot k} - \bar{y}_{\cdot\cdot t}) \pm q_{\alpha,p,(p-1)(p-2)} \sqrt{\frac{s^2}{p}} \qquad (12.3.7)$$

Multiple Comparison tests for $H_0 : \tau_k = \tau_t$:

Decide $\tau_k \neq \tau_t$ if $|\bar{y}_{\cdot\cdot k} - \bar{y}_{\cdot\cdot t}| > q_{\alpha,p,(p-1)(p-2)} \sqrt{\frac{s^2}{p}}$

Example 12.3.1 Refer to the tire wear data in Example 12.1.1 which comes from an LSD with $p = 4$. We let $y_{ij(k)}$ represent the tire wear for the ith car (row), jth position (column), and kth brand (treatment) where brand A, B, C, D are numerically coded 1, 2, 3, 4, respectively. We first find the ANOVA table and the f—tests. Routine calculations yield the following values:

$$y_{\cdot\cdot\cdot} = 395 \qquad \sum_{i=1}^{4} \sum_{j=1}^{4} y_{ij(k)}^2 = 10{,}685$$

$$y_{1\cdot\cdot} = 113 \qquad y_{2\cdot\cdot} = 91 \qquad y_{3\cdot\cdot} = 108 \qquad y_{4\cdot\cdot} = 83$$

$$y_{\cdot 1\cdot} = 121 \qquad y_{\cdot 2\cdot} = 124 \qquad y_{\cdot 3\cdot} = 71 \qquad y_{\cdot 4\cdot} = 79$$

$$y_{\cdot\cdot 1} = 103 \qquad y_{\cdot\cdot 2} = 84 \qquad y_{\cdot\cdot 3} = 119 \qquad y_{\cdot\cdot 4} = 89.$$

Using these values and the computational formulas (12.3.5), it is easy to obtain the analysis of variance table below.

Source	df	SS	MS	f
Car (Row)	3	149.1875	49.729	11.5
Position (Column)	3	573.1875	191.062	44.3
Brand (Treatment)	3	185.1875	61.729	14.3
Error	6	25.875	4.312	
Total	15	933.4375		

Since $f_{.05,3,6} = 4.76$ we note that each of the three f—tests result in rejecting H_0. As a consequence, there is a significant difference between the mean tire wear for at least two brands, at least two cars, and at least two positions. To determine which brands are significantly different we use the Tukey multiple comparison. From (12.3.7) we decide that the kth and tth brands are significantly different if

$$|\bar{y}_{\cdot\cdot k} - \bar{y}_{\cdot\cdot t}| > q_{.05,4,6}\sqrt{\frac{s^2}{p}} = 4.90\sqrt{\frac{4.312}{4}} = 5.09.$$

The results of this Tukey procedure are given in tabular form.

Brand	B	D	A	C
$\bar{y}_{\cdot\cdot k}$	21.00	22.25	25.75	29.75

Since smaller tire wear values are better, it follows that brand C is significantly worse than brand B and D. Purely for illustration purposes we find a 95% confidence interval for $\tau_1 - \tau_2$. Using (12.3.6) the interval is

$$(\bar{y}_{\cdot\cdot 1} - \bar{y}_{\cdot\cdot 2}) \pm t_{\frac{\alpha}{2},\,(p-1)(p-2)}\sqrt{\frac{2s^2}{p}}, \text{ which is,}$$

$$(25.75 - 21.00) \pm 2.447\sqrt{\frac{2(4.312)}{4}}, \text{ which yields,}$$

$$4.75 \pm 2.447\ (1.468) \quad \text{or} \quad (1.16,\ 8.34).$$

Since this confidence interval contains only positive values it indicates that brands A and B are significantly different based on a .05 level t—test. Note that this differs from the conclusion reached by the Tukey procedure but is not surprising because the Tukey procedure has less power in order to protect the overall error level for all $\binom{4}{2} = 6$ comparisons made.

SAS Usage

The analysis of a Latin square can be done using SAS. The SAS commands and printout that perform the calculations we just completed are given in Table 12.3.2. The reader is invited to compare the results.

Table 12.3.2 SAS Commands and Printout for Latin Square

SAS Commands

```
OPTIONS LS = 80;
DATA TXLS;
INPUT CAR POSITION BRAND Y @@;
CARDS;
1  1  2  30  1  2  1  36  1  3  3  25  1  4  4  22
2  1  4  24  2  2  3  34  2  3  1  18  2  4  2  15
3  1  1  35  3  2  4  30  3  3  2  15  3  4  3  28
4  1  3  32  4  2  2  24  4  3  4  13  4  4  1  14
PROC GLM;  CLASS  CAR  POSITION  BRAND;
MODEL Y = CAR POSITION BRAND;
ESTIMATE 'B1 VS B2' BRAND 1  -1  0  0  ;
MEANS BRAND / TUKEY;
```

SAS Printout

```
General Linear Models Procedure
   Class Level Information

Class      Levels     Values

CAR             4     1 2 3 4

POSITION        4     1 2 3 4

BRAND           4     1 2 3 4

Number of observations in data set = 16
```

```
                         General Linear Models Procedure

Dependent Variable: Y
                                        Sum of            Mean
Source                  DF             Squares          Square   F Value     Pr > F

Model                    9        907.56250000    100.84027778     23.38     0.0005

Error                    6         25.87500000      4.31250000

Corrected Total         15        933.43750000

                  R-Square                C.V.        Root MSE                Y Mean

                  0.972280            8.411771       2.0766560             24.687500

Source                  DF           Type I SS     Mean Square   F Value     Pr > F

CAR                      3        149.18750000     49.72916667     11.53     0.0067
POSITION                 3        573.18750000    191.06250000     44.30     0.0002
BRAND                    3        185.18750000     61.72916667     14.31     0.0038

Source                  DF         Type III SS     Mean Square   F Value     Pr > F

CAR                      3        149.18750000     49.72916667     11.53     0.0067
POSITION                 3        573.18750000    191.06250000     44.30     0.0002
BRAND                    3        185.18750000     61.72916667     14.31     0.0038

                                        T for H0:    Pr > |T|    Std Error of
Parameter                  Estimate    Parameter=0                 Estimate

B1 VS B2                 4.75000000           3.23     0.0178      1.46841752
```

```
                         General Linear Models Procedure

              Tukey's Studentized Range (HSD) Test for variable: Y

          NOTE: This test controls the type I experimentwise error rate, but
                generally has a higher type II error rate than REGWQ.

                      Alpha= 0.05  df= 6  MSE= 4.3125
                  Critical Value of Studentized Range= 4.896
                    Minimum Significant Difference= 5.0832

          Means with the same letter are not significantly different.

                Tukey Grouping              Mean      N  BRAND

                             A            29.750      4  3
                             A
                      B      A            25.750      4  1
                      B
                      B                   22.250      4  4
                      B
                      B                   21.000      4  2
```

12.4 Reparametrization of an LSD

It is easy to reparametrize the less than full rank traditional LSD model (12.1.1) into a full rank regression model. One can then use regression techniques to analyze data from an LSD. As an example, consider the 3×3 LSD illustrations given early in Section 12.1 and 12.2. The reparametrized or regression form of the model for this $p = 3$ case is as follows:

$$Y = \alpha_0 + \alpha_1 x_1 + \alpha_2 x_2 + \alpha_3 x_3 + \alpha_4 x_4 + \alpha_5 x_5 + \alpha_6 x_6 + \varepsilon \qquad (12.4.1)$$

where

$$x_1 = \begin{cases} 1, & \text{if experimental unit is in row 1} \\ 0, & \text{if not} \end{cases}$$

$$x_2 = \begin{cases} 1, & \text{if experimental unit is in row 2} \\ 0, & \text{if not} \end{cases}$$

$$x_3 = \begin{cases} 1, & \text{if experimental unit is in column 1} \\ 0, & \text{if not} \end{cases}$$

$$x_4 = \begin{cases} 1, & \text{if experimental unit is in column 2} \\ 0, & \text{if not} \end{cases}$$

$$x_5 = \begin{cases} 1, & \text{if treatment 1 applied} \\ 0, & \text{if not} \end{cases}$$

$$x_6 = \begin{cases} 1, & \text{if treatment 2 applied} \\ 0, & \text{if not.} \end{cases}$$

Using this system of (0,1) coding the correspondence between the parameters of the traditional and reparametrized model is:

$$\alpha_0 = \mu + \rho_3 + \gamma_3 + \tau_3, \quad \alpha_1 = \rho_1 - \rho_3, \quad \alpha_2 = \rho_2 - \rho_3,$$

$$\alpha_3 = \gamma_1 - \gamma_3, \quad \alpha_4 = \gamma_2 - \gamma_3, \quad \alpha_5 = \tau_1 - \tau_3, \quad \alpha_6 = \tau_2 - \tau_3.$$

In matrix form this reparametrized model is

$$\underset{\mathbf{Y}}{\begin{bmatrix} Y_{11(1)} \\ Y_{12(2)} \\ Y_{13(3)} \\ Y_{21(2)} \\ Y_{22(3)} \\ Y_{23(1)} \\ Y_{31(3)} \\ Y_{32(1)} \\ Y_{33(2)} \end{bmatrix}} = \underset{\boldsymbol{\mathcal{X}}}{\begin{bmatrix} 1 & 1 & 0 & 1 & 0 & 1 & 0 \\ 1 & 1 & 0 & 0 & 1 & 0 & 1 \\ 1 & 1 & 0 & 0 & 0 & 0 & 0 \\ 1 & 0 & 1 & 1 & 0 & 0 & 1 \\ 1 & 0 & 1 & 0 & 1 & 0 & 0 \\ 1 & 0 & 1 & 0 & 0 & 1 & 0 \\ 1 & 0 & 0 & 1 & 0 & 0 & 0 \\ 1 & 0 & 0 & 0 & 1 & 1 & 0 \\ 1 & 0 & 0 & 0 & 0 & 0 & 1 \end{bmatrix}} + \underset{\boldsymbol{\alpha}}{\begin{bmatrix} \alpha_0 \\ \alpha_1 \\ \alpha_2 \\ \alpha_3 \\ \alpha_4 \\ \alpha_5 \\ \alpha_6 \end{bmatrix}} + \underset{\boldsymbol{\varepsilon}}{\begin{bmatrix} \varepsilon_{11(1)} \\ \varepsilon_{12(2)} \\ \varepsilon_{13(3)} \\ \varepsilon_{21(2)} \\ \varepsilon_{22(3)} \\ \varepsilon_{23(1)} \\ \varepsilon_{31(3)} \\ \varepsilon_{32(1)} \\ \varepsilon_{33(2)} \end{bmatrix}}$$

The normal equations are $\boldsymbol{\mathcal{X}}'\boldsymbol{\mathcal{X}}\mathbf{a} = \boldsymbol{\mathcal{X}}'\mathbf{y}$ and the unique solution is $\mathbf{a} = (\boldsymbol{\mathcal{X}}'\boldsymbol{\mathcal{X}})^{-1}\boldsymbol{\mathcal{X}}'\mathbf{y}$ for which

$$(\boldsymbol{\mathcal{X}}'\boldsymbol{\mathcal{X}})^{-1} = \begin{bmatrix} \frac{7}{9} & -\frac{1}{3} & -\frac{1}{3} & -\frac{1}{3} & -\frac{1}{3} & -\frac{1}{3} & -\frac{1}{3} \\ -\frac{1}{3} & \frac{2}{3} & \frac{1}{3} & 0 & 0 & 0 & 0 \\ -\frac{1}{3} & \frac{1}{3} & \frac{2}{3} & 0 & 0 & 0 & 0 \\ -\frac{1}{3} & 0 & 0 & \frac{2}{3} & \frac{1}{3} & 0 & 0 \\ -\frac{1}{3} & 0 & 0 & \frac{1}{3} & \frac{2}{3} & 0 & 0 \\ -\frac{1}{3} & 0 & 0 & 0 & 0 & \frac{2}{3} & \frac{1}{3} \\ -\frac{1}{3} & 0 & 0 & 0 & 0 & \frac{1}{3} & \frac{2}{3} \end{bmatrix} \qquad (12.4.2)$$

Note that the $(\mathbf{X}'\mathbf{X})^{-1}$ matrix has six 2×2 submatrices of all zeros. This shows that the correlations between the estimators for parameters of rows and columns, rows and treatments, and columns and treatments are all zero, implying that the LSD is **orthogonal**. It should be noted that the configuration features exhibited in $(\mathbf{X}'\mathbf{X})^{-1}$ matrix (12.4.2) are those associated with a general p rows, p columns and p treatments LSD. As a consequence of this orthogonality property, adjusted and unadjusted sums of squares are the same and the sums of squares "add up," i.e., SSt = SSro + SSco + SStr + SSe.

To complete the usual tests and confidence intervals for a Latin square we use the relationship between the parameters of the reparametrized and traditional models. We illustrate a few of these ideas for a Latin square with $p = 3$ and the reparametrized model (12.4.1).

The usual tests of interest are of the form:

$$H_0 : \rho_1 = \rho_2 = \rho_3 \quad \text{is equivalent to} \quad H_0 : \alpha_1 = \alpha_2 = 0,$$

$$H_0 : \gamma_1 = \gamma_2 = \gamma_3 \quad \text{is equivalent to} \quad H_0 : \alpha_3 = \alpha_4 = 0, \text{ and}$$

$$H_0 : \tau_1 = \tau_2 = \tau_3 \quad \text{is equivalent to} \quad H_0 : \alpha_5 = \alpha_6 = 0.$$

To find the sum of squares for the ANOVA table we first recall that the adjusted and unadjusted sum of squares are equal. The treatment sum of squares (unadjusted) is the regression sum of squares for the model $Y = \alpha_0^* + \alpha_5^* x_5 + \alpha_6^* x_6 + \varepsilon^*$. In terms of notation introduced in Chapters 10 and 11, this can be stated symbolically as SStr = SS(μ,τ). Similarly, SSro = SS(μ,ρ) and SSco = SS(μ,γ). SSe can be found by subtraction.

Finding tests or confidence intervals for functions of parameters is routine. For example, for the Latin square with $p = 3$ a confidence interval for $\tau_1 + \tau_2 - 2\tau_3$ can be found by finding a confidence interval for $\alpha_5 + \alpha_6$ in the reparametrized model (12.4.1). Other confidence intervals or multiple comparisons can be obtained in a similar manner.

In closing this section we note that the reparametrized approach is useful when a general purpose linear models computer program is not available, but a regression program is. Also the regression approach is particularly useful in analyzing data from an incomplete Latin square design such as in a missing data case.

12.5 Expected Mean Squares, Restricted LSD Model

Our first goal is to find the theoretical expected mean squares associated with the ANOVA table for an LSD. Recall from Theorem 7.7.1 the formula $E(MS) = \sigma^2 + W/q$, where W is the expression for the sum of squares when the Y's are replaced by their expected values, and q is the degrees of freedom for the mean square. To find E(MSTR) we recall that $SSTR = p \sum_{k=1}^{p} (\bar{Y}_{..k} - \bar{Y}_{...})^2$; therefore,

$W = p \sum_{k=1}^{p} [E(\bar{Y}_{..k}) - E(\bar{Y}_{...})]^2$. For a LSD, $E(\bar{Y}_{...}) = \sum_{i=1}^{p} \sum_{j=1}^{p} Y_{ij(k)}/p^2 = \mu + \bar{\rho} + \bar{\gamma} + \bar{\tau}$, where $\bar{\rho} = \sum_{i=1}^{p} \rho_i/p$, $\bar{\gamma} = \sum_{j=1}^{p} \gamma_j/p$ and $\bar{\tau} = \sum_{k=1}^{p} \tau_k/p$. Similarly, $E(\bar{Y}_{..k}) = \mu + \bar{\rho} + \bar{\gamma} + \tau_k$; therefore, $W = p \sum_{k=1}^{p} (\tau_k - \bar{\tau})^2$. We then obtain

$$E(MSTR) = \sigma^2 + p \sum_{k=1}^{p} (\tau_k - \bar{\tau})^2/(p-1). \qquad (12.5.1)$$

In a completely analogous manner we can obtain expressions for the expected mean square for rows and columns. These values are summarized in the following table.

Table 12.5.1 Theoretical ANOVA for LSD with E(MS)

Source	df	MS	E(MS)
Rows	p−1	MSRO	$\sigma^2 + p\sum_{i=1}^{p}(\rho_i-\bar{\rho})^2/(p-1)$
Columns	p−1	MSCO	$\sigma^2 + p\sum_{j=1}^{p}(\gamma_j-\bar{\gamma})^2/(p-1)$
Treatments	p−1	MSTR	$\sigma^2 + p\sum_{k=1}^{p}(\tau_k-\bar{\tau})^2/(p-1)$
Error	(p−1)(p−2)	MSE	σ^2
Total	p^2-1		

As we have observed before, E(MSTR) = E(MSE) when H_0 : $\tau_1 = \cdots = \tau_k$ is true. That is, the expected value of the numerator of the F statistic used for testing H_0 equals the expected value of the denominator when H_0 is true. Analogous statements can be made about E(MSRO) and E(MSCO).

The Restricted LSD Model

Some authors include the usual side conditions as part of the model. We have referred to this as the **restricted model**. For the LSD the restricted model is:

$$\begin{cases} Y_{ij(k)} = \mu + \rho_i + \gamma_j + \tau_k + \varepsilon_{ij(k)}; \quad i = 1,\cdots,p \,;\, j = 1,\cdots,p \\ \qquad\qquad\qquad\qquad\qquad\qquad\qquad k = 1,\cdots,p \\ \sum_{i=1}^{p} \rho_i = 0, \quad \sum_{j=1}^{p} \gamma_j = 0, \quad \sum_{k=1}^{p} \tau_k = 0\ . \end{cases} \tag{12.5.2}$$

As we have noted before, for the restricted model all parameters are individually estimable. From (12.2.3) we observe that the BLUE for μ, ρ_i, γ_j and τ_k are $\bar{y}_{...}$, $\bar{y}_{i..} - \bar{y}_{...}$, $\bar{y}_{.j.} - \bar{y}_{...}$, and $\bar{y}_{..k} - \bar{y}_{...}$, respectively. In this restricted case the hypothesis of no treatment effect becomes $H_0 : \tau_1 = \cdots = \tau_k = 0$. Also, E(MSTR) becomes

$\sigma^2 + p \sum_{k=1}^{p} \tau_k^2/(p-1)$.

It should be noted that there is no difference in the usual analysis (i.e., f—tests, inferences for contrasts) for an LSD based on the unrestricted model versus the restricted model. We should also note that many linear model computer programs, including SAS, do not use the restricted model (12.5.2). This is a major reason why we have chosen to focus on the unrestricted model in most of our presentation.

12.6 Design Considerations

The Latin square offers certain advantages as a design. Among these are:

- The experimental error can possibly be reduced by incorporating **two** blocking factors into the design. Effective blocking results in more powerful tests and tighter confidence intervals.

- The analysis of the data is fairly simple and can easily be done even without a computer.

- The design is orthogonal.

- The treatment means are "comparable."

- This three-factor design only requires p^2 observations rather than the p^3 observations required if all possible row, column, treatment combinations are present.

There are also some disadvantages to the LSD. Included are:

- The number of levels of all three factors must be the same.

- Not all row, column, treatment combinations are present in the design. Some of those missing may be important possibilities as they might be the optimal conditions.

- The error degrees of freedom is small when $p \leq 4$.

- The model is an additive one, namely, the effects of rows, columns and treatments are simply added together. There is no easy way to check for relationships (interaction) between the factors — a concept we will consider in the next chapter.

In summary, the LSD is a very useful design for problems involving three factors. This is especially true when the experimenter can only afford a few observations.

The LSD is often used in drug studies. In some problems a subject is given several drugs, and we might be concerned about possible order effects of the drugs. An LSD can be used to help "even out" the order effect of the drug. As an example, consider the case where there are four drugs A, B, C, D and each subject will use each drug in some order. If we use an LSD with $p = 4$ where the treatments are the drugs, the rows are the patients, and the columns are the order in which the drug is taken, then the resulting structure could be as follows:

		Order for Drug			
		1	2	3	4
	1	A	B	D	C
Subject	2	C	A	B	D
	3	D	C	A	B
	4	B	D	C	A

Notice that the LSD design ensures that each drug is used in each order exactly once. We should point out, however, that the LSD does not allow for representation of all 4! = 24 different orders for the drugs. Designs that are used to incorporate order effects are often called **crossover designs**. More will be discussed about these designs in a later chapter and, as we will see then, the Latin square is very useful in developing effective crossover designs.

Earlier in this section we mentioned that when p is small the degrees of freedom for error for a Latin square is small. One remedy for this problem is to replicate the Latin square. For example, if $p = 2$ a design in which the Latin square is replicated three times would appear as follows:

Replication 1

		Column 1	Column 2
Row	1	A	B
	2	B	A

Replication 2

		Column 1	Column 2
Row	1	A	B
	2	B	A

Replication 3

		Column 1	Column 2
Row	1	A	B
	2	B	A

As an illustration, consider a study to compare two varieties of wheat (A,B). In a replicated LSD our rows could represent a fertilizer level (low, high), our columns could represent the pesticide level (none, some), and the replication could represent different fields. The model for this replicated Latin square is discussed in the exercises and, as we discover there, the degrees of freedom for error is increased.

An extension of a Latin square to four factors can be accomplished by using a **Graeco—Latin square design**. In such a design there are four factors, three blocking factors (row, column, and layer) and a treatment factor. All factors have p levels. Each treatment occurs once in every row, column, and layer. Each layer occurs once in every row and column. An example of a Graeco—Latin square design with $p = 4$, treatments (A,B,C,D) and layers $(\alpha,\beta,\gamma,\delta)$ is given below.

		Column			
		1	2	3	4
Row	1	α A	β B	γ C	δ D
	2	β C	α D	δ A	γ B
	3	δ B	γ A	β D	α C
	4	γ D	δ C	α B	β A

Chapter 12 Exercises

Application Exercises

12–1 Starting with the summary totals given in Example 12.3.1, use the computational formulas (12.3.5) to arrive at the numerical entries of the ANOVA table displayed in this example.

12–2 A study is completed to find the work schedule that leads to the best job satisfaction for a group of technicians. Four schedules were used in the study:

A: 4–day week, day shift; B: 4–day week, evening shift;
C: 5–day week, day shift; D: 5–day week, evening shift.

Four technicians were available for the study and each used each work schedule for a week. To help control possible order effects an LSD was employed. The data given in the table represents a job satisfaction score obtained from a questionnaire administered at the end of each week. Higher values indicate more satisfaction.

		Week			
		1	2	3	4
Technician	1	18 (B)	7 (D)	13 (A)	10 (C)
	2	8 (C)	15 (A)	6 (D)	11 (B)
	3	18 (A)	10 (C)	10 (B)	5 (D)
	4	7 (D)	13 (B)	10 (C)	13 (A)

(a) Find the ANOVA table for these data.

(b) Report and interpret the outcome of each f–test using a 0.05 level.
(c) Use a Tukey multiple comparison, at level 0.05, to separate the means for the work schedules. Which schedule(s) lead to the highest satisfaction.
(d) Use a t–test at level 0.05 to test an appropriate contrast comparing day to night shift.
(e) Repeat part (d) using a contrast that compares 4–day to 5–day work weeks.
(f) Analyze the residuals for these data.

12–3 Four varieties of soybeans (V1, V2, V3, V4) are compared for yield in a study. Four fields were available for the study and each was divided into four relatively homogeneous plots of the same size. Four different methods for preventing insects and weeds were used: A: none; B: pesticide spray; C: weed spray; D: both pesticide and weed spray. An LSD was employed so that only four combinations of preventing method and variety were needed at each field. The response variable corresponds to the yield in the plot in bushels. The variety is given in parentheses.

Preventing Method

Field	A	B	C	D
1	29 (V1)	35 (V2)	42 (V3)	43 (V4)
2	30 (V2)	37 (V3)	39 (V4)	42 (V1)
3	37 (V3)	44 (V4)	43 (V1)	47 (V2)
4	32 (V4)	32 (V1)	38 (V2)	43 (V3)

(a) Find the ANOVA table.
(b) Report and interpret the outcome of each f–test using a 0.05 level.
(c) Using a 0.05 level, perform the Tukey multiple comparison for the varieties.
(d) Repeat part (c) for the preventing methods. Is pesticide or weed spray more effective?
(e) Analyze the residuals for these data.

12–4 Hicks (1973) reports a research study at Purdue University on metal–removal rate in which five electrode shapes A, B, C, D, and E were studied. The removal was accomplished by an electric discharge between the electrode and the material being cut. For this experiment five holes were cut in five workpieces, and the order of electrodes was arranged so that only one electrode shape was used in the same position on each of the five workpieces. Thus, the design was a Latin square design, with workpieces (strips) and positions on the strip as restrictions on the randomization. The times in hours necessary to cut the holes were recorded as follows:

		Position				
		1	2	3	4	5
Strip	I	A (3.5)	B (2.1)	C (2.5)	D (3.5)	E (2.4)
	II	E (2.6)	A (3.3)	B (2.1)	C (2.5)	D (2.7)
	III	D (2.9)	E (2.6)	A (3.5)	B (2.7)	C (2.9)
	IV	C (2.5)	D (2.9)	E (3.0)	A (3.3)	B (2.3)
	V	B (2.1)	C (2.3)	D (3.7)	E (3.2)	A (3.5)

(a) Find the ANOVA table for these data.
(b) Report and interpret the outcome of each f–test using a 0.05 level.
(c) Using a 0.05 level, perform the Tukey multiple comparison procedure on the shapes.

12–5 Refer to the data in Table 12.1.1 which were analyzed in Example 12.3.1. Use the reparametrized regression approach to repeat the analysis for these data. Calculate both $SS(\mu,\tau)$ and $SS(\tau \mid \mu,\rho,\gamma)$ to show that the unadjusted and adjusted treatment sums of squares are the same. Would $SS(\tau \mid \mu,\rho)$ also have the same value?

12–6 (Missing Data in a LSD) Suppose that the observation in week 4 for technician 4 is missing in Exercise 12–2. (This could happen due to illness.) We can analyze the resulting data using a linear models program or by reparametrizing the model and using a regression program. (This is similar to the situation of missing observations for a RCBD as described in Chapter 11.)

(a) Using a linear models program such as SAS, find the ANOVA table and perform each f–test using a 0.05 level.

(Be sure to use the adjusted sums of squares for the tests.)

(b) Perform a Bonferroni multiple comparison on the work schedules. Use a 0.05 level. (If SAS is available the LSMEANS command may be used.)

(c) Using a reparametrized regression approach, repeat parts (a) and (b).

12–7 (Youden Square) A Youden square design is an incomplete Latin square design in which the number of columns is less then the number of rows (and treatments). As an illustration, eliminate the last week of the data in Exercise 12–2 and the result is a Youden square. Using either a linear models approach or a reparametrized regression approach, find the ANOVA table and perform each f–test using a 0.05 level. For which of the three factors (rows, columns, treatments) are the adjusted and unadjusted sums of squares identical?

Theoretical Exercises

12–8 Refer to the illustration given at the outset of Section 12.2. Derive the normal equations for this 3×3 Latin square design thereby verifying (12.2.2) for the case when $p = 3$.

12–9 Verify that in a Latin square design all treatment, row, and column contrasts are estimable.

12–10 Verify the computational formulas for the sums of squares for an LSD given in (12.3.5).

12–11 Refer to Exercise 12–4. Assuming the restricted LSD model for the analysis, calculate the BLUES for μ, τ_A, ρ_I, and γ_1. Then compare the estimate for $E(Y_{I1(A)})$ with $y_{I1(A)} = 3.5$.

12–12 Derive (a) E(MSRO) and (b) E(MSCO) for the restricted $p \times p$ Latin square design model.

12–13 The cell means model for a Latin square is

$$Y_{ij(k)} = \mu_{ij(k)} + \varepsilon_{ij(k)};\ i = 1,\cdots,p;\ j = 1,\cdots,p;\ k = 1,\cdots,p.$$

(a) Show that this is a full rank linear model.

(b) Find the least squares estimator for $\mu_{ij(k)}$.

(c) Find a formula for SSe and compare it to the SSe for the traditional model for an LSD.

(d) Can we test the hypothesis of no treatment effect in terms of the $\mu_{ij(k)}$'s? Explain.

12–14 Explain why all parameters of the Latin square model are estimable in the restricted case.

12–15 If one or more cells are missing in an LSD, but all contrasts are still estimable, show that $SSe = SSt - SS(\mu,\rho) - SS(\gamma | \mu,\rho) - SS(\tau | \mu,\rho,\gamma)$.

12–16 (Power of Test) The power of a Latin square can be determined using the Pearson–Hartley Table in the Appendix. To use these tables for a Latin square we let $\nu_1 = p-1$, $\nu_2 = (p-1)(p-2)$ and $\phi^2 = \frac{1}{\sigma^2} \sum_{k=1}^{p} (\tau_k - \bar{\tau})^2$. Find the power for an LSD when $p = 4$, $\sigma^2 = 2$, $\tau_1 = 6$, $\tau_2 = 8$, $\tau_3 = 8$ and $\tau_4 = 10$.

12–17 (Graeco–Latin Square) Consider a Graeco–Latin square design with p rows, columns, treatments and layers as described in Section 12.6. A model for such a design is $Y_{ij(k\ell)} = \mu + \rho_i + \gamma_j + \tau_k + \omega_\ell + \varepsilon_{ij(k\ell)}$; $i, j, k, \ell = 1, \cdots, p$; where $Y_{ij(k\ell)}$ is the observation on the ith row, jth column, kth treatment, and ℓth layer. Note that the k and ℓ subscripts are placed in parentheses since their values depend on i and j.

(a) Recognizing that the ANOVA table for a Graeco–Latin square is similar to a Latin square, find the form of ANOVA table for a Graeco–Latin square.

(b) Perform an analysis of the following data obtained from a Graeco–Latin square design with treatments A, B, C, D and layers α, β, γ, δ.

Column

		1	2	3	4
	1	14 (Dγ)	14 (Aβ)	13 (Cα)	14 (Bδ)
Row	2	21 (Cβ)	17 (Bγ)	22 (Dδ)	16 (Aα)
	3	20 (Bα)	19 (Cδ)	20 (Aγ)	19 (Dβ)
	4	17 (Aδ)	11 (Dα)	19 (Bβ)	18 (Cγ)

12–18 (Replicated Latin Square) Consider a Latin square design with p rows, columns, and treatments. Suppose that this same design is replicated q times so that the levels of the rows and columns are the same for each replication. An illustration of this situation with $p = 2$ and $q = 3$ is given in Section 12.6. A model for this replicated LSD is $Y_{ij(k)\ell} = \mu + \rho_i + \gamma_j + \tau_k + \omega_\ell + \varepsilon_{ij(k)\ell}$; where $Y_{ij(k)\ell}$ is the observation in the ith row, jth column, kth treatment in the ℓth replicated Latin square.

(a) Find the form of the ANOVA table for this replicated Latin square design.

(b) Explain why the model and ANOVA table would change if the levels of the rows were different in each replication. (This could occur if the row factor is subject and different subjects are used for each replication.)

CHAPTER 13

FACTORIAL EXPERIMENTS WITH TWO FACTORS

13.1 Introduction

In the previous chapters we considered the basic designs of experiments that involve just one factor of interest. For example, in the RCBD the factor of interest is the treatment factor, with blocks simply being incorporated into the model to potentially reduce the experimental error. Our goal now is to consider experiments in which there is more than one factor of interest. In an experiment aimed at studying crop yield we might be interested in the effect on yield due to two factors: fertilizer level and amount of spray for weeds. In another study we might be interested in determining the effect of two factors, brand of gasoline and octane level, on the gas mileage for cars. The focus for this chapter will be on designs for experiments with two factors of interest, while the next chapter focuses on experiments with more than two such factors.

When there are two factors of interest in an experiment we need to be concerned about the relationship between the levels of the factors. Two factors A and B are said to be **crossed** if every level of factor A occurs with every level of factor B, and vice versa. The various combinations of the levels of the factors (i.e., the treatments) are sometimes called the **treatment combinations**. In an experiment with factor A having two levels (L and H) and factor B having two levels (L and H), when the factors are crossed the treatment combinations would be as follows:

Treatment	1	2	3	4
Level of A	L	L	H	H
Level of B	L	H	L	H.

It is important to note that not all factors are crossed. For example, suppose we are considering an experiment involving horses and our two factors are sex of the horse (female and male), and animal (horse 1, horse 2, etc.) Since a horse can only be one sex the factors sex and horse are not crossed. (Actually horse is **nested** in sex — a concept we deal with in a later chapter.) Factorial experiments involve crossed factors. We formalize our understanding of a factorial experiment with a definition.

Definition 13.1.1 A **factorial experiment** is one involving two or more factors of interest. For each factor all levels of the factor occur with all levels of every other factor in the experiment. (That is, the factors are crossed.) If the number of replications is the same for each treatment combination, then the experiment is called a **complete factorial** or a **balanced factorial.**

Example 13.1.1 As an illustration of a factorial experiment consider a study aimed at determining the effect of engine speed and engine operating temperature on the lifetime of an electric motor. For the study there are two levels of operating temperature, which we label as low and high. For operating speed there are three levels labeled as low, medium, and high. Since there are two levels of temperature (factor A), three levels of speed (factor B), and in a factorial experiment the factors need to be crossed, there will be $2(3) = 6$ treatment combinations. Eighteen different motors were available for the study, and they were randomly assigned to the six different treatment groups (i.e., a completely randomized design). The resulting data is the time to failure for the motors, which is given in the Table 13.1.1. We note that this experiment is a two—factor factorial experiment with a CRD used to allocate the experimental units (motors) to the six treatment groups. Since there are two levels of the first factor and three for the second, we label this experiment as a 2×3 factorial experiment.

Table 13.1.1 Time to Failure for Electric Motors

		Operating Speed		
		Low	Medium	High
Temperature	Low	67 74 67	67 70 74	46 54 52
	High	82 78 86	75 84 82	64 61 57

We make several observations from the preceding example. First, factorial experiments may be classified by the number of factors and by the number of levels for each factor. For example, a $3 \times 4 \times 2$ factorial experiment is one involving three factors with 3, 4, and 2 levels, respectively. Second, the design used in a factorial experiment is determined by how the experimental units are assigned to the treatment combinations. In the example we used a CRD, but we could have a two–factor factorial experiment in which an RCBD is used, and the resulting data structure would be quite different.

We will first choose to consider the simplest of all factorial experiments, namely the two–factor factorial experiment with the design being a CRD. The next few sections will consider this situation in detail. In the next chapter we consider two–factor factorial experiments with other designs, and factorial experiments with more than two factors of interest.

13.2 Model for Two–Factor Factorial and Interaction

Our goal in this section is to develop a useful model for a balanced two–factor factorial experiment randomized within a CRD. The two factors will be labeled as A and B, where A has "a" levels and B has "b" levels, and there are $r \geq 2$ observations (experimental units) in each treatment (combinations of the levels of A and B). We let Y_{ijk} represent the response to the kth experimental unit to which the ith level of A and jth level of B has been applied.

One model that can be considered is the **main effects model**. In this model we let $\alpha_1,\ldots,\alpha_a$ represent the effects for factor A, and $\beta_1,\ldots,\beta_b$ represent the effects for factor B. These are called the **main effects** and the main effects model is:

$$Y_{ijk} = \mu+\alpha_i + \beta_j+\varepsilon_{ijk} \ ; \ i=1,\cdots,a; \ j=1,\cdots,b; \ k = 1,\cdots,r \ ; \quad (13.2.1)$$

where μ is the overall mean. We assume that the ε_{ijk}'s are uncorrelated with mean zero and variance σ^2. With this model we note that the cell means are given by $\mu_{ij} = \mu + \alpha_i + \beta_j$, and Var $(Y_{ijk}) = \sigma^2$ for all i, j, k. Since μ_{ij} is found by taking the overall mean and adding the effect for A and the effect for B, this model is often called an **additive model**.

The additive model is appropriate for some experiments but not for all. If we examine the cell means for the model we notice a definite pattern. When $a = 2$ and $b = 3$ the pattern for cell means is as follows:

Cell Means

		Level of B		
		1	2	3
Level of A	1	$\mu+\alpha_1+\beta_1$	$\mu+\alpha_1+\beta_2$	$\mu+\alpha_1+\beta_3$
	2	$\mu+\alpha_2+\beta_1$	$\mu+\alpha_2+\beta_2$	$\mu+\alpha_2+\beta_3$

With this pattern notice that the difference between the mean of levels 1 and 2 of factor A is the same within each level of factor B. That is, $\mu_{11} - \mu_{21} = \mu_{12} - \mu_{22} = \mu_{13} - \mu_{23} = \alpha_1 - \alpha_2$. Notice that a similar statement can be made concerning the differences between the levels of B within a level of A.

We now turn to a numerical illustration to explore situations where this constant difference in theoretical cell means holds and where it does not in order to recognize from data when the main effects model is applicable. Let us assume that the numerical values of the underlying means for two experiments are as given in Table 13.2.1 below. In Setting I, the difference between the two levels of A is the same for each of the three levels of B. This cannot be said for Setting II; rather, for this theoretical setting the differences in A level values varies according to the level of B. In conclusion, the main effects model (13.2.1) is an appropriate model for Setting I since its underlying theoretical values conform to the $\mu + \alpha_i + \beta_j$ pattern. However, (13.2.1) does not explain the cell mean values in Setting II and therefore it behooves us to consider another model. To do so requires us to discuss an important concept associated with factorial experiments called **interaction**.

Table 13.2.1 Theoretical Cell Means

Setting I: Main Effects Model, Additive

		Level of B			
		1	2	3	Row Mean
Level of A	1	12	14	10	12
	2	16	18	14	16
Column Mean		14	16	12	

Setting II: Not Additive

		Level of B			
		1	2	3	Row Mean
Level of A	1	15	19	11	15
	2	17	11	11	13
Column Mean		16	15	11	

Interaction

For two crossed factors A and B we say there is **no interaction** between the factors if the relationship between means for the levels of factor A is the same for each level of B. The same holds true if we reverse the roles of A and B. There is **interaction** between the factors if this relationship varies with the level of B. One can detect the presence of interaction by inspecting the cell means. For the cell means in Setting I of Table 13.2.1 we have previously observed that the difference between the means for levels one and two of factor A is four units within each level of B. Since the relationship between the levels of A is the same for every level of B, there is no interaction between A and B. For the cell means in Setting II of Table 13.2.1 there clearly is interaction between the two factors.

For the no interaction case the additive model (13.2.1) is appropriate; however, when interaction is present we need a more complicated model. To account for its presence it is necessary to add an interaction term, denoted by $(\alpha\beta)_{ij}$, which represents the effect of interaction between the ith level of A and jth level of B. Then the **non–additive model** (model with interaction) for a two–factor factorial within a CRD is:

$$Y_{ijk} = \mu + \alpha_i + \beta_j + (\alpha\beta)_{ij} + \varepsilon_{ijk}; \quad i = 1,\cdots,a;$$

$$j = 1,\cdots, b; \quad k = 1,\cdots, r; \qquad (13.2.2)$$

where, as before, the ε_{ijk}'s are uncorrelated random variables with mean zero and variance σ^2. This is the model that we will use most of the time.

We close this section by illustrating a graphical approach for checking for interaction. We can plot the cell means as follows for the values from Table 13.2.1.

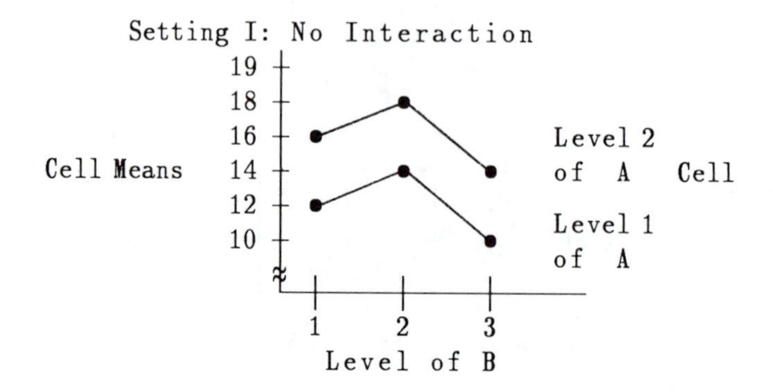

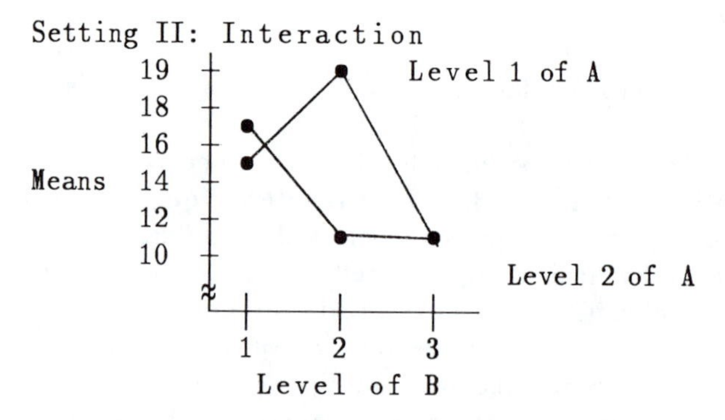

In the case of no interaction the line segments in the plot are equal-distant (parallel), which is characteristic of no interaction. In the interaction case the line segments are not parallel. In Setting II we have a "crossing" interaction as the line segments actually cross. Another form of interaction is a "spread" interaction in which the segments spread apart, as illustrated in the following plot.

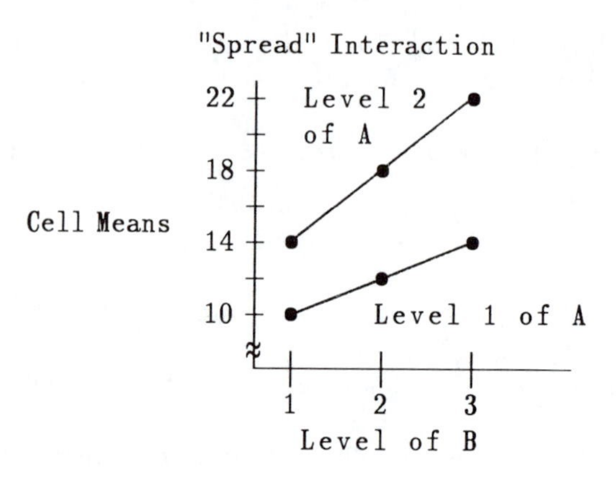

13.3 Least Squares Results

Our purpose in this section is to consider the normal equations for a balanced two–factor factorial experiment within a CRD and to find a solution to these equations. We also determine what is estimable, find the BLUE for estimable functions, and obtain the error sum of squares.

We first note that our model (13.2.2) is a linear model and can be written in the form $\mathbf{Y} = \mathbf{X}\boldsymbol{\beta} + \boldsymbol{\varepsilon}$. As an illustration, consider the case where $a = 2$, $b = 3$, and $r = 2$, with data given in the table.

		Level of B		
		1	2	3
Level of A	1	Y_{111} Y_{112}	Y_{121} Y_{122}	Y_{131} Y_{132}
	2	Y_{211} Y_{212}	Y_{221} Y_{222}	Y_{231} Y_{232}

In the matrix representation we have

$$\underset{\mathbf{Y}}{\begin{bmatrix} Y_{111} \\ Y_{112} \\ Y_{121} \\ Y_{122} \\ Y_{131} \\ Y_{132} \\ Y_{211} \\ Y_{212} \\ Y_{221} \\ Y_{222} \\ Y_{231} \\ Y_{232} \end{bmatrix}} = \underset{\mathbf{X}}{\begin{bmatrix} 1 & 1 & 0 & 1 & 0 & 0 & 1 & 0 & 0 & 0 & 0 & 0 \\ 1 & 1 & 0 & 1 & 0 & 0 & 1 & 0 & 0 & 0 & 0 & 0 \\ 1 & 1 & 0 & 0 & 1 & 0 & 0 & 1 & 0 & 0 & 0 & 0 \\ 1 & 1 & 0 & 0 & 1 & 0 & 0 & 1 & 0 & 0 & 0 & 0 \\ 1 & 1 & 0 & 0 & 0 & 1 & 0 & 0 & 1 & 0 & 0 & 0 \\ 1 & 1 & 0 & 0 & 0 & 1 & 0 & 0 & 1 & 0 & 0 & 0 \\ 1 & 0 & 1 & 1 & 0 & 0 & 0 & 0 & 0 & 1 & 0 & 0 \\ 1 & 0 & 1 & 1 & 0 & 0 & 0 & 0 & 0 & 1 & 0 & 0 \\ 1 & 0 & 1 & 0 & 1 & 0 & 0 & 0 & 0 & 0 & 1 & 0 \\ 1 & 0 & 1 & 0 & 1 & 0 & 0 & 0 & 0 & 0 & 1 & 0 \\ 1 & 0 & 1 & 0 & 0 & 1 & 0 & 0 & 0 & 0 & 0 & 1 \\ 1 & 0 & 1 & 0 & 0 & 1 & 0 & 0 & 0 & 0 & 0 & 1 \end{bmatrix}} \underset{\boldsymbol{\beta}}{\begin{bmatrix} \mu \\ \alpha_1 \\ \alpha_2 \\ \beta_1 \\ \beta_2 \\ \beta_3 \\ (\alpha\beta)_{11} \\ (\alpha\beta)_{12} \\ (\alpha\beta)_{13} \\ (\alpha\beta)_{21} \\ (\alpha\beta)_{22} \\ (\alpha\beta)_{23} \end{bmatrix}} + \underset{\boldsymbol{\varepsilon}}{\begin{bmatrix} \varepsilon_{111} \\ \varepsilon_{112} \\ \varepsilon_{121} \\ \varepsilon_{122} \\ \varepsilon_{131} \\ \varepsilon_{132} \\ \varepsilon_{211} \\ \varepsilon_{212} \\ \varepsilon_{221} \\ \varepsilon_{222} \\ \varepsilon_{231} \\ \varepsilon_{232} \end{bmatrix}}$$

Notice that $\mathbf{X}$ is not full rank, as there are a number of relationships between the columns. The sum of columns two and three (α_1 and α_2 columns) equal column one, as does the sum of columns four, five, and six (β_1, β_2, β_3 columns). The sum of columns seven, eight, and nine ($(\alpha\beta)_{11}$, $(\alpha\beta)_{12}$, $(\alpha\beta)_{13}$) equals column two, and the sum of columns ten, eleven, and twelve ($(\alpha\beta)_{21}$, $(\alpha\beta)_{22}$, $(\alpha\beta)_{23}$) equals column three (the difference between columns one and two). Finally, the sum of columns seven and ten equals column four, and the sum of columns eight and eleven equals column five. As a result, the rank of **X** is $1 + 1 + 2 + 2 = 6$, which is clearly less than full rank. After some reflection we observe that the linearly independent columns in **X** will in general be column one, a—1 columns associated with the α_i's, b—1 columns associated with the β_j's and (a—1)(b—1) columns associated with the interaction terms. It follows that the rank of **X** is $1 + a - 1 + b - 1 + (a-1)(b-1) = ab$. Since the **X** matrix is of order $abr \times (a+1)(b+1)$, the matrix is rank deficient by $(a+1)(b+1) - ab = a + b + 1$.

To find a solution to the normal equations in the general case, we consider the sample form of the model (12.2.3) which is

$$y_{ijk} = m + a_i + b_j + (ab)_{ij} + e_{ijk};$$

$$i = 1,\cdots,a; \quad j = 1,\cdots, b; \quad k = 1,\cdots, r.$$

The normal equations $\mathbf{X}'\mathbf{X}\,\mathbf{b} = \mathbf{X}'\mathbf{y}$ "expand" to the following set of algebraic equations:

$$abrm + rb \sum_{i=1}^{a} a_i + ra \sum_{j=1}^{b} b_j + r \sum_{i=1}^{a} \sum_{j=1}^{b} (ab)_{ij} = y_{\cdots}$$

$$rbm + rba_i + r \sum_{j=1}^{b} b_j + r \sum_{j=1}^{b} (ab)_{ij} = y_{i\cdot\cdot}$$

$$\text{(for } i = 1,\cdots, a\text{; a separate equations)} \qquad (13.3.1)$$

$$ram + r \sum_{i=1}^{a} a_i + rab_j + r \sum_{i=1}^{a} (ab)_{ij} = y_{\cdot j\cdot}$$

$$\text{(for } j = 1,\cdots, b\text{; b separate equations)}$$

$$rm + ra_i + rb_j + r\,(ab)_{ij} = y_{ij\cdot}$$

$$\text{(for } i = 1,\cdots, a;\ j = 1,\cdots, b\text{; ab separate equations).}$$

Here $y_{\cdots}$, $y_{i\cdot\cdot}$, $y_{\cdot j\cdot}$, $y_{ij\cdot}$ are defined in the usual manner. For example, $y_{i\cdot\cdot} = \sum_{j=1}^{b} \sum_{k=1}^{r} y_{ijk}$ with corresponding sample mean given by $\bar{y}_{i\cdot\cdot} = y_{i\cdot\cdot}/br$.

To solve the normal equations, we utilize side conditions. Since $\mathbf{X}$ is rank deficient by $a + b + 1$, we need that many linearly independent side conditions to solve the equations. Side conditions that greatly simplify the algebra are

$$\sum_{i=1}^{a} a_i = 0, \quad \sum_{j=1}^{b} b_j = 0$$
$$\sum_{i=1}^{a} (ab)_{ij} = 0 \text{ for all } j, \quad \sum_{j=1}^{b} (ab)_{ij} = 0 \text{ for all } i. \tag{13.3.2}$$

In the exercises it is shown that (13.3.2) corresponds to $a + b + 1$ linearly independent estimable functions. With these side conditions we obtain the solutions set

$$\begin{aligned} m &= \bar{y}_{\cdot\cdot\cdot} \\ a_i &= \bar{y}_{i\cdot\cdot} - \bar{y}_{\cdot\cdot\cdot} \\ b_j &= \bar{y}_{\cdot j\cdot} - \bar{y}_{\cdot\cdot\cdot} \\ (ab)_{ij} &= \bar{y}_{ij\cdot} - \bar{y}_{i\cdot\cdot} - \bar{y}_{\cdot j\cdot} + \bar{y}_{\cdot\cdot\cdot} \ . \end{aligned} \tag{13.3.3}$$

We can use this solutions set to find a formula for SSe. First note that using (13.3.3) we obtain

$$\hat{y}_{ijk} = m + a_i + b_j + (ab)_{ij} = \bar{y}_{ij\cdot} \ ,$$

which we recall is invariant to the solution to the normal equations. Our formula for SSe is

$$SSe = \sum_{i=1}^{a} \sum_{j=1}^{b} \sum_{k=1}^{r} (y_{ijk} - \bar{y}_{ij\cdot})^2, \tag{13.3.4}$$

also invariant to the solution of the normal equations. The degrees of freedom for SSe is $n - \text{rank } \mathbf{X} = abr - ab = ab(r-1)$. It follows that the mean square error is given by

$$s^2 = SSe/\left[ab(r-1)\right], \tag{13.3.5}$$

which is an unbiased estimate for σ^2.

Having obtained a solution to the normal equations we turn our attention to estimating quantities. We first need to determine what is estimable. We know that the entries in $\mathbf{X\beta}$ are estimable; therefore, $\mu + \alpha_i + \beta_j + (\alpha\beta)_{ij}$ is estimable for all i, j. Further, linear combinations of these quantities are also estimable.

Since $\mu_{ij} = \mu + \alpha_i + \beta_j + (\alpha\beta)_{ij}$, the (i,j) cell mean is estimable, we next find its least squares estimate and its standard error. The least square estimate (BLUE) is $m + a_i + b_j + (ab)_{ij} = \bar{y}_{ij\cdot}$ which has variance given by $\mathrm{Var}(\bar{Y}_{ij\cdot}) = \mathrm{Var}\left[\frac{1}{r}\sum_{k=1}^{r} Y_{ijk}\right] = \frac{1}{r^2}\sum_{k=1}^{r}\mathrm{Var}(Y_{ijk}) = \frac{1}{r^2}(r\sigma^2) = \sigma^2/r$. The standard error is found by replacing σ^2 by s^2 and taking the square root. These results are summarized as follows:

Estimate of Cell Means μ_{ij} in a Balanced

Two–Factor Factorial, CRD (13.3.6)

BLUE: $\bar{y}_{ij\cdot}$ Standard Error: $s/\sqrt{r}$

We next consider what else is estimable in the model (13.2.2). It is not surprising that none of μ, α_i, β_j, $(\alpha\beta)_{ij}$ are individually estimable. What is surprising, however, is that contrasts in the α_i's or β_j's are also not estimable. For example, the contrast $\sum_{i=1}^{a} c_i\alpha_i$, where $\sum_{i=1}^{a} c_i = 0$, is not estimable since it cannot be written as a linear combination of the elements in $\mathbf{X\beta}$. This is a serious deficiency in our model! It is clear that we have an interest in contrasts in the α_i's, for this is how we would study the effects of factor A. We conclude that our model is not sufficient to allow us to perform desired inferences; therefore, the model must be modified before proceeding further.

There are at least three approaches to consider in solving the model deficiency problem. First, we could choose to make inferences only based on appropriate functions of cell means. Since cell means and linear combinations of them are estimable we would have no difficulties. Unfortunately, some inferences are very complex with this approach, so we prefer another method. A second approach is used by Graybill (1976) and Myers and Minton (1991) and is based on reparametrization of the model. An easier approach that is commonly used is to add side conditions to the model to allow for interpretation of model parameters. The reader will recognize this feature to be the defining difference between the restricted and unrestricted models of earlier chapters. We will proceed with this approach.

To make further progress we will adopt the side conditions in (13.3.2) as inclusion in our model. Then our model for a balanced two-factor factorial with a CRD will henceforth be as follows:

$$Y_{ijk} = \mu + \alpha_i + \beta_j + (\alpha\beta)_{ij} + \varepsilon_{ijk};$$

$$i = 1,\cdots,a; \quad j = 1,\cdots,b; \quad k = 1,\cdots,r;$$

$$\sum_{i=1}^{a} \alpha_i = 0 = \sum_{j=1}^{b} \beta_j, \quad \sum_{i=1}^{a} (\alpha\beta)_{ij} = 0 \text{ for all } j \tag{13.3.7}$$

$$\sum_{j=1}^{b} (\alpha\beta)_{ij} = 0 \text{ for all } i.$$

With this model we obtain the unique solutions to the normal equations given by (13.3.3). Further, since there is a unique solution, we note that μ, α_i, β_j, $(\alpha\beta)_{ij}$ are individually estimable and linear combinations of these parameters are therefore also estimable. It follows that contrasts in the α_i's or β_j's are estimable.

We find the BLUE and standard error for a contrast for factor A. The contrast is $\sum_{i=1}^{a} c_i\alpha_i$, $\sum_{i=1}^{a} c_i = 0$, and using the solution to the normal equations given by (13.3.3) the BLUE for this contrast is

$$\sum_{i=1}^{a} c_i (\bar{y}_{i\cdot\cdot} - \bar{y}_{\cdot\cdot\cdot}) = \sum_{i=1}^{a} c_i \bar{y}_{i\cdot\cdot} .$$

The variance is easily found since the Y_{ijk}'s are uncorrelated and so are the $\bar{Y}_{i\cdot\cdot}$'s. Computations yield

$$\mathrm{Var}\left[\sum_{i=1}^{a} c_i \bar{Y}_{i\cdot\cdot}\right] = \sum_{i=1}^{a} c_i^2 \ \mathrm{Var}(\bar{Y}_{i\cdot\cdot}) = \sum_{i=1}^{a} c_i^2 \sigma^2/\mathrm{br}.$$

The standard error is $\sqrt{\frac{s^2}{br} \sum_{i=1}^{a} c_i^2}$. The formulas for a contrast for B are found in the same manner.

BLUE for the Balanced Two-Factor Factorial, CRD

Function	Blue	Standard Error
Contrast for A:		
$\sum_{i=1}^{a} c_i \alpha_i , \ \sum_{i=1}^{a} c_i = 0$	$\sum_{i=1}^{a} c_i \bar{y}_{i\cdot\cdot}$	$\sqrt{\frac{s^2}{br} \sum_{i=1}^{a} c_i^2}$
Contrast for B:		
$\sum_{j=1}^{b} c_j \beta_j , \ \sum_{j=1}^{b} c_j = 0$	$\sum_{j=1}^{b} c_j \bar{y}_{\cdot j \cdot}$	$\sqrt{\frac{s^2}{ar} \sum_{j=1}^{b} c_j^2}$

(13.3.8)

Before closing this section with an illustration, we should make a few comments about the meaning of the model parameters. To illustrate the situation, let us consider the difference $\alpha_1 - \alpha_2$. We might think of this difference as simply the difference in the effect of levels one and two of factor A, but matters are a bit more complicated than that. Within the first level of factor B the difference in the effect of levels one and two of factor A is measured by $\mu_{11} - \mu_{21} = \alpha_1 - \alpha_2 + (\alpha\beta)_{11} - (\alpha\beta)_{21}$. Note that the interaction terms are involved. In general, the difference in the effect of levels one and two of factor A for the jth level of B is $\mu_{1j} - \mu_{2j} = \alpha_1 - \alpha_2 + (\alpha\beta)_{1j} - (\alpha\beta)_{2j}$. To isolate $\alpha_1 - \alpha_2$ we "average" over B by considering

$$\frac{1}{b}\sum_{j=1}^{b}(\mu_{1j}-\mu_{2j})=\frac{1}{b}\sum_{j=1}^{b}\Big(\alpha_1-\alpha_2+(\alpha\beta)_{1j}-(\alpha\beta)_{2j}\Big)$$

$$=\frac{1}{b}(b\alpha_1-b\alpha_2+0-0)=\alpha_1-\alpha_2.$$

Note that the side conditions are important here. Our meaning for $\alpha_1 - \alpha_2$ is now clear. It measures the difference in the effect of levels one and two of factor A **averaged over the levels of factor B**. Similarly, β_1 β_2 measures the difference in the effect of levels one and two of factor B **averaged over the levels of factor A**.

We should also point out that other side conditions could be used for the model. For example, as we shall observe later in this chapter, SAS uses the solution to the normal equations that corresponds to the side conditions: $\alpha_a = 0$, $\beta_b = 0$, $(\alpha\beta)_{a1} = \cdots = (\alpha\beta)_{ab} = 0$, $(\alpha\beta)_{1b} = \cdots = (\alpha\beta)_{ab} = 0$. Which side conditions are used does not affect SSe, or other sums of squares obtained using the principle of conditional error. Different side conditions do, however, affect the interpretation of the parameters.

Example 13.3.1 Refer to the balanced two–factor factorial with a CRD as described in Table 13.1.1. In these data factor A (temperature) has $a = 2$ levels, factor B (operating speed) has $b = 3$ levels, and there are $r = 3$ replications. Routine calculations yield the error sum of squares, $SSe = \sum_{i=1}^{2}\sum_{j=1}^{3}\sum_{k=1}^{2}(y_{ijk} - \bar{y}_{ij\cdot})^2 = 193.3333$. The error mean square is $s^2 = SSe/ab(r-1) = 193.333/(2)(3)(2) = 16.1111$. Using (13.3.8) the BLUE for $\alpha_2 - \alpha_1$ is $\bar{y}_{2\cdot\cdot} - \bar{y}_{1\cdot\cdot} = 74.33 - 63.44 = 9.89$. The standard error is $\sqrt{\frac{s^2}{br}\sum_{i=1}^{a}c_i^2} = \sqrt{\frac{16.1111}{(3)(2)}\Big(1+1\Big)} = 2.317$. The data suggests that level 2 of A (high temperature) has a mean time to failure that is 9.89 units higher than the mean time for level 1 of A (low temperature), when averaged over the levels of B. In the next section we determine if this is a statistically "significant" result.

13.4 Inferences for Two—Factor Factorials

Our goal in this section is to present tests, confidence intervals and multiple comparisons for the balanced two—factor factorial experiment within a CRD. To perform these inferences we will use the model (13.3.7) and assume the **normal** linear model. That is, the ε_{ijk} in the model are independent $N(0,\sigma^2)$ random variables. As a consequence, the Y_{ijk}'s are assumed to be independent $N(\mu + \alpha_i + \beta_j + (\alpha\beta)_{ij}, \sigma^2)$ random variables. We first find the f—tests and the ANOVA table.

Derivations of F—statistics

First consider testing for the presence of interaction. Our hypothesis of interest is

$$H_0 : (\alpha\beta)_{ij} = 0 \text{ for all } i,\ j, \tag{13.4.1}$$

which states there is **no** interaction. To use the principle of conditional error we need to write this hypothesis in the general linear hypothesis form $H_0 : \mathbf{L}\boldsymbol{\beta} = \mathbf{0}$. Since the $(\alpha\beta)_{ij}$'s are estimable, and because of the side conditions on the $(\alpha\beta)_{ij}$'s in (13.3.7) we can write (13.4.1) in the form $\mathbf{L}\boldsymbol{\beta} = \mathbf{0}$ by using

$$[\mathbf{L}\boldsymbol{\beta}]' = \left[(\alpha\beta)_{11}, \cdots, (\alpha\beta)_{1b-1}, \cdots, (\alpha\beta)_{a-1,1}, \cdots, (\alpha\beta)_{a-1,b-1}\right].$$

Note that $\mathbf{L}\boldsymbol{\beta}$ contains (a–1)(b–1) linearly independent estimable functions $\big($i.e., $\mathbf{L}$ has rank (a–1)(b–1)$\big)$. Proceeding with the principle of conditional error we first note that the full model (13.3.7) has $SSe = \sum_{i=1}^{a} \sum_{j=1}^{b} \sum_{k=1}^{r} (y_{ijk} - \bar{y}_{ij\cdot})^2$. The "reduced" model is similar to the model for a RCBD and is given by

$$Y_{ijk} = \mu^* + \alpha_i^* + \beta_j^* + \varepsilon_{ijk}^*. \tag{13.4.2}$$

In the exercises it is verified that the error sum of squares for this reduced model is

$$SSe^* = \sum_{i=1}^{a} \sum_{j=1}^{b} \sum_{k=1}^{r} (y_{ijk} - \bar{y}_{i\cdot\cdot} - \bar{y}_{\cdot j\cdot} + \bar{y}_{\cdot\cdot\cdot})^2.$$

The **interaction sum of squares**, denoted by SSab, is then given by

$$SSab = SSe^{*} - SSe$$

$$= r \sum_{i=1}^{a} \sum_{j=1}^{b} (\bar{y}_{ij\cdot} - \bar{y}_{i\cdot\cdot} - \bar{y}_{\cdot j\cdot} + \bar{y}_{\cdot\cdot\cdot})^2 \qquad (13.4.3)$$

$$= \sum_{i=1}^{a} \sum_{j=1}^{b} \frac{y_{ij\cdot}^2}{r} - \sum_{i=1}^{a} \frac{y_{i\cdot\cdot}^2}{br} - \sum_{j=1}^{b} \frac{y_{\cdot j\cdot}^2}{ar} + \frac{y_{\cdot\cdot\cdot}^2}{abr}.$$

The degrees of freedom for the interaction sum of squares is (a–1) (b–1) and the mean square interaction is MSab = SSab/[(a–1)(b–1)]. The f–test for $H_0 : (\alpha\beta)_{ij} = 0$ for all i, j is

$$f = \frac{(SSe^{*} - SSe)/[(a-1)(b-1)]}{SSe/[ab(r-1)]} = \frac{MSab}{MSe}.$$

We reject H_0 and conclude there is significant interaction if f = $MSab/MSe > f_{\alpha,(a-1)(b-1),\ ab(r-1)}$.

Next, we turn our attention to testing for main effects. For factor A, the hypothesis of interest is

$$H_0 : \alpha_i = 0 \quad \text{for } i = 1, \cdots, a, \qquad (13.4.4)$$

which, stated in words, is "no factor A effect." This hypothesis can be written as a general linear hypothesis $\mathbf{L\beta} = \mathbf{0}$, where the entries in $\mathbf{L\beta}$ are the $a - 1$ linearly independent estimable functions $\alpha_1, \cdots, \alpha_{a-1}$. To find the form of the f–test we impose the hypothesis (13.4.4) and obtain the "reduced" model

$$Y_{ijk} = \mu^{**} + \beta_j^{**} + (\alpha\beta)_{ij}^{**} + \varepsilon_{ijk}^{**}. \qquad (13.4.5)$$

The error sum of squares of this model is

$$SSe^{**} = \sum_{i=1}^{a} \sum_{j=1}^{b} \sum_{k=1}^{r} (y_{ijk} - \bar{y}_{ij\cdot} - \bar{y}_{i\cdot\cdot} + \bar{y}_{\cdot\cdot\cdot})^2.$$

The sum of squares for factor A, denoted by SSa, is given by

$$\begin{aligned} SSa &= SSe^{**} - SSe \\ &= br \sum_{i=1}^{a} (\bar{y}_{i\cdot\cdot} - \bar{y}_{\cdot\cdot\cdot})^2 = \sum_{i=1}^{a} \frac{y_{i\cdot\cdot}^2}{br} - \frac{y_{\cdot\cdot\cdot}^2}{abr} . \end{aligned} \quad (13.4.6)$$

The degrees of freedom for SSa is $a-1$ and the mean square for factor A is $MSa = SSa/(a-1)$. The f—test for $H_0 : \alpha_i = 0$ for all i is

$$f = \frac{(SSe^{**} - SSe)/(a-1)}{SSe/[ab(r-1)]} = \frac{MSa}{MSe} ,$$ and

we reject H_0 and conclude there is a significant factor A effect if $f > f_{\alpha,a-1,\ ab(r-1)}$.

In a similar manner we can obtain a test for factor B, namely a test for $H_0 : \beta_j = 0$ for $j = 1, \cdots, b$. The sum of squares for factor B is SSb, where

$$SSb = ar \sum_{j=1}^{b} (\bar{y}_{\cdot j\cdot} - \bar{y}_{\cdot\cdot\cdot})^2 = \sum_{j=1}^{b} \frac{y_{\cdot j\cdot}^2}{ar} - \frac{y_{\cdot\cdot\cdot}^2}{abr} . \quad (13.4.7)$$

SSb has $b-1$ degrees of freedom, the mean square is $MSb = SSb/(b-1)$ and the f—test is based on rejecting H_0 if $f = MSb/MSe > f_{\alpha,b-1,ab(r-1)}$.

We now can display the ANOVA table.

Table 13.4.1 ANOVA Table for Two—Factor Factorial

Source	df	SS	MS	f
A	a–1	SSa	MSa=SSa/(a–1)	MSa/MSe
B	b–1	SSb	MSb=SSb/(b–1)	MSb/MSe
AB	(a–1)(b–1)	SSab	MSab=SSab/(a–1)(b–1)	MSab/MSe
Error	ab(r–1)	SSe	MSe=SSe/[ab(r–1)]	
Total	abr – 1	SSt		

Formulas for SSa, SSb, SSab and SSt are given by (13.4.6), (13.4.7), (13.4.3), (13.4.8), and SSe = SSt – SSa – SSb – SSab follows from (13.4.9).

Note that the total sum of squares, SSt, is given by

$$SSt = \sum_{i=1}^{a} \sum_{j=1}^{b} \sum_{k=1}^{r} (y_{ijk} - \bar{y}_{...})^2 = \sum_{i=1}^{a} \sum_{j=1}^{b} \sum_{k=1}^{r} y_{ijk}^2 - \frac{y_{...}^2}{abr} \quad (13.4.8)$$

and has abr—1 degrees of freedom. Also note that in the exercises it is verified that

$$SSt = SSa + SSb + SSab + SSe. \quad (13.4.9)$$

Example 13.4.1 Refer to the time to failure data for electric motors given in Table 13.1.1. Letting factor A represent temperature and factor B operating speed, we note that $a = 2$, $b = 3$, and $r = 3$. Routine calculations (see the exercises) yield the following ANOVA table.

Table 13.4.2 ANOVA Table for Time to Failure Data

Source	df	SS	MS	f
A (temperature)	1	533.556	533.556	33.12
B (speed)	2	1573.778	786.889	48.84
AB (interaction)	2	7.111	3.556	0.22
Error	12	193.333	16.111	
Total	17	2307.778		

The 0.05 level test for $H_0 : (\alpha\beta)_{ij} = 0$ for all i, j fails to reject H_0 since $f = 0.22$ is smaller than $f_{.05,2,12} = 3.89$. This suggests that there is no significant interaction between temperature and speed. The 0.05 level test for $H_0 : \alpha_i = 0$ for all i rejects H_0 since $f = 33.12 > f_{.05,1,12} = 4.75$. We conclude that temperature is a significant factor. Similarly, the 0.05 level test for $H_0 : \beta_j = 0$ for all j rejects H_0 since $f = 48.84 > f_{.05,2,12} = 3.89$.

In the previous example we **first** tested for interaction. In general this is the best practice. If there is no significant interaction, then the tests for the main effects are easily interpreted. For example, if there is no interaction between A and B, and if the f—test for factor A rejects, then we can conclude that there are significant differences between at least two levels of A. Further, these differences should be similar at each level of B. It makes sense in this case to perform a multiple comparison to determine which levels of A are significantly different.

If there is significant interaction between A and B, then the tests for main effects are much more difficult to interpret. In many cases we would simply choose not to perform these tests on the main effects. It is important to realize that interaction indicates that the relationship between the levels of A vary with the level of B. Therefore, in the presence of interaction, if the f—test for the main effect for A rejects, then we can only conclude that there are significant differences between at least two levels of A **when averaged over the levels of factor B**. In many problems this is not very useful information. A more common approach is to omit the tests for main effects when interaction is present. Instead a user might want to perform multiple comparisons on the "ab" treatments (cell means). Alternatively, a user might want to compare the levels of factor A for a specified level of factor B. In summary, when interaction is found to be present the user should consider alternatives to testing for main effects, and perform an analysis on some or all of the cell means that helps answer a question of interest to the experimenter such as determining "best treatment combination."

Multiple Comparisons

Multiple comparisons can be performed in a variety of ways in a two factor factorial experiment. First, we consider multiple comparisons for the main effect for factor A. Since for the normal linear model the random variables $\bar{Y}_{1\cdot\cdot},\cdots,\bar{Y}_{a\cdot\cdot}$ are independent normal random variables with the same variance, it follows from material in Chapter 9 that the Tukey procedure is applicable. The Tukey simultaneous confidence intervals with overall confidence level $100(1-\alpha)\%$ for the pairwise contrasts $\alpha_i - \alpha_\ell$ are:

$$(\bar{y}_{i\cdot\cdot}-\bar{y}_{\ell\cdot\cdot}) \pm q_{\alpha,a,ab(r-1)}\sqrt{\frac{s^2}{br}} \quad \text{for } 1 \leq i < \ell \leq a. \qquad (13.4.10)$$

Obviously, the formula (13.4.10) can be easily modified to yield multiple comparisons for factor B. An an illustration, consider the time to failure for electric motor data analyzed in Example 13.4.1. We find the simultaneous confidence intervals for the pairwise contrasts for the speed factor. We note that $s^2 = 16.111$, $\bar{y}_{\cdot 1\cdot} = 75.667$, $\bar{y}_{\cdot 2\cdot} = 75.333$, $\bar{y}_{\cdot 3\cdot} = 55.667$, $a = 2$, $b = 3$, $r = 3$, and the form of the intervals will be $(\bar{y}_{\cdot j\cdot} - \bar{y}_{\cdot t\cdot}) \pm q_{\alpha,b,ab(r-1)}\sqrt{\frac{s^2}{ar}}$. Using the Studentized range value $q_{.05,3,12} = 3.77$, the intervals are $(\bar{y}_{\cdot j\cdot} - \bar{y}_{\cdot t\cdot}) \pm$

$3.77\sqrt{16.111/(2)(3)}$ or $(\bar{y}_{\cdot j \cdot} - \bar{y}_{\cdot t \cdot}) \pm 6.18$. This results in the following intervals:

Contrast	$\beta_1 - \beta_2$	$\beta_1 - \beta_3$	$\beta_2 - \beta_3$
95% Interval	.33 ± 6.18	20 ± 6.18	19.67 ± 6.18

From these intervals we observe that levels 1 and 2 (low and medium) of speed are not significantly different, but clearly levels 1 and 3, and levels 2 and 3 are significantly different. It follows that we can conclude that high speed results in a significantly smaller mean time to failure for these motors.

When interaction is present we will often prefer to perform multiple comparisons on the cell means. Since for the normal linear model the "ab" cell means $\bar{Y}_{ij\cdot}$ are independent normals with the same variance, the Tukey procedure can be applied. The $100(1-\alpha)\%$ simultaneous confidence intervals for the cell means $\mu_{ij} - \mu_{\ell m}$ are:

$$(\bar{y}_{ij\cdot} - \bar{y}_{\ell m\cdot}) \pm q_{\alpha,ab,ab(r-1)} \sqrt{\frac{s^2}{r}} \quad \text{for all } a,b. \tag{13.4.11}$$

In some cases we could be interested in finding simultaneous confidence intervals comparing the levels of A within a fixed, say, jth level of B. Such an analysis would involve comparing the means $\mu_{1j}, \cdots, \mu_{aj}$. The Tukey simultaneous confidence interval for the pairwise difference for these means $\mu_{ij} - \mu_{\ell j}$ are:

$$(\bar{y}_{ij\cdot} - \bar{y}_{\ell j\cdot}) \pm q_{\alpha,a,ab(r-1)} \sqrt{\frac{s^2}{r}} \tag{13.4.12}$$

$$\text{for all } 1 \leq i < \ell < a.$$

We could use other types of multiple comparisons, such as the Duncan or Bonferroni, and the formulas for these procedures follow in a natural way. The Tukey procedure has been presented since it is easy to use, provides simultaneous confidence intervals, and controls the overall type one error level to α or less.

Confidence Intervals and Tests

Finding confidence intervals and t—tests for estimable functions are routinely found using the ideas from Section 5.3, and the BLUE and standard errors given in (13.3.6) and (13.3.8). Confidence intervals for various quantities are summarized in Table 13.4.3. Corresponding t—tests can be found in an analogous manner.

Table 13.4.3 100(1–α)% Confidence Intervals For Two–Factor Factorial

Function	Interval
Contrast for A	
$\sum_{i=1}^{a} c_i\alpha_i, \ \sum_{i=1}^{a} c_i = 0$	$\sum_{i=1}^{a} c_i\bar{y}_{i\cdot\cdot} \pm t_{\frac{\alpha}{2},ab(r-1)}\sqrt{\frac{s^2}{br}\sum_{i=1}^{a} c_i^2}$
Contrast for B	
$\sum_{j=1}^{b} c_j\beta_j, \ \sum_{j=1}^{b} c_j = 0$	$\sum_{j=1}^{b} c_j\bar{y}_{\cdot j\cdot} \pm t_{\frac{\alpha}{2},ab(r-1)}\sqrt{\frac{s^2}{ar}\sum_{j=1}^{b} c_j^2}$
Cell Mean μ_{ij}	$\bar{y}_{ij\cdot} \pm t_{\frac{\alpha}{2},ab(r-1)}\sqrt{\frac{s^2}{r}}$

(13.4.13)

Example 13.4.2 Neter, Wasserman, and Kutner (1985) consider data from a study investigating the effect of a drug compound in providing relief for people with hay fever. Two active ingredients (factor A and B) were each varied at three levels (low, medium and high) to yield nine treatments. A total of 36 subjects were available for the study and four were assigned to each treatment group. The data representing hours of relief are given in Table 13.4.4. To analyze these data we first obtain the ANOVA table. One can use hand calculations or use the computer. The SAS commands and resulting output for these data are given in Table 13.4.5.

Table 13.4.4 Hours Relief from Hay Fever

		Level of B					
		Low(1)		Medium (2)		High (3)	
Level of A	Low (1)	2.4	2.7	4.6	4.2	4.8	4.5
		2.3	2.5	4.9	4.7	4.4	4.6
	Medium (2)	5.8	5.2	8.9	9.1	9.1	9.3
		5.5	5.3	8.7	9.0	8.7	9.4
	High (3)	6.1	5.7	9.9	10.5	13.5	13.0
		5.9	6.2	10.6	10.1	13.3	13.2

Table 13.4.5 SAS Commands and ANOVA for Hay Fever Data

SAS Commands

```
OPTIONS LS = 80;
DATA TWOFF;
INPUT A B RELIEF @@;
CARDS;
1 1 2.4 1 1 2.7 1 1 2.3 1 1 2.5
        Etc.
3 3 13.5 3 3 13.0 3 3 13.3 3 3 13.2
PROC GLM; CLASS A B;
MODEL RELIEF = A B A*B;
MEANS A/TUKEY;
MEANS A*B;
```

SAS Printout

General Linear Models Procedure

Dependent Variable: RELIEF

Source	DF	Sum of Squares	Mean Square	F Value	Pr > F
Model	8	373.10500000	46.63812500	774.91	0.0001
Error	27	1.62500000	0.06018519		
Corrected Total	35	374.73000000			

R-Square	C.V.	Root MSE	RELIEF Mean
0.995664	3.415221	0.2453267	7.1833333

Source	DF	Type I SS	Mean Square	F Value	Pr > F
A	2	220.02000000	110.01000000	1827.86	0.0001
B	2	123.66000000	61.83000000	1027.33	0.0001
A*B	4	29.42500000	7.35625000	122.23	0.0001

Source	DF	Type III SS	Mean Square	F Value	Pr > F
A	2	220.02000000	110.01000000	1827.86	0.0001
B	2	123.66000000	61.83000000	1027.33	0.0001
A*B	4	29.42500000	7.35625000	122.23	0.0001

General Linear Models Procedure

Tukey's Studentized Range (HSD) Test for variable: RELIEF

NOTE: This test controls the type I experimentwise error rate, but generally has a higher type II error rate than REGWQ.

Alpha= 0.05 df= 27 MSE= 0.060185
Critical Value of Studentized Range= 3.506
Minimum Significant Difference= 0.2483

Means with the same letter are not significantly different.

Tukey Grouping	Mean	N	A
A	9.8333	12	3
B	7.8333	12	2
C	3.8833	12	1

General Linear Models Procedure

Level of A	Level of B	N	RELIEF Mean	RELIEF SD
1	1	4	2.4750000	0.17078251
1	2	4	4.6000000	0.29439203
1	3	4	4.5750000	0.17078251
2	1	4	5.4500000	0.26457513
2	2	4	8.9250000	0.17078251
2	3	4	9.1250000	0.30956959
3	1	4	5.9750000	0.22173558
3	2	4	10.2750000	0.33040379
3	3	4	13.2500000	0.20816660

Observe from the SAS Commands that the interaction term on SAS is represented in the model statement as A*B. Also note that the Tukey procedure can be requested for main effects in the usual manner, and that the cell means can be requested by using the command MEANS A*B; .

Using the printout we first consider the f—test for interaction. From the printout we note that for this test $f = 122.23$ and $p = .0001$. With such a small p—value we reject H_0 and declare that there is a significant interaction. Because of this significant interaction we will choose to ignore the tests for main effects and the multiple comparison given for factor A. Instead we will perform the Tukey multiple comparison for the nine cell means. Using (13.4.11) we note that two cell means will be declared significantly different at the 0.05 level if the absolute difference between the means exceeds

$$q_{\alpha,ab,ab(r-1)} \sqrt{\frac{s^2}{r}} = q_{.05,9,27} \sqrt{\frac{s^2}{4}} = 4.72 \sqrt{\frac{.0602}{4}} = .58.$$

The results of the Tukey procedure on the cell means are:

Level of A	1	1	1	2	3	2	2	3	3
Level of B	1	3	2	1	1	2	3	2	3
$\bar{y}_{ij\cdot}$	2.47	4.58	4.60	5.45	5.98	8.92	9.12	10.28	13.25

From the results we observe that the mean hours of relief is longest when both levels of A and B are at the high level. One final note is that the Type I and III sum of squares are identical. As we will show later, this follows since the factors A, B, and AB are orthogonal.

13.5 Reparametrized Model

We can reparametrize the model (13.3.7) for a two—factor factorial into a full rank regression model and then use regression techniques to perform the analysis. Because the side conditions are part of the model, we need to proceed differently than before by using 1, 0, —1 coding. As an illustration consider a two—factor factorial with $a = 2$ and $b = 3$. The reparametrized, regression form of the model is:

$$Y = \gamma_0 + \overbrace{\gamma_1 x_1}^{A} + \overbrace{\gamma_2 x_2 + \gamma_3 x_3}^{B} + \overbrace{\gamma_4 x_1 x_2 + \gamma_5 x_1 x_3}^{AB} + \varepsilon,$$

$$\text{where} \quad x_1 = \begin{cases} 1 & \text{if first level of A} \\ -1 & \text{if second level of A} \end{cases}$$

$$\text{where} \quad x_2 = \begin{cases} 1 & \text{if first level of B} \\ 0 & \text{otherwise} \\ -1 & \text{if third (last) level of B} \end{cases} \tag{13.5.1}$$

$$\text{where} \quad x_3 = \begin{cases} 1 & \text{if second level of B} \\ 0 & \text{otherwise} \\ -1 & \text{if third (last) level of B .} \end{cases}$$

Using this system of coding the correspondence between the parameters of the traditional model (13.3.7) and this reparametrized model is: $\mu = \gamma_0$, $\alpha_1 = \gamma_1$, $\beta_1 = \gamma_2$, $\beta_2 = \gamma_3$, $(\alpha\beta)_{11} = \gamma_4$, $(\alpha\beta)_{12} = \gamma_5$. Because of the side conditions note that $\alpha_2 = -\alpha_1 = -\gamma_1$ and $\beta_3 = -(\beta_1 + \beta_2) = -\gamma_2 - \gamma_3$. Note that the use of 1, 0, −1 coding yields a simple relationship between the parameters of the traditional and reparametrized models. Also note that the interaction variables in the reparametrized model are simply all possible products of a factor A variable (x_1 in this problem) with a factor B variable (x_2 and x_3 in this problem).

It is routine to use regression procedures to carry out tests and to find confidence intervals. For example, a test of no factor B effect, $H_0 : \beta_i = 0$, for $i = 1, 2, 3$ is equivalent to testing $H_0 : \gamma_2 = \gamma_3 = 0$. To find a confidence interval for $\beta_3 - \beta_1$ is equivalent to finding a confidence interval for $-(\gamma_2 + \gamma_3) - \gamma_2 = -(2\gamma_2 + \gamma_3)$.

The ANOVA sums of squares can be found using the full model for the error sum of squares, SSe, and appropriate reduced models. For example, from the terms associated with the factors as noted in (13.5.1), the main effects or "no interaction" model is given by

$$Y = \gamma_0^* + \gamma_1^* x_1 + \gamma_2^* x_2 + \gamma_3^* x_3 + \varepsilon^* .$$

On fitting this reduced model to the data we get SSe^* and SSr^*, the conditional error and regression sums of squares. Using SS( ) notation, we denote $SSr^* = SS(\alpha,\beta)$. With this notation and applying the principal of conditional error we arrive at

$$SSab = SS(\alpha\beta \mid \alpha, \beta) = SS(\alpha, \beta, \alpha\beta) - SS(\alpha, \beta) = SSe^* - SSe.$$

Similarly SSa $= SS(\alpha | \beta, \alpha\beta) = SS(\alpha, \beta, \alpha\beta) - SS(\beta, \alpha\beta)$ and SSb $= SS(\beta | \alpha, \alpha\beta) = SS(\alpha, \beta, \alpha\beta) - SS(\alpha, \alpha\beta)$. As a numerical illustration, consider the data in Table 13.1.1 which was analyzed in Example 13.4.1. Since for these data a = 2 and b = 3 we can use as our full model the reparametrized model (13.5.1) from which we get the model sum of squares $SS(\alpha, \beta, \alpha\beta) = 2114.445$ and the error sum of squares SSe = 193.333. On fitting the above reduced model we obtain, as noted,

$SS(\alpha,\beta) = 2107.334$ and $SSe^{*} = 200.444$. These figures yield

$$SSab = 2114.445 - 2107.334 = 200.444 - 193.333 = 7.111.$$

In like manner, SSa = 2114.445 − 1580.889 = 533.556 and SSb = 2114.445 − 540.667 = 1573.778. It is now a routine matter to perform the various f—tests.

The fact that in the balanced two—factor factorial model factors A, B and the interaction term are orthogonal is easy to observe by inspecting the $(\boldsymbol{X}'\boldsymbol{X})^{-1}$ matrix of the reparametrized model. For the two—factor factorial problem considered in the previous paragraph where a = 2, b = 3 and r = 3 the $(\boldsymbol{X}'\boldsymbol{X})^{-1}$ matrix is given by

$$
\begin{array}{c|cccccc}
 & \mu & A & \multicolumn{2}{c}{B} & \multicolumn{2}{c}{AB} \\
\hline
\mu & .056 & 0 & 0 & 0 & 0 & 0 \\
A & 0 & .056 & 0 & 0 & 0 & 0 \\
B & 0 & 0 & .111 & -.056 & 0 & 0 \\
 & 0 & 0 & -.056 & .111 & 0 & 0 \\
AB & 0 & 0 & 0 & 0 & .111 & -.056 \\
 & 0 & 0 & 0 & 0 & -.056 & .111
\end{array}
$$

Notice that this matrix has the familiar block diagonal form which implies orthogonality. For example, note that the entries relating factor A to factor B are all zero, which implies that factors A and B are orthogonal.

Purely Orthogonal Coding

The x_i coding used in this section's demonstration model

(13.5.1) can, of course, be done in various ways. If the primary objective of an experiment is to calculate the analysis of variance using regression, the **method of orthogonal contrasts** using orthogonal polynomial coding can conveniently be used. (See Table 4 of Appendix B.) The reader can refer to Sections 6.5 and 6.6 for the elements of this approach. Applied to our current problem, the "new" (13.5.1) becomes

$$Y = \gamma_0 + \overbrace{\gamma_1 x_1}^{A} + \overbrace{\gamma_2 x_2 + \gamma_3 x_3}^{B} + \overbrace{\gamma_4 x_1 x_2 + \gamma_5 x_1 x_3}^{AB} + \varepsilon,$$

where
$$x_1 = \begin{cases} -1 & \text{for the first level of A} \\ 1 & \text{for the second level of A} \end{cases}$$

$$x_2 = \begin{cases} -1 & \text{for the first level of B} \\ 0 & \text{for the second level of B} \\ 1 & \text{for the third level of B} \end{cases} \qquad (13.5.2)$$

$$x_3 = \begin{cases} 1 & \text{for the first level of B} \\ -2 & \text{for the second level of B.} \\ 1 & \text{for the third level of B} \end{cases}$$

This x_i coding will result in a $\boldsymbol{X}$ matrix of orthogonal columns and a diagonal $(\boldsymbol{X}'\boldsymbol{X})^{-1}$ matrix so that there is an **independent** sum of squares associated with **each** of the parameters, γ_i, i = 1, 2, 3, 4, 5.

As before, on fitting the full model (13.5.2) we obtain the error sum of squares SSe as well as the model or regression sum of squares SSr. For a convenient description of the process using various reduced models we now adopt the regression coefficients SS( ) notation introduced in Section 6.4. Then we write

$$SSr = SS(\gamma_0,\gamma_1,\gamma_2,\gamma_3,\gamma_4,\gamma_5) = SS(\gamma_1) + SS(\gamma_2) + SS(\gamma_3) + SS(\gamma_4) + SS(\gamma_5),$$

with the partitioning of SSr the result of the complete orthogonal coding. Hence it follows that the interaction sum of squares notationally is

$$SSab = SS(\gamma_4, \gamma_5 | \gamma_0, \gamma_1, \gamma_2, \gamma_3) = SS(\gamma_0, \gamma_1, \gamma_2, \gamma_3, \gamma_4, \gamma_5) - SS(\gamma_0, \gamma_1, \gamma_2, \gamma_3) = SS(\gamma_4) + SS(\gamma_5).$$

Also, for the main effects sums of squares we compute

$$SSa = SS(\gamma_1 | \gamma_0, \gamma_2, \gamma_3, \gamma_4, \gamma_5) = SS(\gamma_1) \text{ and}$$
$$SSb = SS(\gamma_2, \gamma_3 | \gamma_0, \gamma_1, \gamma_4, \gamma_5) = SS(\gamma_2) + SS(\gamma_3).$$

In the exercises, the student will be asked to rework the numerical illustration of this section using regression model (13.5.2).

Nature of Response

The idea of using orthogonal polynomial coding in order to analyze the nature of the response to a quantitative type variable was first introduced in Section 6.6. There we find that if the levels of a quantitative variable are equally spaced and equally replicated, an orthogonal polynomial table can be used to "break out" the polynomial effects of the variable on the response. As we have seen outlined above, even if the variables are qualitative rather than quantitative, we can still use purely orthogonal coding for calculating the ANOVA. However, in the qualitative case, the terms such as linear effect, quadratic effect, and cubic effect do not have a ready mathematical interpretation.

To continue this topic with reference to the data in Table 13.1.1, instead of low and high temperatures and low, medium and high speeds, let us assume temperatures of 10° C and 30° C degrees and speeds of 700, 900, and 1100 revolutions per minute. Then in the preceding example, $SS(\gamma_1)$ is said to be the linear effect of A (temperature) which we can denote by A_L, $SS(\gamma_2)$ is the linear effect of B (speed) denoted by B_L and $SS(\gamma_3)$ is the quadratic effect of speed denoted by B_Q. Since SSb = $SS(\gamma_2) + SS(\gamma_3)$, we see that this procedure results in a partitioning of the B main effect sum of squares into mathematical concepts easily interpretable. For the partitioning of the ANOVA sums of squares into single degree of freedom polynomial effects and its interpretation, see "An Experimental Design Example" near the close of Chapter 6. If it is desirable to study the polynomial effects of a quantitative variable that is not equally spaced and/or equally replicated, Table 4 cannot be used as the basis for coding. Rather, polynomials must be fitted to the data in the direct way as detailed in Section 6.6. There are some problems in the exercises which ask for the implementation of regression analysis using orthogonal polynomial coding.

13.6 Expected Mean Squares

Our next task is to find the theoretical expected mean squares associated with the ANOVA table for a balanced two–factor factorial with a CRD. From Theorem 7.7.1 we know that the expected mean square is of the form $\sigma^2 + W/q$, where W is the expression for the sum of squares when the Y's are replaced by their expected values, and q is the degrees of freedom. For the model (13.3.7) and the ANOVA Table 13.4.1 we find E(MSA) as follows. First, after some routine algebra we obtain $E(\bar{Y}_{...}) = \mu$ and $E(\bar{Y}_{i..}) = \mu + \alpha_i$; therefore, $W = br \sum_{c=1}^{a} (\mu+\alpha_i - \mu)^2 = br \sum_{i=1}^{a} \alpha_i^2$. It follows that $E(MSA) = \sigma^2 + br \sum_{i=1}^{a} \alpha_i^2/(a-1)$. In a similar manner we can find the other expected mean squares and these are summarized in the following table.

Table 13.6.1 Theoretical ANOVA for Two—Factor Factorial with E(MS)

Source	df	MS	E(MS)
A	a–1	MSA	$\sigma^2+br\sum_{i=1}^{a}\alpha_i^2/(a-1)$
B	b–1	MSB	$\sigma^2+ar\sum_{j=1}^{b}\beta_j^2/(b-1)$
AB	(a–1)(b–1)	MSAB	$\sigma^2+r\sum_{i=1}^{a}\sum_{j=1}^{b}(\alpha\beta)_{ij}^2/(a-1)(b-1)$
Error	ab(r–1)	MSE	σ^2
Total	abr – 1		

Notice that when $H_0 : (\alpha\beta)_{ij} = 0$ is true, E(MSAB) = E(MSE); therefore, as is always the case, the expected value of the numerator of the F statistic used for testing H_0 equals the expected value of the denominator when H_0 is true. This is a confirmation on the appropriateness of the form of the f—test. Analogous statements can be made in the tests for A and for B.

13.7 Special Cases for Factorials

One Observation Per Cell

The first special case we consider is the two—factor factorial when there is exactly one observation in each cell (r = 1). This case presents special problems if one tries to use the traditional model (13.3.7) and the usual ANOVA Table 13.4.1. When r = 1 the degrees of freedom for error is $ab(r-1) = 0$, that is, there is no error term. That makes it impossible to perform the usual inference procedures.

A common approach for the analysis when $r = 1$ is to drop the interaction term from the model and base the analysis on the model

$$Y_{ij} = \mu + \alpha_i + \beta_j + \varepsilon_{ij}; \quad i = 1,\cdots,a; \quad j = 1,\cdots,b.$$

With this model the error sum of squares has degrees of freedom $(a-1)(b-1)$ and is equal to the sum of squares for interaction found in the traditional ANOVA Table 13.4.1. One can perform inferences on the main effects, but this model does not yield a test for interaction. The analysis for the data is identical to an RCBD where factor A represents "treatments" and factor B, "blocks." It is important to note, however, that an RCBD and a factorial are different designs due to the difference in randomization.

Tukey (1949) described a procedure for testing for interaction in a factorial experiment when $r = 1$. His test is based upon a quantity having a single degree of freedom which measures the interaction effect. We label this value by SSint, where

$$\text{SSint} = ab\left[\sum_{i=1}^{a} \sum_{j=1}^{b} (\bar{y}_{i\cdot} - \bar{y}_{\cdot\cdot})(\bar{y}_{\cdot j} - \bar{y}_{\cdot\cdot})\, y_{ij}\right]^2 / [\text{SSa SSb}].$$

Tukey's procedure is based upon rejecting H_0: No interaction if

$$f = \frac{\text{SSint}}{(\text{SSe} - \text{SSint})/[(a-1)(b-1)-1]} \geq f_{\alpha,1,(a-1)(b-1)-1}.$$

Unbalanced Case (Unequal Cell Sizes)

Our next special case for a two–factor factorial is when the cell sample sizes are different. We refer to this as the **unbalanced** case. If we let n_{ij} be the number of observations in cell (i,j), then we consider the case where each $n_{ij} \geq 1$ (i.e., no empty cells), but at least two n_{ij} are different. An illustration of this case is given later in this section in Example 13.7.1.

We encounter a number of difficulties in the unbalanced case. First, most of the computing formulas given thus far in this chapter no longer hold. As an illustration, if n is the total number of observations, then the degrees of freedom for error is $n - ab$, which does not equal $ab(r-1)$. Second, in the unbalanced case, the design is no longer orthogonal and, as a consequence, adjusted and unadjusted sums of squares are not the same. For the f–tests we require the adjusted sums of squares obtained by using the principle of conditional error. For example, when testing H_0 : No interaction, the f—test will be based upon the sum of squares for interaction adjusted for A and B, namely $SS(\alpha\beta | \alpha, \beta)$. The f statistic will then be given by

$$f = \frac{SS(\alpha\beta | \alpha,\beta)/(a-1)(b-1)}{SSe/(n-ab)} .$$

Similar statements can be made in the tests for A and for B.

Another major difficulty encountered in the unbalanced case is that the sample means for the levels of A, $\bar{y}_{i\cdot\cdot}$'s, are no longer comparable. That is, $\bar{Y}_{i\cdot\cdot} - \bar{Y}_{\ell\cdot\cdot}$ is no longer an unbiased estimator for $\alpha_i - \alpha_\ell$. To observe why this occurs we note that $E\left[\bar{Y}_{i\cdot\cdot}\right] = \mu + \alpha_i + \sum_{j=1}^{b} n_{ij}[\beta_j + (\alpha\beta)_{ij}] / \sum_{j=1}^{b} n_{ij}$. It follows that $E(\bar{Y}_{i\cdot\cdot} - \bar{Y}_{\ell\cdot\cdot}) \neq \alpha_i - \alpha_\ell$, as the expression also involves the effects for B and for interaction. One ramification of this fact is that the Tukey multiple comparison given in (13.4.10), which is based upon the $\bar{y}_{i\cdot\cdot}$'s, is **no** longer valid. It should be clear to the reader that completely analogous statements can be made concerning factor B.

In order to compare the levels of factor A in the unbalanced case we find a new measure that is based upon the mean of the cell means. For the ith level of factor A, the mean of the cell means is $\frac{1}{b}\sum_{j=1}^{b} \bar{y}_{ij\cdot}$. Note that this mean of cell means is a "weighted mean" of the observation in the ith level of A:

$$\frac{1}{b}\sum_{j=1}^{b} \bar{y}_{ij\cdot} = \frac{1}{b}\sum_{j=1}^{b} \sum_{k=1}^{n_{ij}} y_{ijk}/n_{ij} .$$

Routine calculations can be used to show that $E\left[\frac{1}{b}\sum_{j=1}^{b} \bar{Y}_{ij\cdot}\right] = \mu + \alpha_i$ and from this it follows that an unbiased estimator for $\alpha_i - \alpha_\ell$ is the difference between the mean of the cell means, that is,

$$\widehat{\alpha_i - \alpha_\ell} = \frac{1}{b}\sum_{j=1}^{b} \bar{Y}_{ij\cdot} - \frac{1}{b}\sum_{j=1}^{b} \bar{Y}_{\ell j\cdot} . \tag{13.7.1}$$

It can be shown that (13.7.1) provides the BLUE for $\alpha_i - \alpha_\ell$. A confidence interval for $\alpha_i - \alpha_\ell$ can now be found. First,

$$\begin{aligned}\operatorname{Var}\left(\widehat{\alpha_i - \alpha_\ell}\right) &= \operatorname{Var}\left[\frac{1}{b}\sum_{j=1}^{b}\left(\bar{Y}_{ij.} - \bar{Y}_{\ell j.}\right)\right] \\ &= \frac{1}{b^2}\sum_{j=1}^{b}\left[\operatorname{Var}\left(\bar{Y}_{ij.}\right) + \operatorname{Var}\left(\bar{Y}_{\ell j.}\right)\right] \\ &= \frac{1}{b^2}\sum_{j=1}^{b}\left(\sigma^2/n_{ij} + \sigma^2/n_{\ell j}\right).\end{aligned}$$

The standard error is $\sqrt{\frac{s^2}{b^2}\sum_{j=1}^{b}\left(\frac{1}{n_{ij}} + \frac{1}{n_{\ell j}}\right)}$; therefore, the form of a $100(1-\alpha)\%$ confidence interval for $\alpha_i - \alpha_\ell$ is

$$\frac{1}{b}\sum_{j=1}^{b}(\bar{y}_{ij.} - \bar{y}_{\ell j.}) \pm t_{\frac{\alpha}{2},n-ab}\sqrt{\frac{s^2}{b^2}\sum_{j=1}^{b}\left(\frac{1}{n_{ij}} + \frac{1}{n_{\ell j}}\right)}. \tag{13.7.2}$$

This formula can be used to find confidence intervals for all pairwise differences $\alpha_i - \alpha_\ell$. By using $t_{\frac{\alpha}{2},n-ab}$ in (13.7.2) we obtain a Fisher's LSD procedure, while using $t_{\frac{\alpha}{2c},n-ab}$ in (13.7.2), where $c = a(a-1)/2$, provides a Bonferroni procedure. This procedure is illustrated in the following example.

Example 13.7.1 A study is completed to determine the effect of advertising medium (radio, television) and price (58¢, 60¢, 65¢) on the sale of soup. Eighteen stores with similar characteristics were selected for the study and the stores were randomly assigned to the six treatments (combinations of advertising medium and price) groups. Due to geographic constraints on advertising, four stores were assigned to the (radio, 60¢) treatment, and only two stores were assigned to the (television, 65¢) treatment. Sales were recorded for one week at each store, and the can sales (in thousands) are given in Table 13.7.1.

Table 13.7.1 Sales for Soup

		Medium: Radio		Medium: Television		$\bar{y}_{i\cdot\cdot}$	$\frac{1}{b}\sum_{j=1}^{b}\bar{y}_{ij\cdot}$
Price	58¢	44	47	52	56	51.667	51.667
		50		61			
	60¢	45	49	62	57	50.286	51.000
		42	48	49			
	65¢	33	36	40	43	38.600	39.083
		41					

This illustrates an unbalanced, two—factor factorial experiment within a CRD. The SAS printout in Table 13.7.2, to follow, can be used to analyze these data. First note from the printout that the unadjusted sum of squares for PRICE is 554.538, which is different from the adjusted sum of squares, 524.011. As pointed out earlier, this is a consequence of the design not being orthogonal. The f—tests yield the following results: there is no significant interaction between price and medium (p = .5678), there is a significant difference between the two mediums (p = .0018), and there is a significant difference between at least two levels of price (p = .0006). Using the output from the ESTIMATE statement on SAS we can find a 95% confidence interval for the noted significant difference in the effects of the two mediums. The computed confidence interval for $\beta_{TV} - \beta_{Radio}$ turns out to be 8.056 ± $t_{.025,12}(2.021)$, or (3.652, 12.460). This interval could also be found using (13.7.2). To compare the levels of price we perform a multiple comparison. Note that the printout for the Tukey procedure is based upon the $\bar{y}_{i\cdot\cdot}$, which is not valid. As a consequence, this Tukey procedure yields inappropriate results. The LSMEANS yields the mean of the cell means, $\frac{1}{b}\sum_{j=1}^{b}\bar{y}_{ij\cdot}$. These values and the p—values on the printout can be used to perform a Bonferroni multiple comparison. Since there are c = 3 pairwise comparisons, we declare two levels of price different if the p—value is $\frac{\alpha}{c} = \frac{.05}{(3)} = .0167$ or less. We find that the price of 65¢ leads to significantly smaller sales than either 58¢ or 60¢. There is no significant difference between the sales for the prices of 58¢ and 60¢.

Table 13.7.2 SAS Printout for Soup Sales

SAS Commands

```
options ls = 80;
data tx2ffs;
input price medium sales @@;
cards;
58  1  44    58  1  47    58  1  50
60  1  45    60  1  49    60  1  42    60  1  48
65  1  33    65  1  36    65  1  41
58  2  52    58  2  56    58  2  61
60  2  62    60  2  57    60  2  49
65  2  40    65  2  43
proc glm;  class price medium;
model sales = price medium price*medium;
estimate 'Radio vs TV' medium -1 1;
means price / tukey lines;
lsmeans price / pdiff;
```

SAS Printout

```
                    General Linear Models Procedure

Dependent Variable: SALES
                                    Sum of           Mean
Source                  DF         Squares         Square    F Value       Pr > F

Model                    5    884.66666667   176.93333333      10.02       0.0006

Error                   12    211.83333333    17.65277778

Corrected Total         17   1096.50000000

                  R-Square            C.V.       Root MSE               SALES Mean

                  0.806810        8.845307      4.2015209                47.500000

Source                  DF       Type I SS    Mean Square    F Value       Pr > F

PRICE                    2    554.53809524   277.26904762      15.71       0.0004
MEDIUM                   1    309.17226075   309.17226075      17.51       0.0013
PRICE*MEDIUM             2     20.95631068    10.47815534       0.59       0.5678

Source                  DF     Type III SS    Mean Square    F Value       Pr > F

PRICE                    2    524.01132686   262.00566343      14.84       0.0006
MEDIUM                   1    280.33333333   280.33333333      15.88       0.0018
PRICE*MEDIUM             2     20.95631068    10.47815534       0.59       0.5678

                                       T for H0:    Pr > |T|   Std Error of
Parameter                Estimate    Parameter=0                 Estimate

Radio vs TV            8.05555556           3.99      0.0018     2.02145768
```

```
                    General Linear Models Procedure

         Tukey's Studentized Range (HSD) Test for variable: SALES

      NOTE: This test controls the type I experimentwise error rate, but
            generally has a higher type II error rate than REGWQ.

                 Alpha= 0.05  df= 12  MSE= 17.65278
              Critical Value of Studentized Range= 3.773
                Minimum Significant Difference= 6.5326
                  WARNING: Cell sizes are not equal.
                 Harmonic Mean of cell sizes= 5.88785

      Means with the same letter are not significantly different.

             Tukey Grouping              Mean      N  PRICE

                          A            51.667      6  58
                          A
                          A            50.286      7  60

                          B            38.600      5  65


                          General Linear Models Procedure
                               Least Squares Means

             PRICE            SALES    Pr > |T| H0: LSMEAN(i)=LSMEAN(j)
                             LSMEAN    i/j      1         2         3

             58          51.6666667    1        .     0.7814    0.0004
             60          51.0000000    2   0.7814         .     0.0005
             65          39.0833333    3   0.0004    0.0005         .

   NOTE: To ensure overall protection level, only probabilities associated with
         pre-planned comparisons should be used.
```

Problems With Empty Cells

Another special case for a factorial is when one or more cells are empty, that is, some $n_{ij} = 0$. Empty cells cause a number of problems or changes. One is that the degrees of freedom for interaction no longer equals (a—1)(b—1). In most cases the degrees of freedom for interaction is decreased by one for each empty cell. A serious problem is that there is more than one way to find "adjusted" sums of squares for factors. We will not elaborate on this fact here but the interested reader can find more details in Searle (1987). An additional difficulty is that the least squares means on SAS for a main effect are not estimable for one or more levels. This makes it difficult to perform multiple comparisons.

For a problem with empty cells there are several ways to proceed with the analysis. Searle (1987) describes these approaches in great detail. We mention two possibilities here. A first approach is to use the **cell means model**

$$Y_{ijk} = \mu_{ij} + \varepsilon_{ijk}, \quad i = 1,\cdots, a; \quad j = 1,\cdots, b;$$

$$k = 1,\cdots,n_{ij}\ . \qquad (13.7.3)$$

In this model the μ_{ij}'s are estimable for each cell where $n_{ij} \geq 1$. Hypotheses can be formulated using μ_{ij}'s and tested in the usual manner. This approach is particularly useful for comparing the cell means, but formulating more complicated hypotheses such as one for interaction between factors A and B can be difficult. As an illustration, consider a 2×3 factorial experiment with one missing cell as described in the following table of cell means.

		B		
		1	2	3
A	1	μ_{11}	μ_{12}	μ_{13}
	2	missing	μ_{22}	μ_{23}

The differences $\mu_{12} - \mu_{13}$ and $\mu_{22} - \mu_{23}$ can be compared to check for interaction. Specifically, a test for interaction is $H_0 : \mu_{12} - \mu_{13} = \mu_{22} - \mu_{23}$ or equivalently, $H_0 : (\mu_{12} - \mu_{13}) - (\mu_{22} - \mu_{23}) = 0$. Hypotheses for factor A and factor B can be written in a similar manner.

Next, we describe an alternative approach that can be used when there are empty cells. We first test for interaction between A and B. Even when there are empty cells the adjusted sum of squares for interaction can be easily found (e.g., using SAS or by reparametrization). (We are assuming that there are enough non–empty cells to allow an analysis.) If the interaction is significant, then an analysis of the cell means can be performed. If the interaction is not significant, then eliminate the interaction term from the model and base the analysis on the main effects model $Y_{ijk} = \mu + \alpha_i + \beta_j + \varepsilon_{ijk}$. The empty cells cause no problems for this model, and tests for main effects can be performed in the usual manner. Multiple comparisons can be performed using the LSD or Bonferroni procedure (e.g., use the least square means on SAS). It should be noted that this approach is easy to perform. Also note that the mean for the empty cell, $\mu + \alpha_i + \beta_j$, can be estimated using this approach. One drawback is that the tests for main effects are conditional since they depend on the outcome of the test for interaction. Though this makes the level of the tests **approximately** α, this approach is still a simple way to handle problems involving empty cells.

13.8 Assumptions, Design Considerations

Model Assumptions

To perform tests and confidence intervals requires assuming the normal linear model. For the two–factor factorial this requires that the observations Y_{ijk} (or equivalently the ε_{ijk}) are normally distributed, have equal variance and are independent. These assumptions can be checked by using the residuals in a manner identical to that used in Section 7.9. An analysis of the residuals can also be used to detect possible outliers in the data.

Design Considerations

The major advantage of a factorial experiment is the ability to consider the effects of two or more factors at the same time. It can be less expensive to study factors in the same experiment rather than studying them separately. But more importantly, we can detect relationships between the factors when we study them together.

In some cases we can detect differences in a factorial experiment that might never be found in "one factor at a time" experiments. As an illustration, consider a balanced two–factor factorial with sample cell means, $\bar{y}_{ij\cdot}$, as given in Table 13.8.1.

Table 13.8.1 Sample Cell Means

		B	
		1	2
A	1	10	20
	2	20	10

There appears to be an interaction between factors A and B, which could be detected with the analysis of a factorial experiment. If factor A is analyzed by itself no differences will be found. The same is true for factor B. In other words, single factor at a time experimentation would yield no differences, whereas the two–factor factorial approach could detect differences between the cell means.

Factorial experiments provide a simple way to study the effects of two or more factors. It also provides an opportunity to check for the relationship between the factors, that is, to check for interaction. In this chapter we concentrated on factorial experiments having only two factors and with a CRD. This is the simplest case. In the next chapter we consider more complex factorial experiments.

Chapter 13 Exercises

Application Exercises

13–1 An experiment was conducted to determine the effect of two factors on the consistency of pancakes. Factor one was supplement with levels none and some. Factor two was level of whey (none, low, medium, high). Pancake mixes were made with the combination of supplement and whey and the pancakes were judged for consistency by a panel of judges. The rating scale was 1–6 (6 is best) and the average of the ratings for the judges was recorded. A CRD was used for the experiment with three batches of pancakes made for each combination. The data obtained are listed in the table.

	Level of Whey			
Supplement	None	Low	Medium	High
None	3.3 3.2 3.1	3.8 3.7 3.6	5.0 5.3 4.8	5.4 5.6 5.3
Some	4.4 4.5 4.3	4.6 4.5 4.8	4.5 4.8 4.8	4.6 4.7 5.1

(a) Find the ANOVA table for these data.
(b) Perform appropriate f–tests using a 0.05 level and report the outcomes.
(c) Use a Tukey multiple comparison procedure to determine which combination of supplement and whey yields the best consistency. Use a 0.05 level.
(d) Find a 95% confidence interval for the difference between the cell means $\mu_{14} - \mu_{24}$.
(e) Use the residuals to check the model assumptions.

13–2 A study was completed to determine how assembly line speed and number of rest breaks for employees affect the number of defects in a production process. The three set levels of assembly line speed were 20, 25, and 30 items per hour. The number of breaks given the workers per four hour period were respectively, 1, 2, and 3. The same group of workers and the same assembly line were used for a one–week period with each of the nine combinations of speed and breaks. A CRD design was used and each of the nine treatments were replicated twice. The response variable was the number of defects per 1000 produced items. The table contains the data from the study.

		Number of Breaks					
		1		2		3	
	20	5	7	8	6	2	4
Line Speed	25	9	12	10	13	6	3
	22	22	17	15	19	12	10

(a) Find the ANOVA table for these data.
(b) Perform and interpret the outcome of appropriate f–tests using a 0.05 level.
(c) Perform the Tukey multiple comparison using a 0.05 level for the levels of speed. Repeat for the levels of breaks.
(d) Perform a test for linear and for quadratic effects of line speed on the response variable. Use a 0.05 level for each test.

13–3 The effects of variety of wheat and pesticide level were investigated in an experiment. Three varieties (A, B, and C) and three levels of pesticide (none, low, heavy) were used. Eighteen plots of equal size were available for the experiment and each combination of variety and pesticide was assigned at random to two plots. The yield in bushels is recorded for each plot in the table.

		Pesticide Level					
		None		Low		Heavy	
	A	115	101	120	127	136	130
Variety	B	96	94	113	108	117	124
	C	98	109	110	122	130	128

(a) Find the ANOVA table.
(b) Perform and interpret appropriate f–tests using a 0.05 level.
(c) Perform a Tukey multiple comparison for the variety factor using a 0.05 level.
(d) Use the residuals to check the normality assumption and to check for possible outliers.

13–4 The strength of a produced metal item was thought to be related to the temperature and humidity used in the production process. An experiment was completed to determine the optimal levels of temperature (500, 600, 700, and 800 degrees Centigrade) and humidity (40% and 60%) which were used in a factorial experiment with a CRD. Three replications were used for each treatment. The data obtained on the response variable, strength, is listed in the table.

		Temperature			
		500	600	700	800
Humidity	40%	64 57 58	68 71 65	67 73 68	57 59 63
	60%	59 65 60	74 66 70	73 69 67	65 67 62

(a) Find the ANOVA table for these data using regression with orthogonal polynomial coding for the independent variables humidity and temperature.
(b) Perform and interpret the outcomes of appropriate f–tests using a 0.05 level.
(c) Perform appropriate Tukey multiple comparison(s) using a 0.05 level to determine the optimal levels of temperature and humidity.
(d) Determine the nature of the response with respect to temperature. Test for linear, quadratic, and cubic effects using a 0.05 level.

13–5 Refer to the data table in Exercise 13–4. Suppose the lowest level of temperature in the experiment had been 400 rather than 500 with the responses as given in the table. With this change in level, calculate the ANOVA table and test for linear, quadratic, and cubic effects at the 0.05 risk level.

13–6 A company performed a study to determine how a store

manager's years of experience and education level affect the profitability of the store. Nineteen stores were selected for the study and the characteristics of the store manager were recorded as well as a measure of the store's profitability. The results are given in the following table. (Note that this problem is essentially a **two–way** problem as the data are grouped into cells by two variables. Data from a two–way can be analyzed in the same manner as a two–factor factorial with a CRD.)

Manager's Years of Experience

Education	5–10	11–15	16 or more
Bachelor	10, 18, 12, 17	17, 15	24, 17, 25
Master	16, 8, 13	19, 18, 11	18, 22, 25, 17

(a) Find the ANOVA table for the data.
(b) Perform and interpret the outcome of appropriate f–tests using a 0.05 level.
(c) Perform a Bonferroni multiple comparison on the factor years of experience using a 0.05 level.
(d) Find a 95% confidence interval for the difference between the means of the two education levels.

13–7 A drug company performed a small pilot study to determine the effect of two drugs on decreasing blood pressure in high blood pressure patients. For drug A three levels were used (labeled as low, medium, and high), while drug B was either not used or used (absent, present). Two patients were assigned to the six treatments, but some patients dropped out of the study. The percentage decrease in blood pressure was recorded for each patient. The results are listed in the table.

Drug A

Drug B	L	M	H
Absent	20, 12	18, 26	5
Present	25, 30		10, 17

(a) Perform a test for interaction using a 0.05 level.
(b) Delete the interaction term and find the ANOVA table for the main effects model.
(c) Using the main effects model, perform the f–tests for the main effects. Use a 0.05 level.
(d) Using the main effects model estimate the mean for the empty cell. Find a 95% confidence interval for this mean.
(e) Using the main effects model, perform a Bonferroni multiple comparison on the levels of drug A. Use a 0.05 level.
(f) Using a cell means model approach, within the low level of A test $H_0 : \mu_{11} = \mu_{12}$.
(g) Use SAS to fit the full model including main effects and an interaction term. Find the LSMEANS for factor A. Explain why the least squares mean for the second level is not estimable.

13–8 Bradu and Gabriel (1978) considered data from an experiment to determine the effects of temperature and pressure on the compressibility of natural rubber. One observation was taken at each combination of temperature (four levels) and pressure (six levels). The data are listed in the table.

		Pressure (kg/cm^2)					
		500	400	300	200	100	0
Temperature (Centigrade)	0	137	178	219	263	307	357
	10	197	239	282	328	376	427
	20	256	301	346	394	444	498
	30	286	330	377	426	477	532

(a) Find the ANOVA table for these data.
(b) Perform and interpret the outcome of appropriate f–tests using a 0.05 level.
(c) Perform a Tukey multiple comparison using a 0.05 level on the levels of temperature.
(d) Perform Tukey's test for additivity using a 0.05 level.
(e) Since both factors are quantitative the data can be fit with a regression model. Let x_1 represent temperatures and

x_2 pressure. Fit the following regression models:

i) $Y = \beta_0 + \beta_1 x_1 + \beta_2 x_2 + \varepsilon$

ii) $Y = \beta_0 + \beta_1 x_1 + \beta_2 x_2 + \beta_{12} x_1 x_2 + \beta_{11} x_1^2 + \beta_{22} x_2^2 + \varepsilon.$

Use each model to predict Y when temperature is 15° and pressure is 350.

13–9 Refer to Example 13.4.1. Verify the entries in the ANOVA table given in Table 13.4.2.

13–10 Refer to the Hay Fever Data given in Table 13.4.4 and analyzed in Example 13.4.2.

(a) Perform a Tukey multiple comparison using a 0.05 level comparing the levels of factor A for the medium level of B.

(b) Plot the cell means to display the pattern of interaction between factors A and B.

(c) The variability within a treatment can be compared using the following approach. Find the standard deviation for each cell and label these values as s_{ij}. Place the value $\ln s_{ij}$ in each cell and analyze the resulting data as a two–factor factorial with one observation in each cell. Tests for factors A and B compare the variability between the levels of these two factors. (The purpose for taking logs is to have better conformity with the normality assumption.) Perform this test for the Hay Fever data.

13–11 Refer to the Sales of Soup Data in Table 13.7.1 analyzed in Example 13.7.1.

(a) Perform a Bonferroni multiple comparison using a 0.05 level on the levels of price for the television medium.

(b) Use the LSMEANS statement in SAS to find least squares means for the cell means. Use the p–values from the output to perform a Bonferroni multiple comparison using a 0.05 level on the levels of price for the radio medium.

13–12 Use regression model (13.5.2) with its orthogonal polynomial coding to redo the numerical illustration presented in Section 13.5. Obtain the ANOVA sums of squares in two ways as follows:

(a) Use the principle of conditional error as outlined in the text.

(b) Calculate $SSa = SS(\gamma_0, \gamma_1)$, $SSb = SS(\gamma_0, \gamma_2, \gamma_3)$, and $SSab = SS(\gamma_0, \gamma_4, \gamma_5)$.

Notice that these formulas "work" because of the "independence" generated by purely orthogonal coding.

Theoretical Exercises

13–13 Show that the side conditions given by (13.3.2) correspond to $a + b + 1$ linearly independent conditions.

13–14 SAS uses the side conditions $\alpha_a = 0$, $\beta_b = 0$, $(\alpha\beta)_{a1} = \cdots = (\alpha\beta)_{ab} = 0$, $(\alpha\beta)_{1b} = \cdots = (\alpha\beta)_{ab} = 0$ to solve the normal equations (13.3.1).

(a) Find the solution using these side conditions and compare the results with those obtained in (13.3.3).

(b) Find the form of $\hat{y}_{ijk}$ using SAS's side conditions.

13–15 Refer to Settings I and II in Table 13.2.1. Use the factorial model (13.3.7) to explain the numerical theoretical cell mean values as presented in the table by choosing appropriate values for the 12 components of the model. Begin by assuming $\mu = 14$ for the value of the overall mean in both settings and then assign values to α_i, β_j and $(\alpha\beta)_{ij}$, $i = 1, 2$; $j = 1, 2, 3$ for each of the two settings. Hint: Observe the model side conditions.

13–16 (a) Explain why μ, α_i, β_j, and $(\alpha\beta)_{ij}$ are each estimable in model (13.3.7).

(b) Find the BLUE for α_i and find the form of a 95% confidence interval for it.

13–17 Refer to the reduced model given in (13.4.2). Verify the formula given for SSe^* for this model.

13–18 Refer to the reduced model given in (13.4.5). Verify the formula given for SSe^{**} for this model. Hint: Be sure to use the side conditions on the $(\alpha\beta)_{ij}$'s.

13–19 Verify that for the ANOVA table 13.4.1, SSt = SSa + SSb + SSab + SSe.

13–20 Verify the forms of the confidence intervals for the estimable functions given in Table 13.4.3.

13–21 In Table 13.4.3 we find confidence interval formulas listed for main effects contrasts. In like manner, write down the formula for a 100(1–α)% confidence interval for a linear combination of cell means, $\sum_{i=1}^{a} \sum_{j=1}^{b} c_{ij} \mu_{ij}$.

13–22 For a balanced two–factor factorial find the form of the t–test for testing $H_0 : \mu_{11} + \mu_{12} = \mu_{21} + \mu_{22}$.

13–23 Consider a two–factor factorial with $a = 3 = b$. Using 1, 0, –1 coding explain how to reparametrize the traditional model into a full rank regression model. Find the correspondence between the parameters of the two models.

13–24 Derive the expression for E(MSAB) given in Table 13.6.1.

13–25 Consider an unbalanced two–factor factorial with no empty cells.

(a) Show that $E\left[\frac{1}{b} \sum_{j=1}^{b} \bar{Y}_{ij\cdot}\right] = \mu + \alpha_i$.

(b) Verify that (13.7.1) yields the BLUE for $\alpha_i - \alpha_\ell$.

13–26 Consider an unbalanced two–factor factorial with no empty cells.

(a) Find the form of a confidence interval for the contrast $\sum_{i=1}^{a} c_i \alpha_i$, where $\sum_{i=1}^{a} c_i = 0$.

(b) Find the form of a confidence interval for $\beta_j - \beta_t$.

13–27 A two–factor factorial has missing cells as given in the table.

Are quantities such as $\alpha_1 - \alpha_2$ and $\beta_1 - \beta_2$ estimable? Explain.

		B 1	2	3	4
A	1	* *		* *	
	2		* *		* *

13–28 (Power) In Exercise 7–27 we calculated the power of the f–test for a CRD using the Pearson–Hartley tables given in the Appendix. In a similar manner we can calculate the power for the f–test for factor A for a balanced two–factor factorial with a CRD by using $\emptyset^2 = br \sum_{i=1}^{a} \alpha_i^2 / (a\sigma^2)$. Note that $\nu_1 = a - 1$ and $\nu_2 = ab(r-1)$. Find the power of the f–test for this design at level 0.05 when $a = 3$, $b = 2$, $r = 4$, $\alpha_1 = -4$, $\alpha_2 = 1$, $\alpha_3 = 3$, and $\sigma^2 = 16$.

13–29 (Sample Size) Use the Pearson–Hartley power tables and the formula for $\emptyset^2$ given in Exercise 13–28 to find the sample size for the following balanced two–factor factorials with a CRD.

(a) What sample size r is needed to have a power of at least 0.90 for a 0.05 level test when $a = 3$, $b = 2$, $\sigma^2 = 16$, $\alpha_1 = -4$, $\alpha_2 = 1$, and $\alpha_3 = 3$?

(b) If the difference between the means for two levels of A is at least D, then $\emptyset^2 \geq brD^2/(2a\,\sigma^2)$. If $\Delta = D/\sigma$, then $\emptyset^2 \geq br\,\Delta^2/(2a)$. Find the sample size r required for a 0.05 level test for factor A to have power of 0.95 or greater when $a = 3$, $b = 4$, and $\Delta = 1.5$.

CHAPTER 14

OTHER FACTORIAL EXPERIMENTS

14.1 Factorial Experiments with Three or More Factors

Our purpose in this chapter is to study more complicated factorial experiments than those considered in the previous chapter. In this section we extend factorial experiments to three or more factors. In the next section we consider factorial experiments replicated in designs other than a CRD. Finally, in later sections we look at some special cases for factorial experiments.

To illustrate the modifications needed to analyze data from a factorial experiment with more than two factors we consider a three–factor factorial using a CRD. Label the three factors as A, B, C and let a, b, c represent the number of levels of these factors. We present the balanced case in which there are r observations in each of the abc cells. With a CRD there are rabc experimental units, and they are randomly assigned to the abc treatments with the only restriction being that r are assigned to each treatment. This yields an **a × b × c factorial experiment replicated in a CRD.**

As an example of such an experiment, consider a drug that reduces the anxiety level in patients. The drug is made of three active ingredients, which we label as A, B, C. To determine the effects of these three ingredients, two levels (low, high) of each are used in a 2 × 2 × 2 factorial experiment. Twenty–four subjects, who have high anxiety levels, are randomly assigned to the eight treatments, three per treatment (r = 3). Each subject uses a particular drug compound and, after a specified period, a measure of the subject's anxiety level is recorded. The measure is on a forty–point scale, with high values indicating high anxiety. The data are given in Table 14.1.1.

Table 14.1.1 Anxiety Levels with Drugs

		B			
		Low		High	
A	C	Low	High	Low	High
Low		15	14	32	28
		12	18	31	21
		17	18	27	23
High		21	24	26	39
		13	29	29	30
		17	26	27	34

The analysis of data from a balanced $a \times b \times c$ factorial experiment replicated in a CRD can be based upon the following model. We let $Y_{ijk\ell}$ represent the response to the ℓth experimental unit to which the ith level of A, jth level of B and kth level of C has been applied. The model is:

$$Y_{ijk\ell} = \mu+\alpha_i+\beta_j+\gamma_k+(\alpha\beta)_{ij}+(\alpha\gamma)_{ik}+(\beta\gamma)_{jk}+(\alpha\beta\gamma)_{ijk}+\varepsilon_{ijk\ell} \qquad (14.1.1)$$

$$i = 1,\cdots,a;\ \ j = 1,\cdots,b;\ \ k = 1,\cdots,c;\ \ \ell = 1,\cdots,r$$

$$\sum_i \alpha_i = \sum_j \beta_j = \sum_k \gamma_k = 0$$

$$\sum_i (\alpha\beta)_{ij} = \sum_j (\alpha\beta)_{ij} = \sum_i (\alpha\gamma)_{ik} = \sum_k (\alpha\gamma)_{ik} = \sum_j (\beta\gamma)_{jk} = \sum_k (\beta\gamma)_{jk} = 0,$$

and

$$\sum_i(\alpha\beta\gamma)_{ijk} = \sum_j(\alpha\beta\gamma)_{ijk} = \sum_k(\alpha\beta\gamma)_{ijk} = 0.$$

The main effects α_i, β_j, γ_k represent the effects due to the levels of A, B, and C. There are three two–factor interaction terms with, for example, $(\beta\gamma)_{jk}$ measuring the interaction between the jth level of B and kth level of C. These two–factor interactions have the same interpretation as given in the previous chapter. The three–factor interaction term $(\alpha\beta\gamma)_{ijk}$ measures the interaction between all three factors. Specifically, it measures the variation in the AB interaction over the levels of C, or equivalently the BC interaction over the levels of A, or the AC interaction over the levels of B.

The analysis for this three–factor factorial follows in a manner that is analogous to the two–factor case. The ANOVA table is given as follows.

Table 14.1.2 ANOVA Table for $a \times b \times c$ Factorial with a CRD

Source	df	SS	MS	f
A	a–1	SSa	MSa	MSa/MSe
B	b–1	SSb	MSb	MSb/MSe
C	c–1	SSc	MSc	MSc/MSe
AB	(a–1)(b–1)	SSab	MSab	MSab/MSe
AC	(a–1)(c–1)	SSac	MSac	MSac/MSe
BC	(b–1)(c–1)	SSbc	MSbc	MSbc/MSe
ABC	(a–1)(b–1)(c–1)	SSabc	MSabc	MSabc/MSe
Error	abc(r–1)	SSe	MSe = s^2	
Total	rabc–1	SSt		

Using the "dot" notation in the usual way the formulas for these sum of squares are as follows:

$$SSa = bcr \sum_{i=1}^{a} (\bar{y}_{i\cdot\cdot\cdot} - \bar{y}_{\cdot\cdot\cdot\cdot})^2$$

$$SSb = acr \sum_{j=1}^{b} (\bar{y}_{\cdot j\cdot\cdot} - \bar{y}_{\cdot\cdot\cdot\cdot})^2$$

$$SSc = abr \sum_{k=1}^{c} (\bar{y}_{\cdot\cdot k\cdot} - \bar{y}_{\cdot\cdot\cdot\cdot})^2$$

$$SSab = cr \sum_{i} \sum_{j} (\bar{y}_{ij\cdot\cdot} - \bar{y}_{i\cdot\cdot\cdot} - \bar{y}_{\cdot j\cdot\cdot} + \bar{y}_{\cdot\cdot\cdot\cdot})^2 \qquad (14.1.2)$$

$$SSac = br \sum_{i} \sum_{k} (\bar{y}_{i\cdot k\cdot} - \bar{y}_{i\cdot\cdot\cdot} - \bar{y}_{\cdot\cdot k\cdot} + \bar{y}_{\cdot\cdot\cdot\cdot})^2$$

$$SSbc = ar \sum_{j} \sum_{k} (\bar{y}_{\cdot jk\cdot} - \bar{y}_{\cdot j\cdot\cdot} - \bar{y}_{\cdot\cdot k\cdot} + \bar{y}_{\cdot\cdot\cdot\cdot})^2$$

$$SSabc = r \sum_{i} \sum_{j} \sum_{k} (\bar{y}_{ijk\cdot} - \bar{y}_{ij\cdot\cdot} - \bar{y}_{i\cdot k\cdot} - \bar{y}_{\cdot jk\cdot} + \bar{y}_{i\cdot\cdot\cdot} + \bar{y}_{\cdot j\cdot\cdot} + \bar{y}_{\cdot\cdot k\cdot} - \bar{y}_{\cdot\cdot\cdot\cdot})^2$$

$$SSt = \sum_{i} \sum_{j} \sum_{k} \sum_{\ell} (y_{ijk\ell} - \bar{y}_{\cdot\cdot\cdot\cdot})^2$$

$$SSe = \sum_{i} \sum_{j} \sum_{k} \sum_{\ell} (y_{ijk\ell} - \bar{y}_{ijk\cdot})^2.$$

The "hand" computing formulas for the above sums of squares are easily deducible from those given in Section 13.4 for the two–factor case.

Using the ANOVA table the f–tests can be carried out in the usual manner. Typically, the interactions should be tested first. If no interactions are significant, then tests and multiple comparisons on the main effects are appropriate. If, however, one or more interactions are significant, then one usually should not try to study the effects of the involved factors separately. For example, if the ABC interaction is present, then one cannot separate the effects of these three factors. An analysis of all the cell means would be appropriate in such a case. As another example, suppose that the ANOVA indicates AB interaction is present and no other interaction is significant. In this case one should not separate the effects of the factors A and B, but one can study factor C separately. One approach is to analyze the data as a two–factor study where the factors are C and the "ab" combinations of the levels of factors A and B.

Multiple comparisons or confidence intervals for contrasts can be applied in the usual manner. For example, the Tukey multiple comparison on the levels of factor A is based upon deciding that $\alpha_i \neq \alpha_t$ if

$$|\bar{y}_{i\cdot\cdot\cdot} - \bar{y}_{t\cdot\cdot\cdot}| \geq q_{\alpha,a,abc(r-1)} \sqrt{\frac{s^2}{rbc}} . \quad (14.1.3)$$

The $100(1-\alpha)\%$ confidence interval for the contrast $\sum_{i=1}^{a} c_i\alpha_i$, $\sum_{i=1}^{a} c_i = 0$ is

$$\sum_{i=1}^{a} c_i\bar{y}_{i\cdot\cdot\cdot} \pm t_{\frac{\alpha}{2},abc(r-1)} \sqrt{\frac{s^2}{rbc} \sum_{i=1}^{a} c_i^2} . \quad (14.1.4)$$

Similar formulas can be given for factors B and C.

These calculations can be completed using a standard computer package such as SAS as illustrated in the next example.

Example 14.1.1 Refer to the anxiety data given in Table 14.1.1. The SAS printout in Table 14.1.3 contains an analysis of this $2 \times 2 \times 2$ factorial experiment with $r = 3$ replications. Using a 0.05 level, the f—tests show no significant three—factor interaction and no significant AB and BC interactions. There is a significant AC interaction indicating that we should not separate the effects of factors A and C. Since none of the interactions involving factor B is significant it is appropriate to examine the main effect for factor B. Note that the f—test for B is significant, which indicates a significant difference between the levels of factor B. Using the printout we note that the 95% confidence interval for $\beta_2 - \beta_1$ is $10.25 \pm t_{.025,16}(1.2638)$ or (7.57, 12.93), which indicates that the first (low) level of B leads to lower mean anxiety. To study factors A and C we can consider a Tukey multiple comparison based upon combination of the four AC means, namely $\mu_{i\cdot k}$, $i = 1, 2$; $k = 1, 2$. We decide that two of these means are significantly different

if $|\bar{y}_{i\cdot k\cdot} - \bar{y}_{u\cdot v\cdot}|$ exceeds $q_{.05,4,16}\sqrt{\frac{s^2}{rb}} = 4.05\sqrt{\frac{9.583}{6}} = 5.12.$
This results in the following outcome presented in tabular form:

Levels of A, C:	1,2	2,1	1,1	2,2
Mean	20.33	22.17	22.33	30.33

(The means 20.33, 22.17 and 22.33 are joined by a single line printed beneath them.)

From this analysis we note that the smallest anxiety levels can be found when the drug is based upon B at the low level, and at least one of A or C at the low level. Finally, the analysis of the residuals on the printout indicates no outliers and that the normal distribution assumption is reasonable.

Table 14.1.3 SAS Printout for Anxiety Data

SAS Commands

```
OPTIONS LS = 80;
DATA TXTFF;
INPUT  Y A B C @@;
CARDS;
15  1  1  1        12  1  1  1        17  1  1  1
14  1  1  2        18  1  1  2        18  1  1  2
32  1  2  1        31  1  2  1        27  1  2  1
28  1  2  2        21  1  2  2        23  1  2  2
21  2  1  1        13  2  1  1        17  2  1  1
24  2  1  2        29  2  1  2        26  2  1  2
26  2  2  1        29  2  2  1        27  2  2  1
39  2  2  2        30  2  2  2        34  2  2  2
PROC  GLM;  CLASS A B C;
MODEL Y = A | B | C;
ESTIMATE 'B1 VS B2' B -1 1;
OUTPUT OUT = NEW P = YHAT R = RESID;
PROC  UNIVARIATE PLOT NORMAL; VAR RESID;
```

SAS Printout

```
                    General Linear Models Procedure

Dependent Variable: Y
                                   Sum of              Mean
Source                   DF       Squares            Square    F Value      Pr > F

Model                     7  1046.6250000       149.5178571      15.60      0.0001

Error                    16   153.3333333         9.5833333

Corrected Total          23  1199.9583333

               R-Square              C.V.          Root MSE                 Y Mean

               0.872218          13.01168         3.0956959              23.791667

Source                   DF     Type I SS       Mean Square    F Value      Pr > F

A                         1  145.04166667      145.04166667      15.13      0.0013
B                         1  630.37500000      630.37500000      65.78      0.0001
A*B                       1    7.04166667        7.04166667       0.73      0.4040
C                         1   57.04166667       57.04166667       5.95      0.0267
A*C                       1  155.04166667      155.04166667      16.18      0.0010
B*C                       1   40.04166667       40.04166667       4.18      0.0578
A*B*C                     1   12.04166667       12.04166667       1.26      0.2789

Source                   DF   Type III SS       Mean Square    F Value      Pr > F

A                         1  145.04166667      145.04166667      15.13      0.0013
B                         1  630.37500000      630.37500000      65.78      0.0001
A*B                       1    7.04166667        7.04166667       0.73      0.4040
C                         1   57.04166667       57.04166667       5.95      0.0267
A*C                       1  155.04166667      155.04166667      16.18      0.0010
B*C                       1   40.04166667       40.04166667       4.18      0.0578
A*B*C                     1   12.04166667       12.04166667       1.26      0.2789

                                      T for H0:    Pr > |T|    Std Error of
Parameter               Estimate    Parameter=0                  Estimate

B1 V2 B2              10.2500000           8.11      0.0001     1.26381257



                              Moments

               N                 24  Sum Wgts         24
               Mean               0  Sum               0
               Std Dev     2.581989  Variance   6.666667
               Skewness    0.072255  Kurtosis   -0.89018
               USS         153.3333  CSS        153.3333
               CV                 .  Std Mean   0.527046
               T:Mean=0           0  Pr>|T|       1.0000
               Num ^= 0          24  Num > 0          12
               M(Sign)            0  Pr>=|M|      1.0000
               Sgn Rank           3  Pr>=|S|      0.9338
               W:Normal    0.966546  Pr<W         0.5857
```

```
                    Quantiles(Def=5)

100% Max    4.666667           99%   4.666667
 75% Q3     1.833333           95%          4
 50% Med    -0.16667           90%          4
 25% Q1         -2.5           10%         -3
  0% Min    -4.33333            5%         -4
                                1%   -4.33333

Range              9
Q3-Q1       4.333333
Mode        1.333333


                        Extremes

   Lowest      Obs     Highest      Obs
-4.33333(       23)  2.333333(        3)
      -4(       14)  2.666667(       17)
      -3(       11)         4(       10)
      -3(        9)         4(       13)
-2.66667(        2)  4.666667(       22)

Stem Leaf                     #             Boxplot
   4 007                      3                |
   3                                           |
   2 037                      3                |
   1 0337                     4            +-----+
   0 03                       2            |  +  |
  -0 333                      3            *-----*
  -1 30                       2            |     |
  -2 773                      3            +-----+
  -3 00                       2                |
  -4 30                       2                |
     ----+----+----+----+
```

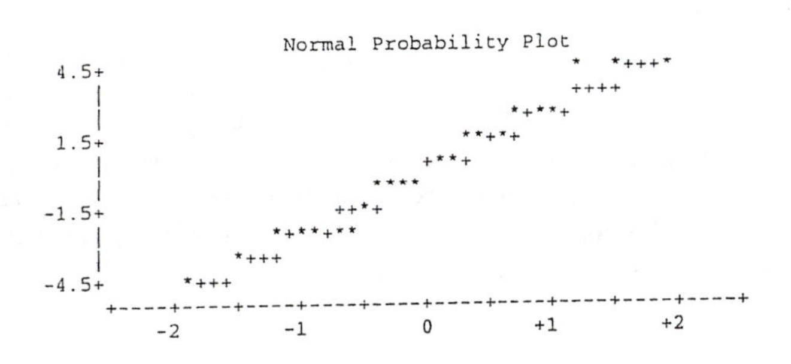

Factorial experiments with more than three factors can be handled in a similar manner. For example, in a four factor factorial with factors A, B, C, D the complete model would contain four main effects, six two–factor interactions (AB, AC, AD, BC, BD, CD), four three–factor interactions (ABC, ABD, ACD, BCD) and a four–factor interaction

ABCD. The four–factor interaction has (a–1)(b–1)(c–1)(d–1) degrees of freedom. More details on this four–factor factorial can be found in the exercises.

It should be pointed out that a factorial experiment requires many observations and a complex model. A four–factor factorial with three levels for each factor requires $3 \times 3 \times 3 \times 3 = 81$ different treatments. In most investigations, having 81 different treatment conditions is too many due to time and cost considerations. Later in this chapter we discuss ways to reduce the number of treatments in such problems.

14.2 Factorial Experiments with Other Designs

Up to this point all of our factorial illustrations have involved the CRD. In this section we consider using other designs with factorial experiments.

First we consider using an RCBD in a factorial experiment. As an illustration consider a 3×2 factorial experiment replicated within an RCBD with $r = 2$ blocks. The layout data from such an experiment could appear as follows:

y_{211}	y_{121}	y_{321}
y_{311}	y_{111}	y_{221}

Block 1

y_{112}	y_{212}	y_{312}
y_{222}	y_{322}	y_{122}

Block 2

where y_{ijk} represents the response associated with the experimental unit in the kth block to the ith level of A and the jth level of B. Note that in this example there are six treatments (combinations of the levels of factor A and B) and these are randomly assigned to the six positions in each block. To see the factorial elements involved, the table below is a rearrangement of the previous "field" layout data.

	A: 1		2		3	
	B1	B2	B1	B2	B1	B2
Block 1	y_{111}	y_{121}	y_{211}	y_{221}	y_{311}	y_{321}
Block 2	y_{112}	y_{122}	y_{212}	y_{222}	y_{312}	y_{322}

.

The model we could use for this 3×2 factorial replicated in an RCBD with $r = 2$ blocks is

$$Y_{ijk} = \mu + \alpha_i + \beta_j + (\alpha\beta)_{ij} + \rho_k + \varepsilon_{ijk} \quad i = 1,2,3; \quad j = 1,2; \quad k = 1,2$$

$$\sum_{i=1}^{3} \alpha_i = \sum_{j=1}^{2} \beta_j = \sum_{i=1}^{3} (\alpha\beta)_{ij} = \sum_{j=1}^{2} (\alpha\beta)_{ij} = 0, \ i = 1,2,3; \ j = 1,2$$

$$\sum_{k=1}^{2} \rho_k = 0.$$

The ANOVA table would appear as

Source	df	SS	MS	f
Blocks	1	SSbl	MSbl	MSbl/MSe
Treatments	5	SStr	MStr	MStr/MSe
A	2	SSa	MSa	MSa/MSe
B	1	SSb	MSb	MSb/MSe
AB	2	SSab	MSab	MSab/MSe
Error	5	SSe	MSe	
Total	11	SSt		

In general, the ANOVA table from **a × b factorial experiment replicated in an RCBD with r blocks** is given in Table 14.2.1.

Table 14.2.1 ANOVA Table for an a × b Factorial Within an RCBD

Source	df	SS	MS	f
Blocks	r–1	SSbl	MSbl	MSbl/MSe
A	a–1	SSa	MSa	MSa/MSe
B	b–1	SSb	MSb	MSb/MSe
AB	(a–1)(b–1)	SSab	MSab	MSab/MSe
Error	(r–1)(ab–1)	SSe	MSe	
Total	abr–1	SSt		

Using this ANOVA table will allow us to test a variety of hypotheses. For example, a test for $H_0 : \beta_j = 0$ (No B effects) rejects H_0 if $f = MSb/MSe \geq f_{\alpha,b-1,(r-1)(ab-1)}$. A Tukey multiple comparison on the levels of factor B decides that $\beta_j \neq \beta_t$ if

$$|\bar{y}_{\cdot j \cdot} - \bar{y}_{\cdot t \cdot}| \geq q_{\alpha,b,(r-1)(ab-1)} \sqrt{\frac{s^2}{ar}} .$$

The next example provides an illustration of these calculations using SAS.

Example 14.2.1 An experiment was conducted to determine the effectiveness of a gasoline additive on gas mileage for cars. Three brands of gasoline (1, 2, 3) were used for the experiment. With each brand the additive was absent (N) and present (Y). Hence we have six treatments. Eight cars were used for the experiment and each car was used with each treatment in randomized order. The response variable is the miles per gallon over a specified driving course.

Table 14.2.2 Gasoline Additive Data

	Brand	1		2		3	
	Additive	N	Y	N	Y	N	Y
Car (Block)	1	24.5	27.2	28.1	30.3	26.1	26.8
	2	26.0	29.3	29.3	32.1	25.3	27.1
	3	22.1	23.2	24.3	28.4	23.0	26.9
	4	25.6	29.3	29.8	31.4	26.3	28.1
	5	30.3	33.4	33.7	36.0	33.4	37.6
	6	34.2	35.9	38.3	40.2	34.3	35.4
	7	26.2	29.3	29.4	32.8	26.9	28.4
	8	24.1	28.7	25.5	29.6	25.8	29.2

We observe that the data in Table 14.2.2 comes from a 2 × 3 factorial experiment replicated in an RCBD with eight blocks. Letting α represent additive, β represent brands and ρ represent blocks (cars) the model we use is $Y_{ijk} = \mu + \alpha_i + \beta_j + (\alpha\beta)_{ij} + \rho_k + \varepsilon_{ijk}$; $i = 1, 2$; $j = 1, 2, 3$; $k = 1, \cdots 8$; with the usual side conditions.

Table 14.2.3 contains the SAS printout for these gasoline additive data. From this printout we observe that at the 0.05 level there is no significant interaction between additive and brand. Each of the main effects, cars (blocks), additive and brand is significant. The means on the printout for the two additive groups indicate that the presence of the additive significantly increases gas mileage. Also, the Tukey multiple comparison for brands indicates that brand 2 has significantly higher gas mileage than brands 1 and 3.

Table 14.2.3 SAS Printout for Gasoline Additive Data

SAS Commands

```
OPTIONS LS = 80;
DATA FRCBD;
INPUT CAR ADDIT BRAND MILEAGE @@;
CARDS;
1 1 1 24.5  1 2 1 27.2  1 1 2 28.1  1 2 2 30.3  1 1 3 26.1  1 2 3 26.8
2 1 1 26.0  2 2 1 29.3  2 1 2 29.3  2 2 2 32.1  2 1 3 25.3  2 2 3 27.1
3 1 1 22.1  3 2 1 23.2  3 1 2 24.3  3 2 2 28.4  3 1 3 23.0  3 2 3 26.9
4 1 1 25.6  4 2 1 29.3  4 1 2 29.8  4 2 2 31.4  4 1 3 26.3  4 2 3 28.1
5 1 1 30.3  5 2 1 33.4  5 1 2 33.7  5 2 2 36.0  5 1 3 33.4  5 2 3 37.6
6 1 1 34.2  6 2 1 35.9  6 1 2 38.3  6 2 2 40.2  6 1 3 34.3  6 2 3 35.4
7 1 1 26.2  7 2 1 29.3  7 1 2 29.4  7 2 2 32.8  7 1 3 26.9  7 2 3 28.4
8 1 1 24.1  8 2 1 28.7  8 1 2 25.5  8 2 2 29.6  8 1 3 35.8  8 2 3 29.2
PROC GLM;  CLASS CAR ADDIT BRAND;
MODEL MILEAGE = CAR ADDIT BRAND ADDIT*BRAND;
MEANS ADDIT;
MEANS BRAND / TUKEY;
```

SAS Printout

```
                         General Linear Models Procedure

Dependent Variable: MILEAGE
                                    Sum of           Mean
Source                  DF         Squares         Square    F Value      Pr > F

Model                   12     807.49000000    67.29083333      49.70      0.0001

Error                   35      47.38812500     1.35394643

Corrected Total         47     854.87812500

                  R-Square            C.V.       Root MSE          MILEAGE Mean

                  0.944567        3.963694      1.1635920             29.356250


Source                  DF       Type I SS    Mean Square    F Value      Pr > F

CAR                      7    635.46312500    90.78044643      67.05      0.0001
ADDIT                    1     85.60020833    85.60020833      63.22      0.0001
BRAND                    2     85.57625000    42.78812500      31.60      0.0001
ADDIT*BRAND              2      0.85041667     0.42520833       0.31      0.7325

Source                  DF     Type III SS    Mean Square    F Value      Pr > F

CAR                      7    635.46312500    90.78044643      67.05      0.0001
ADDIT                    1     85.60020833    85.60020833      63.22      0.0001
BRAND                    2     85.57625000    42.78812500      31.60      0.0001
ADDIT*BRAND              2      0.85041667     0.42520833       0.31      0.7325
```

```
                General Linear Models Procedure

     Level of          -----------MILEAGE-----------
     ADDIT        N        Mean                SD

     1           24     28.0208333        4.16036988
     2           24     30.6916667        4.01723641
```

```
                General Linear Models Procedure

  Tukey's Studentized Range (HSD) Test for variable: MILEAGE

 NOTE: This test controls the type I experimentwise error rate, but
       generally has a higher type II error rate than REGWQ.

              Alpha= 0.05  df= 35  MSE= 1.353946
          Critical Value of Studentized Range= 3.461
            Minimum Significant Difference= 1.0068

 Means with the same letter are not significantly different.

        Tukey Grouping              Mean        N  BRAND

                     A           31.2000       16  2

                     B           28.7875       16  3
                     B
                     B           28.0812       16  1
```

Factorial and Latin Square

A Latin square design can also be used with a two factor factorial experiment. If factors A and B have a and b levels, respectively, then these "ab" treatments can be used in a Latin square with "ab" rows and columns. As an illustration, consider factors A and B each having two levels with the four treatments labeled as A1B1, A1B2, A2B1, A2B2. A particular Latin square with four rows, four columns and these four treatments is as follows:

	Columns			
Rows	1	2	3	4
1	A1B1	A1B2	A2B1	A2B2
2	A2B1	A1B1	A2B2	A1B2
3	A1B2	A2B2	A1B1	A2B1
4	A2B2	A2B1	A1B2	A1B1

Letting $Y_{ijk\ell}$ represent the response at the ith level of factor A, jth level of factor B, kth row and ℓth column, the model that can be used is $Y_{ijk\ell} = \mu + \alpha_i + \beta_j + (\alpha\beta)_{ij} + \rho_k + \gamma_\ell + \varepsilon_{ijk\ell}$; i = 1, 2; j = 1, 2; k = 1,···,4; ℓ = 1,···,4; with the usual side conditions. Of course only certain combinations of values of (i, j, k, ℓ) are present.

The ANOVA table for an a × b factorial experiment replicated in a Latin square with "ab" rows and columns is given in Table 14.2.4.

Table 14.2.4 ANOVA Table for a × b Factorial in a Latin Square

Source	df	SS	MS	f
A	a–1	SSa	MSa	MSa/MSe
B	b–1	SSb	MSb	MSb/MSe
AB	(a–1)(b–1)	SSab	MSab	MSab/MSe
Row	ab–1	SSr	MSr	MSr/MSe
Column	ab–1	SSc	MSc	MSc/MSe
Error	$a^2b^2-3ab+2$	SSe	MSe	
Total	a^2b^2-1	SSt		

An example of this two–factor factorial replicated in a Latin square design is given in the exercises.

It should be pointed out that a Latin square design enables us to make comparisons with fewer observations. For example, the 2 × 2 factorial replicated in a Latin square with four rows and columns requires 16 observations. However, only certain combinations of rows, columns and treatments are present. If we desired to have all combinations of rows, columns, levels of factor A, and levels of factor B, present, then we would need a minimum of 4 × 4 × 2 × 2 = 64 observations. Clearly, the Latin square approach requires less data. Of course with a Latin square design we are unable to test for some of the interaction terms.

In closing this section we note that our examples have involved two–factor factorial experiments in a RCBD or a Latin square. We can consider factorial experiments in a RCBD or Latin square when there are more than two factors. The extension follows in a natural manner.

14.3 Special Factorial Experiments — 2^k

There are special cases of factorial experiments that occur frequently in applications. Some of these have special structures and formulas that make them especially easy to use. In this section we consider one of these classes of special factorial experiments, namely the 2^k factorials.

In a **2^k factorial experiment** there are $k \geq 2$ factors each having two levels. For convenience, we label the two levels as low (L or —) and high (H or +). All factors are crossed and there are $r \geq 1$ observations at each combination of the factors. A factorial arrangement is used to assign the treatments to the experimental units. As an illustration, when $k = 2$ the 2^2 factorial experiment has four treatments which are of the form (A, B): (L, L), (H, L), (L, H), (H, H). Notice that in a 2^k factorial experiment there are 2^k different treatments.

It is helpful to use the **Yates notation** to denote the various treatments. In this notation the factors are denoted by the capital letters A, B, C,···, the lowercase letters a, b, c,···, are used to denote the high level of a factor. The absence of a letter indicates a factor at the low level. For example, in a 2^3 factorial the symbol "ac" represents the treatment where factors A and C are at the high level, and factor B is at the low level. The symbol (1) is used to represent the treatment where all factors are at the low level. With this notation the four treatments for a 2^2 are (1), a, b, ab. The eight treatments for a 2^3 are (1), a, b, c, ab, ac, bc, abc.

There is a nice geometric representation that can be used for 2^2 and 2^3 factorial experiments. As can be observed from Figure 14.3.1, the treatments of a 2^2 are the vertices of a square and those for a 2^3 are the vertices of a cube.

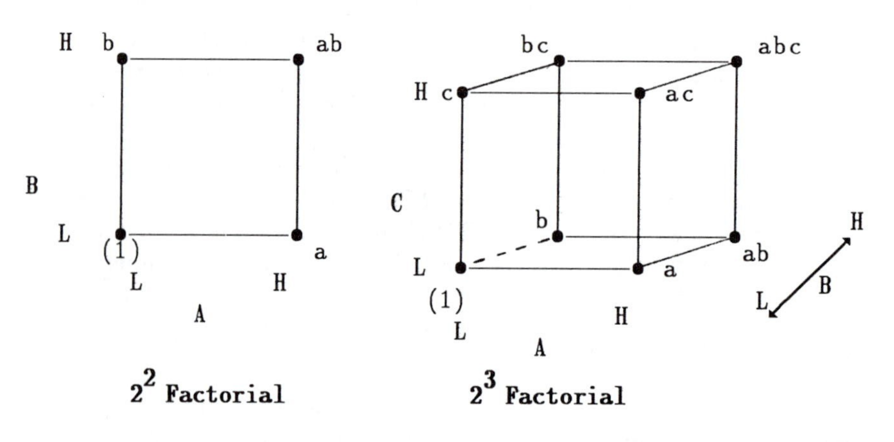

Figure 14.3.1 Geometric Representation of a 2^2 and 2^3

There are easy to use formulas for 2^k factorials that can be applied to find estimates and sums of squares for contrasts. We now let (1), a, b,$\cdots$, represent the **total** of the observations at the indicated treatment. For a 2^2 factorial three contrasts of interest are:

$$\hat{\ell}_A = ab + a - b - (1)$$
$$\hat{\ell}_B = ab + b - a - (1) \qquad (14.3.1)$$
$$\hat{\ell}_{AB} = ab - b - a + (1).$$

From Figure 14.3.1 we note that the contrast for the main effect A is simply the difference between the total for the observations at the high level of A and the total for the observations at the low level of A. To find the AB (interaction) contrast we see that [ab — b] measures the difference between the two levels of A at the high level of B, while [a — (1)] measures this same difference at the low level of B. The difference between these two quantities is a natural measure for interaction. Finally, observe that these three contrasts in (14.3.1) are orthogonal.

The **effect** for a factor is defined to be the change in response produced by changing the level of the factor from low to high, averaged over the levels of the other factor(s). For a 2^2 factorial with r replications at each treatment these effects are as follows:

$$\text{Effect of A: } \hat{\ell}_A / 2r$$

$$\text{Effect of B: } \hat{\ell}_B / 2r \qquad (14.3.2)$$

$$\text{Effect of AB: } \hat{\ell}_{AB} / 2r.$$

Notice that the effect of A is actually the difference between the means of the observations at the high and low levels of A, that is $\bar{y}_{H\cdot\cdot} - \bar{y}_{L\cdot\cdot}$.

The sums of squares for these contrasts can be found in the usual manner from Equation (8.2.1) which yields

$$SSa = \hat{\ell}_A^2 / 4r, \quad SSb = \hat{\ell}_B^2 / 4r, \quad SSab = \hat{\ell}_{AB}^2 / 4r. \qquad (14.3.3)$$

Since these contrasts are orthogonal, we know that the treatment sum of squares, SStr, can be written as SStr = SSa + SSb + SSab. Also, the above formulas for SSa, SSb, and SSab yield the same result as the sums of squares for these factors found in Chapter 13 for a factorial experiment.

To extend the formulas for contrasts, effects and sum of squares to a general 2^k factorial we utilize a coefficient or sign table. In this table we indicate a low level for a treatment by " — " or (—1) and the high level by " + " or (+1). We illustrate this coefficient table for a 2^2 factorial in Table 14.3.2.

Table 14.3.1 Coefficients for Effects in a 2^2 Factorial

	Effects		
Treatment	A	B	AB
(1)	—	—	+
a	+	—	—
b	—	+	—
ab	+	+	+

The main effects columns (A and B) are found by using a "—" or "+" to indicate the level of the factor for the specified treatment. The interaction column is found by taking "products" of the main effects columns. From this table we can easily find the contrasts for A, B, and AB. For example, the contrast is found for A by using the signs in the A column as the coefficients for the treatment totals. That is,

$$\hat{\ell}_A = -(1) + a - b + ab \text{ , and similarly } \hat{\ell}_{AB} = (1) - a - b + ab.$$

This coefficient table can be easily generalized to any 2^k factorial. The contrasts (in total) can be found using the coefficients and the following formulas can be used to find the effects and the sums of squares:

$$\text{Effect} = \text{contrast}/r2^{k-1} \qquad (14.3.4)$$
$$\text{Sum of Squares} = (\text{contrast})^2/r2^k.$$

We illustrate these ideas with a 2^3 factorial with r = 3 replications.

Example 14.3.1 Consider the anxiety data given in Table 14.1.1. These data are obtained from a 2^3 factorial with r = 3 replications. We use the coefficient table to find the contrasts, effects and sums of squares. The coefficient table for these data is given in Table 14.3.2.

Table 14.3.2 Coefficient Table for 2^3 Anxiety Data

		Effect						
Treatment	Total	A	B	C	AB	AC	BC	ABC
(1)	44	-	-	-	+	+	+	-
a	51	+	-	-	-	-	+	+
b	90	-	+	-	-	+	-	+
c	50	-	-	+	+	-	-	+
ab	82	+	+	-	+	-	-	-
ac	79	+	-	+	-	+	-	-
bc	72	-	+	+	-	-	+	-
abc	103	+	+	+	+	+	+	+
Contrast		59	123	37	−13	61	−31	17
Effect		4.92	10.25	3.08	−1.08	5.08	−2.58	1.42
Sum of Squares		145.04	630.38	57.04	7.04	155.04	40.04	12.04

After finding the contrasts, the effects and sums of squares are found using (14.3.4). For example, the contrast for A is $\hat{\ell}_A = -(1) + a - b$ $c + ab + ac - bc + abc = 59$, the effect for A $= \hat{\ell}_A / r2^{k-1} =$ $59/(3)(4) = 4.92$, $SSa = (\hat{\ell}_A)^2 / r2^k = (59)^2/(3)(8) = 145.04$. Other values can be found in the same manner. Notice that the sums of squares values in Table 14.4.2 agree with the values given in the SAS printout in Table 14.1.3. One nice feature of the table is that all of the sum of squares have one degree of freedom; therefore, the values are "comparable." We can quickly see that factor B has the largest effect on the response (largest sum of squares), while the AB interaction has the least effect. These effects can be displayed visually via a main effects plot or an interaction plot. The plots for B and for AC are displayed in Figure 14.3.2.

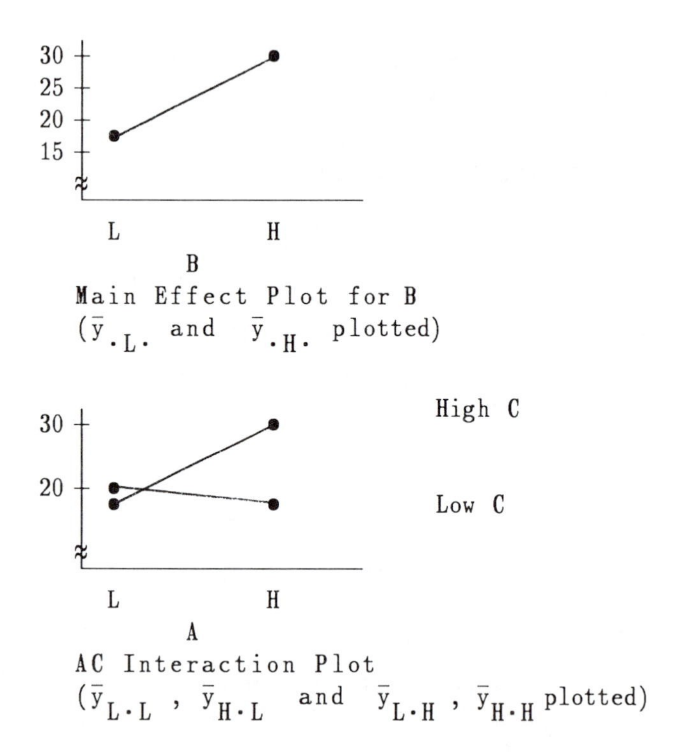

Figure 14.3.2 Main Effect and Interaction Plot

It should be clear that these ideas can be easily extended to 2^4, 2^5, and higher order factorials. There is another algorithm that can be used to find contrasts that is called the **Yates algorithm**. We shall not describe it here, but the interested reader can find a description of it in Hicks (1973) or in Walpole and Myers (1993).

Regression Example

Historically, 2^k factorials have been among the most commonly used experimental plans in industry. A contributing factor to its usage in the pre–computer age was the ease of computation as demonstrated in the last example. Now, with computer programs available, we no longer need to be concerned about the complexity of the computing formulas. To close our discussion of 2^k factorials we illustrate the use of this section's (–1,1) "orthogonal coding" as it applies to a problem in regression. The advantage of using a regression model is that it enables us to predict (or estimate) the response to values of the factors not used in the experiment.

Let us continue to focus on a 2^3 factorial with factors A, B, C.

The regression model for such an experiment is

$$Y = \gamma_0 + \gamma_1 x_1 + \gamma_2 x_2 + \gamma_3 x_3 + \gamma_{12} x_1 x_2 + \gamma_{13} x_1 x_3 + \gamma_{23} x_2 x_3 + \gamma_{123} x_1 x_2 x_3 + \varepsilon.$$

The x_1, x_2, x_3 terms are the main effects terms; the x_1x_2, x_1x_3, x_2x_3 are the two–factor interaction terms; the $x_1x_2x_3$ provides the three–factor interaction term. If we use (–1, 1) coding we let

$$x_1 = \begin{cases} -1, \text{ if A is low} \\ 1, \text{ if A is high} \end{cases} \qquad x_2 = \begin{cases} -1, \text{ if B is low} \\ 1, \text{ if B is high} \end{cases}$$

$$x_3 = \begin{cases} -1, \text{ if C is low} \\ 1, \text{ if C is high} \end{cases}.$$

The next example uses this coding scheme for the SAS printout.

Example 14.3.2 Refer to the data given in Table 14.1.1, obtained in a 2^3 factorial experiment. There we are studying the effects of three ingredients in forming a drug to reduce anxiety levels in patients. For **the purpose of prediction**, we give the numerical levels of the three factors: A (low is 10, high is 20); B (low is 50, high is 90); C (low is 2, high is 3). Table 14.3.3 provides the SAS GLM printout for the regression analysis of these data. The estimating equation is

$$\hat{y} = 23.79 + 2.46\, x_1 + 5.12\, x_2 + 1.54\, x_3 - 0.54\, x_1x_2 + 2.54\, x_1x_3 - 1.29\, x_2x_3 + 0.71\, x_1x_2x_3.$$

We can use this equation to estimate the anxiety level when the drug consists of A = 15, B = 80, and C = 3. For these values we note that $x_1 = 0$, $x_2 = 0.5$, and $x_3 = 1$. Our predicted value is $\hat{y} =$ 27.25. Notice that numerical values given in Table 14.3.3 match the results for the traditional analysis displayed in Table 14.1.3 and the sums of squares values in Table 14.3.2.

A final note about the SAS GLM printout in Table 14.3.3. The (–1,1) coding of the independent variables is an illustration of the "purely orthogonal" coding discussed in Section 13.5 where each parameter has an independent sum of squares associated with it. In using SAS GLM for regression analysis, these sums of squares are printed out for us from which an experimental design ANOVA can easily be constructed. Most regression programs, however, do not provide this information directly in their printouts but all give a t—value for testing individual parameters. These t—values can be used to obtain the ith parameter sum of squares according to the formula

$$SS(\gamma_i) = t_i^2 s^2, \text{ for } i = 1, 2, 3, \ldots \quad (14.3.5)$$

The above formula follows from (6.6.3).

Table 14.3.3 SAS Printout for Regression Model for Anxiety Data

SAS Commands

```
OPTIONS LS = 80;
DATA TXTFF;
INPUT Y A B C @@;
x1 = 0;  x2 = 0;  x3 = 0;
if a = 1 then x1 = -1;  if a = 2 then x1 = 1;
if b = 1 then x2 = -1;  if b = 2 then x2 = 1;
if c = 1 then x3 = -1;  if c = 2 then x3 = 1;
x12 = x1*x2;  x13 = x1*x3;  x23 = x2*x3;  x123 = x1*x2*x3;
CARDS;
15  1  1  1     12  1  1  1     17  1  1  1
14  1  1  2     18  1  1  2     18  1  1  2
32  1  2  1     31  1  2  1     27  1  2  1
28  1  2  2     21  1  2  2     23  1  2  2
21  2  1  1     13  2  1  1     17  2  1  1
24  2  1  2     29  2  1  2     26  2  1  2
26  2  2  1     29  2  2  1     27  2  2  1
39  2  2  2     30  2  2  2     34  2  2  2
proc glm; model y = x1 x2 x3 x12 x13 x23 x123;
```

SAS Printout

General Linear Models Procedure

Dependent Variable: Y

Source	DF	Sum of Squares	Mean Square	F Value	Pr > F
Model	7	1046.6250000	149.5178571	15.60	0.0001
Error	16	153.3333333	9.5833333		
Corrected Total	23	1199.9583333			

R-Square	C.V.	Root MSE	Y Mean
0.872218	13.01168	3.0956959	23.791667

Source	DF	Type I SS	Mean Square	F Value	Pr > F
X1	1	145.04166667	145.04166667	15.13	0.0013
X2	1	630.37500000	630.37500000	65.78	0.0001
X3	1	57.04166667	57.04166667	5.95	0.0267
X12	1	7.04166667	7.04166667	0.73	0.4040
X13	1	155.04166667	155.04166667	16.18	0.0010
X23	1	40.04166667	40.04166667	4.18	0.0578
X123	1	12.04166667	12.04166667	1.26	0.2789

Source	DF	Type III SS	Mean Square	F Value	Pr > F
X1	1	145.04166667	145.04166667	15.13	0.0013
X2	1	630.37500000	630.37500000	65.78	0.0001
X3	1	57.04166667	57.04166667	5.95	0.0267
X12	1	7.04166667	7.04166667	0.73	0.4040
X13	1	155.04166667	155.04166667	16.18	0.0010
X23	1	40.04166667	40.04166667	4.18	0.0578
X123	1	12.04166667	12.04166667	1.26	0.2789

Parameter	Estimate	T for H0: Parameter=0	Pr > |T|	Std Error of Estimate
INTERCEPT	23.79166667	37.65	0.0001	0.63190629
X1	2.45833333	3.89	0.0013	0.63190629
X2	5.12500000	8.11	0.0001	0.63190629
X3	1.54166667	2.44	0.0267	0.63190629
X12	-0.54166667	-0.86	0.4040	0.63190629
X13	2.54166667	4.02	0.0010	0.63190629
X23	-1.29166667	-2.04	0.0578	0.63190629
X123	0.70833333	1.12	0.2789	0.63190629

14.4 Quantitative Factors, 3^k Factorial

When the levels of a factor are quantitative in nature, one can study the type of relationship (e.g., linear, quadratic) that exists between the response variable and the factor. Our purpose in this section is to illustrate how to detect such polynomial relationships using the 3^k class of factorial experiments.

The regression example in the last section dealt with three quantitative factors in a 2^k ($k = 3$) factorial experiment. Since there are just two levels of each factor in this case, a significant main effect is interpreted as a significant linear relationship between that factor and the response variable. In order to check for curvature in the relationship, we need three or more levels and a 3^k factorial is one such experimental plan.

In a **3^k factorial experiment** there are $k \geq 2$ factors each having three levels. In the balanced case there are $r \geq 1$ replications at each of the 3^k treatments. When the levels are numerical we can label them as low (L), medium (M), and high (H). For $k = 2$ the $3^2 = 9$ treatments are displayed below. (The dots correspond to the treatments.)

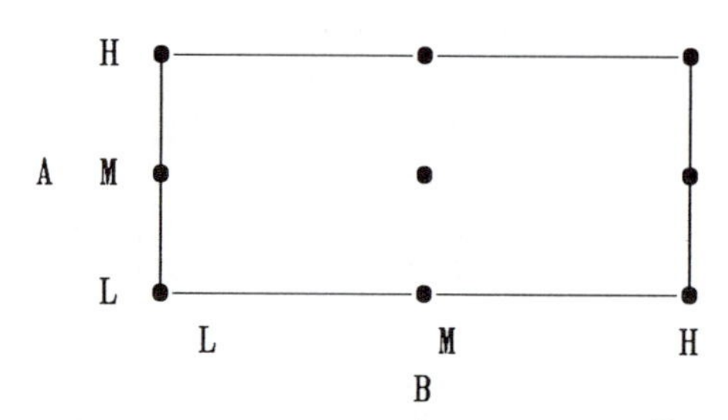

Figure 14.4.1 Treatments for a 3^2

We can study the nature of the relationship between the response and the factors in a 3^k by using orthogonal polynomials or by regression techniques. As an illustration consider a 3^2 factorial experiment with factors A and B within a CRD with r replications. The traditional model for this experiment is

$$Y_{ijk} = \mu + \alpha_i + \beta_j + (\alpha\beta)_{ij} + \varepsilon_{ijk}; \quad i = 1, 2, 3; \quad j = 1, 2, 3;$$

$$k = 1, \cdots, r; \text{ with the usual side conditions.}$$

If the levels of factor A are quantitative and equally spaced, we can use the orthogonal polynomial approach in Chapter 8 to study the nature of the response. For three levels of the factor the linear contrast is $\psi_L = \alpha_3 - \alpha_1$ and the quadratic contrast is $\psi_Q = \alpha_1 - 2\alpha_2 + \alpha_3$. These two orthogonal contrasts can be used to test for linear and quadratic relationships, and can be used to partition the sum of squares for factor A. Note that we can perform these tests for factor A regardless of whether factor B has quantitative levels or not.

We illustrate these ideas in the next example.

Example 14.4.1 A company was developing a soap product for cleaning clothes. The product was made of two major ingredients which we label as A and B. To determine the relationship between the levels of these factors and the cleaning ability of the soap a 3^2 factorial experiment was completed. The levels for A and B are equally spaced and for convenience are labelled as L, M, H. (For factor A, L represents 10% of the product, M is 15% and H is 20%. For factor B, L represents 20%, M is 30% and H is 40%. The remainder of the soap product is filler and other non—active ingredients.) For each combination of the levels of A and B the soap product was produced and used on two batches of clothes. Each batch was judged for overall cleanliness on a 35—point scale. (Seven judges each rated the batch on a 5—point scale and the total score was the sum of the ratings.) The results are given in Table 14.4.1.

Table 14.4.1 Soap Product Data

		B		
		L(20%)	M(30%)	H(40%)
	L (10%)	7 10	21 17	13 15
A	M (15%)	15 17	32 30	25 28
	H (20%)	10 5	19 22	16 14

Table 14.4.2 contains an analysis for these data using SAS GLM. Notice that there is no significant interaction, and the main effects for A and B are significant. The linear and quadratic contrasts are considered for both A and B. Both A and B have significant quadratic effects, indicating curvature in the relationship between each factor and the cleaning ability of the soap product. The curvatures differ in nature, however, as evidenced by the linear effect for B but not for A. Prototypes of these two curvatures are given in Figure 14.4.2.

Table 14.4.2 SAS Printout for Soap Data

SAS Commands

```
OPTIONS LS = 80;
DATA TXTKS;
INPUT A B Y @@;
CARDS;
1  1 7    1  1 10   1  2 21   1  2 17   1  3 13   1  3 15
2  1 15   2  1 17   2  2 32   2  2 30   2  3 25   2  3 28
3  1 10   3  1 5    3  2 19   3  2 22   3  3 16   3  3 14
PROC GLM; CLASS A B;
MODEL Y = A B A*B;
CONTRAST 'LINEAR A' A -1 0 1;
CONTRAST 'QUADRATIC A' A 1 -2 1;
CONTRAST 'LINEAR B' B -1 0 1;
CONTRAST 'QUADRATIC B' B 1 -2 1;
```

SAS Printout

```
                         General Linear Models Procedure

Dependent Variable: Y
                                     Sum of              Mean
Source                  DF          Squares            Square    F Value       Pr > F

Model                    8     952.44444444      119.05555556      25.51       0.0001

Error                    9      42.00000000        4.66666667

Corrected Total         17     994.44444444

                  R-Square             C.V.          Root MSE                  Y Mean

                  0.957765         12.30520         2.1602469               17.555556


Source                  DF        Type I SS       Mean Square    F Value       Pr > F

A                        2     434.77777778      217.38888889      46.58       0.0001
B                        2     502.11111111      251.05555556      53.80       0.0001
A*B                      4      15.55555556        3.88888889       0.83       0.5368

Source                  DF      Type III SS       Mean Square    F Value       Pr > F

A                        2     434.77777778      217.38888889      46.58       0.0001
B                        2     502.11111111      251.05555556      53.80       0.0001
A*B                      4      15.55555556        3.88888889       0.83       0.5368

Contrast                DF      Contrast SS       Mean Square    F Value       Pr > F

LINEAR A                 1       0.75000000        0.75000000       0.16       0.6979
QUADRATIC A              1     434.02777778      434.02777778      93.01       0.0001
LINEAR B                 1     184.08333333      184.08333333      39.45       0.0001
QUADRATIC B              1     318.02777778      318.02777778      68.15       0.0001
```

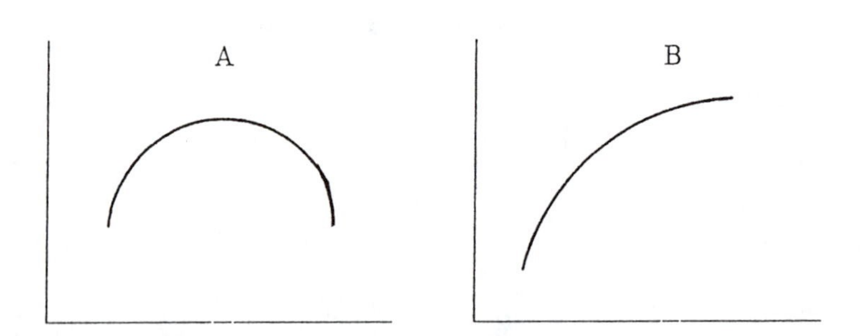

Figure 14.4.2 Response Curves from Example 14.4.1

In the Table 14.4.2 computer printout, we again see the partitioning of sums of squares into single degree of freedom sums of squares characteristic of orthogonal contrasts. (See Chapter 8.) We highlight that property in Table 14.4.3.

Table 14.4.3 The Partitioning into Polynomial Effects

Effect	df	SS	Effect	df	SS
A	2	434.778	B	2	502.111
A_L	1	0.750	B_L	1	184.083
A_Q	1	434.028	B_Q	1	318.028

In the exercises the student is asked to analyze these data using regression approaches and also to compute the sums of squares "by hand" using a table of orthogonal polynomial coefficients.

In closing this section on quantitative type factors, we again emphasize that in a 2^k factorial only linear relationships can be studied. A 3^k factorial allows us to investigate for quadratic effects as well as linear effects. It should be noted, however, that a 3^k requires many more treatments than a 2^k. For example, when k = 4, a 2^4 requires 16 treatments while a 3^k requires 81. Also, by increasing the number of levels of a factor, higher order polynomial effects can be studied. For an illustration, see Example 8.3.1 where 4 levels are used. However, it should be noted that it often is difficult to interpret polynomial relationships higher than degree 2 or 3.

14.5 Fractional Factorials, Confounded

In some cases a factorial experiment can require more treatments or observations than is feasible in terms of time and money. For example, a 2^5 factorial requires 32 treatments, and if each observation is very expensive to obtain, then using 32 treatments might be too costly. Further to run an experiment with 32 treatments might require more time than can be allowed. A solution to the problem is to use only a subset of all possible 2^5 treatments. This leads to a **fractional factorial experiment**, namely a factorial experiment in which only some of the possible treatments are included.

Some issues arise when utilizing a fractional factorial. First, does it matter which of the treatments we choose to use? Second, what information do we lose by only using a fraction of the treatments? We address each of these issues in the ensuing discussion.

To illustrate these concerns we consider a $\frac{1}{2}$ of a 2^3 fractional factorial. Since a 2^3 factorial has 8 treatments, a $\frac{1}{2}$ of a 2^3 could consist of the treatments a b c abc, or it could consist of the treatments b c ab ac. Is one set of treatments better than the other? We explore this by looking at the coefficient table of main and interaction effects together with an added intercept (I) column. The coefficient table for the $\frac{1}{2}$ of a 2^3 with treatments a b c abc is given as follows using (± 1 coding):

	Effect							
Treatment	I	A	B	C	AB	AC	BC	ABC
a	1	1	−1	−1	−1	−1	1	1
b	1	−1	1	−1	−1	1	−1	1
c	1	−1	−1	1	1	−1	−1	1
abc	1	1	1	1	1	1	1	1

From this table we note that the columns for I and ABC are identical. As a consequence we cannot distinguish between these two effects in this experiment. They are what is called **confounded**. Two effects are said to be **confounded** if you cannot distinguish between the two effects, that is, you cannot separate these two effects. Effects that are confounded are called **aliases** of each other. If two effects have the same values in the coefficient table, then the two effects are confounded. Similarly, if the entries for two effects are the negative of each other, then the effects are confounded. From the coefficient table we see that the pairs A and BC, B and AC, C and AB and I and ABC are confounded. When effects are confounded only one of the effects can appear in a model. For example, for the $\frac{1}{2}$ of a 2^3 with treatments a b c abc the model can include the intercept and the main effects A, B, C, but no interaction terms.

We note that in this $\frac{1}{2}$ of a 2^3, each effect is confounded with one other effect. We can, therefore, only include half of the effects in the model. This is what we lose by taking only a fraction of the treatments, namely, only a fraction of the effects can be studied. Note that these facts remain true even if there are replications in the experiment.

Now turn to the question concerning which four treatments to use in a $\frac{1}{2}$ of a 2^3. Suppose we consider the treatments b c ab ac. The coefficient table for these treatments is as follows:

Treatment	Efect I	A	B	C	AB	AC	BC	ABC
b	1	−1	1	−1	−1	1	−1	1
c	1	−1	−1	1	1	−1	−1	1
ab	1	1	1	−1	1	−1	−1	−1
ac	1	1	−1	1	−1	1	−1	−1

The alias pairs for this experiment are: I and BC, A and ABC, B and C, and AB and AC. The model we use for this experiment can include one effect from each alias pair, therefore, only one of the main effects B and C can be included in the model. This is a very poor design since we cannot separate the effects for B and C. This illustrates that it makes a difference which half of the treatments are used.

We should point out that the aliases obtained with the treatments a b c abc are identical to those found with the other half, namely (1) ab ac bc. This always occurs for any $\frac{1}{2}$ of a 2^k. The treatment set containing the treatment (1) is called the **principal part**.

Since it makes a difference which fraction of the treatments we choose, we next describe a procedure for finding a "good" set of treatments for a $\frac{1}{2}$ of a 2^k. The effect that is confounded with the intercept (I) is called the **defining contrast**. We can write this defining contrast as $A^{\alpha_1} B^{\alpha_2} C^{\alpha_3} D^{\alpha_4} \cdots$, where α_i is a zero or one. For example, in a $\frac{1}{2}$ of a 2^3 if the defining contrast is BC, then $\alpha_1 = 0$, and $\alpha_2 = 1 = \alpha_3$. We now calculate for each possible treatment the quantity $L = \alpha_1 x_1 + \alpha_2 x_2 + \cdots$, where

$$x_i = \begin{cases} 1 & \text{if the ith factor is high} \\ 0 & \text{if the ith factor is low} \end{cases}.$$

Next we calculate $S = L \pmod 2$ for each possible treatment. This divides the treatments into two groups of equal size based upon the value of S. We select one of two groups of treatments for the $\frac{1}{2}$ of a 2^k. An illustration will help.

Example 14.5.1 We aim to select the treatments for a $\frac{1}{2}$ of a 2^3. We choose the effect ABC to be the defining contrast (alias of I) so $L = x_1 + x_2 + x_3$. We calculate L and $S = L \pmod 2$ for each treatment.

Treatment	(1)	a	b	c	ab	ac	bc	abc
L	0	1	1	1	2	2	2	3
S = L(mod 2)	0	1	1	1	0	0	0	1

Using the two values of S, the two possible $\frac{1}{2}$ of a 2^3 fractional factorial experiments consist of the treatments (1) ab ac bc (principal part), or equivalently a b c abc.

We can use the defining contrast to find the alias structure. The alias for an effect is found by multiplying the symbols for the effect by the symbols for the defining contrast and taking the answer mod 2. As an illustration, in a $\frac{1}{2}$ of a 2^3 with defining contrast ABC the alias structure is:

Effect	Effect (Defining Contrast) Mod 2	Alias Pair
A	$A(ABC) = A^2BC = BC$	A, BC
B	$B(ABC) = AB^2C = AC$	B, AC
C	$C(ABC) = ABC^2 = AB$	C, AB
I	$I(ABC) = ABC$	I, ABC

This matches what we found earlier in this section. We consider the analysis of a fractional factorial in the following example.

Example 14.5.2 A manufacturer of plastic parts wanted to complete a study aimed at identifying factors that affect the strength of the plastic. Three possible factors were identified that are part of the molding process. The factors were: A (mold temperature), B (time in mold), and C (thickness of mold). It was decided to use a less expensive experiment to obtain a preliminary analysis. Two levels (low and high) were chosen for each factor and a $\frac{1}{2}$ of a 2^3 fractional factorial design was employed. Three replications were taken at each of the four treatments. The data measuring the strength of the plastic follows:

<table>
<tr><td></td><td></td><td colspan="2">C = L</td><td colspan="2">C = H</td></tr>
<tr><td></td><td></td><td colspan="2">B</td><td colspan="2">B</td></tr>
<tr><td></td><td></td><td>L</td><td>H</td><td>L</td><td>H</td></tr>
<tr><td rowspan="2">A</td><td>L</td><td>38 42 29</td><td></td><td></td><td>46 55 50</td></tr>
<tr><td>H</td><td></td><td>27 37 34</td><td>24 22 15</td><td></td></tr>
</table>

Table 14.5.1 contains an analysis of these data using SAS. Note that the treatments used were (1) ab ac bc; which is a design considered earlier in this section. We can use the main effects model $Y_{ijk} = \mu + \alpha_i + \beta_j + \gamma_k + \varepsilon_{ijk}$ for the analysis. No interaction can be included in the model since they are confounded with main effects. From the SAS printout we note that factors A and B are significant at the 0.05 level, and factor C is not. Further, by inspecting the "ESTIMATE" statements we note that the plastic is stronger when A is at the low level and B is at the high level. In fact, the low level of A has mean strength that is 16.8 units more than the mean for the high level of A. If further experimentation is to be performed, then it is clear that the focus should be on factor A at the low level, and factor B at the high level.

Table 14.5.1 SAS Printout for Plastic Data

SAS Commands

```
options LS = 80;
data tx2kff;
input a b c y @@;
cards;
1  1 1   38      1  1 1   42      1  1 1   29
1  2 2   55      1  2 2   50      1  2 2   46
2  2 1   37      2  2 1   34      2  2 1   27
2  1 2   22      2  1 2   24      2  1 2   15
proc glm;
class a b c;
model y = a b c;
estimate 'a L vs H' a -1 1;
estimate 'b L vs H' b -1 1;
estimate 'c L vs H' c -1 1;
```

SAS Printout

```
                    General Linear Models Procedure

Dependent Variable: Y
                                   Sum of            Mean
Source                  DF        Squares          Square   F Value     Pr > F

Model                    3   1372.2500000     457.4166667     16.14     0.0009

Error                    8    226.6666667      28.3333333

Corrected Total         11   1598.9166667

              R-Square               C.V.        Root MSE                Y Mean

              0.858237           15.24460       5.3229065             34.916667


Source                  DF      Type I SS     Mean Square   F Value     Pr > F

A                        1   850.08333333    850.08333333     30.00     0.0006
B                        1   520.08333333    520.08333333     18.36     0.0027
C                        1     2.08333333      2.08333333      0.07     0.7931

Source                  DF    Type III SS     Mean Square   F Value     Pr > F

A                        1   850.08333333    850.08333333     30.00     0.0006
B                        1   520.08333333    520.08333333     18.36     0.0027
C                        1     2.08333333      2.08333333      0.07     0.7931

                                         T for H0:    Pr > |T|    Std Error of
Parameter                 Estimate      Parameter=0                 Estimate

a L vs H               -16.8333333           -5.48      0.0006      3.07318149
b L vs H                13.1666667            4.28      0.0027      3.07318149
c L vs H                 0.8333333            0.27      0.7931      3.07318149
```

Blocking

Sometimes we can afford to run all of the treatments in a 2^k factorial, but we cannot run them all on the same day or at the same factory. This leads us to running the treatments in blocks (e.g., day, factory). Questions that now arise are which treatments should occur in which blocks, and what do we lose by using blocks? We investigate these issues with an illustration.

Suppose that we want to run a 2^3 factorial in two blocks, with treatments a b c abc in block one and (1) ab ac bc in block two. To investigate any confounding we display the coefficient table. From the coefficient table we observe that ABC and blocks are confounded. By running this experiment in two blocks we have lost the ability to include the interaction term ABC in the model. This is generally not a serious loss.

Treatment		Effect I	A	B	C	AB	AC	BC	ABC	Block
Block one	a	1	1	−1	−1	−1	−1	1	1	1
	b	1	−1	1	−1	−1	1	−1	1	1
	c	1	−1	−1	1	1	−1	−1	1	1
	abc	1	1	1	1	1	1	1	1	1
Block two	(1)	1	−1	−1	−1	1	1	1	−1	−1
	ab	1	1	1	−1	1	−1	−1	−1	−1
	ac	1	1	−1	1	−1	1	−1	−1	−1
	bc	1	−1	1	1	−1	−1	1	−1	−1

It is interesting to note the relationship between blocking in a factorial and fractional factorials. Notice in the previous illustration that the treatments used in block one, namely a b c abc, are precisely the ones obtained for a $\frac{1}{2}$ of a 2^3 factorial using the defining contrast ABC (see Example 14.5.1). Further notice that the effect confounded with the blocks is the defining contrast term. This is the general pattern. To run a 2^k in two blocks we select a defining contrast, which will be confounded with blocks. We find the treatments for a $\frac{1}{2}$ of a 2^k factorial associated with this defining contrast. These treatments are placed in block one and the other half of the treatments is placed in block two.

We have concentrated in this section on $\frac{1}{2}$ of a 2^k fractional factorials. One can also consider $\frac{1}{4}$ of a 2^k or $\frac{1}{8}$ of a 2^k or in general a $1/2^p$ of a 2^k (sometimes written as 2^{k-p}). We will not consider these nor 3^k fractional factorial experiments here. The interested reader can consult Montgomery (1991) for further details.

Chapter 14 Exercises

Application Exercises

14–1 A study was completed to determine the effect of three factors on the performance in an elementary statistics course. The three factors were: class format (lecture only, lecture with discussion), review sessions before exams (yes, no), and homework handed in (yes, no). The combinations of these factors led to eight different classes. Thirty–two students of fairly equal abilities were randomly assigned to the eight classes with four students assigned to each class. Each class used the same exams, and the response variable was the course percentage for each student. The results are as follows:

		Homework Handed In			
		YES		NO	
	Review Sessions	Yes	No	Yes	No
Format	Lecture Only	92 86 69 76	74 63 58 89	50 54 65 79	74 48 65 40
	Lecture and Discussion	94 75 93 76	84 62 88 70	77 82 59 59	64 52 79 45

(a) Find the ANOVA table for these data. Perform and interpret each f–test using a 0.05 level.

(b) Find a 95% confidence interval for the difference between the means for the two levels of homework.

(c) Use residuals to check the model assumptions.

14–2 In a factorial experiment with a CRD the effect of each factor on the chemical yield of a process was considered. Each factor had two levels with three replications for each treatment. Unfortunately, one observation was lost during the experiment. The results are as follows:

	C1		C2	
	B1	B2	B1	B2
A1	227 223 232	262 260 271	249 241 256	263 231 246
A2	225 231	243 325 251	212 220 208	246 258 260

(a) Find the ANOVA table for this data. Perform and interpret each f–test using a 0.05 level.

(b) Use the residuals to check the model assumptions. Identify the outlier in the data.

(c) Eliminate the outlier and rerun the analysis. Note the effect that this has on s^2 and the various f–tests.

(d) (SAS Required) Find the least squares mean for the eight cell means. Use a Bonferroni multiple comparison with $\alpha = 0.10$ to determine differences among these means. Why might a 0.10 level be more appropriate in this case than the usual 0.05?

14–3 The yield for three varieties of wheat was compared in an experiment. Three fields were used in the experiment and each field was divided into six plots of equal size. Another factor used was spray for weeds, with levels yes and no. Six treatments were formed with the combinations of varieties and spray. Each treatment was randomly assigned to a plot in a field. The response variable, bushels of wheat, is recorded as follows:

	Variety	1		2		3	
	Spray	N	Y	N	Y	N	Y
Field	1	53	72	47	69	63	84
	2	48	63	39	62	60	75
	3	59	77	46	67	68	89

(a) Find the ANOVA table for the data. Perform and interpret the outcome of each f–test using a 0.05 level.

(b) Using a 0.05 level perform the Tukey multiple comparison on the three varieties.

14–4 A manufacturing firm performed a study aimed at increasing the productivity of four operators of a drill press. Two factors considered were number of rest breaks per four hour shift (1 and 2), and amount of supervision (minimal, close). This led to four treatment conditions: 1 (1 break, minimal), 2 (1 break, close), 3 (2 breaks, minimal), 4 (2 breaks, close). A Latin square design was used in which each operator used each treatment for a full week and a measure of productivity was recorded. The results are as follows with the treatment number given in parentheses.

	Operator	1	2	3	4
Week	1	19 (1)	16 (2)	22 (3)	25 (4)
	2	25 (3)	17 (1)	26 (4)	15 (2)
	3	20 (2)	24 (4)	14 (1)	19 (3)
	4	27 (4)	20 (3)	16 (2)	14 (1)

(a) Write down an appropriate model for the analysis of these data using the effect terms: week, operator, number of breaks, supervision, and the interaction between breaks and supervision.
(b) Find the ANOVA table for these data. Perform and interpret the f–tests using a 0.05 level.
(c) Perform a Tukey multiple comparison on the four treatments using a 0.05 level.

14–5 Refer to Exercise 14–1 which contains data from a 2^3 factorial experiment. Using the special formulas for a 2^3 factorial, find the coefficient table and use it to find the effect and sum of squares for each term in the model.

14–6 Refer to the soap product data in Table 14.4.1 which was analyzed in Example 14.4.1. Notice that both factors are quantitative and the levels are equally spaced.

(a) Let x_1 represent the actual level of factor A (10, 15 or 20) and let x_2 represent the actual level of factor B (20, 30, 40). Use regression procedures to fit the second degree model $Y = \beta_0 + \beta_1 x_1 + \beta_2 x_2 + \beta_{12} x_1 x_2 + \beta_{11} x_1^2 + \beta_{22} x_2^2 + \varepsilon$.
(b) Using the model in part (a) predict the value of Y when A is 18% and B is 25%.
(c) (SAS required) Use the contour option in the plot procedure of SAS to obtain a contour plot for y based on the regression equation in part (a). Use x_1 values from 5 to 25 and x_2 values from 15 to 45. Using the plot, what values of x_1 and x_2 tend to give the best cleaning values?
(d) Because of the equal spacing and equal replication, we can use regression with orthogonal polynomial coding in order to analyze these data. Let L, M and H values of each factor be represented by –1, 0, 1 and 1, –2, 1 for linear and quadratic effects, respectively. Using a **regression program**, calculate the ANOVA table for these data and compare your results with those displayed in Table 14.4.2.
(e) Patterned after Table 14.3.2 and using orthogonal polynomial coding, construct a table of coefficients for the purpose of calculating the effects sums of squares for the

Table 14.4.1 data. Then use this table to obtain the sums of squares. Hint: For effect headings use A_L, A_Q, B_L, B_Q, A_LB_L, A_LB_Q, A_QB_L, A_QB_Q and note that the divisor in the sum of squares formula, $(\text{contrast})^2/\text{divisor}$, varies according to the coefficients used to calculate the contrast total.

14–7 Johnson and Leone (1977) report an experiment aimed at improving the tensile strength (Y) of a rubber compound. The following four factors were considered for the experiment each having two levels: A (location), B (mixing method), C (curing method), D (test stage strategy). The experimenters decided to use a 1/2 of a 2^4 fractional factorial with a CRD for the experiment. The data obtained are as follows:

		A_1		A_2	
		B_1	B_2	B_1	B_2
C_1	D_1	3400			2850
	D_2		3350	3200	
C_2	D_1		3150	3050	
	D_2	3250			3200

(a) Using a main effects model for the data, perform the f–tests with a 0.05 level. Which main effects seems most important?

(b) Find the alias structure for this experiment. Hint: Use the coefficient table to determine the defining contrast.

(c) Consider a model containing the "intercept," the main effects, and the three interactions AB, AC, AD. This is a **saturated model** (i.e., no degrees of freedom for error). Find the sum of squares for each term. We cannot perform f–tests (no SSe term), but since each sum of squares has one degree of freedom we can compare these values. Which terms in the model seem to be the most important?

14–8 Consider a 1/2 of a 2^3 fractional factorial experiment.

(a) Find the treatments (principal part) that should be used if the defining contrast is AB.

(b) Find the alias structure for this experiment.
(c) What model could be used with the data obtained from this experiment?
(d) Compare the treatments chosen for this experiment with the ones chosen for the 1/2 of a 2^3 in Example 14.5.1. Which is the better experiment? Why?

14–9 Find the treatments required for a 1/2 of a 2^4 with defining contrast ABCD. Compare the treatments obtained with those used in Exercise 14–7.

14–10 Refer to the gasoline additive data in Example 14.2.1. Find a 95% confidence interval for the difference between the mean gas mileage for brand 2 versus brand 1.

14–11 Refer to the soap product data analyzed in Example 14.4.1. Perform a Tukey multiple comparison on the levels of factor A using a 0.05 level.

14–12 Refer to the Anxiety Data given in Table 14.1.1 and analyzed in Examples 14.1.1 and 14.3.1.

(a) Verify the values given for the effects and sum of squares in Table 14.3.2.
(b) Display the main effects plot for A.
(c) Display the interaction plot for BC.

Theoretical Exercises

14–13 Consider a balanced three–factor factorial replicated in a CRD.

(a) Verify the formula given in (14.1.3) for the Tukey multiple comparison.
(b) Verify the formula given in (14.1.4) for a confidence interval for a contrast in factor A.
(c) Find the form of a t–test for a contrast in factor B.
(d) Find a formula for the Tukey multiple comparison for the "abc" cell means.

14–14 Consider a balanced four–factor factorial replicated in a CRD. Label the factors as A, B, C, D with levels a, b, c, d, and r replications.

(a) Write down the model for this experiment.

(b) Write down the ANOVA table for this experiment in a form similar to Table 14.1.2.

14–15 Consider a 3 × 2 factorial experiment replicated in an RCBD with 2 blocks.

(a) Write down the sample form of the model.
(b) Find the **X** matrix for the model and identify its rank.
(c) Find the normal equations for the model.
(d) Impose the usual side conditions to find a solution to these normal equations.
(e) Find a formula for SSe.
(f) Find a formula for the sum of squares for blocks.

14–16 Consider an a × b factorial experiment replicated in an RCBD with r blocks.

(a) Derive the formula for a Tukey multiple comparison for the levels of factor A.
(b) If the interaction between A and B is significant, then we could perform a multiple comparison on the "ab" means, $\mu_{ij\cdot}$. Find the form of the Tukey multiple comparison for these means.

14–17 Consider a a × b factorial experiment replicated in a Latin square with p = ab rows and columns.

(a) Show that $\bar{Y}_{i\cdot\cdot\cdot} - \bar{Y}_{t\cdot\cdot\cdot}$ is an unbiased estimator for $\alpha_i - \alpha_t$.
(b) Explain why the degrees of freedom for error is $a^2b^2 - 3ab + 2$.
(c) Find the form of a 95% confidence interval for $\alpha_i - \alpha_t$. (Assume normal distributions.)
(d) Find the form of the Tukey multiple comparison for factor A.

14–18 Show that the three contrasts in (14.3.1) are orthogonal.

14–19 Consider the usual model for a 2^2 factorial experiment replicated within a CRD. Find an algebraic relationship between the effect for A and the least squares estimate for $\alpha_H - \alpha_L$.

14–20 Use (8.2.1) to verify the formulas for the sums of squares of the contrasts given in (14.3.3).

14–21 Verify that the formula for SSa given in (14.3.3) equals the value for SSa for a 2 × 2 factorial given in Chapter 13.

14–22 Consider a 2^2 factorial experiment with quantitative factors. Let x_1 and x_2 represent the values for factor A and B, respectively. Use the **X** matrix to explain why the x_1^2 and x_2^2 terms cannot be used in the regression model

$$Y = \beta_0 + \beta_1 x_1 + \beta_2 x_2 + \beta_{12} x_1 x_2 + \beta_{11} x_1^2 + \beta_{22} x_2^2 + \varepsilon.$$

14–23 Use the coefficient table to explain why the treatments (1) a c ac would be a poor choice for the treatments in a 1/2 of a 2^3 fractional factorial.

14–24 An experiment is called **resolution III** if no main effects are confounded with other main effects, but main effects are confounded with two–factor interactions. In **resolution IV** experiments main effects are not confounded with main effects or two–factor interactions; however, at least one of the two–factor interactions is confounded with another two–factor interaction. An experiment is called **resolution V** if no main effect or two–factor interaction is confounded with any other main effect or two–factor interaction.

(a) What is the resolution of the experiment in Example 14.5.1?

(b) What is the resolution of the experiment in Exercise 14.7?

CHAPTER 15

ANALYSIS OF COVARIANCE

15.1 Introduction

In the designs considered thus far we have studied how factors that are controlled by the experimenter affect a response variable. In some problems, however, the response variable is also affected by one or more quantitative variables, called **concomitant variables (or covariates)**, that the experimenter is unable to control at a constant level throughout the experiment. Since erroneous conclusions regarding factor effects can result because of relationships between the response variable and concomitant variables, it is important to remove the effects that these concomitant variables have on the response variable. As an illustration consider an experiment aimed at comparing the yield for three varieties of corn planted in a number of different fields. The response variable, yield, depends on the treatment (variety), but also depends on other uncontrolled variables such as rainfall, which could serve as a covariate in the experiment.

In this chapter we present a method called the **analysis of covariance**. This method incorporates the covariates into the model along with the controlled factors in order to reduce the experimental error. It also provides as an end—product treatment effects **adjusted** for the effects of the covariate(s). The linear model associated with **covariance analysis** depends upon the number of covariates present and upon the particular experimental design used in allocating treatments to experimental units. If there is only one covariate and the relationship between this variable and the response variable is assumed to be first degree linear, the analysis is called **simple covariance**; if of higher degree, **curvilinear covariance**. If two or more covariates are incorporated in the experiment, the analysis is called **multiple covariance**.

Our approach will be to illustrate the methods of analysis of covariance in the case that involves a single factor, one covariate having a linear relationship with the response variable, and a CRD. We refer to this as the simple covariance replicated in a CRD. As an example of such a case consider an experiment aimed at comparing the effectiveness of three instructors teaching the same course. These instructors use varying teaching methods in the course. Thirty students are randomly assigned to the three instructors (treatments), and at the end of the course their score, on a common comprehensive final exam is recorded. To adjust for the fact that the students might have had differing academic abilities, it was decided to record each student's grade point average (gpa). Since gpa is an uncontrolled, quantitative variable, it serves as a covariate in this problem. The data are given in Table 15.1.1.

Table 15.1.1 Test Scores and GPA

Instructor					
1		2		3	
Score	GPA	Score	GPA	Score	GPA
75	3.0	91	3.5	93	3.8
78	2.9	84	3.2	75	2.3
85	3.0	80	2.5	69	1.4
85	3.8	78	2.7	89	3.5
93	3.8	87	3.8	77	1.7
69	2.4	65	2.5	98	3.0
83	3.9	91	3.1	94	2.6
85	3.1	95	3.5	84	2.9
72	3.0	81	3.0	83	2.1
87	3.3	62	2.3	85	3.6

We note from the data that both the scores and gpa values vary from instructor to instructor. Our goal is to compare the student performances for the three instructors (treatments), while adjusting for the effect of the differing gpa (covariate).

We describe a model useful for analyzing data from a simple covariance replicated in a CRD. The model for p treatments, and n observations with n_i experimental units assigned to treatment i is:

$$Y_{ij} = \mu + \tau_i + \beta x_{ij} + \varepsilon_{ij} \tag{15.1.1}$$

$$i = 1,2,\cdots,p \; ; \quad j = 1,2,\cdots,n_i \; ;$$

where

Y_{ij} is the response to the jth experimental unit of the ith treatment,

μ is a nonestimable parameter,

τ_i is the effect of the ith treatment,

β is the common slope of the "treatment lines,"

x_{ij} is the covariate value corresponding to the jth experimental unit of the ith treatment, and

ε_{ij} is the random error associated with the jth experimental unit of the ith treatment with $E(\varepsilon_{ij}) = 0$ and $Var(\varepsilon_{ij}) = \sigma^2$ for every i and j. Furthermore, it is assumed that the ε_{ij} are uncorrelated. In order to carry out statistical inference procedures it will be necessary for us to further assume that the ε_{ij} are normally distributed.

From model (15.1.1) we note that $E(Y_{ij}) = \mu + \tau_i + \beta x_{ij}$ and $Var(Y_{ij}) = \sigma^2$. If we graph $E(Y_{ij})$ versus x_{ij} we see that we have p lines with equal slopes. Figure 15.1.1 provides this graph for a case when $p = 3$, $\beta > 0$, and $\tau_1 < \tau_2 < \tau_3$. The fact that we have ideally required equal slopes in this model is an assumption that we will need to check because this is the situation whereby the pictured relationship among the τ_i holds for all values in the experimental domain of the covariate, x_{ij}.

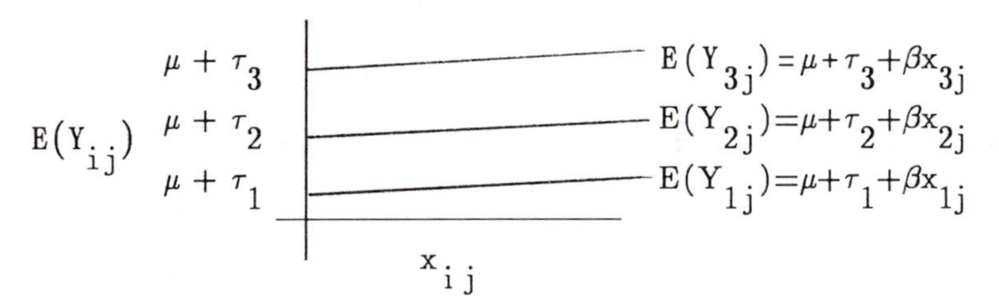

Figure 15.1.1 Relationship Between Mean and Covariate

In the next section we use the model to perform statistical inferences. Of particular interest will be a test for "no difference between the treatments," namely a test for $H_0 : \tau_1 = \cdots = \tau_p$. Also of interest will be a test for "no covariate effect" on the response variable, that is $H_0 : \beta = 0$.

15.2 Inferences for a Simple Covariance Model

Our goal in this section is to obtain tests and confidence intervals for the analysis of covariance model (15.1.1). We first note that this model can be written in the form $\mathbf{Y} = \mathbf{X}\boldsymbol{\beta} + \boldsymbol{\varepsilon}$ and is, therefore, a linear model. As an illustration when $p = 3$, $n_1 = n_2 = n_3 = 2$ we have

$\mathbf{Y}' = (Y_{11}, Y_{12}, Y_{21}, Y_{22}, Y_{31}, Y_{32})$, $\boldsymbol{\beta}' = (\mu, \tau_1, \tau_2, \tau_3, \beta)$ and

$$\mathbf{X} = \begin{bmatrix} 1 & 1 & 0 & 0 & x_{11} \\ 1 & 1 & 0 & 0 & x_{12} \\ 1 & 0 & 1 & 0 & x_{21} \\ 1 & 0 & 1 & 0 & x_{22} \\ 1 & 0 & 0 & 1 & x_{31} \\ 1 & 0 & 0 & 1 & x_{32} \end{bmatrix}$$

In this case $\mathbf{X}$ has rank 1 + (3—1) + 1 = 4. It is easy to observe that in general the $\mathbf{X}$ matrix is less than full rank and has rank equal to 1 + (p—1) + 1 = p + 1. Letting $\mathbf{b}' = (m, t_1, \cdots, t_p, b)$ the normal equations $\mathbf{X}'\mathbf{X}\mathbf{b} = \mathbf{X}'\mathbf{y}$ for model (15.1.1) are of the form:

$$\begin{aligned} &nm + \sum_{i=1}^{p} n_i t_i + bx_{..} = y_{..} \\ &n_i m + n_i t_i + b\, x_{i.} = y_{i.}\ , \quad i = 1, \cdots, p \qquad (15.2.1) \\ &mx_{..} + \sum_{i=1}^{p} t_i x_i + b \sum_{i=1}^{p} \sum_{j=1}^{n_i} x_{ij}^2 = \sum_{i=1}^{p} \sum_{j=1}^{n_i} x_{ij}\, y_{ij}\ . \end{aligned}$$

Since $\mathbf{X}$ is rank deficient by one we add one side condition to find a solution. Using the side condition $\sum_{i=1}^{p} n_i t_i = 0$ we obtain the following solution:

$$b = \frac{\sum_{i=1}^{p} \sum_{j=1}^{n_i} (y_{ij} - \bar{y}_{i.})(x_{ij} - \bar{x}_{i.})}{\sum_{i=1}^{p} \sum_{j=1}^{n_i} (x_{ij} - \bar{x}_{i.})^2}$$

$$m = \bar{y}_{..} - b\bar{x}_{..} \qquad (15.2.2)$$

$$t_i = \bar{y}_{i.} - \bar{y}_{..} - b\,(\bar{x}_{i.} - \bar{x}_{..})\ .$$

We can find $\hat{y}_{ij} = m + t_i + bx_{ij}$, and then the error sum of squares is $SSe = \sum_{i=1}^{p} \sum_{j=1}^{n_i} (y_{ij} - \hat{y}_{ij})^2$. The degrees of freedom for SSe is n—rank $\mathbf{X}$ = n—(p+1) = n—p—1. The mean square error is then

$$MSe = s^2 = SSe/(n-p-1). \tag{15.2.3}$$

Before proceeding to find tests and confidence intervals we need to determine which quantities are estimable. We know that the entries in $\mathbf{X}\boldsymbol{\beta}$ are estimable; therefore, $\mu + \tau_i + \beta x_{ij}$ is estimable for all i, j. It is easy to show that β, $\mu + \tau_i$ and contrasts in the τ_i's are all estimable. The least squares estimates and Blue for β and $\tau_i - \tau_k$ involve b from (15.2.2) and are:

Estimable Function	Blue
β	b
$\tau_i - \tau_k$	$(\bar{y}_{i\cdot} - \bar{y}_{k\cdot}) - b(\bar{x}_{i\cdot} - \bar{x}_{k\cdot})$.

(15.2.4)

If we assume a normal distribution then we can obtain the customary tests and confidence intervals. Confidence intervals are found in the usual manner. For example, if $Se(\hat{\beta})$ and $Se(\widehat{\tau_i - \tau_k})$ represent the standard errors for $\hat{\beta}$ and $\widehat{\tau_i - \tau_k}$, respectively, then confidence intervals for these quantities are given by

Estimable Function	Confidence Interval
β	$b \pm t_{\frac{\alpha}{2}, n-p-1} Se(\hat{\beta})$
$\tau_i - \tau_k$	$(\bar{y}_{i\cdot} - \bar{y}_{k\cdot}) - b(\bar{x}_{i\cdot} - \bar{x}_{k\cdot}) \pm t_{\frac{\alpha}{2}, n-p-1} Se(\widehat{\tau_i - \tau_k})$

(15.2.5)

Note that b is given in (15.2.2), and the standard error for an estimable function $\boldsymbol{\ell}'\boldsymbol{\beta}$ is $\sqrt{s^2 \boldsymbol{\ell}' (\mathbf{X}'\mathbf{X})^{-}\boldsymbol{\ell}}$, see (5.3.5).

We now turn our attention to finding f—tests and the ANOVA table for an analysis of covariance. The hypotheses of primary interest are $H_0 : \beta = 0$, a general linear hypothesis involving one estimable function, and $H_0 : \tau_1 = \cdots = \tau_p$, a general linear hypothesis involving $p - 1$ linearly independent estimable functions. These hypotheses suggest the following reduced models and associated error sum of squares.

$$Y_{ij} = \mu^* + \tau_i^* + \varepsilon_{ij}^*, \quad SSe^*$$

$$Y_{ij} = \mu^{**} + \beta^{**} x_{ij} + \varepsilon_{ij}^{**}, \quad SSe^{**}.$$

Recalling that the full model (15.1.1) has SSe and s^2 with $n - p - 1$ degrees of freedom, we find the following **adjusted** sum of squares for covariance and treatments, respectively:

$$SScov = SSe^* - SSe \quad \text{with } 1 \text{ df}, \tag{15.2.6}$$
$$SStr = SSe^{**} - SSe \quad \text{with } p{-}1 \text{ df}.$$

The f—test for $H_0 : \beta = 0$ rejects H_0 if $f = SScov/s^2 \geq f_{\alpha,1,n-p-1}$, and the f—test for $H_0 : \tau_1 = \cdots = \tau_p$ rejects H_0 if $f = [SStr/(p-1)]/s^2 \geq f_{\alpha,p-1,n-p-1}$. We summarize these values in an ANOVA table. As usual the total sum of squares is

$$SSt = \sum_{i=1}^{p} \sum_{j=1}^{n_i} (y_{ij} - \bar{y}_{..})^2.$$

Table 15.2.1 ANOVA Table for Analysis of Covariance

Source	df	SS	MS	f
Treatments	p–1	SStr	MStr	MStr/MSe
Covariate	1	SScov	MScov	MScov/MSe
Error	n–p–1	SSe	MSe	
Total	n–1			

It should be noted that our design is not orthogonal. As a consequence, the adjusted sum of squares used in Table 15.2.1 do not generally equal the unadjusted sum of squares. Further, the sum of squares in the ANOVA table do not "add—up," that is, SSt $\neq$ SStr + SScov + SSe.

We close this section with an example that demonstrates the methods obtained in this section.

Example 15.2.1 Refer to the data in Table 15.1.1 which provides test scores for students of three instructors. We use an analysis of covariance for these data based upon p = 3 treatments (instructors) and a covariate (gpa). Some summary calculations are:

Instructor	1	2	3
Mean Score $\bar{y}_{i\cdot}$	81.2	81.4	84.7
Mean GPA $\bar{x}_{i\cdot}$	3.22	3.01	2.69

Since the mean gpa are different, especially for instructor three, it is appropriate to use analysis of covariance methods to adjust for the effect of gpa on the test scores. Table 15.2.2 contains a printout from SAS for these data. Note that in the program statements for the PROC GLM procedure the class statement includes the factor treatment, but not the covariate. (In general, on SAS the class statement contains the names of factors with specified levels, but does not contain regression type factors such as covariates.) Using $\alpha = .05$ from the printout we note that H_0 : "No covariate (gpa) effect" is rejected since the p—value = .0001 < α, and H_0 : "No treatment (instructor) effect" is also rejected since p = .0149 < α. We also note that SStr (unadj) = 77.267 $\neq$ SStr (adj) = 410.052, as we expect in a non—orthogonal design. A solution to the normal equations on the printout provides the least squares estimate for the covariate b = 10.98. This coefficient indicates that there is a positive linear relationship between gpa and test score. Using the estimate for $\tau_2 - \tau_1$ and the standard error on the printout, and noting that $t_{.025,26} = 2.056$, a 95% confidence interval for $\tau_2 - \tau_1$ is 2.506 ± 2.056 (2.904) or (—3.46, 8.48). This suggests that there is no significant difference between the mean scores of the students of instructors one and two, when adjusted for gpa.

Table 15.2.2 SAS Printout for Test Score Data

SAS Program

```
OPTIONS LS = 80;
DATA TXCOV;
INPUT INST SCORE GPA @@;
CARDS;
1 75 3.0    1 78 2.9    1 85 3.0    1 85 3.8    1 93 3.8
1 69 2.4    1 83 3.9    1 85 3.1    1 72 3.0    1 87 3.3
2 91 3.5    2 84 3.2    2 80 2.5    2 78 2.7    2 87 3.8
2 65 2.5    2 91 3.1    2 95 3.5    2 81 3.0    2 62 2.3
3 93 3.8    3 75 2.3    3 69 1.4    3 89 3.5    3 77 1.7
3 98 3.0    3 94 2.6    3 84 2.9    3 83 2.1    3 85 3.6
PROC GLM; CLASS INST;
MODEL SCORE = INST GPA / SOLUTION;
ESTIMATE 'INST2 VS INST1' INST -1 1 0;
```

SAS Printout

```
                       General Linear Models Procedure

Dependent Variable: SCORE
                                   Sum of              Mean
Source                   DF       Squares            Square   F Value      Pr > F

Model                     3  1325.6166860       441.8722287     10.70      0.0001

Error                    26  1073.7499807        41.2980762

Corrected Total          29  2399.3666667

                   R-Square          C.V.          Root MSE            SCORE Mean

                   0.552486      7.795824         6.4263579             82.433333


Source                   DF     Type I SS       Mean Square   F Value      Pr > F

INST                      2    77.2666667        38.6333333      0.94      0.4052
GPA                       1  1248.3500193      1248.3500193     30.23      0.0001

Source                   DF   Type III SS       Mean Square   F Value      Pr > F

INST                      2   410.0520909       205.0260454      4.96      0.0149
GPA                       1  1248.3500193      1248.3500193     30.23      0.0001

                                         T for H0:     Pr > |T|    Std Error of
Parameter                    Estimate    Parameter=0                 Estimate

INSTR2 VS INST1            2.50586247           0.86     0.3961      2.90439546


                                         T for H0:     Pr > |T|    Std Error of
Parameter                    Estimate    Parameter=0                 Estimate

INTERCEPT                55.16299981 B          9.60     0.0001      5.74384838
INST         1           -9.31955766 B         -3.04     0.0053      3.06268118
             2           -6.81369519 B         -2.31     0.0288      2.94415505
             3            0.00000000 B           .         .          .
GPA                      10.98029747            5.50     0.0001      1.99715067

NOTE: The X'X matrix has been found to be singular and a generalized inverse
      was used to solve the normal equations.   Estimates followed by the
      letter 'B' are biased, and are not unique estimators of the parameters.
```

Our analysis of covariance model (15.1.1) assumes equal slopes for the p treatments. In the previous example, this assumes that the relationship between gpa and test score is the **same** for each instructor. This is a stringent assumption and one that we should check. If this assumption is not reasonable, then one might say that "all bets are off" concerning the interpretation of test results using model (15.1.1).

Covariance Regression

A regression model can be used to analyze data obtained in an analysis of covariance problem. For example, we consider the regression approach to simple covariance as applied to Table 15.1.1 data and analyzed in Example 15.2.1. With $p = 3$, the regression version of model (15.2.1) in this example becomes

$$Y = \alpha_0 + \alpha_1 x_1 + \alpha_2 x_2 + \alpha_3 x_3 + \varepsilon, \tag{15.2.7}$$

where $x_1 = \begin{cases} 1, & \text{if Instructor 1} \\ 0, & \text{if not} \end{cases}$, $x_2 = \begin{cases} 1, & \text{if Instructor 2} \\ 0, & \text{if not} \end{cases}$,
$x_3 = x_{ij}$, the gpa value.

Keeping in mind that α_1 and α_2 are treatment parameters and $\alpha_3 \equiv \beta$ is the covariate parameter, we can obtain the analysis of variance sums of squares as directed by SS( ) notation as follows:

$$SSr = SS(\alpha_1, \alpha_2, \alpha_3) \equiv SS(\tau,\beta) = 1325.617, \quad SSe = 1073.750$$

$$SStr = SS(\tau|\beta) = SS(\tau,\beta) - SS(\beta) = 1325.617 - 915.565 = 410.052$$

$$SScov = SS(\beta|\tau) = SS(\tau,\beta) - SS(\tau) = 1325.617 - 77.267 = 1248.350.$$

These results displayed in an ANOVA table are:

Table 15.2.3 ANOVA for Test Score Data

Source	df	SS	MS	f
Treatments (Instructors)	2	410.052	205.026	4.96
Covariate (gpa)	1	1248.350	1248.350	30.23
Error	26	1073.750	41.298	
Total	29			

Beyond the ANOVA, additional tests and estimates concerning treatments can be made using regression model (15.2.7) by observing the relationships between it and traditional model (15.1.1). In this $p = 3$ case, the correspondences are $\alpha_0 = \mu + \tau_3$, $\alpha_1 = \tau_1 - \tau_3$, $\alpha_2 = \tau_2 - \tau_3$ and $\alpha_3 = \beta$.

15.3 Testing for Equal Slopes

As pointed out in the previous section, we assumed that the linear relationship between the covariate and the response variable had the same slope for each treatment. Since this is an important assumption, we should check its appropriateness **before** completing the analysis of the data. Our purpose in this section is to provide a test for the equality of the slopes.

To obtain the desired test we introduce a more general model that allows the slope to vary with the treatment:

$$Y_{ij} = \mu^* + \tau_i^* + \beta_i^* x_{ij} + \varepsilon_{ij}^* \; ; i = 1,\cdots,p; \; j = 1,\cdots,n_i \, . \qquad (15.3.1)$$

(The stars on the parameters point out the fact that the parameter values associated with the models (15.3.1) and (15.1.1) are not the same.)

The hypothesis for the equality of the slopes is $H_0 : \beta_1^* = \cdots = \beta_p^*$. It is easy to show that the model (15.3.1) is a linear model, the **X** matrix has rank $2p$, and that the β_i^* are estimable. The equal slope hypothesis can be written as $H_0 : \beta_1^* - \beta_p^* = 0, \beta_2^* - \beta_p^* = 0, \cdots, \beta_{p-1}^* - \beta_p^* = 0$, which is a general linear hypothesis involving $p - 1$ linearly independent estimable functions. To test this hypothesis, note that model (15.3.1), with error sum of squares denoted by SSe^* has $n - 2p$ degrees of freedom. The equal slopes model (15.1.1) for this test has error sum of squares SSe as given in the ANOVA Table 15.2.1. Using the principle of conditional error, the f—test for testing $H_0 : \beta_1^* = \cdots = \beta_p^*$ is based upon rejecting H_0 if

$$f = \frac{(SSe - SSe^*)/(p-1)}{SSe^*/(n-2p)} \geq f_{\alpha,p-1,n-2p} . \tag{15.3.2}$$

While it is routine to conceptualize the f—test in (15.3.2) it is not obvious as to how we can calculate the f—value. The solution to this problem is provided by the way in which we handled the question of additivity in factorial experiments. Just as in the analysis of factorials, where we check for parallelism in the response to main effects, here we can use interaction terms to investigate the tenability of parallel treatment lines. Let us illustrate by again turning to the example in the last section within the framework of regression analysis.

Illustrating Equality of Slopes Using Covariance Regression

Refer to the regression model (15.2.7) which is the equal slopes model when $p = 3$. If we extend this equal slopes model by adding treatments, covariate interaction terms, we obtain the model

$$Y = \alpha_0^* + \alpha_1^* x_1 + \alpha_2^* x_2 + \alpha_3^* x_3 + \alpha_{13}^* x_1 x_3 + \alpha_{23}^* x_2 x_3 + \varepsilon^* , \tag{15.3.3}$$

where x_1, x_2 and x_3 are defined as in (15.2.7). This extended model represents the full ranked reparametrized version of (15.3.1) with $p = 3$ which allows for unequal slopes. By comparing the functional forms of $E(Y)$ for the three "treatment lines" implicit in model (15.3.3), it can be verified that the slopes of all three are theoretically the same when $\alpha_{13}^* = \alpha_{23}^* = 0$.

We now illustrate the test for equality of slopes in Example 15.2.1 using the f—test (15.3.2). Recall from the previous section that when model (15.2.7), or equivalently model (15.1.1), was used, we obtained $SSe = 1073.750$. On fitting (15.3.3) to the data we get $SSe^* = 909.021$. Hence

$$f = \frac{(SSe - SSe^*)/2}{SSe^*/24} = \frac{(1073.750 - 909.021)/2}{909.021/24} = 2.17.$$

Since $2.17 < f_{.05,2,24} = 3.40$, we fail to reject the null hypothesis H_0 : Three slopes equal. As a result, we have insufficient evidence to conclude that the three treatment lines are not parallel at the 0.05 risk level. This then gives validity to the results concerning treatment comparisons discussed and illustrated in Example 15.2.1.

Recommendation

The recommendation for an analysis of covariance problem is to first test for equal slopes (i.e., covariate, treatment interaction). If the equal slope assumption is reasonable (i.e., fail to reject the test), then use the equal slope model (15.1.1) and the associated procedures described in Section 15.2. If the slopes are unequal, then a comparison of the treatments is difficult to interpret. To observe why this is so, see Figure 15.3.1 which gives an illustration of unequal slopes when p = 2.

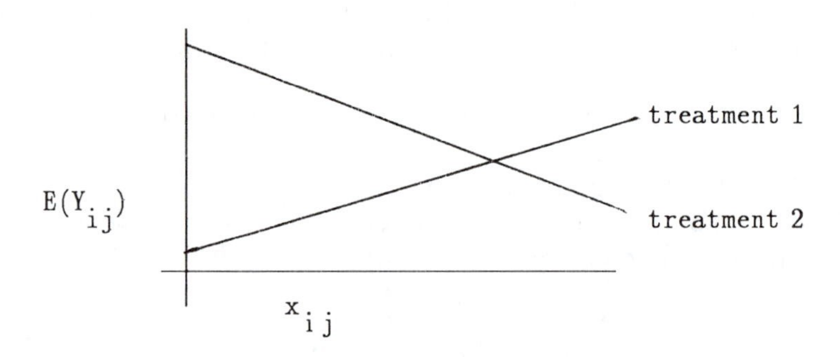

Figure 15.3.1 Unequal Slopes in Analysis of Covariance

From the diagram we note that when the slopes are unequal the relationship between the two treatments varies with the value of the covariate. In such a case the experimenter may be interested in making treatment comparisons at specified covariate values.

SAS Illustration

We close this section with an illustration of testing for equal slopes using SAS. Again refer to the test score data given in Table 15.1.1. We aim to test $H_0 : \beta_1^* = \beta_2^* = \beta_3^*$. To complete the calculations we can use the following SAS commands:

```
PROC GLM; CLASS INST;
MODEL SCORE = INST GPA INST * GPA; .
```

Recall the test for equal slopes corresponds to the presence of interaction between the treatment and covariate factors. Table 15.3.1 contains the SAS printout resulting from these commands. From the printout, the f—test for equal slopes (or equivalently the INST * GPA term) is f = 2.17 with p = 0.1355. As before, we fail to reject $H_0 : \beta_1^* = \beta_2^* = \beta_3^*$

at the 0.05 level; therefore, the equal slopes assumption is reasonable. Note that the values for f—tests for testing treatments and covariates are different on the printout in Table 15.3.1 than the values obtained with the equal slopes model (see Table 15.2.2). Since the equal slopes assumption is reasonable, we use the results from the equal slopes model to test for treatments and the covariate.

Table 15.3.1 SAS Printout for Testing Equal Slopes

General Linear Models Procedure

Dependent Variable: SCORE

Source	DF	Sum of Squares	Mean Square	F Value	Pr > F
Model	5	1490.3457850	298.0691570	7.87	0.0002
Error	24	909.0208816	37.8758701		
Corrected Total	29	2399.3666667			

	R-Square	C.V.	Root MSE		SCORE Mean
	0.621141	7.465836	6.1543375		82.433333

Source	DF	Type I SS	Mean Square	F Value	Pr > F
INST	2	77.2666667	38.6333333	1.02	0.3757
GPA	1	1248.3500193	1248.3500193	32.96	0.0001
GPA*INST	2	164.7290991	82.3645495	2.17	0.1355

Source	DF	Type III SS	Mean Square	F Value	Pr > F
INST	2	255.6922919	127.8461459	3.38	0.0511
GPA	1	1305.7108571	1305.7108571	34.47	0.0001
GPA*INST	2	164.7290991	82.3645495	2.17	0.1355

15.4 Multiple Comparisons, Adjusted Means

In this section we discuss procedures for comparing the treatments when the assumption of equal slopes is reasonable. We first note that the sample means, $\bar{y}_{i\cdot}$, are not "comparable" as in general $E(\bar{Y}_{i\cdot} - \bar{Y}_{k\cdot}) \neq \tau_i - \tau_k$. Recall from (15.2.4) that the Blue for $\tau_i - \tau_k$ is $\bar{y}_{i\cdot} - \bar{y}_{k\cdot} - b(\bar{x}_{i\cdot} - \bar{x}_{k\cdot})$. A Bonferroni multiple comparison for the treatments can be obtained from (15.2.5) by deciding that

$$\tau_i \neq \tau_k \text{ if } |\bar{y}_{i\cdot} - \bar{y}_{k\cdot} - b(\bar{x}_{i\cdot} - \bar{x}_{k\cdot})| \geq t_{\frac{\alpha}{2\binom{p}{2}}, n-p-1} \widehat{Se(\tau_i - \tau_k)}$$

$$\text{for } 1 \leq i < k \leq p. \tag{15.4.1}$$

Another approach that can be used to compare the treatments is to use adjusted means. The **adjusted mean** for treatment i, denoted by am_i, is given by

$$am_i = \bar{y}_{i\cdot} - b(\bar{x}_{i\cdot} - \bar{x}_{\cdot\cdot}) . \tag{15.4.2}$$

The adjusted mean adjusts the treatments means for effect of the covariate. In the exercises it is shown that

$$E(AM_i) = \mu + \tau_i + \beta \bar{x}_{\cdot\cdot} . \tag{15.4.3}$$

This illustrates that the adjusted mean measures the ith treatment average at the average value of the covariate. It is obvious that the difference between the adjusted means, $AM_i - AM_k$, is unbiased for $\tau_i - \tau_k$.

We can use the adjusted means to perform a Bonferroni multiple comparison for the treatments. In the exercises it is shown that the standard error for $AM_i - AM_k$, denoted by Se $(AM_i - AM_k)$ is given by:

$$Se(AM_i - AM_k) = \sqrt{s^2\left[\frac{1}{n_i} + \frac{1}{n_k} + (\bar{x}_{i\cdot} - \bar{x}_{k\cdot})^2 \Big/ \sum_{i=1}^{p} \sum_{j=1}^{n_i} (x_{ij} - \bar{x}_{i\cdot})^2\right]} \tag{15.4.4}$$

The Bonferroni multiple comparison for the treatments, which is equivalent to (15.4.1), is based upon deciding that $\tau_i \neq \tau_k$ if

$$|am_i - am_k| \geq t_{\frac{\alpha}{2\binom{p}{2}}, n-p-1} Se(AM_i - AM_k). \tag{15.4.5}$$

On SAS we can calculate the adjusted means by using the least squares means. We illustrate this in the next example.

Example 15.4.1 Refer to the test score data in Table 15.1.1 which was analyzed using the SAS printout in Table 15.2.2. We first calculate the adjusted means. Some useful values are: $\bar{x}_{..} = 2.973$, and

Treatment	1	2	3
$\bar{y}_{i.}$	81.2	81.4	84.7
$\bar{x}_{i.}$	3.22	3.01	2.69

From the printout we note that the estimated slope for the covariate is $b = 10.9803$. The adjusted means are then

$$am_1 = \bar{y}_{1.} - b(\bar{x}_{1.} - \bar{x}_{..})$$

$$= 81.2 - 10.9803\ (3.22 - 2.973) = 78.49,$$

and similarly $am_2 = 80.99$ and $am_3 = 87.81$.

Notice that the adjusted mean for instructor (treatment) one is smaller than the sample mean, $\bar{y}_{1.}$. Since the mean gpa (covariate) for students with instructor one is larger than the mean gpa for all students, and since there is a positive relationship (slope) between test scores and gpa, our **adjusted** mean needs to be smaller than the usual sample mean.

As mentioned previously, the least squares means from SAS equals the adjusted means. The LSMEANS option on SAS can be used to provide an LSD or Bonferroni multiple comparison for the treatments. Table 15.4.1 contains the printout for the command: LSMEANS INST/ PDIFF;. For a Bonferroni multiple comparison with level 0.05 we decide that two treatments are significantly different if the p—value is $.05/3 = .0167$ or less. From Table 15.4.1 we obtain the following outcome:

Instructor	1	2	3
Adjusted Mean	78.492	80.997	87.811 .

Table 15.4.1 Least Squares Means for Test Score Data

```
                General Linear Models Procedure
                      Least Squares Means

INST           SCORE    Pr > |T|  H0: LSMEAN(i)=LSMEAN(j)
              LSMEAN    i/j       1         2         3

1         78.4915266    1       .      0.3961    0.0053
2         80.9973891    2  0.3961         .      0.0288
3         87.8110843    3  0.0053    0.0288         .

NOTE: To ensure overall protection level, only probabilities associated with
      pre-planned comparisons should be used.
```

15.5 Other Covariance Models

In the previous sections of this chapter we focused on the situation involving one covariate, a CRD, and a linear relationship between the covariate and the response variable. Obviously there are other situations encountered in practice. In this section we briefly describe some of these situations.

First we concentrate on other analysis of covariance situations involving a CRD. One such case is curvilinear covariance which involves a single covariate x, but the relationship between response variable Y and x could be more complicated than linear. For example, if the relationship is quadratic, an appropriate model, which assumes "parallel" treatment curves, would be:

$$Y_{ij} = \mu + \tau_i + \beta_1 x_{ij} + \beta_2 x_{ij}^2 + \varepsilon_{ij} ; \qquad (15.5.1)$$

$$i = 1, \cdots, p ; \;\; j = 1, \cdots, n_i .$$

Another case could involve a CRD with more than one covariate, the multiple covariance situation. The model involving a CRD and two covariates x_1 and x_2, each having a linear relationship with the response variable, would be

$$Y_{ij} = \mu + \tau_i + \beta_1 x_{1ij} + \beta_2 x_{2ij} + \varepsilon_{ij} ; \qquad (15.5.2)$$

$$i = 1, \cdots, p\ ;\ \ j = 1, \cdots, n_i\ .$$

Notice that this model assumes that the "slope" for each covariate is the same for each treatment which results in assuming that the treatment curves are equidistant.

Finally, we can use analysis of covariance with designs other than a CRD. As an illustration, consider an RCBD with p treatments, r blocks and a covariate x. If there is a linear relationship between the covariate and the response variable, then an appropriate model would be

$$Y_{ij} = \mu + \tau_i + \rho_j + \beta x_{ij} + \varepsilon_{ij}\ ;\ i = 1, \cdots, p;\ \ j = 1, \cdots, r. \quad (15.5.3)$$

In this model τ_i represents the treatment parameters, ρ_j represents the block parameters, and β is the common slope parameter. With this model we have

$$E(Y_{ij}) = \mu + \tau_i + \rho_j + \beta\, x_{ij}\ ,$$

which is essentially the equation of pr lines with common slope β and intercept $\mu + \tau_i + \rho_j$.

There clearly are other situations that can arise for analysis of covariance. Our purpose in this section has been to briefly present some of the possible situations, with the understanding that the reader can extend these ideas to other situations in a natural manner.

15.6 Design Considerations

The major purpose of an analysis of covariance is to adjust for the effect of the covariate and as a consequence reduce the experimental error. As we know, a reduced error can lead to more power in tests and shorter confidence intervals. To illustrate this reduction, consider the test scores data from Table 15.1.1 analyzed as an analysis of covariance in Example 15.2.1. From this analysis the experimental error is $s^2 = 41.298$ and the p—value for the f—test for testing the equality of the treatments is p = .0149. Table 15.6.1 provides the output for these same data analyzed as a CRD **without** the covariate. Note that without the adjustment of the covariate, the experimental error is $s^2 = 86.004$, and the p—value for the f—test for treatments is p = .6428. Clearly, having the covariate in the analysis is advantageous. In these illustrations there is a significant difference between the treatments in the analysis of covariance, but no significant difference in the CRD analysis without the covariate.

Analysis of covariance is an alternative to blocking. When blocking, we group the experimental units into blocks usually based upon the "value" of the blocking variable. If there are r blocks, then we can allow for only r different values of the blocking variable. Of course the purpose of blocking is to reduce the experimental error. In analysis of covariance instead of forming blocks we use a covariate to help reduce the experimental error. An advantage of using a covariate is that we can allow for a different value of the covariate variable for each observation.

Table 15.6.1 SAS Printout for Test Scores Data Without the Covariate

General Linear Models Procedure

Dependent Variable: SCORE

Source	DF	Sum of Squares	Mean Square	F Value	Pr > F
Model	2	77.26666667	38.63333333	0.45	0.6428
Error	27	2322.10000000	86.00370370		
Corrected Total	29	2399.36666667			

R-Square	C.V.	Root MSE	SCORE Mean
0.032203	11.25008	9.2738182	82.433333

Source	DF	Type I SS	Mean Square	F Value	Pr > F
INST	2	77.26666667	38.63333333	0.45	0.6428

Source	DF	Type III SS	Mean Square	F Value	Pr > F
INST	2	77.26666667	38.63333333	0.45	0.6428

Chapter 15 Exercises

Application Exercises

15–1 A small college desires to compare the salaries of faculty in three areas: science, humanities, and business. Several faculty, all of whom have doctorates, are selected from each area, and their salaries, in thousands of dollars, are recorded. Since years of experience is expected to affect salary that is also recorded. The results are given as follows:

Science		Humanities		Business	
Salary	Years Exp.	Salary	Years Exp.	Salary	Years Exp.
35	2	68	28	46	5
47	7	54	17	39	1
65	22	38	6	47	7
51	14	59	19	63	18
45	4	47	10	68	22
		36	5		
		32	4		

(a) Consider an analysis of covariance with three treatments (areas), covariate (years of experience), and response variable (salary). Use a 0.05 level to test the equal slope assumption.

(b) Using a 0.05 level, test for differences between the salaries for the three areas.

(c) Find a 95% confidence interval for the common slope.

(d) Use the Bonferroni procedure to find 95% simultaneous confidence intervals for the pairwise difference in the salaries for the three areas.

15–2 A large convenience store chain wanted to compare the effect of two advertising methods (newspaper, radio) on the sales in stores. For comparison purposes a control group (no advertising) was added to the study. Fifteen stores in different regions of the country were selected for the study and five were randomly assigned to each advertising method (newspaper, radio, none). Sales in thousands were recorded during the week before the advertising, label as x, and during the week after the advertising, label as y. The results are as follows:

Advertising Methods

Newspaper			Radio			None		
Store	x	y	Store	x	y	Store	x	y
1	34	39	1	27	30	1	22	24
2	26	30	2	34	35	2	31	29
3	18	25	3	22	24	3	34	35
4	15	23	4	33	37	4	28	24
5	31	36	5	29	33	5	19	21

(a) Using an analysis of covariance model, test for equal slopes. Use a 0.05 level. Here y is the response variable and x is the covariate.

(b) Using a 0.05 level, test for differences between the three advertising methods.
(c) Perform a Bonferroni multiple comparison for the advertising methods. Use a 0.05 level.
(d) An alternative to analysis of covariance for these data is to take the difference d = y − x. We can then analyze these differences using the traditional model for a CRD. Perform the f–test for treatments based upon the differences and a CRD model. Use a 0.05 level. Follow the f–test with a Tukey multiple comparison using a 0.05 level. Compare the results with those obtained in parts (b) and (c).

15–3 Two drugs (A and B) and a placebo (control) were compared for their effects for testing allergies. Six people suffering from allergies were available for the study. Each person used each drug and placebo during three different one–week periods. The order the person used the drugs and placebo was randomized. People had the severity of their allergy rated on a twenty–point scale before taking the drug, label as x, and after the drug, label as y. The results are as follows:

Treatment

Drug A			Drug B			Placebo		
Person	x	y	Person	x	y	Person	x	y
1	12	16	1	10	18	1	15	13
2	10	11	2	12	17	2	12	10
3	8	9	3	8	15	3	19	18
4	4	7	4	6	14	4	12	14
5	7	12	5	13	19	5	9	12
6	3	17	6	5	12	6	17	16

(a) Consider an analysis of covariance model with three treatments, blocks (person), covariate (x), and response variable (y). Check the equal slopes assumption using a 0.05 level.
(b) Find the ANOVA table for the equal slope model. Use a 0.05 level to test for treatments, blocks, and covariate.
(c) Perform a Bonferroni multiple comparison for the treatments using a 0.05 level.
(d) If we ignore the effect of the covariate we can use the RCBD model. Using this model perform the f–test for treatments. Compare the results with those obtained in part (b).

15–4 Refer to the data in Exercise 15–1. Use reparametrized regression models to answer the same questions given in Exercise 15–1.

15–5 Refer to the SAS Printout in Table 15.2.2 that provides an analysis for the test score data in Table 15.1.1. Find and interpret a 95% confidence interval for the slope.

15–6 Refer to the Salary Discrimination Problem in Section 2 of Chapter 6. The question raised concerning equality in male and female salary structures and the data set given in Table 6.2.1 places us in a multiple covariance setting. Here the "treatments" are sex and the amount of education and work experience are the two covariates.

(a) From an analysis of covariance point of view, the model used in the text assumed the two treatment planes to be parallel. Test this assumption at the 0.05 significance level.

(b) If the test in (a) confirms the validity of model (6.2.1), reformat the regression ANOVA displayed in Example 6.3.1 to a covariance ANOVA, combining covariates sums of squares as one source of variation. What conclusions do you draw from your table?

15–7 In Example 15.2.1 a confidence interval for $\tau_2 - \tau_1$ is given. Use regression model (15.2.7) to construct the same interval. Also find a 95% confidence interval for $\tau_3 - \tau_1$, using this regression model.

Theoretical Exercises

15–8 Verify that (15.2.1) constitute the normal equations for the analysis of covariance model (15.1.1.)

15–9 Show that (15.2.2) provides a solution to the normal equations in (15.2.1.)

15–10 Show that for the simple analysis of covariance model (15.1.1) the following are estimable:

(i) β (ii) $\mu + \tau_1$ (iii) treatment contrasts.

15–11 Show that for the simple analysis of covariance model (15.1.1) SSt = SStrt (adj) + SScov (unadjusted) + SSe.

15–12 Consider the non–equal slope covariance model (15.3.1).

(a) Show that the **X** matrix has rank 2p.
(b) Show that the β_i are estimable.

15–13 Find the relationship between the parameters of the non–equal slopes covariance model (15.3.1) and the reparametrized regression model (15.3.3).

15–14 Verify the expression for $E(AM_i)$ given in (15.4.3).

15–15 Derive the formula (15.4.4) for the standard error of the difference between two adjusted means.

15–16 Refer to the extended regression model (15.3.3) with x_1, x_2, and x_3 defined as in (15.2.7). Verify that the slopes of the three treatment lines theoretically have a common value when $\alpha^*_{13} = \alpha^*_{23} = 0$.

15–17 In Section 15.5, other traditional covariance models are highlighted. For each of these models write down corresponding regression models together with their extended (interaction) models used for checking "parallelism" in the treatment effects. Include the definitions for the independent x_i variables.

(a) Model (15.5.1) with $p = 2$.
(b) Model (15.5.2) with $p = 3$.
(c) Model (15.5.3) with $p = 3$, $r = 2$.

CHAPTER 16

RANDOM AND MIXED MODELS

16.1 Random Effects

In previous chapters, the experimental design models that we considered assumed that the effects for factors were fixed constants (except for the random error term). In these models the levels of the factors were predetermined by the experimenter. Such factors are called **fixed** and the associated effects are called **fixed effects**. For example, in an experiment aimed at comparing three brands of gasoline the factor brand is fixed since the levels of the factor are predetermined by the experimenter. Notice that our interest is only in these three brands and not in other possible brands, which is a characteristic of a fixed factor.

In some experiments the assumption that the levels of the factors are predetermined or controlled is not realistic or desired. For example, in an experiment comparing brands of gasolines (treatments) with cars used as blocks, it may be more appropriate to assume that the cars (blocks) were randomly selected from a population. When the levels of a factor are selected at random from a population, then the factor is called a **random factor**. The effects for a random factor are then random variables.

Our main purpose in this chapter is to consider experimental design models that involve random factors. There are three possible cases. A **fixed effects** model (Model I) is one in which all factors are fixed, that is, the levels of the factors are predetermined and the model effects (except for ε) are assumed to be unknown constants. All of the models we have considered thus far fit into this case. A **random effects model** (Model II) is one in which all of the factors are random. Except for the overall mean, μ, all of the terms in the model are assumed to be random variables. A **mixed effects model** (Model III) is one in which at least one factor is fixed and at least one is random.

To see how these new ideas impact on experimental design concepts, we first illustrate with a simple random effects model. Consider a CRD with p treatments and n_i experimental units assigned to treatment i. We assume that the treatment factor is random, that is, the p levels are randomly chosen from a large population. Since the treatment factor is random, the treatment effects, τ_i, will be random variables. The model and assumptions for this case are:

$$Y_{ij} = \mu + \tau_i + \varepsilon_{ij}\ ;\ \ i = 1,\cdots, p;\ \ j = 1,\cdots, n_i\ ,$$

where the τ_i and ε_{ij} are uncorrelated random variables,

$$E(\tau_i) = 0, \ Var(\tau_i) = \sigma_\tau^2 \ \text{for all } i, \tag{16.1.1}$$

$$E(\varepsilon_{ij}) = 0, \ \text{and} \ Var(\varepsilon_{ij}) = \sigma_\varepsilon^2 \ \text{for all } i,j.$$

To perform inferences we will require the normality assumptions, which can be written as: τ_i are independent $N(0,\sigma_\tau^2)$, ε_{ij} are independent $N(0,\sigma_\varepsilon^2)$, and the τ_i and ε_{ij} are independent. With these normality assumptions we note that Y_{ij} is $N(\mu,\sigma_\tau^2 + \sigma_\varepsilon^2)$. We also note that in random effects models some observations are dependent, while others might be independent. In model (16.1.1) observations within a treatment are dependent, while observations in different treatments are independent. That is, $Cov(Y_{ij}, Y_{ik}) \neq 0$ for $j \neq k$, and $Cov(Y_{ij}, Y_{km}) = 0$ for $i \neq k$.

The inferences that are made in model (16.1.1) involve the parameters σ_τ^2 and σ_ε^2, which are sometimes called the **variance components** in the model. Since hypotheses usually involve parameters, and since the τ_i's are random variables, the hypothesis we have used in the past for treatments ($H_0 : \tau_1 = \cdots = \tau_p$) is no longer appropriate. Instead, for the random factor treatments we test $H_0 : \sigma_\tau^2 = 0$. If this hypothesis is true, then the variation in the observations, Y_{ij} is due only to the random error term and the treatments have no effect on the response variable. If $\sigma_\tau^2 > 0$, then treatments have an effect on the observations. Notice that multiple comparisons on the levels of treatments would be inappropriate, since the levels are not fixed.

We close this section with a numerical example involving a random factor. The analysis of these data will be completed later in the exercises.

Example 16.1.1 It is desired to determine if large differences exist in the grading of term papers by English professors. From a large group of English professors, four are randomly selected. Twenty—four students who received the same grade in freshmen English were asked to write a term paper on the same topic. The 24 papers were randomly assigned to the four professors, six papers per professor. The grades for the term papers on a 0—100 scale are given in Table 16.1.1.

Table 16.1.1 Grades for Term Papers

Professor	1	89	98	80	82	88	97
	2	94	80	80	89	78	77
	3	99	91	69	81	84	78
	4	72	77	61	87	70	75

The experiment utilizes a CRD, and it is appropriate to consider the treatment factor (professor) as random. To determine grading differences, our primary interest will be on the variance for the treatments, σ_τ^2. A large σ_τ^2 implies that there is a large difference between the grades assigned by the various professors.

16.2 Mixed Effects Models

Our goal in this section is to introduce a mixed effects model and to discuss how the assumptions of the model affect the statistical inference procedures. To illustrate these ideas we will consider an RCBD in which the treatments are fixed and the blocks are random. This leads to a mixed model, since at least one factor is fixed and at least one is random.

For a mixed model RCBD having p fixed treatments and r random blocks the model we use has the familiar functional form

$$Y_{ij} = \mu + \tau_i + \beta_j + \varepsilon_{ij}; \quad i = 1,\cdots,p; \quad j = 1,\cdots, r.$$

In this model, as before, μ and $\tau_1,\cdots,\tau_p$ are unknown parameters (fixed constants). However, since blocks are random, we assume that the β_j and ε_{ij} are uncorrelated random variables, where $E(\beta_j) = 0$, $Var(\beta_j) = \sigma_\beta^2$ for all j, and $E(\varepsilon_{ij}) = 0$, $Var(\varepsilon_{ij}) = \sigma_\varepsilon^2$ for all i, j.

To perform statistical inference we add the normality assumptions. With these additional assumptions the β_j's are independent $N(0, \sigma_\beta^2)$, the ε_{ij}'s are independent $N(0, \sigma_\varepsilon^2)$ random variables, and the β_j's and ε_{ij}'s are independent of each other. It follows that the Y_{ij}'s have a $N(\mu + \tau_i, \sigma_\beta^2 + \sigma_\varepsilon^2)$ distribution. We note that observations in different blocks are independent, while observations within the same block are dependent whenever $\sigma_\beta^2 \neq 0$. In fact the covariance for observations within the same block is:

$$\mathrm{Cov}(Y_{ij}, Y_{tj}) = E\{[Y_{ij} - E(Y_{ij})]\,[Y_{tj} - E(Y_{tj})]\}$$

$$= E\{[\mu + \tau_i + \beta_j + \varepsilon_{ij} - (\mu+\tau_i)]\,[\mu + \tau_t + \beta_j + \varepsilon_{tj} - (\mu + \tau_t)]\}$$

$$= E[(\beta_j + \varepsilon_{ij})\,(\beta_j + \varepsilon_{tj})] = E[\beta_j^2] + E(\beta_j)\,E(\varepsilon_{tj}) + E(\varepsilon_{ij})\,E\,(\beta_j)$$

$$+ E(\varepsilon_{ij})\,E\,(\varepsilon_{tj})$$

$$= \sigma_\beta^2 + 0 + 0 + 0 = \sigma_\beta^2.$$

For the RCBD with fixed treatments and random blocks the tests of interest are as follows. The hypothesis "no treatment effect" is represented by the hypothesis $H_0 : \tau_1 = \cdots = \tau_p$, and "no block effect" is represented by the hypothesis $H_0 : \sigma_\beta^2 = 0$. Confidence intervals can be found for treatment contrasts and for the "variance components" σ_β^2 and σ_ε^2. Multiple comparisons can be performed on the fixed effects, treatments; however, since the β_j's are random variables, it is inappropriate to consider multiple comparisons or confidence intervals on the blocks effects, β_j.

Now that we have introduced random effects and mixed effects models, we should consider how the analysis of data associated with these models differs from the analysis for fixed effects models. We mention some of the differences in this section and illustrate them more fully in the next section. Some ramifications of random and mixed effects models in comparison to fixed effects models are listed below.

1. (Sum of Squares) The sum of squares for a factor remains the same whether the factor is fixed or random. The same is true for the error and total sum of squares. As a consequence, formulas for sums of squares can be found for a random or mixed effects model in the same manner as a fixed effects model.

2. (Hypotheses) Hypotheses for fixed effects factors involve the model parameters (e.g. $H_0 : \tau_1 = \cdots = \tau_p$), while hypotheses for

random effects factors involve the variance component (e.g. H_0 : $\sigma_\beta^2 = 0$).

3. (Estimation, Confidence Intervals, Multiple Comparisons) For fixed effects factors it is appropriate to estimate the parameters associated with the factor, to find confidence intervals for estimable functions involving these parameters, and to perform multiple comparisons on the levels of the factor. For random effects factors (say, factor B with model effects $\beta_1, \cdots, \beta_r$) the situation is different. There is little interest in estimating the β_j's since these are random variables rather than parameters. Similarly, confidence intervals or multiple comparisons on the β_j's are also inappropriate. A confidence interval for the variance component, σ_β^2, could be of interest.

4. (Distribution of the Observations) A major change resulting from the presence of one or more random effects factor is that the distribution of the Y observations is different than the distribution in the fixed effects case. The presence of random effects generally causes some observations to be dependent, while in the fixed effects case all observations are independent. Also, the mean and variance of the observations are generally different for a random or mixed model than for a fixed effects model.

5. (Distribution of Sum of Squares, E(MS), F) The presence of a random effects factor may alter the distribution of a sum of squares. If so, the form of the expected mean squares, and as a consequence, the form of the f—test may change.

In the remainder of this chapter we will continue to explore modifications resulting from the presence of random effects factors in a model. First, however, we will introduce another concept not seen before, that of **nesting**.

16.3 Introduction to Nested Designs—Fixed Case

In this section we discuss the concept of nesting in a design. In addition, we will use designs with nesting to illustrate the differences in the statistical analysis for the fixed effects case versus the mixed effects case.

To understand the concept of nesting we contrast it with the concept of crossed factors. Recall that two factors A and B are **crossed** if each level of factor A occurs with each level of factor B, and vice versa. The two factors are **nested** (or **hierarchical**) if each level of one factor occurs with only **one** level of the other factor. More specifically, factor **B is nested within factor A** if each level of factor B occurs with only one level of factor A. An easy example of nesting is when factor A is gender (female, male) and factor B is person. Clearly each person (level of B) occurs with only one level of gender (level of A); therefore, person is nested within gender.

To further understand the difference between crossed and nested factors we present an illustration. Suppose we have an experiment which consists of 3 machines (levels of factor A), each operated by the **same** 2 operators (levels of factor B). Further suppose two observations are obtained for each machine–operator combination. The model for this experiment is that of a factorial within a CRD. The 12 observations in this experiment can be classified in the following two–way table.

		A — Machines	
	1	2	3
Operators B (Mary) 1	y_{111} y_{112}	y_{211} y_{212}	y_{311} y_{312}
Operators B (Sam) 2	y_{121} y_{122}	y_{221} y_{222}	y_{321} y_{322}

For convenience we have named the two operators Mary and Sam. Since each operator (level of B) uses each machine (level of A) and vice versa, factors A and B are crossed.

Now suppose we have an experiment which consists of 3 machines (levels of factor A) and 6 operators (levels of factor B). Further suppose each operator uses only one machine with two operators to a machine and two observations per operator. The 12 observations for this experiment can be represented by the branching scheme given in Figure 16.3.1. The operators are labelled as Mary, Sam, Pete, Joe, Lucy, and Sue. Notice that each operator (level of B) uses only one machine (level of A); therefore, factor B is nested within factor A. This is an illustration of a **balanced** nested design in that the number of levels of B associated with each level of A is the same, and the number of observations associated with each level of B within each level of A is the same.

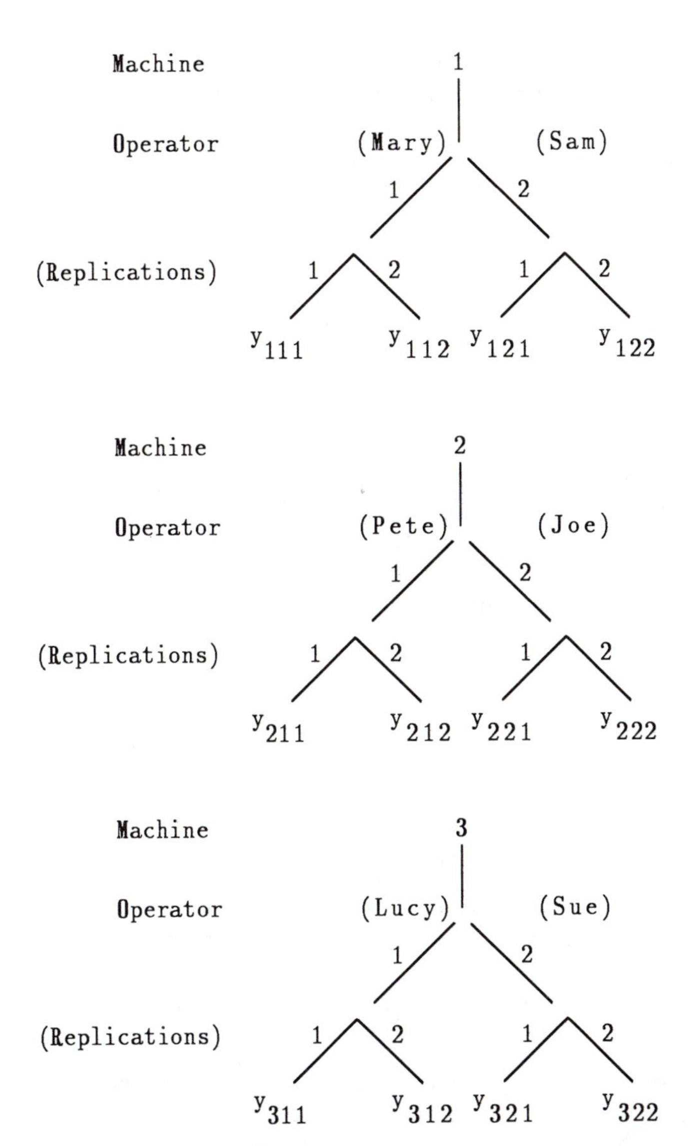

Figure 16.3.1 Observations in Nested Experiment

Our goal now is to describe an appropriate model for a balanced nested design with factor A having a levels, and factor B having b levels nested within each level of A. We assume c replications at each combination of A and B. For the present we will consider both factors A and B to be fixed. (Later in the chapter we will consider factor B to be random, which is the more common case.) In writing the model we use parentheses to identify a nested factor. The model is then

$$Y_{ijk} = \mu + \alpha_i + \beta_{j(i)} + \varepsilon_{k(ij)}$$

$$i = 1,2,\cdots,a;\ j = 1,2,\cdots,b;\ k = 1,2,\cdots,c; \tag{16.3.1}$$

$$\sum_{i=1}^{a} \alpha_i = 0 \ \text{ and } \ \sum_{j=1}^{b} \beta_{j(i)} = 0,\ \ i = 1,2,\cdots,a;$$

where

Y_{ijk} is the response to the kth observation of the jth level of B within the ith level of A.

μ is an overall mean response.

α_i is the effect of the ith level of A.

$\beta_{j(i)}$ is the effect of the jth level of B within the ith level of A.

$\varepsilon_{k(ij)}$ is the random error associated with the kth observation within the jth level of B within the ith level of A. Furthermore, we assume $E(\varepsilon_{k(ij)}) = 0$ and $Var(\varepsilon_{k(ij)}) = \sigma^2$ for every i, j, and k.

We have used the restricted form of the model to ensure that linear contrasts in the α_i's are estimable.

Next we consider the analysis associated with the balanced two–factor nested design, namely model (16.3.1). When $a = 3$, $b = 2$, $c = 2$, the model in matrix form is given by:

$$\underset{\mathbf{Y}}{\begin{bmatrix} Y_{111} \\ Y_{112} \\ Y_{121} \\ Y_{122} \\ Y_{211} \\ Y_{212} \\ Y_{221} \\ Y_{222} \\ Y_{311} \\ Y_{312} \\ Y_{321} \\ Y_{322} \end{bmatrix}} = \underset{\mathbf{X}}{\begin{bmatrix} 1 & 1 & 0 & 0 & 1 & 0 & 0 & 0 & 0 & 0 \\ 1 & 1 & 0 & 0 & 1 & 0 & 0 & 0 & 0 & 0 \\ 1 & 1 & 0 & 0 & 0 & 1 & 0 & 0 & 0 & 0 \\ 1 & 1 & 0 & 0 & 0 & 1 & 0 & 0 & 0 & 0 \\ 1 & 0 & 1 & 0 & 0 & 0 & 1 & 0 & 0 & 0 \\ 1 & 0 & 1 & 0 & 0 & 0 & 1 & 0 & 0 & 0 \\ 1 & 0 & 1 & 0 & 0 & 0 & 0 & 1 & 0 & 0 \\ 1 & 0 & 1 & 0 & 0 & 0 & 0 & 1 & 0 & 0 \\ 1 & 0 & 0 & 1 & 0 & 0 & 0 & 0 & 1 & 0 \\ 1 & 0 & 0 & 1 & 0 & 0 & 0 & 0 & 1 & 0 \\ 1 & 0 & 0 & 1 & 0 & 0 & 0 & 0 & 0 & 1 \\ 1 & 0 & 0 & 1 & 0 & 0 & 0 & 0 & 0 & 1 \end{bmatrix}} \underset{\boldsymbol{\beta}}{\begin{bmatrix} \mu \\ \alpha_1 \\ \alpha_2 \\ \alpha_3 \\ \beta_{1(1)} \\ \beta_{2(1)} \\ \beta_{1(2)} \\ \beta_{2(2)} \\ \beta_{1(3)} \\ \beta_{2(3)} \end{bmatrix}} + \underset{\boldsymbol{\varepsilon}}{\begin{bmatrix} \varepsilon_{1(11)} \\ \varepsilon_{2(11)} \\ \varepsilon_{1(12)} \\ \varepsilon_{2(12)} \\ \varepsilon_{1(21)} \\ \varepsilon_{2(21)} \\ \varepsilon_{1(22)} \\ \varepsilon_{2(22)} \\ \varepsilon_{1(31)} \\ \varepsilon_{2(31)} \\ \varepsilon_{1(32)} \\ \varepsilon_{2(32)} \end{bmatrix}}$$

Upon inspection of the **X** matrix we find the rank to be $1 + (3-1) + 3\ (2-1) = 6$. In general the rank of **X** for the balanced case is $1 + (a-1) + a(b-1) = ab$.

Using m, a_i and $b_{j(i)}$ for the sample form of the model, the normal equations when $a = 3$, $b = 2$, $c = 2$ are

$$\begin{cases} 12m + 4\sum_{i=1}^{3} a_i + 2\sum_{j=1}^{2} b_{j(1)} + 2\sum_{j=1}^{2} b_{j(2)} + 2\sum_{j=1}^{2} b_{j(3)} = y_{...} \\ 4m + 4a_i + 2\sum_{j=1}^{2} b_{j(i)} = y_{i..}, \quad i = 1,2,3 \\ 2m + 2a_i + 2b_{j(i)} = y_{ij.}, \quad i = 1,2,3; \quad j = 1,2 \end{cases}$$

When we impose the conditions $\sum_{i=1}^{3} a_i = 0$ and $\sum_{j=1}^{2} b_{j(i)} = 0$, $i = 1,2,3$ we obtain the following solution:

$$\begin{cases} m = \bar{y}_{...} \\ a_i = \bar{y}_{i..} - \bar{y}_{...} \\ b_{j(i)} = \bar{y}_{ij.} - \bar{y}_{i..} \end{cases} \qquad (16.3.2)$$

To find the error sum of squares we observe that $\hat{y}_{ijk} = m + a_i + b_{j(i)} = \bar{y}_{ij.}$, so

$$SSe = \sum_{i=1}^{a}\sum_{j=1}^{b}\sum_{k=1}^{c} (y_{ijk} - \bar{y}_{ij.})^2. \qquad (16.3.3)$$

Its degrees of freedom is $n - \text{rank}\,\mathbf{X} = abc - ab = ab(c-1)$.

To find the ANOVA table we note that the hypotheses of interest for the fixed effects case are $H_0 : \alpha_i = 0$ for all i, and $H_0 : \beta_{j(i)} = 0$ for all i, j. In the exercises the principle of conditional error is used to find the following sum of squares

$$SSa = \sum_{i=1}^{a} bc\,(\bar{y}_{i..} - \bar{y}_{...})^2, \quad a - 1 \text{ degrees of freedom;}$$

$$SSb(a) = \sum_{i=1}^{a}\sum_{j=1}^{b} c(\bar{y}_{ij.} - \bar{y}_{i..})^2, \quad a(b-1) \text{ degrees of freedom.} \qquad (16.3.4)$$

The total sum of squares is $SSt = \sum_{i=1}^{a}\sum_{j=1}^{b}\sum_{k=1}^{c} (y_{ijk} - \bar{y}_{...})^2$, and it is easy to show that $SSt = SSa + SSb(a) + SSe$. Notice that we are using the symbol $b(a)$ to represent factor B nested in A. If we assume a normal distribution, then the f—tests are: reject $H_0 : \alpha_i = 0$ for all i if $f = MSa/MSe \geq f_{\alpha,a-1,ab(c-1)}$, and reject $H_0 : \beta_{j(i)} = 0$ for all i,j if $f = MSb(a)/MSe \geq f_{\alpha,a(b-1),ab(c-1)}$.

The expected mean squares for the factors are easily found in the exercises, and these values along with the sum of squares are summarized in the following ANOVA table.

Table 16.3.1 ANOVA Table for Balanced Two Factor Nested Design – Fixed Case

Source	df	SS	MS	f	E(MS)
A	a−1	SSa	MSa	MSa/MSe	$\sigma^2 + \frac{bc}{a-1}\sum_{i=1}^{a} \alpha_i^2$
B(A)	a(b−1)	SSb(a)	MSb(a)	MSb(a)/MSe	$\sigma^2 + \frac{c}{a(b-1)}\sum_{i=1}^{a}\sum_{j=1}^{b} \beta_{j(i)}^2$
Error	ab(c−1)	SSe	MSe		σ^2
Total	abc−1	SSt			

Confidence intervals for contrasts in factor A, $\sum_{i=1}^{a} c_i\alpha_i$, $\sum_{i=1}^{a} c_i = 0$ are easily found and have the form:

$$\sum_{i=1}^{a} c_i\bar{y}_{i\cdot\cdot} \pm t_{\frac{\alpha}{2},ab(c-1)} \sqrt{\frac{s^2}{bc} \sum_{i=1}^{a} c_i^2} . \tag{16.3.5}$$

Similarly, a Tukey multiple comparison for factor A has the form:

$$\text{Decide } \alpha_i \neq \alpha_t \text{ if } |\bar{y}_{i\cdot\cdot} - \bar{y}_{t\cdot\cdot}| \geq q_{\alpha,a,ab(c-1)} \sqrt{\frac{s^2}{bc}} . \tag{16.3.6}$$

An illustration of the analysis of a balanced two—factor nested design in the fixed case is given in the next example.

Example 16.3.1 A company uses three machines (lathes) to shape metal pieces, and an experiment is designed to compare the three machines. Six operators are available to use the machines, but there is insufficient time to have all six use each machine. As a consequence, two operators are assigned to each machine. Each operator uses a machine twice to prepare a metal piece. The response variable recorded is the absolute difference in mm between the actual diameter of the piece and the desired diameter. The results are given in Table 16.3.2.

Table 16.3.2 Operator, Machine, Nested Data

		Machine 1	Machine 2	Machine 3
Operator	1	.62 .34	.21 .32	.73 .64
	2	.56 .41	.14 .23	.52 .59

We observe that the design involves two factors: machine, and operator, which is nested within machine. It is appropriate to consider machine as fixed, since these are the only machines the company has. For the present, we consider operator as a fixed factor. It is important to note that this is appropriate if there are only six operators in the company. We use SAS to carry out the calculations for f—tests and for a Tukey multiple comparison for the machines. The SAS commands and printout are given in Table 16.3.3. Note that in the model statement we indicate that operator is nested within machine by writing "OPER(MACH)". From the printout we see that there is a significant difference between at least two machines (p = .0051). The Tukey multiple comparison indicates that machine 2 has a mean that is significantly different from machines 1 and 3. There is no significant difference between the operators (p = .5876). In the next section we will analyze this same data using the assumption that the operator factor is random.

Table 16.3.3 SAS Commands and Printout for Nested Data — Fixed Model

SAS Commands

```
OPTIONS LS = 80;
DATA TXNESTF;
INPUT OPER MACH Y @@;
CARDS;
1 1 .62  1 1 .34  1 2 .21  1 2 .32  1 3 .73  1 3 .64
2 1 .56  2 1 .41  2 2 .14  2 2 .23  2 3 .52  2 3 .59
PROC GLM;  CLASS OPER MACH;
MODEL Y = MACH OPER (MACH);
MEANS MACH / TUKEY;
```

SAS Printout

```
                     General Linear Models Procedure

Dependent Variable: Y
                                  Sum of            Mean
Source                  DF       Squares          Square   F Value     Pr > F

Model                    5    0.34497500      0.06899500      6.17     0.0233

Error                    6    0.06705000      0.01117500

Corrected Total         11    0.41202500

                  R-Square          C.V.        Root MSE               Y Mean

                  0.837267      23.88969       0.1057119            0.4425000

Source                  DF     Type I SS     Mean Square   F Value     Pr > F

MACH                     2    0.32165000      0.16082500     14.39     0.0051
OPER(MACH)               3    0.02332500      0.00777500      0.70     0.5876

Source                  DF   Type III SS     Mean Square   F Value     Pr > F

MACH                     2    0.32165000      0.16082500     14.39     0.0051
OPER(MACH)               3    0.02332500      0.00777500      0.70     0.5876
```

```
                   General Linear Models Procedure

              Tukey's Studentized Range (HSD) Test for variable: Y

          NOTE: This test controls the type I experimentwise error rate, but
                generally has a higher type II error rate than REGWQ.

                      Alpha= 0.05  df= 6  MSE= 0.011175
                  Critical Value of Studentized Range= 4.339
                    Minimum Significant Difference= 0.2293

          Means with the same letter are not significantly different.

               Tukey Grouping             Mean      N  MACH

                            A          0.62000      4  3
                            A
                            A          0.48250      4  1

                            B          0.22500      4  2
```

16.4 Nested Designs — Mixed Model

In many cases involving two–factor nested designs it is appropriate to assume that the nested factor, say B, is random. If we assume the other factor, say A, is fixed, then we have a mixed effects model. Our purpose in this section is to consider the analysis for this mixed effects model, and to contrast this analysis with the fixed effects case presented in Section 16.3. If we assume c replications at each combination of the levels of A and B, and if we include the normality assumptions needed for statistical inference, then the appropriate model with assumptions is:

$$Y_{ijk} = \mu + \alpha_i + \beta_{j(i)} + \varepsilon_{k(ij)};\ i = 1,\cdots,a;$$

$$j = 1,\cdots,b;\ k = 1,\cdots,c;\ \text{where} \tag{16.4.1}$$

$\sum_{i=1}^{a} \alpha_i = 0$, $\beta_{j(i)}$'s are independent $N(0,\sigma^2_\beta)$ random variables, and $\varepsilon_{k(ij)}$'s are independent $N(0,\sigma^2_\varepsilon)$ random variables. The $\beta_{j(i)}$'s and the $\varepsilon_{k(ij)}$'s are assumed to be independent.

For the mixed effects case the statistical inferences of major interest are tests for:

$$H_0 : \alpha_i = 0,\ \text{for all } i \text{ (i.e. no factor A effect)},$$

$$H_0 : \sigma^2_\beta = 0 \text{ (i.e. no factor B effect)}.$$

Multiple comparisons or confidence intervals for contrasts in factor A may also be of interest.

We need to consider the effect of the assumptions in (16.4.1) on the analysis. We have previously pointed out that the sums of squares and degrees of freedom are the same in the fixed, random, and mixed cases; therefore, the formulas for SSa, SSb(a), SSe, and SSt given in Table 16.3.1 for the fixed case remain valid. These quantities remain the same since sums of squares are found by minimizing certain numerical quantities based upon the observed data. The distributional assumptions of mixed and random effects models do not change the observed data; therefore, the sums of squares are unchanged.

The greatest changes that result from assuming factor B random involve distributional forms. First, we note that Y_{ijk} is distributed as $N(\mu + \alpha_i,\ \sigma^2_\beta + \sigma^2_\varepsilon)$. Next we note that observations involving different

$\beta_{j(i)}$ values are independent, but observations that involve the same $\beta_{j(i)}$ term will be dependent (if $\sigma_\beta^2 > 0$). In fact, the covariance between Y_{ijk} and $Y_{ij\ell}$ is easily found as follows:

$$
\begin{aligned}
\mathrm{Cov}(Y_{ijk}, Y_{ij\ell}) &= E\left\{\left[Y_{ijk} - E(Y_{ijk})\right]\left[Y_{ij\ell} - E(Y_{ij\ell})\right]\right\} \\
&= E\Big\{\left[\mu + \alpha_i + \beta_{j(i)} + \varepsilon_{k(ij)} - (\mu + \alpha_i)\right] \\
&\qquad\qquad \left[\mu + \alpha_i + \beta_{j(i)} + \varepsilon_{\ell(ij)} - (\mu + \alpha_i)\right]\Big\} \\
&= E\left[\beta_{j(i)}^2 + \beta_{j(i)}\,\varepsilon_{\ell(ij)} + \varepsilon_{k(ij)}\,\beta_{j(i)} + \varepsilon_{k(ij)}\,\varepsilon_{\ell(ij)}\right] \\
&= \sigma_\beta^2 + 0 + 0 + 0 + 0 = \sigma_\beta^2.
\end{aligned}
$$

Using this fact it is shown in the exercises that

$$\bar{Y}_{ij\cdot} \text{ is } N(\mu + \alpha_i, (\sigma_\varepsilon^2 + c\,\sigma_\beta^2)/c).$$

Since the $\bar{Y}_{ij\cdot}$'s are independent and since $\bar{Y}_{i\cdot\cdot} = \frac{1}{b}\sum_{j=1}^{b} \bar{Y}_{ij\cdot}$ It follows that $E(\bar{Y}_{i\cdot\cdot}) = \mu + \alpha_i$ and $\mathrm{Var}(\bar{Y}_{i\cdot\cdot}) = \dfrac{\sigma_\varepsilon^2 + c\sigma_\beta^2}{bc}$. Therefore, the $\bar{Y}_{i\cdot\cdot}$'s are independent $N(\mu + \alpha_i, (\sigma_\varepsilon^2 + c\sigma_\beta^2)/bc)$ random variables. In the exercises it is shown that $\bar{Y}_{\cdot\cdot\cdot}$ is $N(\mu, (\sigma_\varepsilon^2 + c\sigma_\beta^2)/abc)$.

How do we obtain tests for H_0 : all $\alpha_i = 0$ and $H_0 : \sigma_\beta^2 = 0$? Since our distribution forms are different than the fixed effects case, we can no longer find f—tests using the principle of conditional error. Instead we will use the expected mean squares to find the appropriate form of the f—test. Recall that in the f—test the E(MS) have the property that E(MS) for the numerator (when H_0 is true) equals the E(MS) of the denominator. We use this property to find the form of f—tests for mixed and random effects models.

Expressions for the E(MS) can be calculated using statistical properties. We illustrate this for MSA. First,

$$SSA = bc \sum_{i=1}^{a} (\bar{Y}_{i..} - \bar{Y}_{...})^2 = bc \sum_{i=1}^{a} \bar{Y}_{i..}^2 - abc\, \bar{Y}_{...}^2 .$$

$$\text{So } E(SSA) = bc \sum_{i=1}^{a} E(\bar{Y}_{i..}^2) - abc\, E(\bar{Y}_{...}^2)$$

$$= bc \sum_{i=1}^{a} \{Var(\bar{Y}_{i..}) + [E(\bar{Y}_{i..})]^2\} - abc\, \{Var(\bar{Y}_{...})]^2 + [E(\bar{Y}_{...}^2)]\}$$

$$= bc\, [\frac{a(\sigma_\varepsilon^2 + c\sigma_\beta^2)}{bc} + \sum_{i=1}^{a} (\mu+\alpha_i)^2] - abc\, [\frac{(\sigma_\varepsilon^2 + c\sigma_\beta^2)}{abc} + \mu^2]$$

$$= (a-1)\, \sigma_\varepsilon^2 + (a-1)\, c\sigma_\beta^2 + bc \sum_{i=1}^{a} \alpha_i^2 .$$

Since MSA = SSA/(a − 1) it follows that

$$E(MSA) = \sigma_\varepsilon^2 + c\sigma_\beta^2 + \frac{bc}{a-1} \sum_{i=1}^{a} \alpha_i^2 . \tag{16.4.2}$$

In a similar manner we can find the other E(MS) values

$$E(MSE) = \sigma_\varepsilon^2,\ E(MSB(A)) = \sigma_\varepsilon^2 + c\sigma_\beta^2 . \tag{16.4.3}$$

Using the principle that the E(MS) of the numerator of the f—test equals the E(MS) of the denominator when H_0 is true, our f—tests are as follows:

Reject $H_0 : \alpha_i = 0$ for all i if $f = MSa/MSb(a) \geq f_{\alpha,a-1,a(b-1)}$,

Reject $H_0 : \sigma_\beta^2 = 0$ if $f = MSb(a)/MSe \geq f_{\alpha,a(b-1),ab(c-1)}$.

Note that the f—test for the factor A does not involve the mean square error. This is a consequence of a mixed or random effects model, namely the denominator for the f—test might not be the mean square error. We summarize our findings in the ANOVA Table 16.4.1.

Table 16.4.1 Table for Balanced Two Factor Nested Design — A fixed, B random

Source	df	SS	MS	f	E(MS)
A	a–1	SSa	MSa	MSa/MSb(a)	$\sigma_\epsilon^2 + c\sigma_\beta^2 + \frac{bc}{a-1} \sum_{i=1}^{a} \alpha_i^2$
B(A)	a(b–1)	SSb(a)	MSb(a)	MSb(a)/MSe	$\sigma_\epsilon^2 + c\sigma_\beta^2$
Error	ab(c–1)	SSe	MSe		σ_ϵ^2
Total	abc–1	SSt			

Formulas for SSa and SSb(a) are given in (16.3.4) and the formula for SSe is given in (16.3.3). Also, SSt = SSa + SSb(a) + SSe.

To find a multiple comparison for the levels of factor A we note that the $\bar{Y}_{i..}$'s are independent $N(\mu + \alpha_i, (\sigma_\varepsilon^2 + c\sigma_\beta^2)/bc)$. Also, MSb(a) is an unbiased estimate for $\sigma_\varepsilon^2 + c\sigma_\beta^2$. A Tukey multiple comparison for the levels of A is based upon the following:

$$\text{Decide } \alpha_i \neq \alpha_t \text{ if } |\bar{y}_{i..} - \bar{y}_{t..}| \geq q_{\alpha,a,a(b-1)} \sqrt{\frac{MSb(a)}{bc}}. \qquad (16.4.4)$$

Observe that the parameters for the Studentized range are a, the number of levels of A, and a(b—1), the degrees of freedom for MSb(a). Note that the mean square used in the Tukey procedure is the same mean square as that used in the denominator of the f—test for the factor. This is often the case. The multiple comparison along with the f—test are illustrated in the next example.

Example 16.4.1 Refer to the data in Table 16.3.2 which involve three machines (lathes) with two operators nested in each machine. In Example 16.3.1 we thought of both machines and operators as fixed. In this example we consider the six operators to be a sample of operators from a population; therefore, it is appropriate to declare the operator factor to be random. This suggests that we want our comparison of the machines to be reflective of the population of operators. The model appropriate for this problem is (16.4.1) and the f—tests and multiple comparisons are given in Table 16.4.1 and decision (16.4.4), respectively. The f—test for $H_0 : \alpha_i = 0$ for all i (no difference between the means for machines) is based upon the statistic $f = MSa/MSb(a) = \frac{.32165/2}{.023325/3}$ = 20.68. We reject H_0 since $f = 20.68 > f_{.05,2,3} = 9.55$. The SAS printout in Table 16.4.2 contains an analysis for these data. Be aware that the statement TEST H = MACH E = OPER(MACH); requests a test for the hypothesis concerning machines using the mean square for OPER(MACH) as the denominator instead of MSe for the f—test. Notice that this f = 20.68 on the printout is different from the f used for the fixed effects case (f = 14.39). To obtain the Tukey multiple comparison we specified in the MEANS statement that error term is OPER(MACH). Note that the outcome of the multiple comparison yields a different result than obtained in the fixed effects case in Table 16.3.3. Finally, the statement RANDOM OPER(MACH) provide the expression for the E(MS). In these expressions Q(MACH) represents the quantity $\frac{bc}{a-1} \sum_{i=1}^{a} \alpha_{i.}^{2}$.

Table 16.4.2
SAS Commands and Printout for Nested Data—Mixed Model

SAS Commands

```
OPTIONS LS = 80;
DATA TXNEST;
INPUT OPER MACH Y @@;
CARDS;
1 1 .62  1 1 .34  1 2 .21  1 2 .32  1 3 .73  1 3 .64
2 1 .56  2 1 .41  2 2 .14  2 2 .23  2 3 .52  2 3 .59
PROC GLM; CLASS OPER MACH;
MODEL Y = MACH OPER(MACH);
TEST H = MACH  E = OPER(MACH);
MEANS MACH / TUKEY E = OPER(MACH);
RANDOM OPER(MACH);
```

SAS Printout

```
                         General Linear Models Procedure

Dependent Variable: Y
                                         Sum of             Mean
Source                      DF          Squares           Square   F Value     Pr > F

Model                        5       0.34497500       0.06899500      6.17     0.0233

Error                        6       0.06705000       0.01117500

Corrected Total             11       0.41202500

                      R-Square             C.V.         Root MSE               Y Mean

                      0.837267         23.88969        0.1057119            0.4425000


Source                      DF        Type I SS      Mean Square   F Value     Pr > F

MACH                         2       0.32165000       0.16082500     14.39     0.0051
OPER(MACH)                   3       0.02332500       0.00777500      0.70     0.5876

Source                      DF      Type III SS      Mean Square   F Value     Pr > F

MACH                         2       0.32165000       0.16082500     14.39     0.0051
OPER(MACH)                   3       0.02332500       0.00777500      0.70     0.5876

Tests of Hypotheses using the Type III MS for OPER(MACH) as an error term

Source                      DF      Type III SS      Mean Square   F Value     Pr > F

MACH                         2       0.32165000       0.16082500     20.68     0.0176
```

SAS Printout

```
                  General Linear Models Procedure

     Tukey's Studentized Range (HSD) Test for variable: Y

NOTE: This test controls the type I experimentwise error rate, but
      generally has a higher type II error rate than REGWQ.

              Alpha= 0.05  df= 3  MSE= 0.007775
          Critical Value of Studentized Range= 5.910
            Minimum Significant Difference= 0.2605

Means with the same letter are not significantly different.

       Tukey Grouping              Mean       N  MACH

                    A           0.62000       4  3
                    A
            B       A           0.48250       4  1
            B
            B                   0.22500       4  2
```

```
                        General Linear Models Procedure

Source          Type III Expected Mean Square

MACH            Var(Error) + 2 Var(OPER(MACH)) + Q(MACH)

OPER(MACH)      Var(Error) + 2 Var(OPER(MACH))
```

In mixed and random effects cases it is clear that we need to find expressions for the E(MS). These are needed to obtain the proper form for the f—tests and to estimate standard errors needed for multiple comparisons or confidence intervals. While we can find these expressions using statistical properties of expectations, this is often quite tedious. Instead we will use an algorithm that is presented in the next section.

16.5 Expected Mean Squares Algorithm

By using an algorithm it is not difficult to find the expected mean squares for a **balanced** design. The algorithm we describe was developed by Bennett and Franklin (1954) and by Cornfield and Tukey (1956).

E(MS) Algorithm

Step 1: **(Model)** Write the model including subscripts and the range for each subscript. Use brackets (parentheses) for nested subscripts.

Step 2: **(Auxillary Table)** Form a two–dimensional table called the Auxillary Table.

a) (Table Heading – Rows) Use the terms of the model, including subscripts, as the row headings.

b) (Table Heading – Columns) Use the subscripts in the model as the column headings. Over each of these subscripts write F if the factor associated with the subscript is fixed or write R if the factor is random.

c) (Table Entries) We now fill in the table.

i) For each row write the number of levels under each subscript, providing the subscript does **not** appear in the row heading.

ii) For each row write a 1 in the columns for those subscripts that are inside the brackets of the row heading.

iii) Fill in the remaining entries of the table with a 0 or 1; use a 0 if the column heading is fixed, and a 1 if it is random.

d) (Finding E(MS) for an Effect)

i) Cover entries in the column which contain non–bracketed subscripts that are associated with the effect. Multiply the remaining numbers in each row. Each of these products is the coefficient of the σ^2 or Q for each row effect. (If the row effect is random, then use σ^2 to represent the variance for the effect. If the row effect is fixed, then use Q to represent some function of the model parameters.)

ii) The E(MS) for an effect is σ_ε^2 plus the sum of the row products in part i (multiplied by the proper σ^2 or Q) for rows that contain **all** of the subscripts of the effect. For the error term $E(MSE) = \sigma_\varepsilon^2$.

As an illustration of this E(MS) algorithm consider the balanced two–factor nested design, with factor A fixed and factor B random and nested within factor A. The model for this design is given in (16.4.1). The Auxillary Table for this model is given in Table 16.5.1. Note that we consider the replication subscript to be random in all cases.

Table 16.5.1 Auxillary Table for Two–Factor Nested–Mixed Model

	a	b	c
	F	R	R
	i	j	k
α_i	0	b	c
$\beta_{j(i)}$	1	1	c

Using this table and the algorithm we find:

$$E(MSE) = \sigma_\varepsilon^2$$

$$E(MSA) = \sigma_\varepsilon^2 + bc\ Q_A + c\sigma_\beta^2$$

$$E(MSB(A)) = \sigma_\varepsilon^2 + c\sigma_\beta^2.$$

With $Q_A = \sum_{i=1}^{a} \alpha_i^2/(a-1)$, we note that these E(MS) quantities match with the ones in Table 16.4.1 and on the SAS Printout in Table 16.4.2.

Our use of this E(MS) algorithm should be clear. By using the algorithm we find the E(MS) for a model, and the expressions for the E(MS) indicate the proper form of the f–tests. We illustrate this with an RCBD, and with some additional examples in the next section and chapter.

RCBD with Random Blocks

Consider an RCBD with p fixed treatments and r random blocks. The appropriate model with assumptions for this design was given in Section 16.2. Recall that with the normality assumption the β_j's are independent $N(0,\sigma_\beta^2)$ and the ε_{ij}'s are independent $N(0,\sigma_\varepsilon^2)$. The analysis of major interest are tests for H_0 : all $\tau_i = 0$ and H_0 : $\sigma_\beta^2 = 0$, and a multiple comparison for the treatments. Our immediate goal is to illustrate how to carry out these procedures.

To find the proper form of the f—tests we find expressions for the E(MS). The Auxillary Table for the mixed model for an RCBD is as follows:

	p F i	r R j
τ_i	0	r
β_j	p	1

Using the E(MS) algorithm we find

$$E(MSTR) = \sigma_\varepsilon^2 + rQ_T, \quad E(MSBL) = \sigma_\varepsilon^2 + p\sigma_\beta^2, \quad \text{and } E(MSE) = \sigma_\varepsilon^2.$$

From these expressions we find that the appropriate form for the f—tests are:

Reject H_0 : all $\tau_i = 0$, for all i, if f = MStr/MSe $\geq f_{\alpha,p-1,(p-1)(r-1)}$,

Reject $H_0 : \sigma_\beta^2 = 0$ if f = MSbl/MSe $\geq f_{\alpha,(r-1),(p-1)(r-1)}$.

Note that these f—tests have the same form as in the fixed effects case presented in Chapter 10.

Next we find the form of a multiple comparison for the treatments. In Section 16.2 we showed that observations within the same block are not necessarily independent. In fact, $Cov(Y_{ij}, Y_{tj}) = \sigma_\beta^2$ for $i \neq t$. It follows that the $\bar{Y}_{i\cdot}$'s are not necessarily independent, and as a consequence, the Tukey procedure presented in Chapter 9 is not appropriate. To obtain a multiple comparison for the treatments we will use the Bonferroni procedure for pairwise treatment contrasts. We first note that $E(\bar{Y}_{i\cdot} - \bar{Y}_{k\cdot}) = \tau_i - \tau_k$, and $Var(\bar{Y}_{i\cdot} - \bar{Y}_{k\cdot}) = \frac{2\sigma_\varepsilon^2}{r}$. Since 2MSe/r is an unbiased estimate for $Var(\bar{Y}_{i\cdot} - \bar{Y}_{k\cdot})$, it follows that a 100(1—α)% confidence interval for $\tau_i - \tau_k$ is

$$(\bar{y}_{i\cdot} - \bar{y}_{k\cdot}) \pm t_{\frac{\alpha}{2},(p-1)(r-1)} \sqrt{\frac{2MSe}{r}}. \tag{16.5.1}$$

The Bonferroni multiple comparison for the treatments is then:

$$\text{Decide } \tau_i \neq \tau_k \text{ if } |\bar{y}_{i\cdot} - \bar{y}_{k\cdot}| \geq t_{\frac{\alpha}{2c},(p-1)(r-1)} \sqrt{\frac{2MSe}{r}}. \quad (16.5.2)$$

where $c = p(p-1)/2$.

16.6 Factorial Experiment — Mixed Model

In this section we study a mixed model in a factorial experiment. Specifically, we consider a balanced two–factor factorial experiment with factors A and B. Factor A is fixed with "a" levels, and factor B is random with "b" levels. If either factor A or B is random, then the interaction term is also random.

There are two alternative models for this case that one encounters in the literature, and we shall present them both. The difference between the two models is in the assumptions used for the random interaction term. The first model, which we call the **standard model**, with the normality assumptions is as follows:

$$Y_{ijk} = \mu + \alpha_i + \beta_j + (\alpha\beta)_{ij} + \varepsilon_{ijk};$$
$$i = 1,\cdots,a; \quad j = 1,\cdots,b; \quad k = 1,\cdots,r;$$
$$\text{where } \sum_{i=1}^{a} \alpha_i = 0,$$
$$\text{the } \beta_j\text{'s are independent } N(0,\sigma^2_\beta), \quad (16.6.1)$$
$$\text{the } (\alpha\beta)_{ij}\text{'s are } N(0, \tfrac{a-1}{a}\sigma^2_{\alpha\beta}),$$
$$\sum_{i=1}^{a} (\alpha\beta)_{ij} = 0 \text{ for all } j; \text{ and}$$
$$\text{the } \varepsilon_{ijk}\text{'s are independent } N(0,\sigma^2_\varepsilon).$$

A few comments are in order concerning this model. First, the model allows for interaction terms in different levels of A within the same level of B to be dependent. In fact, in the exercises it is shown that $\text{Cov}((\alpha\beta)_{ij}, (\alpha\beta)_{tj}) = -\frac{1}{a}\sigma^2_{\alpha\beta}$ for $i \neq t$. Second, the variance for the $(\alpha\beta)_{ij}$'s is assumed to be the constant "(a—1)/a" times $\sigma^2_{\alpha\beta}$ to simplify the expected mean squares, and so the E(MS) values match the ones obtained with the algorithm given in Section 16.5. Third, the summation constraint on the interaction is summed over the fixed factor, but not the random one. Finally, with this model, observations within the same cell are dependent, as are observations within the same level of factor B.

Using the E(MS) algorithm in the previous section we find the following expressions:

$$E(MSA) = \sigma_{\varepsilon}^2 + r\sigma_{\alpha\beta}^2 + \frac{br}{a-1}\sum_{i=1}^{a} \alpha_i^2$$

$$E(MSB) = \sigma_{\varepsilon}^2 + ar\sigma_{\beta}^2$$

$$E(MSAB) = \sigma_{\varepsilon}^2 + r\sigma_{\alpha\beta}^2 \tag{16.6.2}$$

$$E(MSE) = \sigma_{\varepsilon}^2.$$

The appropriate f—tests are:

Reject $H_0 : \sigma_{\alpha\beta}^2 = 0$ if $f = MSab/MSe \geq f_{\alpha,(a-1)(b-1),ab(r-1)}$,
Reject $H_0 : \alpha_i = 0$, for all i, if $f = MSa/MSab \geq f_{\alpha,a-1,(a-1)(b-1)}$,
Reject $H_0 : \sigma_{\beta}^2 = 0$ if $f = MSb/MSe \geq f_{\alpha,b-1,ab(r-1)}$.

The formulas for the sum of squares and degrees of freedom for the various terms are identical to the fixed effects case presented in Chapter 13.

Performing a multiple comparison or obtaining confidence intervals for comparing the levels of factor A can be of interest in many cases. We note that $\bar{Y}_{i..} - \bar{Y}_{t..}$ is an unbiased estimator for $\alpha_i - \alpha_t$. Also, in the exercises it is shown that

$$Var(\bar{Y}_{i..} - \bar{Y}_{t..}) = 2(\sigma_{\varepsilon}^2 + r\sigma_{\alpha\beta}^2)/rb;$$

therefore, the standard error is

$$Se(\widehat{\alpha_i - \alpha_t}) = \sqrt{2\ MSab/rb}.$$

A 100(1—α)% confidence interval for $\alpha_i - \alpha_t$ is

$$(\bar{y}_{i..} - \bar{y}_{t..}) \pm t_{\frac{\alpha}{2},(a-1)(b-1)} \sqrt{2\ MSab/rb}. \tag{16.6.3}$$

Since the $\bar{Y}_{i..}$'s are not independent, a Bonferroni multiple comparison is appropriate. It can be performed on the levels of A by using the $t_{\frac{\alpha}{2c},(a-1)(b-1)}$ value in (16.6.3), where $c = a(a-1)/2$.

The second form of the model for a balanced two–factor factorial with A fixed and B random is now considered. In this form we assume that the interaction terms are uncorrelated with each other. The resulting model with normality assumptions is:

$$\begin{gathered} Y_{ijk} = \mu + \alpha_i + \beta_j + (\alpha\beta)_{ij} + \varepsilon_{ijk}; \\ i = 1,\cdots,a; \;\; j = 1,\cdots,b; \;\; k = 1,\cdots,r; \\ \text{where } \sum_{i=1}^{a} \alpha_i = 0, \\ \text{the } \beta_j\text{'s are independent } N(0,\sigma^2_\beta), \\ \text{the } (\alpha\beta)_{ij}\text{'s are independent } N(0,\sigma^2_{\alpha\beta}), \text{ and} \\ \text{the } \varepsilon_{ijk}\text{'s are independent } N(0,\sigma^2_\varepsilon). \end{gathered} \tag{16.6.4}$$

The independence between the interaction terms leads to altered expressions for the E(MS). The expressions are:

$$\begin{aligned} E(MSA) &= \sigma^2_\varepsilon + r\sigma^2_{\alpha\beta} + \frac{br}{a-1} \sum_{i=1}^{a} \alpha_i^2 \\ E(MSB) &= \sigma^2_\varepsilon + r\,\sigma^2_{\alpha\beta} + ar\,\sigma^2_\beta \\ E(MSAB) &= \sigma^2_\varepsilon + r\,\sigma^2_{\alpha\beta} \\ E(MSE) &= \sigma^2_\varepsilon. \end{aligned} \tag{16.6.5}$$

While the f–tests for H_0 : all $\alpha_i = 0$ and $H_0 : \sigma^2_{\alpha\beta} = 0$ have the same form as in the standard model, the f–test for $H_0 : \sigma^2_\beta = 0$ for this model is based upon $f = MSb/MSab$. The form of the f–test for H_0 : $\sigma^2_\beta = 0$ is the primary difference between the standard model (16.6.1) and our second model (16.6.4).

The choice between the two models is not an easy one. The standard model seems to be used more often in the literature and is most appropriate when it makes sense to assume that the interaction terms are correlated. Interesting, when using the RANDOM statement in SAS one obtains the E(MS) for the second model as given in (16.6.5). For example, if a = 2, b = 2 and r = 6, the SAS statement "RANDOM B A * B;" yields the following output:

General Linear Models Procedure

Source	Type III Expected Mean Square
A	Var(Error) + 6 Var(A*B) + Q(A)
B	Var(Error) + 6 Var(A*B) + 12 Var(B)
A*B	Var(Error) + 6 Var(A*B)

Observe that the E(MS) match formulas (16.6.5).

16.7 Pseudo F—Tests

In the previous sections we found the appropriate form of the f—test by using the expected means squares, that is, so that the E(MS) of the numerator and denominator are equal when H_0 is true. In some cases there is no ANOVA line whose E(MS) equals the E(MS) of the numerator when H_0 is true; therefore, there is not a natural denominator for the f—test. The following is an example of such a case.

Example 16.7.1 Hicks (1973) reports an experiment in which three gate settings were compared in terms of their effect on the dry—film thickness of varnish. Three operators were chosen from a large group of operators and two measurements were made on each operator at each gate setting on two days that were randomly selected in a given month. The results from Hicks are given in Table 16.7.1.

Table 16.7.1 Dry—Film Thickness Experiment

		Day					
		1			2		
		Operator			Operator		
		A	B	C	A	B	C
Gate Setting	2	0.38	0.39	0.45	0.40	0.39	0.41
		0.40	0.41	0.40	0.40	0.43	0.40
	4	0.63	0.72	0.78	0.68	0.77	0.85
		0.59	0.70	0.79	0.66	0.76	0.84
	6	0.76	0.95	1.03	0.86	0.86	1.01
		0.78	0.96	1.06	0.82	0.85	0.98

Clearly there are three factors: day, operator, and gate. All factors are crossed and it seems appropriate to assume that gate is fixed, but day and operator are random. If we let $Y_{ijk\ell}$ represent the ℓth measurement on the ith day, jth operator and kth gate setting, our model is

$$Y_{ijk\ell} = \mu + \alpha_i + \beta_j + \gamma_k + (\alpha\beta)_{ij} + (\alpha\gamma)_{ik} + (\beta\gamma)_{jk}$$
$$+ (\alpha\beta\gamma)_{ijk} + \varepsilon_{ijk\ell};$$

$$i = 1,2; \quad j = 1,2,3; \quad k = 1,2,3; \quad \ell = 1,2;$$

with appropriate assumptions.

Using the algorithm of Section 16.5, we can find expressions for each E(MS). Also, using the formulas in Chapter 14 for the sum of squares and the degrees of freedom for a three–factor factorial, we obtain the quantities given in Table 16.7.2.

Table 16.7.2 Sum of Squares and Expected Mean Squares for Dry–Film Thickness Experiment

Source	df	SS	MS	E(MS)
Day (α_i)	1	0.0010	MSd=0.0010	$\alpha_{\varepsilon}^2 + 6\sigma_{\alpha\beta}^2 + 18\sigma_{\alpha}^2$
Operator (β_j)	2	0.1121	MSo=0.0560	$\alpha_{\varepsilon}^2 + 6\sigma_{\alpha\beta}^2 + 12\sigma_{\beta}^2$
Gate (γ_k)	2	1.5732	MSg=0.7866	$\alpha_{\varepsilon}^2 + 2\sigma_{\alpha\beta\gamma}^2 + 6\sigma_{\alpha\gamma}^2 + 4\sigma_{\beta\gamma}^2 + 12Q_G$
Day * Operator	2	0.0060	MSdo=0.0030	$\sigma_{\varepsilon}^2 + 6\sigma_{\alpha\beta}^2$
Day * Gate	2	0.0113	MSdg=0.0056	$\sigma_{\varepsilon}^2 + 2\sigma_{\alpha\beta\gamma}^2 + 6\sigma_{\alpha\gamma}^2$
Operator * Gate	4	0.0428	MSog=0.0107	$\sigma_{\varepsilon}^2 + 2\sigma_{\alpha\beta\gamma}^2 + 4\sigma_{\beta\gamma}^2$
Day*Oper.*Gate	4	0.0099	MSdog=0.0025	$\sigma_{\varepsilon}^2 + 2\sigma_{\alpha\beta\gamma}^2$
Error	18	0.0059	MSe=0.0003	σ_{ε}^2
Total	35	1.7622		

A hypothesis of particular interest is no difference in gate settings, namely $H_0 : \gamma_k = 0$ for all k (or equivalently $H_0 : Q_G = 0$). The numerator of the f usually would be MSg, and the denominator would be the mean squares which has E(MS) of $\sigma_{\varepsilon}^2 + 2\sigma_{\alpha\beta\gamma}^2 + 6\sigma_{\alpha\gamma}^2 + 4\sigma_{\beta\gamma}^2$. There is no line that has this E(MS); therefore, we need to find a different way to perform the test. Our approach will be to form a pseudo f—test using a procedure suggested by Satterthwaite (1946).

Satterthwaite Approximation

Satterthwaite proposed a procedure for obtaining a test statistic that has an approximate f—distribution. In the procedure the numerator of the f is the mean square of the term associated with the null hypothesis. For convenience label this mean square by MSw, and let u represent its degrees of freedom. The next step is to use some of the mean squares in the problem, labelled as $MS_1,...,MS_k$ with $\nu_1,...,\nu_k$ degrees of freedom, to form a linear combination

$$MSx = a_1 MS_1 + \cdots + a_k MS_k \qquad (16.7.1)$$

that satisfies E(MSW) = E(MSX) when H_0 is true. In (16.7.1), MS_i is the mean square value associated with the ith line of the ANOVA table. The resulting constructed f—test rejects H_0 if $f = MSw/MSx \geq f_{\alpha,u,\nu}$. To find ν, the degrees of freedom of MSx, we use the Satterthwaite approximation

$$\nu = (MSx)^2/[\sum_{i=1}^{k} (a_i MS_i)^2/\nu_i]. \qquad (16.7.2)$$

A test formed in this manner is called a **pseudo f—test**, since f = MSw/MSx has only an approximate f—distribution. Since ν in (16.7.2) might not be an integer, we recommend the value be truncated to an integer to yield a conservative test.

As an illustration of the Satterthwaite procedure let us test for no difference between gate settings in the Dry Film experiment considered in Example 16.7.1. For testing $H_0 : \gamma_k = 0$, for all k, the numerator of the f—test will be MSg. For the denominator we seek a linear combination of mean squares, MSx, so that E(MSX) = E(MSG) when H_0 is true. From Table 16.7.2 we observe that we must have E(MSX) $= \sigma^2_\varepsilon + 2\sigma^2_{\alpha\beta\gamma} + 6\sigma^2_{\alpha\gamma} + 4\sigma^2_{\beta\gamma}$. One such linear combination is MSx = MSdg + MSog — MSdog. Our pseudo f—test is then based upon the statistic

$$\begin{aligned} f &= MSg/(MSdg + MSog - MSdog) \\ &= 0.7866/(0.0056 + 0.0107 - 0.0025) \\ &= 0.7866 / 0.0138 = 57 . \end{aligned}$$

We reject H_0 if $f \geq f_{.05,2,\nu}$, where

$$\nu = (\text{MSdg} + \text{MSog} - \text{MSdog})^2/[(\text{MSdg})^2/2 + (\text{MSog})^2/4 + (\text{MSdog})^2/4]$$

$$= (0.0138)^2/[(0.0056)^2/2 + (0.0107)^2/4 + (0.0025)^2/4]$$

$$= 4.2, \text{ which we truncate to } 4.$$

Our critical point is $f_{.05,2,4} = 6.94$ and since $f = 57 > 6.94$, we reject H_0 and conclude that at least two gate settings have significantly different effects.

One difficulty that can be encountered with the Satterthwaite procedure is that the linear combination (16.7.1) could be negative. Naturally this presents a serious problem in carrying out the f—test. One alternative is to only use **sums** in forming the linear combination MSx, that is, $\text{MSx} = a_1\text{MS}_1 + \cdots + a_k\text{MS}_k$, where each $a_i \geq 0$. In such a case MSx cannot be negative. In fact, we could also **add** one or more mean squares in the numerator. For example, in the Dry Film experiment we could use MSg + MSdog as the numerator of the f—statistic and MSdg + MSog as the denominator. This is appropriate since $E(\text{MSG} + \text{MSDOG}) - E(\text{MSDG} + \text{MSOG}) = 12Q_G$ which is zero when H_0 is true. We would then use the Satterthwaite approximation (16.7.2) to find the degrees of freedom for both the numerator and denominator in the f—ratio.

16.8 Variance Components

In Section 16.1 we introduced the concept of a random effects model, namely a model in which all elements in the model are random except for the constant μ term. Sometimes these models are called variance components models, since the interest in the models is studying the variances (variance components) of the various random variables. Knowing which component in the model contributes most to the variability in the response variable will be of critical interest in some problems. Our goal in this section is to consider examples of such problems, and to illustrate the types of statistical analyses that can be employed.

To more fully appreciate the ideas involved in a variance component model we consider an illustration based upon **three—stage nested sampling**.

Three–Stage Nested Sampling

A grocery store chain noticed considerable variation in the size of grapefruits it purchases. The chain receives shipments of grapefruits from a large number of suppliers, so to better understand the sources of variation in the weight of the grapefruits an experiment was completed. A random sample of three suppliers was selected from the group of all suppliers. For each of the three suppliers a random sample of two shipments was selected, and a random sample of four bags of grapefruits was obtained from each shipment. Each bag contained the same number of grapefruits and the total weight of each bag was recorded. The results are given in Table 16.8.1.

Table 16.8.1 Weights of Grapefruits

	Supplier					
	1		2		3	
Shipment	1	2	1	2	1	2
	12	16	18	15	15	14
	13	19	17	16	10	19
	17	12	14	19	18	13
	14	15	21	23	16	12

For this experiment we observe that there are two factors, supplier and shipment. Both are random factors and shipment is nested within supplier. Notice that this experiment involves three different samples, namely, a sample of suppliers, a sample of shipments within suppliers, and a sample of bags within a shipment. It is natural, therefore, to refer to this experiment as **three–stage nested sampling**. The model that we should employ is similar to the model (16.3.1) that was used for a two–factor experiment with one factor nested within the other factor. We need to modify the assumptions to account for our random factors. If we let Y_{ijk} represent the weight of the kth bag within the jth shipment within the ith supplier, then the appropriate model with assumptions (including normality) is

$$Y_{ijk} = \mu + \alpha_i + \beta_{j(i)} + \varepsilon_{k(ij)};$$

$$i = 1,2,3; \quad j = 1,2; \quad k = 1,\cdots,4; \qquad (16.8.1)$$

The α_i's, $\beta_{j(i)}$'s and $\varepsilon_{k(ij)}$'s

are independent normal random variables with mean zero and variance

σ^2_α, σ^2_β and σ^2_ε, respectively.

With this assumption we note that the Y_{ijk}'s are distributed as $N(\mu, \sigma^2_\alpha + \sigma^2_\beta + \sigma^2_\varepsilon)$. Some issues of interest are the following. Since the variation in the weights of the bags of grapefruit is measured by $\sigma^2_\alpha + \sigma^2_\beta + \sigma^2_\varepsilon$, we want to study these three variance components individually to determine their contribution to the overall variation. A point estimate or confidence interval for the variance component should be of interest. Also, the parameter μ represents the mean weight; therefore, an estimate and confidence interval for μ should be of interest. We need to consider how to find these estimates and confidence intervals for μ and for a variance component.

Estimating Variance Components

To find point estimates for variance components we use the **analysis of variance method**. In this method the expected mean squares are found and are equated to the computed mean squares from the analysis of variance table. These equations are then solved for the various variance components to obtain point estimates. The resulting point estimators are unbiased and, in the balanced care, are often minimum variance.

To illustrate this analysis of variance method, consider the balanced random effects CRD model given in (16.1.1), namely, $Y_{ij} = \mu + \tau_i + \varepsilon_{ij}$; $i = 1,\cdots,p$; $j = 1,\cdots,r$; with the τ_i's and ε_{ij}'s as uncorrelated random variables with means zero and variances σ^2_τ and σ^2_ε. We aim to find point estimators for σ^2_τ and σ^2_ε. Let MSTR and MSE be the random variable forms of the mean squares for treatment and error, respectively. Routine calculations yield the following E(MS):

$$E(MSTR) = \sigma^2_\varepsilon + r\sigma^2_\tau,\ E(MSE) = \sigma^2_\varepsilon.$$

Setting $\sigma^2_\varepsilon + r\sigma^2_\tau = MSTR$ and $\sigma^2_\varepsilon = MSE$ and solving, we obtain the corresponding point estimators, denoted by $\hat{\sigma}^2_\varepsilon$ and $\hat{\sigma}^2_\tau$, as

$$\hat{\sigma}^2_\varepsilon = MSE,\ \hat{\sigma}^2_\tau = (MSTR - MSE)/r. \qquad (16.8.2)$$

To find confidence intervals for variance components we use properties of chi–square random variables to find exact confidence intervals, and the Satterthwaite (1946) approximation to find approximate confidence intervals. The method we use is based upon the following fact. In a balanced variance components model, with the usual normality assumptions, if $\hat{\sigma}_i^2$ is a random variable mean square associated with the ith line in the ANOVA table that has ν_i degrees of freedom and $E(\hat{\sigma}_i^2) = \sigma_i^2$, then $\nu_i \hat{\sigma}_i^2/\sigma_i^2$ is distributed as $\chi^2(\nu_i)$. This fact can be used to find an exact confidence interval for σ_i^2, the E(MS) of the ith ANOVA line. For example, in the random effects CRD model just considered, we know that MSE $= \hat{\sigma}_\varepsilon^2$. Since MSE is an entry in the ANOVA table with $n - p$ degrees of freedom, $(n - p)\mathrm{MSE}/\sigma_\varepsilon^2 \equiv \mathrm{SSE}/\sigma_\varepsilon^2$ is $\chi^2(n - p)$. An exact $100(1-\alpha)\%$ confidence interval for σ_ε^2 is then

$$(\mathrm{SSe}/\chi^2_{\frac{\alpha}{2},n-p}\ ,\ \mathrm{SSe}/\chi^2_{(1-\frac{\alpha}{2}),n-p}).$$

In many cases the estimator for the variance component is a linear combination of two or more variance components. In these cases we use the Satterthwaite approximation. If our unbiased estimator for a variance component σ^2 is of the form $\hat{\sigma}^2 = \sum_{i=1}^{k} a_i \hat{\sigma}_i^2$, then $\nu\, \hat{\sigma}^2/\sigma^2$ is approximately $\chi^2(\nu)$, where ν is the Satterthwaite approximation referred to in (16.7.2). Assuming the σ_i^2 are known, this approximation can be written as

$$\nu = (\sum_{i=1}^{k} a_i \sigma_i^2)^2/[\sum_{i=1}^{k} a_i^2 \sigma_i^4/\nu_i]. \tag{16.8.3}$$

In practice, MS_i, the calculated ith mean square, is substituted for the theoretical σ_i^2. As an illustration of this approach we refer again to the random effects CRD model. In (16.8 2.) we found the unbiased estimate for the variance component σ_τ^2 to be $s_\tau^2 = (\mathrm{MStr} - \mathrm{MSe})/r$. It follows that $\nu[(\text{MSTR} - \text{MSE})/r]/\sigma_\tau^2$ is approximately $\chi^2(\nu)$, where $\nu = (s_\tau^2)^2/[(\text{MStr})^2/(p-1) + (\text{MSe})^2/(n-1)]$. An approximate $100(1-\alpha)\%$ confidence interval for σ_τ^2 is then $(\nu\, s_\tau^2/\chi^2_{\frac{\alpha}{2},\nu},\ \nu\, s_\tau^2/\chi^2_{(1-\frac{\alpha}{2}),\nu})$.

We close this section by illustrating these procedures with a numerical example.

Example 16.8.1 Refer to the weights of grapefruit data given in Table 16.8.1. These data were obtained from a three–stage nested sampling experiment and the appropriate model is given in (16.8.1). Our goal is to perform statistical analysis on the variance components σ_α^2, σ_β^2 and σ_ε^2. We first need the ANOVA table. Using SAS we can find the numerical elements of the ANOVA table. (Formulas for the sum of squares are the same as for the fixed effects case discussed in Section 16.3.) Using the E(MS) algorithm of Section 16.5 we can find expressions for the E(MS). These are given in Table 16.8.2.

Table 16.8.2 ANOVA Table for Grapefruit Data

Source	df	SS	MS	E(MS)
Supplier	2	54.25	27.125	$\sigma_\varepsilon^2 + 4\sigma_\beta^2 + 8\sigma_\alpha^2$
Shipment	3	5.75	1.9167	$\sigma_\varepsilon^2 + 4\sigma_\beta^2$
Error	18	166.5	9.25	σ_ε^2
Total	23	226.5		

For convenience we label suppliers as factor A and shipments as factor B. We first find estimates for the variance components. We set up the equations

$$\text{MSA} = \sigma_\varepsilon^2 + 4\sigma_\beta^2 + 8\sigma_\alpha^2,\quad \text{MSB(A)} = \sigma_\varepsilon^2 + 4\sigma_\beta^2,\quad \text{MSE} = \sigma_\varepsilon^2,$$

and solving we obtain the corresponding point estimators.

$$\hat{\sigma}_\varepsilon^2 = \text{MSE},$$

$$\hat{\sigma}^2_{\alpha} = [\text{MSA} - \text{MSB(A)}]/8$$

$$\hat{\sigma}^2_{\beta} = [\text{MSB(A)} - \text{MSE}]/4.$$

Substituting in the values from the ANOVA Table 16.8.2 we find the point estimates s^2_{ε} and s^2_{α} to be $s^2_{\varepsilon} = 9.25$, $s^2_{\alpha} = 3.15$. This shows that the variation due to suppliers is substantially smaller than the random variation among bags of grapefruits. In the exercises we find a confidence interval for σ^2_{α} using the Satterthwaite approach. Inspecting the E(MS) entries we note that the proper f—test is based upon $f = \text{MSa}/\text{MSb(a)}$. From Table 16.8.2 we note that $f = 27.125/1.9167 = 14.15$. Since $14.15 > f_{.05,2,3} = 9.55$, we have evidence that σ^2_{α} is significantly greater than zero.

There is a disturbing item that arises with this grapefruit data. Note that the estimate for σ^2_{β} is given by $s^2_{\beta} = [\text{MSb(a)} - \text{MSe}]/4 = (1.9167 - 9.25)/4 = -1.83$. This is an absurd estimate since we know that variances cannot be negative. There are several approaches that one can consider. First, we can consider a negative estimate as strong evidence that the variance component is zero. Second, we could choose to use a different method for estimating the variance component that insures a non—negative estimate. Searle (1971) has a good discussion of other approaches that can be used.

To estimate the mean weight, μ, for the grapefruit data we note that $\bar{Y}_{...}$ is an unbiased estimator for μ. In the exercises it is shown that $\text{Var}(\bar{Y}_{...}) = \frac{\sigma^2_{\alpha}}{a} + \frac{\sigma^2_{\beta}}{ab} + \frac{\sigma^2_{\varepsilon}}{abc}$. An unbiased estimate for this variance is MSa/abc; therefore, a confidence interval for μ is $\bar{y}_{...} \pm t_{\frac{\alpha}{2},a-1}\sqrt{\frac{\text{MSa}}{\text{a bc}}}$. The calculations for this confidence interval are left for the exercises.

Chapter 16 Exercises

Application Exercises

16–1 An experiment is conducted to compare the water quality of three creeks in an area. Five water samples are selected from each creek. Each sample is divided into two parts, and the dissolved oxygen content is measured for each part. (Higher dissolved oxygen contents indicate higher water quality.) The results are given as follows:

	Water Sample 1		2		3		4		5	
Creek 1	5.2	5.4	5.6	5.7	5.4	5.4	5.6	5.5	5.8	5.5
Creek 2	5.1	5.3	5.1	5.0	5.3	5.2	5.0	5.0	4.9	5.1
Creek 3	5.9	5.8	5.8	5.8	5.7	5.8	5.8	5.9	5.9	5.9

(a) Write down an appropriate model with assumptions (include normality).
(b) Find the ANOVA table for the data.
(c) Perform the f–test comparing the creeks using a 0.05 level.
(d) Perform a Tukey multiple comparison on the creeks using a 0.05 level.

16–2 Refer to Exercise 10–3 which involves an RCBD design with four varieties of wheat and five fields. Assume that the varieties are fixed, but the fields are randomly selected from a population.

(a) Perform the f–tests for the data using a 0.05 level.
(b) Perform a Bonferroni multiple comparison on the varieties using a 0.05 level.
(c) Estimate the variance components associated with these data.
(d) Using Exercise 16–15 results, find a 95% confidence interval for the mean of the second variety.

16–3 Three cooking temperatures are considered for baking a brand of cake. A sample of four ovens is selected, and cakes are baked on two separate occasions in each of the ovens at each temperature level. A judge rates the quality of the texture of the cake on a twenty–point scale. The data follow:

		Temperature					
		325° F		350° F		375° F	
Oven	1	11	13	18	17	11	12
	2	16	15	19	19	14	16
	3	14	12	17	18	13	11
	4	15	14	20	17	17	14

(a) Write down an appropriate model with assumptions.
(b) Find the ANOVA table.
(c) Perform the f–tests using a 0.05 level.
(d) Find a 95% confidence interval for the difference between means for 350° and 375° F.
(e) Perform a Bonferroni multiple comparison on the temperatures using a 0.05 level.
(f) Estimate the variance component for ovens.

16–4 Refer to the Dry–Film Thickness experiment given in Example 16.7.1.

(a) Verify the sum of squares entries in Table 16.7.2.
(b) Use the E(MS) algorithm to obtain the expressions for the E(MS) given in Table 16.7.2.
(c) Perform an appropriate f–test for each of the effects using a 0.05 level.
(d) Perform an appropriate multiple comparison on the levels of gate. Use a 0.05 level.
(e) Estimate the variance components for operator and day. Which component contributes more to the variation in the observations?

16–5 Refer to the Grapefruit Data analyzed in Example 16.8.1.

(a) Find a 95% confidence interval for σ_α^2.
(b) Find a 95% confidence interval for μ.

16–6 Consider the Grades for Term Papers data given in Table 16.1.1, where professors are the treatments (τ).
(a) Find the ANOVA table for the data.
(b) Find point estimates for the two variance components.

(c) Test $H_0 : \sigma_\tau^2 = 0$ using a 0.05 level.

(d) Find an appropriate 95% confidence interval for σ_τ^2.

Theoretical Exercises

16–7 Consider the random effects CRD model given in (16.1.1).

(a) Show that $Cov(Y_{ij}, Y_{k\ell}) = 0$ for $i \neq k$.

(b) Find $Cov(Y_{ij}, Y_{ik})$, for $j \neq k$.

16–8 In a Latin square with p rows, columns and treatments assume that the rows and columns are random, while the treatments are fixed.

(a) Write down an appropriate model, with assumptions for this situation. Include the normality assumptions.
(b) Find the covariance of observations within the:
i) same row ii) same column iii) same treatment
(c) Which observations are independent?
(d) Find the E(MS) column for this model and describe how to carry out the f–test for treatments.

16–9 Verify that the rank of the **X** matrix for the nested model (16.3.1) is ab.

16–10 Consider a design with fixed factors A and B, and factor B nested within A. Assume the model (16.3.1).

(a) Verify that (16.3.2) provides a solution to the normal equations.
(b) Use the principle of conditional error to verify the formulas for SSa and SSb(a) given in (16.3.4). Find the "hand"–computing formulas for these sums of squares.
(c) Verify that SSt = SSa + SSb(a) + SSe.

16–11 Consider the nested design of Exercise 16–10 and model (16.3.1).

(a) Derive the confidence interval (16.3.5) for a contrast in factor A.
(b) Derive the Tukey multiple comparison given in (16.3.6) for factor A.

(c) Derive the E(MS) and verify the entries in Table 16.3.1. (Do not use the E(MS) algorithm of Section 16.5, rather use statistical properties of expectations for the derivation.)

16–12 Consider a two–factor nested design with the mixed model given in (16.4.1).

(a) Show that $\bar{Y}_{ij\cdot}$ is $N(\mu + \alpha_i, (\sigma_\varepsilon^2 + c\sigma_\beta^2)/c)$.

(b) Show that $\bar{Y}_{\cdot\cdot\cdot}$ is $N(\mu, (\sigma_\varepsilon^2 + c\sigma_\beta^2)/abc)$.

(c) Show that $E(MSB(A)) = \sigma_\varepsilon^2 + c\sigma_\beta^2$.

(d) Show that $E(MSE) = \sigma_\varepsilon^2$.

(e) Show that $E(MSA) = \sigma_\varepsilon^2 + c\sigma_\beta^2 + \frac{bc}{a-1}\sum_{i=1}^{a} \alpha_i^2$.

16–13 Consider a two–factor nested design with the mixed model given in (16.4.1). Find the form of a $100(1-\alpha)\%$ confidence interval for a contrast involving the effects of factor A.

16–14 Refer to the setting of a two–factor nested design with the mixed model given in (16.4.1), and $a = 2$, $b = 3$, $c = 2$.

(a) Write down expressions for E(MSA), E(MSB(A)), and E(MSE).

(b) Suppose that observation y_{112} is missing. Show that the expressions for the E(MS) are:

i) $E(MSA) = \sigma_\varepsilon^2 + \frac{104}{55}\sigma_\beta^2 + 5\alpha_1^2 + \frac{65}{11}\alpha_2^2$

ii) $E(MSB(A)) = \sigma_\varepsilon^2 + \frac{9}{5}\sigma_\beta^2$

iii) $E(MSE) = \sigma_\varepsilon^2$.

16–15 Consider a mixed model RCBD with p fixed treatments and r random blocks as presented in Section 16.5

(a) Show that $\bar{Y}_{i\cdot} - \bar{Y}_{k\cdot}$ is $N(\tau_i - \tau_k, 2\sigma_\varepsilon^2/r)$.

(b) Show that $\bar{Y}_{i\cdot}$ is $N(\mu + \tau_i, (\sigma_\varepsilon^2 + \sigma_\beta^2)/r)$, and $\bar{Y}_{i\cdot}$ is unbiased for the ith treatment mean.

(c) Show that $[(p - 1)MSE + MSBL]/rp$ is an unbiased estimator for the variance of $\bar{Y}_{i.}$.

(d) Using the Satterthwaite Approximation, find the form of an approximate confidence interval for $\mu + \tau_i$.

16–16 Consider an RCBD in which both treatments and blocks are random.

(a) Write down an appropriate model with assumptions.
(b) Using the E(MS) algorithm, find expressions for the E(MS).
(c) Find the form of the f–tests and compare these with the ones found for the mixed model in Section 16.5 and for the fixed effects model in Chapter 10.
(d) Find estimators for the variance components σ_τ^2 , σ_β^2 , and σ_ε^2.
(e) Find the form of an approximate confidence interval for σ_β^2.

16–17 Consider a balanced two–factor factorial experiment with factor A fixed and factor B random. Assume the standard model given in (16.6.1).

(a) Show that $Cov[(\alpha\beta)_{ij}, (\alpha\beta)_{tj}] = -\sigma_{\alpha\beta}^2/a$ for $i \neq t$.
(b) Find $Cov(Y_{ijk}, Y_{rtk})$ for various r, t.
(c) Use the E(MS) algorithm to find the expressions for the E(MS) given in (16.6.2).
(d) Show that $\bar{Y}_{i..} - \bar{Y}_{t..}$ is $N[\alpha_i - \alpha_t, 2(\sigma_\varepsilon^2 + r\sigma_{\alpha\beta}^2)/rb]$.
(e) Find estimators for the variance components.
(f) Using the Satterthwaite Approximation, find the form of the confidence interval for σ_β^2 .
(g) Use properties of expectations to find E(MSA).

16–18 Consider a balanced three–factor factorial experiment with factors A, B, and C.

(a) Assume that factors A and B are fixed and C is random.

Find expressions for the E(MS) and describe the appropriate f–tests for each effect.

(b) Repeat part (a) assuming all three factors are random.

16–19 Refer to the three–stage nested sampling experiment described in Section 16.8.

(a) Show that $\bar{Y}_{...}$ is $N(\mu, \frac{\sigma_\alpha^2}{a} + \frac{\sigma_\beta^2}{ab} + \frac{\sigma_\varepsilon^2}{abc})$.

(b) Show that MSA/abc is unbiased for $Var(\bar{Y}_{...})$.

(c) Find $Cov(Y_{ijk}, Y_{rtk})$ for various r, t.

16–20 Consider a three–stage nested sampling experiment. Let the total cost of sampling be represented by the linear function

$$C = aC_1 + abC_2 + abcC_3$$

where C_1, C_2, and C_3 are the sampling unit costs at the first, second, and third stages, respectively. Using the method of Lagrange multipliers, show that if c = 2, $Var(\bar{Y})$ is minimized when

$$b^2 = \frac{C_1(2\sigma_\beta^2 + \sigma_\varepsilon^2)}{2\sigma_\alpha^2(C_2 + 2C_3)}$$

and

$$a = \frac{C}{C_1 + bC_2 + 2bC_3} .$$

CHAPTER 17

NESTED DESIGNS AND ASSOCIATED TOPICS

17.1 Higher Order Nested Designs

In Section 16.3 we introduced the two–factor nested design with factor B nested within factor A. Naturally we can have more complicated nested designs. In a **p–factor nested design** with factors $A_1, \cdots, A_p$ we have factor A_2 nested within A_1, A_3 nested within $A_2, \cdots$, and A_p nested with A_{p-1}.

As an illustration when $p = 3$ we consider an experiment aimed at comparing the effectiveness of three sprays (factor A) used on apple trees. We decide to choose a random sample of two groves (factor B) of apple trees to be used for each spray. From each of the six groves a random sample of three trees (factor C) is selected. On each tree two large branches are selected, and all of the apples on each branch are inspected. The percentage, Y, of apples without blemishes is recorded for each branch. This is an example of a three–factor nested design as factor B (groves) is nested within factor A (spray), and factor C (tree) is nested within factor B (grove). Data from this experiment are given in Table 17.1.1. It would be appropriate to consider factor A as fixed, and factors B and C as random.

Table 17.1.1 Spray Data, Three–Factor Nested

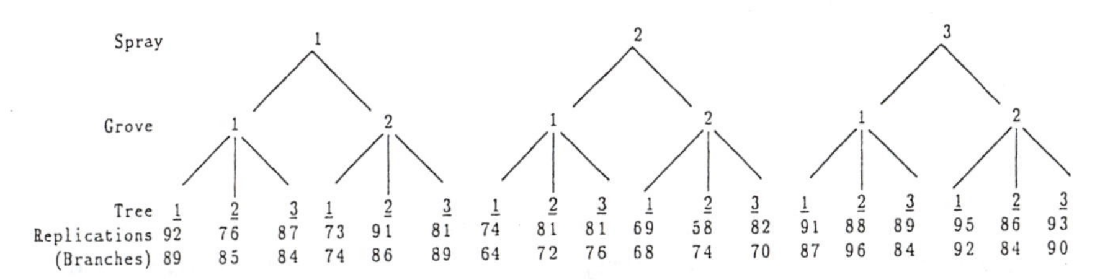

Spray	1						2						3					
Grove	1			2			1			2			1			2		
Tree	1	2	3	1	2	3	1	2	3	1	2	3	1	2	3	1	2	3
Replications	92	76	87	73	91	81	74	81	81	69	58	82	91	88	89	95	86	93
(Branches)	89	85	84	74	86	89	64	72	76	68	74	70	87	96	84	92	84	90

The model we use for a three–factor nested design is

$$Y_{ijk\ell} = \mu + \alpha_i + \beta_{j(i)} + \gamma_{k(ij)} + \varepsilon_{\ell(ijk)};$$

$$i = 1, \cdots, a;\ j = 1, \cdots, b;\ k = 1, \cdots, c; \tag{17.1.1}$$

$\ell = 1, \cdots r;$ where

$Y_{ijk\ell}$ is the ℓth observation on the kth level of C nested in the jth level of B, nested in the ith level of A.

The assumptions for the model (17.1.1) depends upon which factors are fixed and random. We assume that fixed effects sum to zero, and random effects are independent normal random variables with mean zero and common variance at each level.

The analysis for a three–factor nested design is based upon the E(MS). Using the E(MS) algorithm of Chapter 16 we can find expressions for the E(MS) and use these to form appropriate f–tests. The E(MS) are given for two cases of the three–factor nested design. The sum of squares and degrees of freedom are found in the usual manner, and these are summarized in Table 17.1.2.

Table 17.1.2 Balanced Three–Factor Nested Design, SS, E(MS)

Source	df	SS	A,B,C all random E(MS)	A fixed, B and C random E(MS)
A	a−1	$bcr\sum_{i=1}^{a}(\bar{y}_{i...}-\bar{y}_{....})^2$	$\sigma_\varepsilon^2+r\sigma_\gamma^2+cr\sigma_\beta^2+bcr\sigma_\alpha^2$	$\sigma_\varepsilon^2+r\sigma_\gamma^2+cr\sigma_\beta^2+bcrQ_A$
B(A)	a(b−1)	$cr\sum_{i=1}^{a}\sum_{j=1}^{b}(\bar{y}_{ij..}-\bar{y}_{i...})^2$	$\sigma_\varepsilon^2+r\sigma_\gamma^2+cr\sigma_\beta^2$	$\sigma_\varepsilon^2+r\sigma_\gamma^2+cr\sigma_\beta^2$
C(B,A)	ab(c−1)	$r\sum_{i=1}^{a}\sum_{j=1}^{b}\sum_{k=1}^{c}(\bar{y}_{ijk.}-\bar{y}_{ij..})^2$	$\sigma_\varepsilon^2+r\sigma_\gamma^2$	$\sigma_\varepsilon^2+r\sigma_\gamma^2$
Error	abc(r−1)	$\sum_{i=1}^{a}\sum_{j=1}^{b}\sum_{k=1}^{c}\sum_{\ell=1}^{r}(y_{ijk\ell}-\bar{y}_{ijk.})^2$	σ_ε^2	σ_ε^2
Total	abcr−1	$\sum_{i=1}^{a}\sum_{j=1}^{b}\sum_{k=1}^{c}\sum_{\ell=1}^{r}(y_{ijk\ell}-\bar{y}_{....})^2$		

For the data in Table 17.1.1 we have a balanced three–factor nested design with factor A (spray) fixed, and factors B (grove) and C (tree) random. In the model (17.1.1) we assume that $\sum_{i=1}^{a}\alpha_i = 0$, the $\beta_{j(i)}$'s are independent $N(0,\sigma_\beta^2)$, the $\gamma_{k(ij)}$'s are independent $N(0,\sigma_\gamma^2)$, and the $\varepsilon_{\ell(ijk)}$'s are independent $N(0,\sigma_\varepsilon^2)$. With these assumptions the expressions for the E(MS) are given in the last column of Table 17.1.2. The proper f–test for no difference between the sprays ($H_0 : \alpha_i = 0$ for all i) is to reject H_0 if $f = MSa/MSb(a) \geq f_{\alpha,a-1,a(b-1)}$. From the SAS printout for these data given in Table 17.1.3, we note that for this test, $f = 29.64$ ($p = .0106$); therefore, we reject H_0 and conclude that at least two sprays have significantly different effects.

The proper f—test for grove is also given on the printout using the "TEST" statement. Note in the SAS commands that factor C nested within B, which is in turn nested within A is denoted by C(BA).

To determine which levels of spray are significantly different we employ a multiple comparison procedure. Since the $\bar{Y}_{i\cdots}$'s are independent $N(\mu + \alpha_i, (\sigma_\varepsilon^2 + r\sigma_\gamma^2 + cr\sigma_\beta^2)/bcr)$ it follows that a Tukey multiple comparison is based upon deciding $\alpha_i \neq \alpha_t$ if

$$|\bar{y}_{i\cdots} - \bar{y}_{t\cdots}| \geq q_{\alpha,a-1,a(b-1)} \sqrt{MSb(a)/bcr}. \qquad (17.1.2)$$

The results of this procedure are found on the printout and lead us to conclude that sprays 1 and 3 are not significantly different, but both are significantly superior to spray 2.

Table 17.1.3 SAS Printout for Spray Data

SAS Commands

```
options ls = 80;
data txnt3f;
input spray grove tree percent @@;
cards;
1 1 1 92 1 1 1 89 1 1 2 76 1 1 2 85 1 1 3 87 1 1 3 84
1 2 1 73 1 2 1 74 1 2 2 91 1 2 2 86 1 2 3 81 1 2 3 89
2 1 1 74 2 1 1 64 2 1 2 81 2 1 2 72 2 1 3 81 2 1 3 76
2 2 1 69 2 2 1 68 2 2 2 58 2 2 2 74 2 2 3 82 2 2 3 70
3 1 1 91 3 1 1 87 3 1 2 88 3 1 2 96 3 1 3 89 3 1 3 84
3 2 1 95 3 2 1 92 3 2 2 86 3 2 2 84 3 2 3 93 3 2 3 90
proc glm; class spray grove tree;
model percent = spray grove (spray) tree (grove spray);
test h = spray e = grove (spray);
test h = grove (spray) e = tree (grove spray);
means spray / tukey e = grove (spray);
```

SAS printout

```
                    General Linear Models Procedure

Dependent Variable: PERCENT
                                    Sum of              Mean
Source                   DF        Squares            Square   F Value     Pr > F

Model                    17   2593.4722222       152.5571895      5.95     0.0002

Error                    18    461.5000000        25.6388889

Corrected Total          35   3054.9722222

                  R-Square           C.V.          Root MSE         PERCENT Mean

                  0.848935       6.177075         5.0634858            81.972222

Source                   DF      Type I SS       Mean Square   F Value     Pr > F

SPRAY                     2   1836.2222222       918.1111111     35.81     0.0001
GROVE(SPRAY)              3     92.9166667        30.9722222      1.21     0.3353
TREE(SPRAY*GROVE)        12    664.3333333        55.3611111      2.16     0.0680

Source                   DF    Type III SS       Mean Square   F Value     Pr > F

SPRAY                     2   1836.2222222       918.1111111     35.81     0.0001
GROVE(SPRAY)              3     92.9166667        30.9722222      1.21     0.3353
TREE(SPRAY*GROVE)        12    664.3333333        55.3611111      2.16     0.0680

Tests of Hypotheses using the Type III MS for GROVE(SPRAY) as an error term

Source                   DF    Type III SS       Mean Square   F Value     Pr > F

SPRAY                     2   1836.2222222       918.1111111     29.64     0.0106

Tests of Hypotheses using the Type III MS for TREE(SPRAY*GROVE) as an error term

Source                   DF    Type III SS       Mean Square   F Value     Pr > F

GROVE(SPRAY)              3    92.91666667       30.97222222      0.56     0.6518


                    General Linear Models Procedure

            Tukey's Studentized Range (HSD) Test for variable: PERCENT

           NOTE: This test controls the type I experimentwise error rate, but
                 generally has a higher type II error rate than REGWQ.

                      Alpha= 0.05  df= 3  MSE= 30.97222
                   Critical Value of Studentized Range= 5.910
                     Minimum Significant Difference= 9.4941

           Means with the same letter are not significantly different.

               Tukey Grouping              Mean      N  SPRAY

                            A            89.583     12  3
                            A
                            A            83.917     12  1

                            B            72.417     12  2
```

17.2 Designs with Nested and Crossed Factors

In experiments that involve several factors some factors might be crossed and others might be nested. Such designs that involve both crossed and nested factors are often called **nested—factorial experiments**. Our purpose in this section is to illustrate such an experiment and its analysis.

For our illustration consider an experiment aimed at comparing the effectiveness of two drugs for reducing anxiety levels in patients who have high anxiety. It was believed that gender might effect the results; therefore, random samples of five men and five women were selected. Each person used the two drugs and a placebo (control) in random order for two separate periods. An appropriate waiting time was used between the application of the drugs. The response variable was a measure of anxiety level on a forty—point scale. (Higher scores suggest higher anxiety.) The results are given in Table 17.2.1.

Table 17.2.1 Anxiety Data, Nested—Factorial Experiment

Gender	Subject	Drug 1		Drug 2		Drug 3 (control)	
Female	1	28	25	21	19	34	38
	2	31	36	27	31	37	39
	3	15	18	12	17	28	24
	4	29	21	30	22	37	34
	5	18	14	11	13	22	30
Male	1	26	22	18	16	27	24
	2	21	25	17	14	32	34
	3	19	25	12	12	18	22
	4	22	17	18	14	29	32
	5	18	15	10	15	21	26

In this experiment we have three factors: A (drug, 3 fixed levels), B (gender, 2 fixed levels), C (subject, 5 random levels for each gender). Drug and gender are crossed, drug and subject are crossed, but subject is nested within gender.

An appropriate model for these data is:

$$Y_{ijk\ell} = \mu + \alpha_i + \beta_j + \gamma_{k(j)} + (\alpha\beta)_{ij} + (\alpha\gamma)_{ik(j)} + \varepsilon_{\ell(ijk)}, \tag{17.2.1}$$

$$i = 1,2,3 = a;\ j = 1,2 = b;\ k = 1,2,3,4,5 = c;\ \ell = 1,2 = r.$$

$Y_{ijk\ell}$ represents the ℓth observation on the kth subject within the jth gender using the ith drug. Since subject is random, the effects $\gamma_{k(j)}$ and $(\alpha\gamma)_{ik(j)}$ are random variables as are the $\varepsilon_{\ell(ijk)}$'s. To find appropriate f—tests we refer to the E(MS). Using the E(MS) algorithm we obtain the expressions in Table 17.2.2. The sum of squares and degrees of freedom are also given for each effect.

Table 17.2.2 ANOVA Table for Nested—Factorial

Source	df	SS	E(MS)
A	a–1	$bcr\sum_{i=1}^{a}(\bar{y}_{i...}-\bar{y}_{....})^2$	$\sigma_\varepsilon^2+r\sigma_{\alpha\gamma}^2+bcrQ_A$
B	b–1	$acr\sum_{j=1}^{b}(\bar{y}_{.j..}-\bar{y}_{....})^2$	$\sigma_\varepsilon^2+ar\sigma_\gamma^2+acrQ_B$
C(B)	b(c–1)	$ar\sum_{j=1}^{b}\sum_{k=1}^{c}(\bar{y}_{.jk.}-\bar{y}_{.j..})^2$	$\sigma_\varepsilon^2+ar\sigma_\gamma^2$
A*B	(a–1)(b–1)	$cr\sum_{i=1}^{a}\sum_{j=1}^{b}(\bar{y}_{ij..}-\bar{y}_{i...}-\bar{y}_{.j..}+\bar{y}_{....})^2$	$\sigma_\varepsilon^2+r\sigma_{\alpha\gamma}^2+crQ_{AB}$
A*C(B)	(a–1)b(c–1)	$r\sum_{i=1}^{a}\sum_{j=1}^{b}\sum_{k=1}^{c}(\bar{y}_{ijk.}-\bar{y}_{ij..}-\bar{y}_{.jk.}+\bar{y}_{.j..})^2$	$\sigma_\varepsilon^2+r\sigma_{\alpha\gamma}^2$
Error	abc(r–1)	$\sum_{i=1}^{a}\sum_{j=1}^{b}\sum_{k=1}^{c}\sum_{\ell=1}^{r}(y_{ijk\ell}-\bar{y}_{ijk.})^2$	σ_ε^2
Total	abcr–1	$\sum_{i=1}^{a}\sum_{j=1}^{b}\sum_{k=1}^{c}\sum_{\ell=1}^{r}(y_{ijk\ell}-\bar{y}_{....})^2$	

From the E(MS) we note that the f—test for drugs rejects $H_0 : \alpha_i = 0$ for all i if $f = MSa/MSac(b) \geq f_{\alpha,a-1,(a-1)b(c-1)}$. The ANOVA table, with proper f—tests, for the data in Table 17.2.1 is given in Table 17.2.3.

Table 17.2.3 ANOVA Table for Anxiety Data

Source	df	MS	f
A (Drug)	2	723.217	MSa/MSac(b) = 55.8
B (Gender)	1	326.667	MSb/MSc(b) = 2.1
C(B), (Subject)	8	156.533	MSc(b)/MSe = 16.7
A*B	2	17.617	MSab/MSac(b) = 1.4
A*C(B)	16	12.958	MSac(b)/MSe = 1.4
Error	30	9.400	
Total	59		

For the test for drugs, since $f = 55.8 \geq f_{.05,2,16} = 3.63$ we conclude that significant differences exist between at least two of the drugs. In the exercises a multiple comparison is used to determine where the differences exist.

In closing this section we make two important observations. First, the formulas in Table 17.2.2 for the degrees of freedom and sum of squares of the nested–factorial experiment have a distinctive pattern and can easily be generalized to more complicated balanced designs. Second, if there are no replications then the degrees of freedom for error is zero in Table 17.2.2. Our model is then called a **saturated model** since the error sum of squares is zero. Notice that when SSe = 0 we can still carry out some of the f–tests; specifically those that do not involve MSe in the denominator.

17.3 Subsampling

In some experiments replications are obtained by taking several observations on the same experimental units. Observations of this type are referred to as **subsamples** and the process of obtaining the observations is called **subsampling**.

Subsampling frequently occurs with nested designs. For example, in Example 16.3.1 we considered an experiment with three machines and six operators nested within machines. Two operators are assigned to each machine, and each operator uses a machine twice. In this experiment we can consider the machines as treatments and the operators as experimental units. The two readings on each operator can be considered as a subsample. This provides us with another way of looking at this two—factor nested design. That is, we can consider it a CRD with subsampling.

CRD With Subsampling

The presence of subsampling in an experiment provides another source of variation. In a CRD with subsampling the sources of variation are treatments, variation from subsamples, and variation between the experimental units. Let Y_{ijk} represent the response to the kth sample taken from the jth experimental unit to which the ith treatment has been applied. The model (with normality) for a CRD with subsampling may be written as

$$Y_{ijk} = \mu + \alpha_i + \beta_{j(i)} + \varepsilon_{k(ij)};$$

$$i = 1,\cdots,a; \quad j = 1,\cdots,b; \quad k = 1,\cdots,c; \qquad (17.3.1)$$

where $\sum_{i=1}^{a} \alpha_i = 0$. The α_i's represent the treatment effects; $\beta_{j(i)}$ represent the random error associated with the jth experimental unit of the ith treatment; and $\varepsilon_{k(ij)}$ represents the random error associated with the kth sample taken from the jth experimental unit of the ith treatment. We assume that the $\beta_{j(i)}$'s and the $\varepsilon_{k(ij)}$'s are independent normal random variables with means zero and variance σ_β^2 and σ_ε^2, respectively. The $\beta_{j(i)}$ component is the **experimental error**, while $\varepsilon_{k(ij)}$ represents the **sampling error**.

We note that the model (17.3.1) for a CRD with subsampling has the same form and the same assumptions as the mixed model two—factor nested design considered in (16.4.1). As a consequence, the analysis of a CRD with subsampling is identical to the mixed model two—factor nested design. In this way subsampling is simply another way of looking at a nested design.

RCBD With Subsampling

While subsampling often occurs in nested designs, it can also occur in other designs. We consider subsampling in the context of a RCBD. In a RCBD with subsampling we have "a" treatments, b blocks and at each treatment, block combination we have r observations based upon a subsample of size r. For discussion let us assume that blocks are random, and let Y_{ijk} represent the observation on the kth sample in the jth block and ith treatment. The model in this case is

$$Y_{ijk} = \mu + \alpha_i + \beta_j + (\alpha\beta)_{ij} + \varepsilon_{ijk};$$

$$i = 1,\cdots,a; \quad j = 1,\cdots,b; \quad k = 1,\cdots,r. \qquad (17.3.2)$$

The α_i's represent the treatment effects. The β_j's represent the random block effect, while the $(\alpha\beta)_{ij}$'s represent the random error associated with the jth block and ith treatment. The ε_{ijk} represents the random error associated with the kth sample taken from the jth block and ith treatment. Naturally the components β_j, $(\alpha\beta)_{ij}$, and ε_{ijk} are assumed to be random variables. The assumptions we use are the same as those used in model (16.6.1) for the mixed model two—factor factorial. Because models (17.3.2) and (16.6.1) are identical in form and in assumptions, the analysis of a RCBD with subsampling is identical to the mixed model two—factor factorial experiment.

Subsampling is an important concept used in developing a design for an experiment. It provides a method for obtaining additional experimental observations. We have seen, however, that the analysis of a design with subsampling is often identical to the analysis of a design developed from a different point of view.

17.4 Repeated Measures Designs

In experiments involving subjects it is often the case that multiple observations are made upon the same subject. Examples include cases where a subject receives each of the treatments at different times, or cases where measurements are made on the subject at several different times. Designs in which repeated measurements are made on a subject are often called **repeated measures designs**.

We have previously encountered repeated measures designs in this text. In Table 10.2.1 we considered a RCBD in which four workers (subjects) used each of three machines (treatments). In this case workers are repeated over machines; therefore, this is an example of a repeated measures design. In Table 17.2.1 we considered an example involving factors: gender, subject, and drug. In that example each subject used each drug; therefore, the design is a repeated measures design in which subject is repeated over drug.

As pointed out, we have previously analyzed data from repeated measures designs and so we already know how to analyze data from such designs. It is appropriate, however, for us to discuss some of the terminology and special issues that occur in these designs.

We consider first the simplest form of a repeated measures design, that is an RCBD. In this setting we have r subjects (usually randomly chosen), p treatments, and the subjects repeated over the treatments (see Figure 17.4.1).

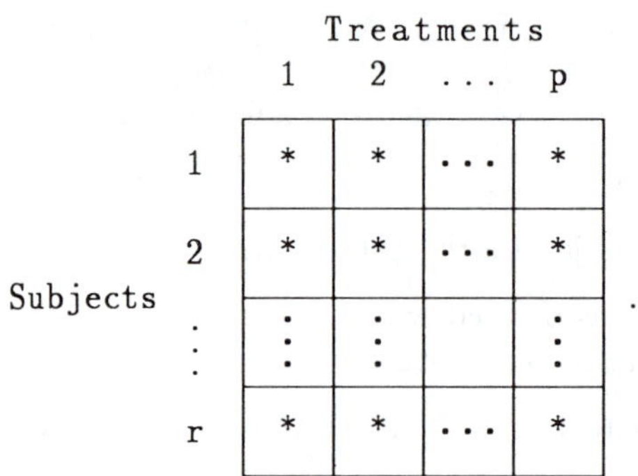

Figure 17.4.1 Repeated Measures RCBD

If subjects are randomly selected and if Y_{ij} represents the observation for subject j using treatment i, then the model is:

$$Y_{ij} = \mu + \tau_i + \beta_j + \varepsilon_{ij} ;$$

$$i = 1,\cdots,p; \quad j = 1,\cdots,r; \qquad (17.4.1)$$

where the β_j's and the ε_{ij}'s are independent normal random variables with mean zero and variances σ^2_β and σ^2_ε, respectively.

We note that this is the mixed model for an RCBD considered in Section 16.5; therefore, the analysis is identical to the one presented in that section.

Note that in this repeated measures design (Figure 17.4.1) the covariance between different observations within the same block (subject) is σ^2_β. This is one of the concerns in a repeated measures design. The model suggests that all observations on the same subject are related in the same manner. The model does not consider the order in which the observations are taken or possible carry—over effects. For example, if the treatments are various drugs, then it is possible that a subject's observation might be different if a drug is taken first versus last (after all of the other drugs). This is one of the disadvantages of a repeated measures design. (See Section 18.2 for designs dealing with carry—over effects.) To counter this shortcoming, however, is the fact that repeated measures designs are very efficient designs. It is generally more efficient to take more than one observation on a subject rather than to take just a single observation. In addition, we can help control the variability in the experiment by blocking on subjects.

In repeated measures designs it is common to report the variation in two parts, namely within subjects variation and between subjects variation. For example, in the RCBD (repeated measures) with model (17.4.1) the sum of squares measuring **within subjects variation** is

$$\text{SSwithin} = \sum_{i=1}^{p} \sum_{j=1}^{r} (y_{ij} - \bar{y}_{\cdot j})^2,$$

with $r(p-1)$ degrees of freedom, while the **between subjects variation** is

$$\text{SSbetween} = p \sum_{j=1}^{r} (\bar{y}_{\cdot j} - \bar{y}_{\cdot\cdot})^2,$$

with r—1 degrees of freedom.

It is easy to observe that SSwithin = SStr + SSe, SSbetween = SSbl, and that SSt = SSbetween + SSwithin. The ANOVA table can be presented as given in Table 17.4.1.

Table 17.4.1 ANOVA Table for Repeated Measures RCBD

Source	df	SS	MS	f
Between subjects	r—1	SSbetween		
Subjects	r—1	SSbl	MSbl	MSbl/MSe
Within subjects	r(p—1)	SSwithin		
Treatments	p—1	SStr	MStr	MStr/MSe
Error	(p—1)(r—1)	SSe	MSe	

As noted before, the f—tests in Table 17.4.1 are identical to those for an RCBD.

We close this section illustrating a more complicated repeated measures design.

Example 17.4.1 A study was completed to determine if the concentration of a drug in a person's blood varies over the three times 1, 3, and 5 hours. A random sample of four patients, who had been using the drug, was selected for each of two randomly selected doctors. For each patient the concentration of the drug was measured 1, 3, and 5 hours after receiving it. The data appears in Table 17.4.2.

Table 17.4.2 Drug Concentration Data

Doctor	Patient	Time 1	Time 3	Time 5
1	1	25	28	24
	2	34	31	29
	3	22	26	25
	4	29	31	28
2	1	24	27	22
	2	31	28	28
	3	29	26	27
	4	21	17	20

The design is clearly a repeated measures design as patients are repeated over time. We consider time as fixed, while doctor and patient are random. Patient is nested within doctor, while doctor and time are crossed. An appropriate model can be written as follows with Y_{ijk} representing the reading on the jth patient of the ith doctor at the kth time.

$$Y_{ijk} = \mu + \alpha_i + \beta_{j(i)} + \gamma_k + (\alpha\gamma)_{ik} + (\gamma\beta)_{kj(i)} + \varepsilon_{j(ik)}$$

$$i = 1,2; \;\; j = 1,2,3,4; \;\; k = 1,2,3.$$

We assume that the α, β, $(\alpha\gamma)$, $(\gamma\beta)$, and ε terms are normal random variables with means zero and appropriate variances. The ANOVA table with E(MS) is given in Table 17.4.3.

Table 17.4.3 ANOVA Table for Concentration Data

Source	df	E(MS)
Between Subjects	7	
Doctor	1	$\sigma_{\varepsilon}^2 + 3\sigma_{\beta}^2 + 12\sigma_{\alpha}^2$
Patients (Doctor)	6	$\sigma_{\varepsilon}^2 + 3\sigma_{\beta}^2$
Within Subjects	16	
Time	2	$\sigma_{\varepsilon}^2 + 4\sigma_{\alpha\gamma}^2 + \sigma_{\beta\gamma}^2 + 8Q_{\gamma}$
Doctor * Time	2	$\sigma_{\varepsilon}^2 + 4\sigma_{\alpha\gamma}^2 + \sigma_{\beta\gamma}^2$
Time * Patients (Doctor)	12	$\sigma_{\varepsilon}^2 + \sigma_{\beta\gamma}^2$

The error term is not listed since the model is saturated and has no degrees of freedom for error. From the E(MS) entries we note that the f–test for time $(H_0 : \gamma_k = 0$ for $k = 1,2,3)$ is based upon f = MStime/MSdoctor * Time. The SAS printout in Table 17.4.4 contains an analysis for these data. From the printout we note that there is no significant difference between the concentrations at the three times.

Table 17.4.4 SAS Printout for Concentration Data

SAS Commands

```
options 1s = 80;
data txrm;
input doctor patient time y @@;
1 1 1 25 1 1 3 28 1 1 5 24
1 2 1 34 1 2 3 31 1 2 5 29
1 3 1 22 1 3 3 26 1 3 5 25
1 4 1 29 1 4 3 31 1 4 5 28
2 1 1 24 2 1 3 27 2 1 5 22
2 2 1 31 2 2 3 28 2 2 5 28
2 3 1 29 2 3 3 26 2 3 5 27
2 4 1 21 2 4 3 17 2 4 5 20
proc glm; class doctor patient time;
model y = doctor patient (doctor) time doctor*time time*patient
(doctor);
test h = time e = doctor*time;
test h = doctor*time e = time*patient (doctor);
```

SAS Printout

```
                    General Linear Models Prccedure
                       Class Level Information

                     Class      Levels     Values

                     DOCTOR          2     1 2

                     PATIENT         4     1 2 3 4

                     TIME            3     1 3 5

                Number of observations in data set = 24



                     General Linear Models Procedure

Dependent Variable: Y
                                       Sum of            Mean
Source                      DF        Squares          Square    F Value      Pr > F

Model                       23   365.33333333     15.88405797          .           .

Error                        0              .               .

Corrected Total             23   365.33333333

                  R-Square               C.V.        Root MSE                 Y Mean

                  1.000000                  0               0              26.333333

Source                      DF      Type I SS     Mean Square    F Value      Pr > F

DOCTOR                       1    42.66666667     42.66666667          .           .
PATIENT(DOCTOR)              6   256.00000000     42.66666667          .           .
TIME                         2    11.08333333      5.54166667          .           .
DOCTOR*TIME                  2    11.08333333      5.54166667          .           .
PATIENT*TIME(DOCTOR)        12    44.50000000      3.70833333          .           .

Source                      DF    Type III SS     Mean Square    F Value      Pr > F

DOCTOR                       1    42.66666667     42.66666667          .           .
PATIENT(DOCTOR)              6   256.00000000     42.66666667          .           .
TIME                         2    11.08333333      5.54166667          .           .
DOCTOR*TIME                  2    11.08333333      5.54166667          .           .
PATIENT*TIME(DOCTOR)        12    44.50000000      3.70833333          .           .

Tests of Hypotheses using the Type III MS for DOCTOR*TIME as an error term

Source                      DF    Type III SS     Mean Square    F Value      Pr > F

TIME                         2    11.08333333      5.54166667       1.00      0.5000

Tests of Hypotheses using the Type III MS for
PATIENT*TIME(DOCTOR) as an error term

Source                      DF    Type III SS     Mean Square    F Value      Pr > F

DOCTOR*TIME                  2    11.08333333      5.54166667       1.49      0.2633
```

Chapter 17 Exercises

Application Exercises

17–1 A study was completed in two states to determine the percentage of acres on farms planted in corn. In each state a random sample of four counties was taken, and in each county a random sample of two townships was selected. In each township five farms were selected and the percentage of acres in corn was recorded. The results are given below.

	State 1							
County	1		2		3		4	
Township	1	2	1	2	1	2	1	2
	70	48	61	76	96	72	78	94
	62	51	77	84	84	43	64	82

	State 2							
County	1		2		3		4	
Township	1	2	1	2	1	2	1	2
	47	71	81	51	34	48	76	37
	59	63	62	46	70	52	48	58

(a) Write down an appropriate model with assumptions for this study.
(b) Using a 0.05 level perform a test comparing the two states.
(c) Estimate the mean for each state with a 95% confidence interval.

17–2 Refer to the nested–factorial experiment of Table 17.2.1 and the model (17.2.1).

(a) Use the algorithm to verify the E(MS) entries in Table 17.2.2.
(b) Use a computer program to verify the mean square entries in ANOVA Table 17.2.3 for these data.
(c) Using a 0.05 level perform an appropriate multiple comparison for the levels of drug.

17–3 Refer to Exercise 16.1. Explain how subsampling plays a role in the experiment.

17–4 An experiment was completed to estimate and compare the mean gas mileage for three brands of cars. For each brand a random sample of five cars was selected. Each car was driven in the same manner over the same driving course three separate times, and the mileage (mgp) was recorded.

The data follow:

		Car 1	Car 2	Car 3	Car 4	Car 5
Brand	1	27.4 26.7 28.1	28.2 27.4 27.6	26.9 27.6 28.2	26.3 26.5 27.0	28.1 26.5 27.3
	2	30.3 29.1 29.4	28.6 29.7 29.2	31.1 30.4 30.7	29.4 29.8 29.4	30.7 29.8 29.4
	3	26.2 27.1 25.4	28.1 27.3 27.0	26.4 25.1 25.6	25.9 24.9 24.8	26.0 24.9 25.2

(a) Describe the design used for this experiment.
(b) Write down an appropriate model for this experiment. Include the normality assumptions.
(c) Find the ANOVA table and perform the f–tests using a 0.05 level.
(d) Estimate the mean gas mileage for each brand using a 95% confidence interval.
(e) Perform a Tukey multiple comparison on the brands using a 0.05 level.
(f) Find an estimate for the variance components. Is there more variability within a car or between cars?

17–5 (Efficiency in Design) The efficiency of a design is often based on the size of the variance of an important estimator. Refer to Exercise 17–4 and let b represent the number of cars used for each brand, and let r represent the subsample size (number of times each car is used). Suppose that we want to improve the efficiency of the design by decreasing the variance of $\bar{Y}_{i\cdot\cdot}$, the estimator for the brand means. Based upon the analysis results obtained in Exercise 17–4 in which $b = 5$ and $r = 3$, which of the following two designs would lead to better efficiency?

i) $b = 6, \ r = 4$
ii) $b = 4, \ r = 6.$

17–6 Refer to the repeated measures design presented in Example 17.4.1.

(a) Using the algorithm verify the E(MS) entries in Table 17.4.3.
(b) Explain how to test for no difference between doctors. Using the values from the SAS printout in Table 17.4.4 carry out this test using a 0.05 level.
(c) Find point estimates for each variance component.

17–7 The effectiveness of a nasal decongestant was studied in an experiment. Two hospital clinics (representing many such hospitals) were chosen for the study. From each clinic a sample of two men and two women with nasal congestion was selected. Each subject was given the decongestant and was monitored for 45 minutes. A measure of relief (10–point scale, 10 representing most relief) from congestion was recorded at 15, 30, and 45 minutes. The results are as follows:

		Clinic 1				Clinic 2			
Gender		F		M		F		M	
Subject		1	2	1	2	1	2	1	2
Time	15	8	9	7	5	7	9	6	4
	30	8	8	4	3	4	6	4	3
	45	6	5	2	2	2	3	2	1

(a) Write down an appropriate model with assumptions.
(b) Find the E(MS) using the algorithm.
(c) Find the ANOVA table.
(d) If possible, perform f–tests for time, gender, and clinic. Use a 0.05 level.

Theoretical Exercises

17–8 Refer to the three–factor nested design with model (17.1.1).

(a) Use the algorithm to verify the E(MS) entries in Table 17.1.2.
(b) Use the principle of conditional error to verify the sum of squares and degrees of freedom given in Table 17.1.2. (Hint: Since the sum of squares is the same for the mixed model as for the fixed model, assume the latter case with appropriate side conditions.)
(c) Show that when A is fixed, and B and C are random that the $\bar{Y}_{i\cdots}$'s are independent $N(\mu + \alpha_i, (\sigma_\varepsilon^2 + r\sigma_\gamma^2 + cr\sigma_\beta^2)/bcr)$. Use this to justify the formula for the Tukey procedure in (17.1.2).

17–9 Consider a balanced random effects three–factor nested design.

(a) Find point estimates for the variance components.
(b) Find the form of a confidence interval for σ_γ^2.

17–10 Consider a balanced four–factor nested design with factor D nested in C, C nested in B, and B nested in A. Assume r replications.

(a) Write down an appropriate model.
(b) Write down the formulas and degrees of freedom for the sums of squares.
(c) For the case when C and D are random, but A and B are fixed, find the E(MS) and the form of the appropriate f–tests.

17–11 A balanced design has four factors A, B, C, D with number of levels a, b, c, d, and r replications. Factor A and D are fixed, while factors B and C are random. Factor C is nested in factors A and B. Factors A and B are crossed, while factor D is crossed with A, B and C.

(a) Write down an appropriate model for this problem. Which terms in the model are random variables?
(b) Write down formulas for the sums of squares and the corresponding degrees of freedom.

(c) Using the algorithm find the E(MS).
(d) Identify the form for the f–tests.
(e) Find point estimators for the variance components σ_β^2 and σ_γ^2.
(f) If $r = 1$, that is, there are no replications, the model becomes saturated. What is the effect on the f–tests?

17–12 Consider a nested–factorial experiment with factors A, B, C, and D such that factor B is nested within factor A and factor D is nested within factor C. Write down the model sums of squares and df. Then compute the E(MS) column assuming r replications and:

(a) all factors random,
(b) all factors fixed,
(c) A and C fixed; B and D random.

17–13 Refer to the balanced CRD with subsampling model (17.3.1). Consider the random variables

$$\bar{Y}_{i\cdot\cdot} = \sum_{j=1}^{b} \sum_{k=1}^{c} Y_{ijk}/bc, \quad i = 1,2,\cdots,a.$$

Show that (a) $E(\bar{Y}_{i\cdot\cdot}) = \mu + \alpha_i$

(b) $Var(\bar{Y}_{i\cdot\cdot}) = (\sigma_\varepsilon^2 + c\sigma_\beta^2)/bc$

(c) $Var(\bar{Y}_{i\cdot\cdot} - \bar{Y}_{\ell\cdot\cdot}) = 2(\sigma_\varepsilon^2 + c\sigma_\beta^2)/bc.$

17–14 Consider an RCBD with subsampling, assuming "p" treatments fixed and "b" blocks random. Let the total cost of sampling be given by

$$c = brc_1 + bc_2$$

where c_1 is the cost per subsample in an experimental unit and c_2 is the cost for each experimental unit. Using the method of Lagrange multipliers, show that for a **fixed** p, the variance of the estimator of the ith treatment mean, $Var(\bar{Y}_{i\cdot\cdot})$, is minimized when

$$r^2 = \frac{c_2 \sigma_\varepsilon^2}{c_1[\sigma_\beta^2 + \frac{(p-1)}{p} \sigma_{\alpha\beta}^2]} \quad \text{and} \quad b = \frac{c}{rc_1 + c_2}.$$

17–15 Consider an experiment consisting of A at 4 levels, B at 5 levels crossed with A, C nested within each A, B combination at 3 levels, and two observations at each level of C. Write the model for this experiment and display the sum of squares and degrees of freedom.

CHAPTER 18

OTHER DESIGNS AND TOPICS

18.1 Split—Plot Designs

Introduction

Our purpose in this section is to introduce another design called a **split—plot design**. Designs in this class are variations of factorial experiments and usually involve mixed models as found in many areas of application. Split—plot designs are useful when the nature of the experiment requires the use of large experimental units for some factors and smaller experimental units for others. These designs are also useful when it is more practical to assign the factors in stages to experimental units of varying sizes.

As an illustration of a split—plot design, consider an experiment involving two factors A and B that are to be assigned to experimental units in two blocks. Suppose that factor A has three levels and factor B has two levels. In a factorial experiment we would require blocks of size six. In a split—plot design, due to constraints in the problem, the blocks are divided into three large experimental units called **whole plots**. The three levels of factor A are randomly assigned to these whole plots. Next each whole plot is subdivided into two smaller experimental units called **subplots**. The two levels of factor B are randomly assigned to these subplots. A possible arrangement of the factors in this split—plot design is given in Figure 18.1.1.

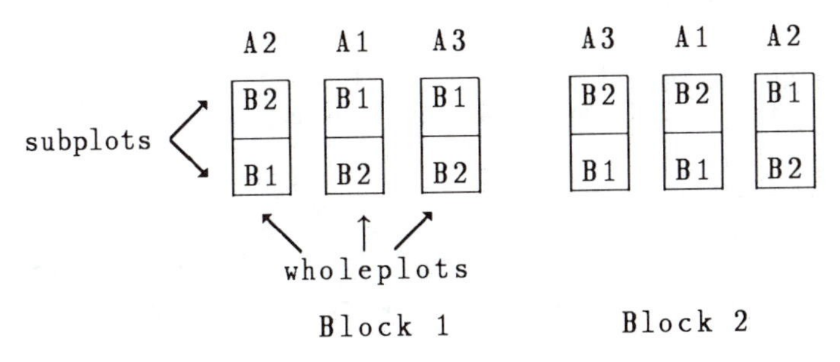

Figure 18.1.1 Split—Plot Design Layout

We make a few observations about the split—plot design given in Figure 18.1.1. First note that the levels of factor A are confounded with the whole plots, that is, we cannot separate their effects. Second, notice that the randomization in a split—plot design is completed in two stages:
1) levels of A are randomly assigned to whole plots and then
2) levels of B are randomly assigned to the subplots in each whole plot.

This two stage randomization and the presence of whole and subplots makes a split–plot design different from a factorial experiment. For purposes of contrast, if a factorial arrangement were used, then the six treatments (A1 B1, A1 B2, A2 B1, A2 B2, A3 B1, A3 B2) would be randomly assigned to the six units in a block. A typical assignment of the six treatments in block one, for example, could be as follows:

A2 B1	A3 B2	A1 B2
A3 B1	A1 B1	A2 B2

As demonstrated, it is how factors A and B are assigned to the experimental units that determines whether a split–plot or a usual factorial analysis is required by the experimenter.

In summary, a two–factor split–plot design has the following characteristics:

1. There are two factors of interest, A and B.
2. The experiment involves whole plots and these plots are divided into subplots.
3. The levels of factor A are randomly assigned to the whole plots.
4. The levels of factor B are randomly assigned to the subplots in each whole plot.

As with a factorial, a split–plot experiment can be incorporated in other basic designs such as a CRD and a Latin square.

A Split–Plot Example

To further illustrate a split–plot design we present an example. The example is an agricultural one, the source of the terms "whole plot" and "subplot."

Example 18.1.1 An experiment was completed to compare the yield of two varieties of wheat. An additional factor to be considered was type of spray for weeds and three different brands were to be considered. Two farms were selected for the study and for each farm three fields were available for planting. It was deemed impractical to use different sprays in a field; therefore, a split–plot design was utilized. At each farm the three sprays were randomly assigned to the fields (whole plots), with the restriction that there be one spray used per field. Each field was divided into two subplots, again with the restriction that each variety is used in exactly one subplot in each field. The response variable is the yield in bushels. The results are given in Table 18.1.1.

Table 18.1.1 Varieties of Wheat Data

	Farm 1			Farm 2		
Field	1	2	3	1	2	3
Spray	2	1	3	3	1	2
	56 (2)*	71 (1)	84 (1)	88 (2)	79 (2)	77 (1)
	64 (1)	66 (2)	82 (2)	97 (1)	83 (1)	73 (2)

*The variety used is given in parentheses.

Note that the data can alternately be displayed as follows:

		Spray		
Farm	Variety	1	2	3
1	1	71	64	84
1	2	66	56	82
2	1	83	77	97
2	2	79	73	88

We observe that a split–plot design has been utilized in this experiment within an RCBD. The farms are the blocks, the fields are the whole plots, factor A is spray and factor B is variety. As mentioned before, the spray factor is confounded with fields (whole plots).

We present the general model for a two–factor split–plot experiment within an RCBD with r random blocks, and factors A and B having a and b fixed levels. Let Y_{ijk} represent the response to the jth level of A and kth level of B in the ith block. The model with normality assumptions is as follows:

$$Y_{ijk} = \mu + \rho_i + \alpha_j + (\rho\alpha)_{ij} + \beta_k + (\alpha\beta)_{jk} + \varepsilon_{ijk};$$

$$i = 1,\cdots,r;\ j = 1,\cdots,a;\ k = 1,\cdots,b; \tag{18.1.1}$$

where $\sum_{j=1}^{a} \alpha_j = \sum_{k=1}^{b} \beta_k = 0$; $\sum_j (\alpha\beta)_{jk} = \sum_k (\alpha\beta)_{jk} = 0$; and $\sum_j (\rho\alpha)_{ij} = 0$. We assume that the ρ_i's, $(\rho\alpha)_{ij}$'s and ε_{ijk}'s are normal random variables with mean zero and variances σ_ρ^2, $\frac{a-1}{a}\sigma_{\rho\alpha}^2$, σ_ε^2, respectively. The ρ_i's and ε_{ijk}'s are independent sets of random variables, and the ρ_i, $(\rho\alpha)_{ij}$, ε_{ijk} are mutually independent of each other.

For reasons that will be explained later, the $(\rho\alpha)_{ij}$ term is often called the **whole plot error**, while ε_{ijk} is called the **subplot error**. With this model we obtain the ANOVA table given in Table 18.1.2. The E(MS) are found using the algorithm.

Table 18.1.2 ANOVA Table for Split—Plot; Factors Fixed, Blocks Random

Source	df	SS	E(MS)
Blocks (R)	r–1	SSbl	$\sigma_\varepsilon^2 + ab\sigma_\rho^2$
A	a–1	SSa	$\sigma_\varepsilon^2 + b\sigma_{\rho\alpha}^2 + rb\ Q_A$
Whole Plot Error (RA)	(r–1)(a–1)	SSra	$\sigma_\varepsilon^2 + b\sigma_{\rho\alpha}^2$
B	b–1	SSb	$\sigma_\varepsilon^2 + ra\ Q_B$
AB	(a–1)(b–1)	SSab	$\sigma_\varepsilon^2 + r\ Q_{AB}$
Subplot Error	a(r–1)(b–1)	SSe	σ_ε^2
Total	rab – 1	SSt	

From the E(MS) we note that the f—test for factor A involves using the whole plot error term RA. That is, we reject $H_0 : \alpha_j = 0$, for all j, if $f = MSa/MSra \geq f_{\alpha,a-1,(r-1)(a-1)}$.

Since factor A is the "whole plot" factor, it is natural to label the RA effect as the whole plot error. The tests for factor B and for the AB interaction involve the subplot error. That is, we reject $H_0 : \beta_k = 0$, for all k, if $f = MSb/MSe \geq f_{\alpha,b-1,a(r-1)(b-1)}$, and we reject $H_0 : (\alpha\beta)_{jk} = 0$, for all j, k, if $f = MSab/MSe \geq f_{\alpha,(a-1)(b-1),a(r-1)(b-1)}$.

Computational formulas for the sum of squares for the split–plot design of Table 18.1.2 are as follows:

$$SSt = \sum_i \sum_j \sum_k y_{ijk}^2 - \frac{y_{...}^2}{rab}, \quad SSbl = \sum_i \frac{y_{i..}^2}{ab} - \frac{y_{...}^2}{rab}$$

$$SSa = \sum_j \frac{y_{.j.}^2}{rb} - \frac{y_{...}^2}{rab}, \quad SSb = \sum_k \frac{y_{..k}^2}{ra} - \frac{y_{...}^2}{rab}$$

$$SSra = \sum_i \sum_j \frac{y_{ij.}^2}{b} - \frac{y_{...}^2}{rab} - SSbl - SSa \qquad (18.1.2)$$

$$SSab = \sum_j \sum_k \frac{y_{.jk}^2}{r} - \frac{y_{...}^2}{rab} - SSa - SSb$$

$$SSe = SSt - SSbl - SSa - SSb - SSra - SSab.$$

Estimates for the variance components are easy to find and are as follows:

Variance Component	Point Estimate
σ_ρ^2	$(MSbl - MSe)/ab$
$\sigma_{\rho\alpha}^2$	$(MSra - MSe)/b$
σ_ε^2	MSe

(18.1.3)

Some comments are in order concerning the split–plot model in (18.1.1). First, the model allows for dependence among various observations. Observations in subplots within the same whole plot are correlated because they have the same ρ_i and $(\rho\alpha)_{ij}$ terms in the model. Further, observations within the same block, but different whole plots, are correlated because they have the same ρ_i term in the model. Observations in different blocks are independent. One can note that this correlation structure makes sense intuitively in the context of the data from Example 18.1.1. Finally, we observe from the E(MS) entries in Table 18.1.2 that the whole plot error, MSra, is theoretically and therefore generally larger than the subplot error, MSe. Further MSra has smaller degrees of freedom than MSe. As a consequence, the test for factor B is generally more efficient than the test for factor A. Because of this fact, it is advantageous to make the factor of major interest the subplot factor.

To find confidence intervals or to perform multiple comparisons on the levels of factors A or B require knowing the point estimators and variances. In the exercises the estimators, variances and standard errors for the differences $\alpha_j - \alpha_\ell$ and $\beta_k - \beta_m$ are obtained. The results are summarized as follows:

Function	Estimator	Variance	Standard Error	
$\alpha_j - \alpha_\ell$	$\bar{Y}_{\cdot j \cdot} - \bar{Y}_{\cdot \ell \cdot}$	$2(b\sigma^2_{\rho\alpha} + \sigma^2_\varepsilon)/rb$	$\sqrt{2\ \text{MSra}/rb}$	
$\beta_k - \beta_m$	$\bar{Y}_{\cdot\cdot k} - \bar{Y}_{\cdot\cdot m}$	$2\sigma^2_\varepsilon/ra$	$\sqrt{2\ \text{MSe}/ra}$	(18.1.4)

The Bonferroni procedure can be used to carry out multiple comparisons for factor A and/or B in the usual manner. For example, for factor A we decide that

$$\alpha_j \neq \alpha_\ell \text{ if } |\bar{y}_{\cdot j \cdot} - \bar{y}_{\cdot \ell \cdot}| \geq t_{\frac{\alpha}{2c},(r-1)(a-1)} \sqrt{\frac{2\text{MSra}}{rb}}, \qquad (18.1.5)$$

where $c = a(a-1)/2$.

Analysis of Split–Plot Data Using SAS

We can now analyze the data in Example 18.1.1. The SAS printout in Table 18.1.3 contains an analysis of the data. The "TEST" statement is used to carry out the test for the whole plot factor (i.e., spray). From the printout we note that there is a significant effect for the main effects spray ($p = .0177$) and variety ($p = .0292$) but there is no significant spray * variety interaction ($p = .9026$). The mean for the two varieties are $\bar{y}_{\cdot\cdot 1} = 79.3$ and $\bar{y}_{\cdot\cdot 2} = 74.0$; therefore, it is clear that variety one has a significantly greater mean yield. The LSMEANS statement with error term specified as the whole plot error (farm * spray) provides a Bonferroni comparison for the levels of spray. Using a 0.05 level, we declare means significantly different if $p \leq .05/3 = .017$. This suggests that spray 3 has significantly higher mean yield than spray 2, but other means are not significantly different.

Table 18.1.3 SAS Printout for Varieties of Wheat Data

SAS Commands

```
options ls = 80;
data txsp;
input farm spray variety y @@;
cards;
1 1 1 71  1 1 2 66  1 2 1 64  1 2 2 56  1 3 1 84  1 3 2 82
2 1 1 83  2 1 2 79  2 2 1 77  2 2 2 73  2 3 1 97  2 3 2 88
proc glm; class farm spray variety;
model y = farm spray farm*spray variety spray*variety;
test h = spray e = farm*spray;
means variety;
lsmeans spray / pdiff e = farm*spray;
```

SAS Printout

General Linear Models Procedure

Dependent Variable: Y

Source	DF	Sum of Squares	Mean Square	F Value	Pr > F
Model	8	1400.1666667	175.0208333	31.82	0.0081
Error	3	16.5000000	5.5000000		
Corrected Total	11	1416.6666667			

R-Square	C.V.	Root MSE	Y Mean
0.988353	3.058967	2.3452079	76.666667

Source	DF	Type I SS	Mean Square	F Value	Pr > F
FARM	1	456.33333333	456.33333333	82.97	0.0028
SPRAY	2	842.16666667	421.08333333	76.56	0.0027
FARM*SPRAY	2	15.16666667	7.58333333	1.38	0.3761
VARIETY	1	85.33333333	85.33333333	15.52	0.0292
SPRAY*VARIETY	2	1.16666667	0.58333333	0.11	0.9026

Source	DF	Type III SS	Mean Square	F Value	Pr > F
FARM	1	456.33333333	456.33333333	82.97	0.0028
SPRAY	2	842.16666667	421.08333333	76.56	0.0027
FARM*SPRAY	2	15.16666667	7.58333333	1.38	0.3761
VARIETY	1	85.33333333	85.33333333	15.52	0.0292
SPRAY*VARIETY	2	1.16666667	0.58333333	0.11	0.9026

Tests of Hypotheses using the Type III MS for FARM*SPRAY as an error term

Source	DF	Type III SS	Mean Square	F Value	Pr > F
SPRAY	2	842.16666667	421.08333333	55.53	0.0177

General Linear Models Procedure

Level of VARIETY	N	--------------Y-------------- Mean	SD
1	6	79.3333333	11.4658914
2	6	74.0000000	11.6103402

General Linear Models Procedure
Least Squares Means

Standard Errors and Probabilities calculated using the Type III MS for
FARM*SPRAY as an Error term

SPRAY	Y LSMEAN	Pr > |T| H0: LSMEAN(i)=LSMEAN(j) i/j	1	2	3
1	74.7500000	1	.	0.0652	0.0217
2	67.5000000	2	0.0652	.	0.0091
3	87.7500000	3	0.0217	0.0091	.

NOTE: To ensure overall protection level, only probabilities associated with pre-planned comparisons should be used.

Other Split–Plot Designs

As noted earlier, other basic designs can be employed to achieve split—plot randomization arrangements. The ANOVA tables for two—factor split—plots replicated in a CRD and Latin square are sketched in Table 18.1.4 below. (See the exercises for others.) Since the corresponding model is "obvious" and the E(MS) column depends on which effects are fixed and which are random, only the ANOVA source and df columns are given. Also, the reader by this time should be familiar with the formula "patterns" used to calculate the ANOVA sums of squares for balanced designs.

Table 18.1.4 ANOVA Tables for Split–Plot Designs

Split–Plot in Balanced CRD (r Replications)

Source	df
A	a − 1
Whole Plot Error	a(r − 1)
B	b − 1
AB	(a − 1)(b − 1)
Subplot Error	a(r − 1)(b − 1)
Total	rab − 1

Split–Plot in Latin–Square Design

Source	df
Rows	a − 1
Columns	a − 1
A	a − 1
Whole Plot Error	(a − 1)(a − 2)
B	b − 1
AB	(a − 1)(b − 1)
Subplot Error	a(a − 1)(b − 1)
Total	$a^2b - 1$

Thus far in this section we have concentrated on split—plot designs involving two factors of interest. We can easily extend the ideas to experiments involving three factors A, B, and C. In such a case we have whole plots divided into subplots which are divided into sub—subplots. Levels of A are assigned to whole plots, levels of B are assigned to subplots and levels of C are assigned to the sub—subplots. Such a design is called a **split—split—plot design**. The ANOVA table (source and df only) for such a design is given in Table 18.1.5.

Table 18.1.5 ANOVA Table for Split—Split Plot Design in an RCBD

Source	df
Blocks	r − 1
A	a − 1
Whole Plot Error	(r − 1)(a − 1)
B	b − 1
AB	(a − 1)(b − 1)
Subplot Error	a(r − 1)(b − 1)
C	c − 1
AC	(a − 1)(c − 1)
BC	(b − 1)(c − 1)
ABC	(a − 1)(b − 1)(c − 1)
Subsubplot Error	ab(r − 1)(c − 1)
Total	rabc − 1

18.2 Crossover Designs

Suppose that we desire to compare the effects of p different treatments and r experimental units are available for the study. The treatments can be different drugs and the subjects can be people or animals that will use the drugs. To control for the variability due to the subjects (experimental units) it is desirable to have each subject use each treatment. One possible design that can be used in this setting is an RCBD. In this design each subject will use each drug and the order in which they use the drugs is randomized.

One difficulty with the RCBD approach is that while the order for using the drugs is randomized, there may be an imbalance in that some drugs will be used first or last more often than others. If there is an order effect for the drugs, then the RCBD approach can lead to inaccurate results. As an illustration, when there are two drugs (A and B) suppose that the randomization in the order leads to the following:

	Subject				
	1	2	3	4	5
Drug Received First	A	A	B	A	A
Drug Received Second	B	B	A	B	B

If the mean response for drug A is larger we cannot be sure whether this occurs because drug A leads to a larger response or because drug A is used first most of the time. Clearly, we need a more appropriate design for this situation.

Our purpose in this section is to introduce a design that will balance out the possible order effects of the drugs (treatments). The design is called a **crossover design** (sometimes called a **change–over** design). In a crossover design the experimental units (subjects) receive more than one (usually all) treatments in a specified sequence. By specifying the sequence we can ensure a reasonable balance for the order in which the treatments are used by the subjects.

One common way to obtain a crossover design is by the use of Latin squares. The sequences specifying the order for the treatments are determined by one or more Latin squares, and the subjects are then randomly assigned to the established sequences. To illustrate, suppose that we have p = 3 treatments labelled by A, B, C. Each subject is to use all three treatments. The possible sequences for using the treatments are established by the following Latin square:

		Order for Treatment		
		1	2	3
	1	A	B	C
Sequence	2	B	C	A
	3	C	A	B

Figure 18.2.1 Treatment Sequences

We now randomly assign the subjects to the three sequences with the restriction that each sequence is used the same number of times. This requires, of course, that the number of subjects be divisible by three.

We present an example using a crossover design. An experiment was completed to compare the effectiveness of two sleep–inducing drugs A and B. A placebo (drug C) was used for a control. Nine subjects were available for the study. The subjects were randomly assigned to the treatment sequences defined by the Latin square in Figure 18.2.1 with three subjects per sequence. The resulting data, number of hours of sleep after using the drug, is given in Table 18.2.1. In the table the drug used is given in parentheses.

Table 18.2.1 Crossover Design — Hours of Sleep

		Subject 1	2	3	4	5	6	7	8	9
Sequence		3	1	2	3	3	1	2	2	1
Order	1	5.2 (C)	7.9 (A)	8.5 (B)	5.5 (C)	7.1 (C)	7.2 (A)	8.0 (B)	7.0 (B)	6.7 (A)
	2	6.7 (A)	8.1 (B)	6.4 (C)	6.9 (A)	8.1 (A)	7.8 (B)	6.1 (C)	5.3 (C)	7.5 (B)
	3	7.4 (B)	6.3 (C)	7.6 (A)	6.8 (B)	8.8 (B)	6.0 (C)	7.4 (A)	6.3 (A)	6.8 (C)

Notice that the crossover design used in the previous table balances the order of the treatments in some ways, but not in other ways. For example, the design ensures that each drug (treatment) is used first, second, and third the same number of times. The design does not, however, balance the order in terms of which drug immediately precedes another drug. In the design, drug B is immediately preceded by drug A, but never by drug C. While drug A is always immediately preceded by drug C. An alternative to the design used in Table 18.2.1 is to use all $3! = 6$ possible sequences of treatment orders. This would require, however, that the number of subjects be divisible by six.

The Analysis of a Crossover Design Model

The model used for a crossover design depends a great deal on the assumption of **carryover** (residual) effects. Carryover effects are the effects of a treatment that continue into a subsequent time period. The best strategy is to run the experiment in such a manner as to avoid, or at least minimize, these carryover effects. In any case it seems advisable to use a model that allows for the possible presence of a treatment sequence or carryover effect. The model that we propose using is based upon $Y_{ijk\ell}$, which represents the response to the ith treatment, jth sequence, kth subject, and ℓth order (period). In this situation subject is nested within sequence. If we have p treatments, a sequences, b subjects nested in each sequence, and c time periods, then our model is:

$$Y_{ijk\ell} = \mu + \tau_i + \alpha_j + \beta_{k(j)} + \gamma_\ell + \varepsilon_{ijk\ell}; \tag{18.2.1}$$

$$i = 1,\cdots,p;\ \ j = 1,\cdots,a;\ \ k = 1,\cdots,b;\ \ \ell = 1,\cdots,c.$$

The assumptions that we use for the model depend upon the situation. Notice that the model has been written to allow for the case when each subject uses each treatment (i.e., p = c as in Figure 18.2.1), but also for the case when each subject uses only some of the treatments (i.e., p > c). In this latter case the sequences could be determined using the ideas of incomplete blocks. (See the exercises for more details.)

For the data in Table 18.2.1 we assume that the subjects are random, so in model (18.2.1) we assume that the $\beta_{k(j)}$ and $\varepsilon_{ijk\ell}$ are independent normal random variables with means zero and variances σ^2_β and σ^2_ε, respectively. The ANOVA table for this case when p = c (each subject uses all treatments) is given in Table 18.2.2. (It is important to note that the design is incomplete in that some combinations of i, j, k, ℓ are not present; therefore, the actual coefficients for the terms in the expressions for the E(MS) will be different than the ones found with the E(MS) algorithm).

Table 18.2.2 ANOVA Table for Crossover Design, Subjects Assumed Random

Source	df	SS	f	E(MS)
Treatments	p–1	SStr	MStr/MSe	$\sigma^2_\varepsilon + bp\,Q_{tr}$
Sequences	a – 1	SSsq	MSsq/MSsb(sq)	$\sigma^2_\varepsilon + p\sigma^2_\beta + p^2Q_{sq}$
Subjects (Seq)	a(b – 1)	SSsb(sq)	MSsb(sq)/MSe	$\sigma^2_\varepsilon + p\sigma^2_\beta$
Order	p – 1	SSor	MSor/MSe	$\sigma^2_\varepsilon + bp\,Q_{or}$
Error	abp–ab–2p+2	SSe		σ^2_ε
Total	abp – 1	SSt		

The form of the f—tests are found by inspecting the expressions for the E(MS).

The test of major interest in Table 18.2.2 is the f—test for the treatments. The test for sequence provides a method for determining if there are carryover effects. A significant sequence effect would suggest the presence of a carryover effect among the treatments. These ideas are illustrated in the following example.

Example 18.2.1 Refer to the Hours of Sleep data in Table 18.2.1. We use the model (18.2.1) in which $p = c = 3$, $a = 3$ and $b = 3$. We assume that subjects are random and other factors are fixed; therefore, the analysis provided in the ANOVA Table 18.2.2 is appropriate. The SAS printout in Table 18.2.3 provides the analysis of these data. From the printout we note that the f—test for sequences is not significant ($f = .08$; $p = .9205$); therefore, there is no evidence of a carryover effect. The f—test for treatments (drugs) is significant ($f = 63.44$; $p = .0001$) indicating differences between at least two drugs. A Bonferroni multiple comparison can be used to determine which drugs are significantly different. From the printout we declare two drugs significantly different if the p—value for the LSMEANS is $\alpha/(\text{number of comparison}) = .05/3 = .017$ or less. It follows that all three drugs are significantly different, with drug B having the highest mean, and drug A the next highest.

Table 18.2.3 SAS Printout for Hours of Sleep Data

SAS Commands

```
options ls = 80;
data txcross;
input subj seq order drug y @@;
cards;
1 3 1 3 5.2    1 3 2 1 6.7    1 3 3 2 7.4
2 1 1 1 7.9    2 1 2 2 8.1    2 1 3 3 6.3
3 2 1 2 8.5    3 2 2 3 6.4    3 2 3 1 7.6
4 3 1 3 5.5    4 3 2 1 6.9    4 3 3 2 6.8
5 3 1 3 7.1    5 3 2 1 8.1    5 3 3 2 8.8
6 1 1 1 7.2    6 1 2 2 7.8    6 1 3 3 6.0
7 2 1 2 8.0    7 2 2 3 6.1    7 2 3 1 7.4
8 2 1 2 7.0    8 2 2 3 5.3    8 2 3 1 6.3
9 1 1 1 6.7    9 1 2 2 7.5    9 1 3 3 6.8
proc glm; class subj seq order drug;
model y = drug seq subj(seq) order;
test h = seq e = subj(seq);
lsmeans drug / pdiff;
```

SAS Printout

General Linear Models Procedure

Dependent Variable: Y

Source	DF	Sum of Squares	Mean Square	F Value	Pr > F
Model	12	21.66666667	1.80555556	17.23	0.0001
Error	14	1.46740741	0.10481481		
Corrected Total	26	23.13407407			

R-Square	C.V.	Root MSE	Y Mean
0.936569	4.615249	0.3237512	7.0148148

Source	DF	Type I SS	Mean Square	F Value	Pr > F
DRUG	2	13.29851852	6.64925926	63.44	0.0001
SEQ	2	0.22740741	0.11370370	1.08	0.3648
SUBJ(SEQ)	6	8.12666667	1.35444444	12.92	0.0001
ORDER	2	0.01407407	0.00703704	0.07	0.9354

Source	DF	Type III SS	Mean Square	F Value	Pr > F
DRUG	2	13.29851852	6.64925926	63.44	0.0001
SEQ	2	0.22740741	0.11370370	1.08	0.3648
SUBJ(SEQ)	6	8.12666667	1.35444444	12.92	0.0001
ORDER	2	0.01407407	0.00703704	0.07	0.9354

Tests of Hypotheses using the Type III MS for SUBJ(SEQ) as an error term

Source	DF	Type III SS	Mean Square	F Value	Pr > F
SEQ	2	0.22740741	0.11370370	0.08	0.9205

General Linear Models Procedure
Least Squares Means

DRUG	Y LSMEAN	Pr > |T| H0: LSMEAN(i)=LSMEAN(j) i/j	1	2	3
1	7.20000000	1	.	0.0023	0.0001
2	7.76666667	2	0.0023	.	0.0001
3	6.07777778	3	0.0001	0.0001	.

NOTE: To ensure overall protection level, only probabilities associated with pre-planned comparisons should be used.

Additional Considerations and Crossover Models

There are other models that are sometimes used for crossover designs. First, if we assume that there is no carryover effect, then we can eliminate the sequence term in the model (18.2.1). If we have a total of r subjects and if $Y_{ij\ell}$ represents the response to the ith treatment, jth subject, and ℓth order (period), then the model is

$$Y_{ij\ell} = \mu + \tau_i + \beta_j + \gamma_\ell + \varepsilon_{ij\ell}; \tag{18.2.2}$$

$$i = 1,\cdots,p; \; j = 1,\cdots,r; \text{ and } \ell = 1,\cdots,c.$$

The difference between models (18.2.1) and (18.2.2) is that the sum of squares for subjects in model (18.2.2) is partitioned into two parts (sequences, subjects nested in sequences) in model (18.2.1).

Another model that is sometimes used for crossover designs relates the response variable to the effect of the treatment applied in the previous period. This enables us to account for the carryover effect from the previous period. Keuhl (1994) provides a detailed discussion on this model and the corresponding analysis.

If carryover effects are expected to be large, then a more complicated crossover design may be utilized. Williams (1949) and John (1971) provide methods for constructing designs that balance out these carryover effects. One design used in this case is a **double crossover design**. In this design each treatment occurs in each order the same number of times, and each treatment immediately follows each of the other treatments the same number of times. An example of a double crossover design for four treatments (A,B,C,D) appears in Figure 18.2.2.

		Order			
		1	2	3	4
Subject	1	A	B	C	D
	2	B	D	A	C
	3	D	C	B	A
	4	C	A	D	B

Figure 18.2.2 Double Crossover Design

In this design each treatment immediately follows each of the other treatments exactly once.

In summary, crossover designs are useful for comparing several treatments when subjects use the treatments over two or more periods of time. Crossover designs are constructed to avoid confounding between the treatments and the order (period) effects. They also enable us to check for the presence of carryover effects between the treatments in the current and previous periods.

18.3 Response Surfaces

The purpose of some experiments is to study the relationship between two or more factors and a response variable Y. In some cases we desire to predict the value of Y for specified values of the factors, while in other cases we may desire to determine the values of the factors that lead to the optimal value of the response variable. Our purpose in this section is to consider these types of problems when the factors are quantitative in nature.

Our situation of interest involves a response variable Y and quantitative factors $x_1, \cdots, x_k$. To study the relationship between Y and the factors, we examine a response surface. The **response surface** is a geometric representation obtained when the response variable Y is plotted against the factors. Algebraically we can express this response surface by $Y = f(x_1, \cdots, x_k)$, where f is some function. Some three–dimensional response surfaces are given in Figure 18.3.1.

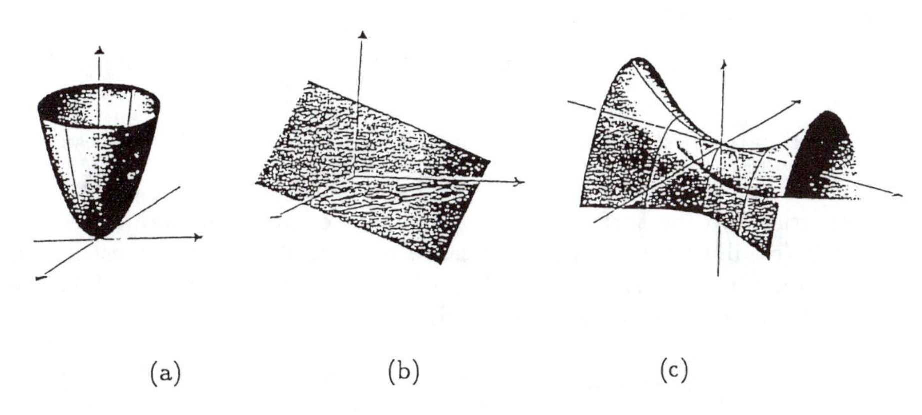

Figure 18.3.1 Some Response Surfaces

If we obtain the equation of the response surface, we can use this equation to predict values of Y or to search for an optimal value. To study these surfaces we use **response surface methodology**. In this methodology we first choose an appropriate design to measure the value of the response at selected values of the factors (sometimes called **design points**). With the data obtained from the designed experiment we select an appropriate model relating Y to the x_i's. Regression techniques are used to estimate the unknown parameters in the model, and the estimated regression equation provides an estimated response surface. Obviously we need to choose a design that lends itself to providing data that can be used to estimate the chosen equation for the response surface.

As an illustration that we will use throughout this section, suppose that we aim to produce a new soap product. We know that our soap product will have two major active ingredients, labelled as A and B. The remainder of the soap will consist of non–active fillers. We believe that the amount of ingredient A should be between 10 and 20 grams out of 100 grams of the product, and the amount of ingredient B should be between 7 and 8 grams. Our task is to determine what the levels of ingredients A and B should be in order to yield the largest value of the response variable Y, a measure of quality of the soap's cleaning ability as elicited on a 0—40 scale. Our goal is to find a response surface representing the relationship between the levels of ingredients A and B and the cleaning ability. We can then study the response surface to find the optimal levels of A and B. We first need to consider appropriate models for a response surface.

The most commonly used response surfaces are based upon low—order polynomial models. When there are k factors $x_1, \cdots, x_k$ the **first—order model** is given by:

$$Y = \beta_0 + \beta_1 x_1 + \cdots + \beta_k x_k + \varepsilon. \tag{18.3.1}$$

Our estimated first—order model $\hat{y} = b_0 + \sum_{i=1}^{k} b_i x_i$ provides a response surface that is a plane (k = 2) or hyperplane (k > 2). There is no "curvature" to the surface and no maximum or minimum value. Figure 18.3.1 (b) illustrates a response surface associated with an estimated first—order model when k = 2 . A more complicated model is the **second—order model** which is given by:

$$Y = \beta_0 + \sum_{i=1}^{k} \beta_i x_i + \sum_{i=1}^{k} \beta_{ii} x_i^2 + \sum_{i<j} \sum \beta_{ij} x_i x_j + \varepsilon. \tag{18.3.2}$$

When k = 2, this model takes the form

$$Y = \beta_0 + \beta_1 x_1 + \beta_2 x_2 + \beta_{11} x_1^2 + \beta_{22} x_2^2 + \beta_{12} x_1 x_2 + \varepsilon.$$

Figures 18.3.1 (a) and (c) are examples of response surfaces associated with this model. Observe that the model allows curvature in the response surface and, consequently, there can be maximum and/or minimum values for y in some cases.

Our purpose in the remainder of this section is to discuss the designs that are useful for estimating first— and second—order models.

First–Order Models

Designs that are used for first—order models are sometimes called **screening designs**. Screening designs are used in the early stages of a product development to help determine what the levels of the factors should be to yield the "best" product. A first—order model is useful in such a case when we are not quite ready to establish the final optimal situation, but want to explore in the simplest way the relationship between the factors and the response variable.

A design that is often used for fitting a first—order model is the 2^k factorial, or perhaps a fraction of a 2^k. In a 2^k design when the factors are numerical it is convenient to code the values of the factors as 1 for high and −1 for low. This provides for an orthogonal regression design and an ease in comparing the coefficients for the various factors. We illustrate the 2^k and the use of a first—order model in the following example.

Example 18.3.1 Refer to the problem described earlier in this section concerning the development of a soap product. Two ingredients, A and B, are to be used as the main active ingredients in the soap. We decide to first use a simple, inexpensive experiment to obtain an idea how the two ingredients relate to the cleaning ability of the soap. A 2^2 with a first—order model should be sufficient. Since we believe that A should be between 10 and 20 grams we choose these two levels coded as −1 (for 10 grams) and 1 (for 20 grams). Similarly we use for factor B −1 for 7 grams and 1 for 8 grams. Three replications will be used for each of the four design points (−1,−1), (−1,1), (1,−1), (1,1). The data obtained are given in Table 18.3.1. The response variable Y measures the cleaning ability.

Table 18.3.1 Soap Data With 2^2 Factorial

		Level of B	
		−1 (7 grams)	1 8 (grams)
Level of A	−1 (10 grams)	15 14 12	12 6 7
	1 (20 grams)	24 20 19	9 5 8

.

We fit a first–order model $Y = \beta_0 + \beta_1 x_1 + \beta_2 x_2 + \varepsilon$. Using regression techniques (see SAS printout in Table 18.3.2) we obtain the fitted model $\hat{y} = 12.583 + 1.583x_1 - 4.75x_2$. We note from the printout that the β_1 coefficient is not significantly different from zero, while β_2 is significantly different. Since x_2 has a significant negative coefficient we should expect a larger mean response (cleaning ability) if we decrease the amount of ingredient B. There is no **significant** evidence supporting the need to increase or decrease the amount of ingredient A, although the regression coefficient on x_1 suggests an increase might help. The results of our screening design suggest that a new experiment should be completed, definitely using a smaller amount of B, and possibly a somewhat large amount of A.

Table 18.3.2 SAS Printout for 2^2 Experiment Soap Data

SAS Commands

```
options ls = 80;
data txres2k;
input x1 x2 y @@;
cards;
-1 -1   15  -1   -1  14  -1 -1    12  -1   1 12    -1    1
6 -1    1  7
1 -1    24   1   -1  20   1 -1    19   1   1 9 1    1    5
1 1     8
proc glm;
model y = x1 x2;
```

SAS Printout

General Linear Models Procedure

Dependent Variable: Y

Source	DF	Sum of Squares	Mean Square	F Value	Pr > F
Model	2	300.83333333	150.41666667	13.53	0.0019
Error	9	100.08333333	11.12037037		
Corrected Total	11	400.91666667			

R-Square	C.V.	Root MSE	Y Mean
0.750364	26.50110	3.3347219	12.583333

Source	DF	Type I SS	Mean Square	F Value	Pr > F
X1	1	30.08333333	30.08333333	2.71	0.1344
X2	1	270.75000000	270.75000000	24.35	0.0008

Source	DF	Type III SS	Mean Square	F Value	Pr > F
X1	1	30.08333333	30.08333333	2.71	0.1344
X2	1	270.75000000	270.75000000	24.35	0.0008

Parameter	Estimate	T for H0: Parameter=0	Pr > |T|	Std Error of Estimate
INTERCEPT	12.58333333	13.07	0.0001	0.96265130
X1	1.58333333	1.64	0.1344	0.96265130
X2	-4.75000000	-4.93	0.0008	0.96265130

Sometimes we can plot the **contours** of a response surface to obtain a better understanding of the relationship between the factors and the response variable. A **contour plot** is a set of lines or curves that identify the values of the factors for which the response variable is constant. As an illustration of a contour plot see Figure 18.3.2. In part (a) of the figure we have a response surface in three dimensions (x_1, x_2, y), while in part (b) we have the corresponding contour plot in two dimensions (x_1, x_2).

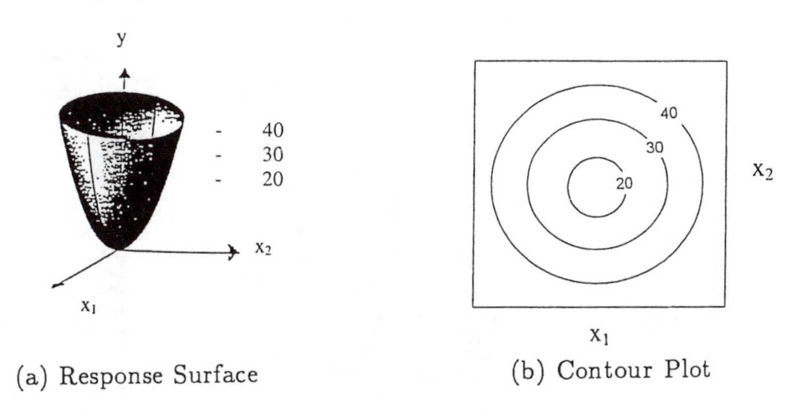

(a) Response Surface (b) Contour Plot

Figure 18.3.2 Response Surface and Contour Plot

SAS can be used to obtain contour plots. Table 18.3.3 contains a contour plot for the response surface (plane) $\hat{y} = 12.583 + 1.583x_1 - 4.75x_2$ calculated in Example 18.3.1. From the contour plot we notice that the response tends to increase most dramatically as we decrease x_2 (level of ingredient B), which matches our conclusion in Example 18.3.1.

Table 18.3.3 Contour Plot for 2^2 Soap Data

SAS Commands

```
options ls = 80;
data txres2k1;
do d1 = -1 to 1 by .01;
do x2 = -1 to 1 by .01;
y = 12.583 + 1.583 * x1 - 4.75 * x2;
output; end; end;
proc plot; plot x2 * x1 = y / contour = 10;
```

SAS Printout

```
                              Contour plot of X2*X1.

     X2 |
   1.00 + ...........................'''''''''''''''''''''''''''''-----------------
        | .....................''''''''''''''''''''''''''''''----------------------
        | ................'''''''''''''''''''''''''''''----------------------------
        | ..........''''''''''''''''''''''''''''''-----------------------------====
        | .....'''''''''''''''''''''''''''''------------------------------=========
   0.75 + '''''''''''''''''''''''''''''-----------------------------===============
        | '''''''''''''''''''''''------------------------------====================
        | ''''''''''''''''''-----------------------------==========================
        | ''''''''''''------------------------------=============================++
        | '''''''-----------------------------==============================+++++++
   0.50 + ''-----------------------------==============================++++++++++++
        | -------------------------==============================++++++++++++++++++
        | --------------------=============================++++++++++++++++++++++++
        | --------------==============================++++++++++++++++++++++++++++O
        | ---------=============================+++++++++++++++++++++++++++++OOOOOO
   0.25 + ----=============================+++++++++++++++++++++++++++++OOOOOOOOOOO
        | ===========================++++++++++++++++++++++++++++++OOOOOOOOOOOOOOOO
        | ======================+++++++++++++++++++++++++++++OOOOOOOOOOOOOOOOOOOOOO
        | ================++++++++++++++++++++++++++++++OOOOOOOOOOOOOOOOOOOOOOOOOOO
        | ===========+++++++++++++++++++++++++++++OOOOOOOOOOOOOOOOOOOOOOOOOOOOOXXXX
   0.00 + =====+++++++++++++++++++++++++++++OOOOOOOOOOOOOOOOOOOOOOOOOOOOOOXXXXXXXXX
        | +++++++++++++++++++++++++++++OOOOOOOOOOOOOOOOOOOOOOOOOOOOOOXXXXXXXXXXXXXX
        | ++++++++++++++++++++++++OOOOOOOOOOOOOOOOOOOOOOOOOOOOOXXXXXXXXXXXXXXXXXXXX
        | ++++++++++++++++++OOOOOOOOOOOOOOOOOOOOOOOOOOOOOOXXXXXXXXXXXXXXXXXXXXXXXXX
        | +++++++++++++OOOOOOOOOOOOOOOOOOOOOOOOOOOOOXXXXXXXXXXXXXXXXXXXXXXXXXXXXXθθ
  -0.25 + +++++++OOOOOOOOOOOOOOOOOOOOOOOOOOOOOOXXXXXXXXXXXXXXXXXXXXXXXXXXXXXθθθθθθθ
        | ++OOOOOOOOOOOOOOOOOOOOOOOOOOOOOXXXXXXXXXXXXXXXXXXXXXXXXXXXXXXθθθθθθθθθθθθ
        | OOOOOOOOOOOOOOOOOOOOOOOOOOXXXXXXXXXXXXXXXXXXXXXXXXXXXXXθθθθθθθθθθθθθθθθθθ
        | OOOOOOOOOOOOOOOOOOOOXXXXXXXXXXXXXXXXXXXXXXXXXXXXXXθθθθθθθθθθθθθθθθθθθθθθθ
        | OOOOOOOOOOOOOOOXXXXXXXXXXXXXXXXXXXXXXXXXXXXXθθθθθθθθθθθθθθθθθθθθθθθθθθθθ⊠
  -0.50 + OOOOOOOOXXXXXXXXXXXXXXXXXXXXXXXXXXXXXXθθθθθθθθθθθθθθθθθθθθθθθθθθθθθ⊠⊠⊠⊠⊠⊠
        | OOOOXXXXXXXXXXXXXXXXXXXXXXXXXXXXXθθθθθθθθθθθθθθθθθθθθθθθθθθθθθ⊠⊠⊠⊠⊠⊠⊠⊠⊠⊠⊠
        | XXXXXXXXXXXXXXXXXXXXXXXXXXXXθθθθθθθθθθθθθθθθθθθθθθθθθθθθθ⊠⊠⊠⊠⊠⊠⊠⊠⊠⊠⊠⊠⊠⊠⊠⊠
        | XXXXXXXXXXXXXXXXXXXXXXθθθθθθθθθθθθθθθθθθθθθθθθθθθθθθ⊠⊠⊠⊠⊠⊠⊠⊠⊠⊠⊠⊠⊠⊠⊠⊠⊠⊠⊠⊠⊠
        | XXXXXXXXXXXXXXXXXθθθθθθθθθθθθθθθθθθθθθθθθθθθθθ⊠⊠⊠⊠⊠⊠⊠⊠⊠⊠⊠⊠⊠⊠⊠⊠⊠⊠⊠⊠⊠⊠⊠⊠⊠⊠⊠
  -0.75 + XXXXXXXXXXXθθθθθθθθθθθθθθθθθθθθθθθθθθθθθθ⊠⊠⊠⊠⊠⊠⊠⊠⊠⊠⊠⊠⊠⊠⊠⊠⊠⊠⊠⊠⊠⊠⊠⊠⊠⊠⊠⊠▩▩▩▩
        | XXXXXθθθθθθθθθθθθθθθθθθθθθθθθθθθθθθ⊠⊠⊠⊠⊠⊠⊠⊠⊠⊠⊠⊠⊠⊠⊠⊠⊠⊠⊠⊠⊠⊠⊠⊠⊠⊠⊠⊠⊠⊠▩▩▩▩▩▩▩▩
        | θθθθθθθθθθθθθθθθθθθθθθθθθθθθθθ⊠⊠⊠⊠⊠⊠⊠⊠⊠⊠⊠⊠⊠⊠⊠⊠⊠⊠⊠⊠⊠⊠⊠⊠⊠⊠⊠⊠⊠▩▩▩▩▩▩▩▩▩▩▩▩▩▩
        | θθθθθθθθθθθθθθθθθθθθθθθθ⊠⊠⊠⊠⊠⊠⊠⊠⊠⊠⊠⊠⊠⊠⊠⊠⊠⊠⊠⊠⊠⊠⊠⊠⊠⊠⊠⊠⊠⊠▩▩▩▩▩▩▩▩▩▩▩▩▩▩▩▩▩▩▩
        | θθθθθθθθθθθθθθθθθθθ⊠⊠⊠⊠⊠⊠⊠⊠⊠⊠⊠⊠⊠⊠⊠⊠⊠⊠⊠⊠⊠⊠⊠⊠⊠⊠⊠⊠⊠▩▩▩▩▩▩▩▩▩▩▩▩▩▩▩▩▩▩▩▩▩▩▩▩▩
  -1.00 + θθθθθθθθθθθθθ⊠⊠⊠⊠⊠⊠⊠⊠⊠⊠⊠⊠⊠⊠⊠⊠⊠⊠⊠⊠⊠⊠⊠⊠⊠⊠⊠⊠⊠⊠▩▩▩▩▩▩▩▩▩▩▩▩▩▩▩▩▩▩▩▩▩▩▩▩▩▩▩▩▩▩
        --+---------+----------+---------+----------+---------+----------+-
        -1.0      -0.7       -0.4      -0.1        0.2       0.5        0.8        1.1

                                        X1
```

Symbol	Y	Symbol	Y	Symbol	Y
.....	6.250 - 7.517	+++++	11.316 - 12.583	▨▨▨▨▨	16.383 - 17.649
'''''	7.517 - 8.783	OOOOO	12.583 - 13.850	▩▩▩▩▩	17.649 - 18.916
-----	8.783 - 10.050	XXXXX	13.850 - 15.116		
=====	10.050 - 11.316	⊖⊖⊖⊖⊖	15.116 - 16.383		

NOTE: 37613 obs hidden.

We have discussed the idea of using the results of a 2^k design to help decide where the next set of design points should be taken. A formal process for deciding how to move the design can be found using the **method of steepest ascent**. While we will not discuss this procedure here, the interested reader can learn more about it in Montgomery (1991).

There are other designs that are sometimes used for fitting first–order models. A **2^k factorial design with a center point** is sometimes used. This design simply augments the design points of a 2^k with a design point at $(0, 0, \cdots, 0)$. Figure 18.3.3 contains the design points for a 2^2 with center point.

		x_2	
x_1	(1,−1)		(1,1)
		(0,0)	
	(−1,−1)		(−1,1)

Figure 18.3.3 2^2 Design With Center Point

The advantage of having a center point is that we can obtain a measure for curvature of the response surface. For a quick check for curvature we can calculate $\bar{y}_F$, the mean of the observations at the four factorial design points $(\pm 1, \pm 1)$, and $\bar{y}_0$ the mean of the observations at the center point. If $\bar{y}_F$ and $\bar{y}_0$ are about the same, then the observations at the center point lie close to the plane determined by the factorial points; therefore, there is no evidence of curvature. A simple measure of curvature is $\bar{y}_F - \bar{y}_0$, with values close to zero suggesting no curvature. For the balanced case (same number of observations at each factorial design point), a sum of squares type measure of curvature, having a single degree of freedom, is given by

$$\text{SScur} = n_F n_0 \, (\bar{y}_F - \bar{y}_0)^2 \, / \, (n_F + n_0), \tag{18.3.3}$$

where n_F represents the total number of observations taken at the factorial design points, and n_0 is the number of observations taken at the center point. A formal f—test for curvature is given in the exercises.

Second–Order Models

In order to find an estimated second—order model (18.3.2) which involves quadratic terms our design will need to have at least three levels for each factor. Recall that with just one factor, an estimated second—order response curve, $\hat{y} = b_0 + b_1 x + b_{11} x^2$, can only be found when there are three or more different values of x. The same is true for higher dimensional problems. We cannot, therefore, fit a second—order model for data from a 2^k design. Even a 2^k with a center point cannot be used to fit a second—order model. (See the exercises.)

One experiment that can be used with second—order models is a **3^k factorial experiment.** In such an experiment there are $k \geq 2$ factors each having three levels. Figure 18.3.4 gives the design points for a 3^2 factorial. We have conveniently labelled the three levels of the factors by L, M, H. Note that a 3^2 factorial experiment involves $3^2 = 9$ different design points (treatments).

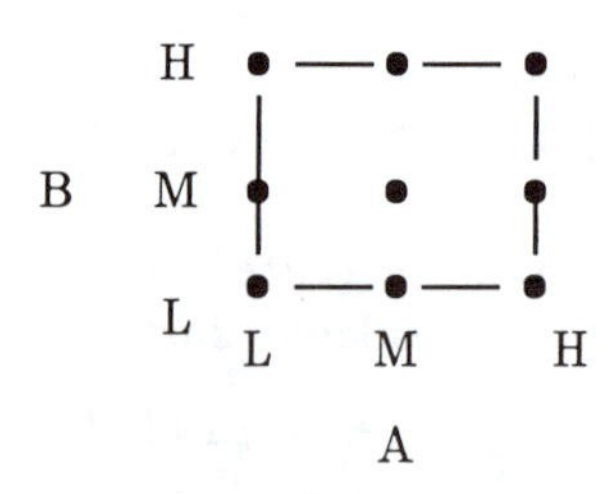

Figure 18.3.4 Representation of a 3^2

The next example illustrates the use of a 3^2 factorial for fitting a second—order model response surface. As mentioned before, the advantage of using a second—order model is that we can sometimes obtain a maximal or minimal value.

Example 18.3.2 We return again to our task of producing a quality soap product as discussed in Example 18.3.1. Our soap product involves main active ingredients A and B, and our goal is to find the levels of A and B that lead to the best soap product, best in terms of cleaning ability.

In Example 18.3.1 we illustrated a 2^2 factorial experiment with levels of A at 10 and 20, and levels of B at 7 and 8. We discovered in the previous example that to improve the cleaning ability we definitely need to decrease the level of B, and perhaps increase somewhat the level of A. To that end we perform a new experiment. Since we want to use a second—order model, hopefully to find a maximal value, we use a 3^2 factorial experiment. The levels we choose for factor A are 12, 17, and 22, while the levels for factor B are 5, 6, and 7. Since these levels are equally spaced we can code the levels as —1, 0, and 1 to simplify comparisons in the regression equation.

Three replications are taken at each of the $3^2 = 9$ design points (treatments) using a CRD. The resulting data are given in Table 18.3.4.

Table 18.3.4 Soap Data With 3^2 Factorial

		B		
		L(5)	M(6)	H(7)
A	L(12)	21, 19, 22	30, 33, 31	18, 21, 20
	M(17)	24, 27, 26	35, 35, 37	22, 24, 26
	H(22)	10, 12, 15	22, 20, 23	15, 12, 14

The SAS printout in Table 18.3.5 provides a fit to the second—order model yielding the response surface $\hat{y} = 35.259 - 4.000x_1 - .222x_2 + .583x_1x_2 - 8.556x_1^2 - 10.222x_2^2$, where x_1 represents the level of A coded as —1,0,1, and x_2 represents the coded level of B. From the printout we note a reasonable fit for this model, with $r^2 = .956$. Both quadratic terms are significant, which suggests the need of using a model allowing for curvature. Table 18.3.6 provides the contour plot for the estimated response surface, and from the plot we observe that a maximal value occurs approximately when $x_1 \approx -.2$ and $x_2 \approx 0$. By taking partial derivatives of the response function and setting them equal to zero, we can obtain the specific value that leads to a maximum. This results in the maximal value occurring when $x_1 = .23$ and $x_2 = -.02$. Recognizing that the levels for A and B are given by $5x_1 + 17$ and $x_2 + 6$, respectively, we find that the maximal cleaning ability occurs when A's level is 15.85 and B's level is 5.98.

Table 18.3.5 SAS Printout for Soap Data: Second–Order Model

SAS Commands

```
options ls = 80;
data tx3k;
input x1 x2 y @@'
x11 = x1*x1;  x22 = x2*x2;  x12 = x1*x2;
cards;
-1  -1  21     -1  -1  19     -1  -1  22
-1   0  30     -1   0  33     -1   0  31
-1   1  18     -1   1  21     -1   1  20
 0  -1  24      0  -1  27      0  -1  26
 0   0  35      0   0  35      0   0  37
 0   1  22      0   1  24      0   1  26
 1  -1  10      1  -1  12      1  -1  15
 1   0  22      1   0  20      1   0  23
 1   1  15      1   1  12      1   1  14
proc glm;
model y = x1 x2 x12 x11 x22;
```

SAS Printout

General Linear Models Procedure

Dependent Variable: Y

Source	DF	Sum of Squares	Mean Square	F Value	Pr > F
Model	5	1359.1203704	271.8240741	91.97	0.0001
Error	21	62.0648148	2.9554674		
Corrected Total	26	1421.1851852			

R-Square	C.V.	Root MSE	Y Mean
0.956329	7.559768	1.7191473	22.740741

Source	DF	Type I SS	Mean Square	F Value	Pr > F
X1	1	288.00000000	288.00000000	97.45	0.0001
X2	1	0.88888889	0.88888889	0.30	0.5892
X12	1	4.08333333	4.08333333	1.38	0.2530
X11	1	439.18518519	439.18518519	148.60	0.0001
X22	1	626.96296296	626.96296296	212.14	0.0001

Source	DF	Type III SS	Mean Square	F Value	Pr > F
X1	1	288.00000000	288.00000000	97.45	0.0001
X2	1	0.88888889	0.88888889	0.30	0.5892
X12	1	4.08333333	4.08333333	1.38	0.2530
X11	1	439.18518519	439.18518519	148.60	0.0001
X22	1	626.96296296	626.96296296	212.14	0.0001

Parameter	Estimate	T for H0: Parameter=0	Pr > |T|	Std Error of Estimate
INTERCEPT	35.25925926	47.66	0.0001	0.73980320
X1	-4.00000000	-9.87	0.0001	0.40520690
X2	-0.22222222	-0.55	0.5892	0.40520690
X12	0.58333333	1.18	0.2530	0.49627507
X11	-8.55555556	-12.19	0.0001	0.70183894
X22	-10.22222222	-14.56	0.0001	0.70183894

Table 18.3.6
SAS Printout for Soap Data: Contour Plot for Second–Order Model

<u>SAS Commands</u>

```
options ls = 80;
data tx3k1;
do x1 = –1 to 1 by .01;
do x2 = –1 to 1 by .01;
y = 35.259 –4.000*x1 – .222*x2 + .583*x1*x2 –8.556*x1*x1 –10.222*x
output; end; end;
proc plot; plot x2*x1 = y / contour = 10;
```

<u>SAS Printout</u>

```
                              Contour plot of X2*X1.

      X2 |
    1.00 + =====++++++++00000000000000000000000000000++++++++=====-----''''...
         | ==+++++++000000000000XXXXXXXXXXXX0000000000000++++++=====-----''''.
         | ++++++000000000XXXXXXXXXXXXXXXXXXXXXXXXX000000000+++++=====-----'''
         | +++00000000XXXXXXXXXXXXXXXXXXXXXXXXXXXXXXXX0000000++++++=====----''
         | +0000000XXXXXXXXXXθθθθθθθθθθθθθθθθθXXXXXXXXXXX0000000+++++====----'
    0.75 + 00000XXXXXXXXXθθθθθθθθθθθθθθθθθθθθθθθθθθθXXXXXXX000000+++++=====---
         | 000XXXXXXXXθθθθθθθθθθθθθθθθθθθθθθθθθθθθθθθθXXXXXXX000000+++++====--
         | 0XXXXXXXθθθθθθθθθθθθ▒▒▒▒▒▒▒▒▒▒▒▒▒▒▒θθθθθθθθθθθXXXXXX00000+++++====-
         | XXXXXXθθθθθθθθθ▒▒▒▒▒▒▒▒▒▒▒▒▒▒▒▒▒▒▒▒▒▒▒θθθθθθθθθXXXXXX000000++++====
         | XXXXθθθθθθθθ▒▒▒▒▒▒▒▒▒▒▒▒▒▒▒▒▒▒▒▒▒▒▒▒▒▒▒▒▒θθθθθθθθXXXXXX00000++++===
    0.50 + XXXθθθθθθθ▒▒▒▒▒▒▒▒▒▒▒▒▒▒▒▒▒▒▒▒▒▒▒▒▒▒▒▒▒▒▒▒▒▒θθθθθθθXXXXXX00000++++==
         | Xθθθθθθθ▒▒▒▒▒▒▒▒▒▒▒▓▓▓▓▓▓▓▓▓▓▓▓▓▒▒▒▒▒▒▒▒▒▒▒▒▒▒θθθθθθXXXXX0000+++++=
         | θθθθθθθ▒▒▒▒▒▒▒▒▒▒▓▓▓▓▓▓▓▓▓▓▓▓▓▓▓▓▓▓▒▒▒▒▒▒▒▒▒▒▒θθθθθθXXXXXX0000++++=
         | θθθθθ▒▒▒▒▒▒▒▒▒▒▓▓▓▓▓▓▓▓▓▓▓▓▓▓▓▓▓▓▓▓▓▓▓▒▒▒▒▒▒▒▒▒θθθθθθXXXXX00000++++
         | θθθθ▒▒▒▒▒▒▒▒▒▓▓▓▓▓▓▓▓▓▓▓▓▓▓▓▓▓▓▓▓▓▓▓▓▓▓▓▒▒▒▒▒▒▒▒▒θθθθθθXXXXX0000++++
    0.25 + θθθ▒▒▒▒▒▒▒▓▓▓▓▓▓▓▓▓▓▓▓▓▓▓▓▓▓▓▓▓▓▓▓▓▓▓▓▓▓▓▓▓▓▒▒▒▒▒▒▒▒θθθθθθXXX00000+++
         | θθθ▒▒▒▒▒▒▓▓▓▓▓▓▓▓▓▓▓▓▓▓▓▓▓▓▓▓▓▓▓▓▓▓▓▓▓▓▓▓▓▓▓▓▒▒▒▒▒▒▒θθθθθθXXXX0000+++
         | θθ▒▒▒▒▒▒▓▓▓▓▓▓▓▓▓▓▓▓▓▓▓▓▓▓▓▓▓▓▓▓▓▓▓▓▓▓▓▓▓▓▓▓▓▓▒▒▒▒▒▒▒θθθθθXXXX0000+++
         | θθ▒▒▒▒▒▒▓▓▓▓▓▓▓▓▓▓▓▓▓▓▓▓▓▓▓▓▓▓▓▓▓▓▓▓▓▓▓▓▓▓▓▓▓▓▓▒▒▒▒▒▒θθθθθθXXX0000+++
         | θθ▒▒▒▒▒▒▓▓▓▓▓▓▓▓▓▓▓▓▓▓▓▓▓▓▓▓▓▓▓▓▓▓▓▓▓▓▓▓▓▓▓▓▓▓▒▒▒▒▒▒▒θθθθθθXXX00000++
    0.00 + θθ▒▒▒▒▒▒▒▓▓▓▓▓▓▓▓▓▓▓▓▓▓▓▓▓▓▓▓▓▓▓▓▓▓▓▓▓▓▓▓▓▓▓▓▓▒▒▒▒▒▒▒θθθθθθXXX00000++
         | θθ▒▒▒▒▒▒▒▓▓▓▓▓▓▓▓▓▓▓▓▓▓▓▓▓▓▓▓▓▓▓▓▓▓▓▓▓▓▓▓▓▓▓▓▓▒▒▒▒▒▒▒θθθθθθXXX0000+++
         | θθ▒▒▒▒▒▒▒▓▓▓▓▓▓▓▓▓▓▓▓▓▓▓▓▓▓▓▓▓▓▓▓▓▓▓▓▓▓▓▓▓▓▓▓▓▒▒▒▒▒▒▒θθθθθXXXX0000+++
         | θθ▒▒▒▒▒▒▒▓▓▓▓▓▓▓▓▓▓▓▓▓▓▓▓▓▓▓▓▓▓▓▓▓▓▓▓▓▓▓▓▓▓▓▓▒▒▒▒▒▒▒θθθθθθXXXX0000+++
         | θθ▒▒▒▒▒▒▒▓▓▓▓▓▓▓▓▓▓▓▓▓▓▓▓▓▓▓▓▓▓▓▓▓▓▓▓▓▓▓▓▓▓▓▓▒▒▒▒▒▒▒θθθθθθXXX00000+++
   -0.25 + θθθ▒▒▒▒▒▒▒▓▓▓▓▓▓▓▓▓▓▓▓▓▓▓▓▓▓▓▓▓▓▓▓▓▓▓▓▓▓▓▓▓▒▒▒▒▒▒▒▒θθθθθθXXXX0000++++
         | θθθθ▒▒▒▒▒▒▒▒▓▓▓▓▓▓▓▓▓▓▓▓▓▓▓▓▓▓▓▓▓▓▓▓▓▓▓▓▓▓▒▒▒▒▒▒▒▒θθθθθθXXXX00000+++=
         | θθθθθ▒▒▒▒▒▒▒▒▒▒▓▓▓▓▓▓▓▓▓▓▓▓▓▓▓▓▓▓▓▓▓▓▒▒▒▒▒▒▒▒▒▒θθθθθθXXXXX0000++++=
         | θθθθθθ▒▒▒▒▒▒▒▒▒▒▒▒▓▓▓▓▓▓▓▓▓▓▓▓▓▓▓▓▒▒▒▒▒▒▒▒▒▒▒▒θθθθθθXXXXX0000++++==
         | θθθθθθθ▒▒▒▒▒▒▒▒▒▒▓▓▓▓▓▓▓▓▓▓▒▒▒▒▒▒▒▒▒▒▒▒▒▒▒▒▒θθθθθθθXXXXX0000++++===
   -0.50 + XXθθθθθθθ▒▒▒▒▒▒▒▒▒▒▒▒▒▒▒▒▒▒▒▒▒▒▒▒▒▒▒▒▒▒▒▒▒θθθθθθθθXXXXX0000+++++===
         | XXXθθθθθθθθθ▒▒▒▒▒▒▒▒▒▒▒▒▒▒▒▒▒▒▒▒▒▒▒▒▒▒▒▒▒θθθθθθθθXXXXX00000++++===-
         | XXXXXθθθθθθθθθ▒▒▒▒▒▒▒▒▒▒▒▒▒▒▒▒▒▒▒▒▒▒▒θθθθθθθθθθXXXXXX00000++++===--
         | XXXXXXθθθθθθθθθθθθθ▒▒▒▒▒▒▒▒▒▒▒▒▒▒θθθθθθθθθθθXXXXXXX000000+++====---
         | 00XXXXXXXθθθθθθθθθθθθθθθθθθθθθθθθθθθθθθθθXXXXXXX000000+++++====----
   -0.75 + 0000XXXXXXXXXθθθθθθθθθθθθθθθθθθθθθθθθXXXXXXXXX000000+++++=====----'
         | 000000XXXXXXXXXXθθθθθθθθθθθθθθθθθXXXXXXXXXXXX0000000++++====----'''
         | ++0000000XXXXXXXXXXXXXXXXXXXXXXXXXXXXXXXX0000000++++++=====----''''
         | ++++000000000XXXXXXXXXXXXXXXXXXXXXXX0000000000++++++=====----''''..
         | ++++++0000000000000XXXXXXXXXX0000000000000++++++++=====-----''''...
   -1.00 + ==++++++++0000000000000000000000000000000++++++++=====----'''''....
         --+---------+---------+---------+---------+---------+---------+---------+-
         -1.0      -0.7      -0.4      -0.1       0.2       0.5       0.8       1.1

                                         X1

Symbol              Y       Symbol              Y       Symbol              Y

.....   12.120 - 14.481     +++++   21.564 - 23.925     ▒▒▒▒▒   31.008 - 33.368
'''''   14.481 - 16.842     00000   23.925 - 26.286     ▓▓▓▓▓   33.368 - 35.729
-----   16.842 - 19.203     XXXXX   26.286 - 28.647
=====   19.203 - 21.564     θθθθθ   28.647 - 31.008

NOTE: 37613 obs hidden.
```

Several comments are in order concerning the 3^k factorial experiment and the second–order model. First, inspecting the contour plot of an estimated second–order model works well when there are only two factors, but is impractical for more than two factors. For second–order models involving more than two factors a special method called a canonical analysis can be employed to study the response surface. See Montgomery (1991) for more details. Second, a 3^k factorial experiment requires many design points (treatments), usually more than practical for an experiment when $k > 2$. For example, when $k = 3$ we require a total of $3^3 = 27$ design points and when $k = 4$ we require 81. We next present a design called a **central composite design** that can be used to fit a second–order model that requires fewer design points than a 3^k factorial when $k > 2$.

Central Composite Design

A central composite design with k factors consists of the following design points: 1) the design points for a 2^k factorial, 2) a center point labelled as $(0,\cdots,0)$, and 3) the $2k$ axial points $(\pm\alpha, 0,\cdots,0)$, $(0, \pm\alpha, 0,\cdots,0)$, $\cdots$, $(0,\cdots,0, \pm\alpha)$ for α a selected constant. Figure 18.3.5 provides a diagram of the design points for the case $k = 3$.

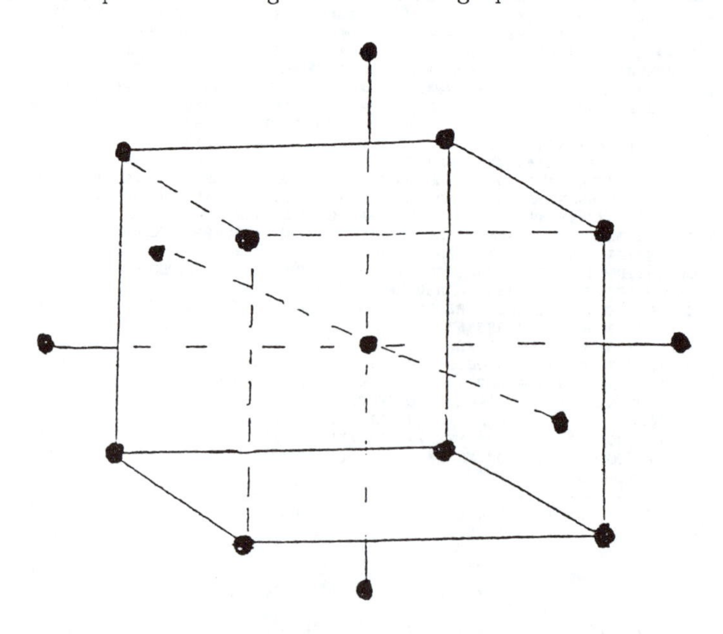

Figure 18.3.5 Central Composite Design for $k = 3$

Usually one observation is taken at the 2^k and the axial points, and $n_0 > 1$ observations are taken at the center point. By having multiple observations at the center point we are able to obtain a measure of variation in the response variable under the same experimental conditions (sometimes called **pure error**).

A major advantage of a central composite design is that it can be used to obtain a second–order model and it has fewer design points than a 3^k design. For example, when $k = 3$ a 3^3 requires 27 design points while a central composite design requires $2^k + 2k + 1 = 15$ design points.

An issue to consider for central composite designs is the choice of α used for the axial points. There are several criteria used to select α, one of which is selecting α so that the design is **rotatable**. A design is called rotatable if the variance of $\hat{Y}$ is only a function of the distance a point is from the center point, that is, the variance does not depend upon the direction. In a rotatable design $Var(\hat{Y})$ is unchanged when the design is rotated about the center point. For a central composite design to be rotatable it is required that $\alpha = n_F^{\frac{1}{4}}$, where n_F is the number of observations in the 2^k portion of the design. (Of course we are assuming the same number of observations are taken at each of the 2^k and axial points.)

More details about central composite designs are available in Montgomery (1991), and in the exercises. In summary, however, a central composite design is useful for fitting second–order models, and for $k \geq 3$ it requires fewer design points than a 3^k factorial.

18.4 Selecting a Design

We conclude this text as we began it, namely discussing how to design an experiment. Now that we are familiar with many of the experimental designs it is useful to discuss the strategy for selecting an appropriate design. As we have discovered, the design for an experiment depends upon many things including: the number of factors, the type of factors (e.g., fixed, random, numerical, categorical), the number of levels of the factors, the relationship between the factors (e.g., nested, crossed), and the randomization procedure.

In choosing a design for an experiment, one usually first selects the factors to be studied. The number of levels of each factor is then determined. As we have discovered, if the factor is numerical in nature, then two levels only allow us to study a linear relationship between the factor and the response variable, while three or more levels allow us to detect curvature in the relationship. Next we consider how the factors are to be related in terms of nested or crossed. As we have seen, this leads us to various treatment arrangements.

Identification of blocking factors is critically important in setting up a design. A single blocking factor can easily be utilized with an RCBD, while two blocking factors can be embodied in a Latin square design — although this latter design is incomplete in nature. If the blocking factor is an uncontrolled, numerical variable, then its effect can be incorporated via analysis of covariance.

Another item affecting the selection of the design is the type of randomization that is desirable and possible. If there is no restriction on the randomization, then a CRD is appropriate whereas designs such as an RCBD, a Latin square, and a split—plot all require various restrictions on the randomization.

Cost is almost always a consideration in selecting a design. Costs increase as treatments and observations increase; therefore, the number of treatments and replications that can be used in an experiment often depends on the funds available for an experiment. Less expensive designs have fewer factors with each factor having a small number of levels, but such designs sometimes provide inadequate information. Alternatively, designs can be made less expensive by using incomplete designs such as a BIB, a Latin square, or a fractional factorial.

Another issue of interest in selecting a design is sample size and allocation of observations. Sample size is related to power of tests and tightness of confidence intervals. One must be certain to use large enough sample sizes to ensure tests powerful enough to detect meaningful differences between treatments. Of course sample size is also directly related to cost.

The preferred allocation of observations to the treatments depends upon the goals for the experiment and the sources and magnitude of variation. As an illustration in this regard, suppose that we are interested in finding a confidence interval for the difference between the mean of the two levels of factor A and that the design involves a random factor B which is nested in factor A. Assume we can afford to take only twelve observations. Is it better to have three levels of factor B at each level of A and two replications, or to have two levels of factor B and three replications? (See Figure 18.4.1.) By comparing the variance of the difference between the means for A we can determine which is a better allocation. Of course the answer depends on the amount of variability associated with the various terms in the model. (It may be necessary to do a preliminary study in order to estimate the magnitude of the variance components involved.)

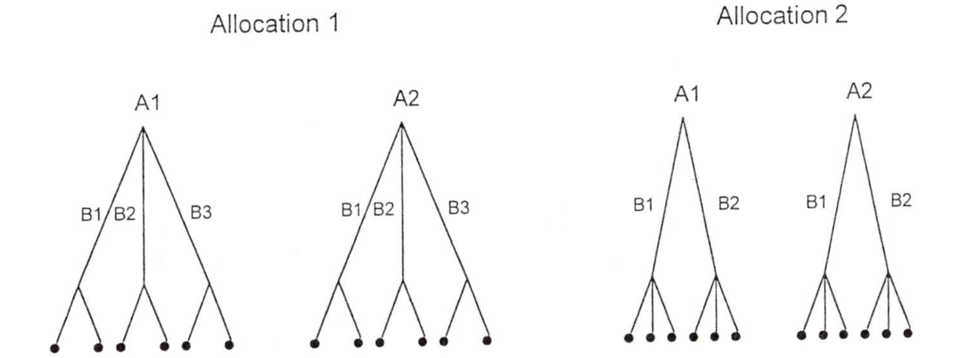

Figure 18.4.1 Possible Allocations

There are approaches to the design of experiments other than the one presented in this text. One prominent one is that promoted by **Genechi Taguchi** in Japan. There are many similarities between the two approaches, but there are some major differences. One major difference is that Taguchi suggests the use of a response variable that measures both the mean and the variability of observations. These measures are called **signal to noise ratios**, and one such measure is

$$SN_T = 10 \log (\bar{y}^2/s^2),$$

where $\bar{y}$ and s^2 are the mean and variance of the observations taken for a treatment. Another characteristic of Taguchi's approach is the use of highly fractional designs, which he labels as L_4, L_8, L_9, L_{12}, L_{16}, L_{18} and so forth. For example, L_4 is a 1/2 of a 2^3 factorial design, while L_9 is a 1/3 of a 3^4 factorial design. In most of Taguchi's designs there is little opportunity to check for interaction between factors.

Taguchi has contributed much to the use of designed experiments in industry. For all practical purposes there is little difference between the designs suggested by Taguchi and those presented in this text. The greatest difference involves Taguchi's use of the signal to noise ratios for a response variable. The interested reader can learn more about the Taguchi approach in Kuehl (1994).

Chapter 18 Exercises

Application Exercises

18–1 Refer to the two–factor split–plot data in a RCBD analyzed in Section 18.1.

(a) Estimate the variance components for these data.
(b) Find a 95% confidence interval for the difference between the means for the two varieties.
(c) Reparametrize model (18.1.1) and then use it and a regression program to reproduce the ANOVA displayed in Table 18.1.3.

18–2 A company that makes frozen pizzas performed an experiment to help determine the desired cooking time and temperature. Three temperatures (400, 425, 450° F) and three cooking times (8, 10, 12 minutes) were selected. Two ovens were randomly selected for the experiment. Ovens were heated to each of the three temperatures (in random order). At each temperature three pizzas were placed in the oven. One of the three (randomly selected) was removed at 8 minutes, another at 10 minutes, and the third one at 12 minutes. An expert rates the quality of the cooking of the pizza on a ten–point scale (10 is best). The data are given as follows:

			Temperature	
		Time 400	425	450
Oven	1	8	4	7	5
		10	4	9	6
		12	3	6	2
	2	8	3	6	5
		10	6	10	7
		12	5	6	3

(a) Write down an appropriate model with assumptions for this experiment.
(b) Find the ANOVA table and perform appropriate f–tests using a 0.05 level.

(c) Perform an appropriate Bonferroni multiple comparison for the times using a 0.05 level.
(d) Estimate the variance components associated with the model.

18–3 A company desired to determine if a drug (label as A) is effective in providing relief from asthma. It was decided to use a placebo (label as P) and to have each subject use both A and P. Ten subjects, suffering from asthma, were available for the study. Five subjects were randomly selected and they used A first for two weeks, then two weeks later, P was used for two weeks. The other five subjects reversed the order of A and P. After the two–week period each subject rated the effectiveness of the treatment on a five–point scale (1 = no effect, 5 = highly effective). The resulting data are as follows:

		Subject				
Period	Drug	1	2	3	4	5
1	A	4	3	3	5	4
2	P	2	1	2	1	1
1	P	2	1	1	2	1
2	A	5	4	5	5	4

(a) Write down an appropriate model with assumptions for this experiment.
(b) Find an ANOVA table for these data.
(c) Using a 0.05 level perform appropriate f–tests. Is there evidence that the drug is effective?
(d) Is there evidence of carryover effects?

18–4 Refer to the crossover design utilized in Table 18.2.1. Suppose that each subject is able to use only two drugs. How would you design the experiment? How many subjects would be required?

18–5 Another model that is sometimes used with a 2^k factorial experiment is a first–order model with interaction. The model, when $k = 2$, has the form $Y = \beta_0 + \beta_1 x_1 + \beta_2 x_2 + \beta_{12} x_1 x_2$

$+ \varepsilon$. Find the estimated model for the 2^2 Soap Data in Table 18.3.1. Interpret the estimated coefficients.

18–6 A restaurant in a city wanted to determine the best advertising strategy by means of radio and newspaper. For the radio they decided to use daily advertisements lasting between 20 and 40 seconds. For the newspaper they decided to use a daily advertisement in size ranging from 20 to 80 square inches. They settled on using a 2^2 factorial with a center point. Twelve weeks were used for the study and a CRD design was used to assign two weeks to each factorial point and four weeks to the center point. Let x_1 represent the radio factor with coding $-1,0,1$ representing 20, 30, and 40 seconds. Similarly, x_2 represents the newspaper factor with $-1,0,1$ representing 20, 50, and 80 square inches. The response variable Y measures the weekly profit of the restaurant minus the advertising costs, in thousands. The data are as follows:

Design Point	(−1,−1)	(−1,1)	(1,−1)	(1,1)	(0,0)
y	9,6	14,18	3,4	11,8	17,18,15,17

(a) Fit a first–order model to these data.
(b) Find the contours for the first–order model.
(c) Using the results of parts (a) and (b), make a statement concerning the proper strategy for advertising. Do you believe the strategy obtained from the first–order model is optimal? Explain why.
(d) Calculate the curvature sum of squares. Do you believe there is evidence of curvature?
(e) Consider the model $Y = \beta_0 + \beta_1 x_1 + \beta_2 x_2 + \beta_{12} x_1 x_2 + \beta_{11} x_1^2 + \varepsilon$.

i) What does the x_1^2 term represent in this model.

Hint: Try adding x_2^2 to the model and inspect the **X** matrix.

ii) Using these data find the estimated form of the model.

iii) Find the contours for the model. Using these contours make a statement concerning the proper advertising strategy. Compare your answer to the one obtained in part (c) and comment on the value of having a center point in the design.

(f) Pure error is defined to be variation in the response variable under the same experimental conditions. If $y_i, \cdots, y_r$ are observations under the same conditions, and $\bar{y}$ is the mean of these observations, then the sum of squares for pure error is $\sum_{i=1}^{r} (y_i - \bar{y})^2$ with $r - 1$ degrees of freedom. For our data we can find sum of squares for pure error at each design point. The overall sum of squares for pure error is then the sum of these five terms, and the degrees of freedom is the sum of the five separate degrees of freedom. Find the overall sum of squares for pure error and its degrees of freedom. Compare these values to the error term in the ANOVA table for the model in part (e).

(g) Using the pure error mean square from part (f) as s^2, and the f–test for curvature given in Exercise 18–17. Carry out the f–test for curvature using a 0.05 level. Compare the f–value for this f–test to the f–value for the f–test of H_0 : $\beta_{11} = 0$ for the model in part (e).

18–7 Suppose there are $k = 4$ factors. Compare the number of design points required for a 3^k factorial and a central composite design.

18–8 Refer to the soap product development considered in Section 18.3. The soap product had two main ingredients A and B, and Examples 18.3.1 and 18.3.2 helped find the optimal levels of A and B. Suppose we want to determine if another ingredient, labelled as C, has an effect on Y, the cleaning ability of the soaps. To find out, a central composite design is employed. From Example 18.3.2 we found that the optimal values for A and B are around 16 and 6. It is believed that the amount of factor C should be between 6 and 10 grams per 100 grams of product. It is decided to center the central composite design at $A = 16$, $B = 6$, $C = 8$. Letting x_1, x_2, and x_3 represent the amount of A, B, and C, respectively, the coded values of the variables are:

		A	B	C
Coded Values	–1	14	5.5	7
	1	18	6.5	9

For the central composite design two observations are taken at each factorial and axial point, and six observations are taken at the center point. The resulting data are as follows:

(x_1,x_2,x_3)	y	(x_1,x_2,x_3)	y
(–1,–1,–1)	34.33	(–1.68,0,0)	33,32
(–1,–1,1)	33,35	(1.68,0,0)	30,33
(–1,1,–1)	30,33	(0, –1.68,0)	30,27
(–1,1,1)	32,32	(0,1.68,0)	25,30
(1,–1,–1)	30,28	(0,0,–1.68)	35,34
(1,–1,1)	29,34	(0,0,1.68)	38,33
(1,1,–1)	33,29	(0,0,0)	37,36,35,38,33,32
(1,1,1)	30,31		

(a) Use these data to fit a second–order model.
(b) Is there any evidence that ingredient C has an effect?
(c) Ignoring ingredient C, fit a second–order model using x_1 and x_2.
(d) Find the contours for the second–order model involving x_1 and x_2. Where does the maximal value occur? Compare your results with those obtained in Example 18.3.2.

18–9 Hicks (1973) reports the results of an experiment in which porosity readings are made on condenser paper obtained from four rolls each taken at random from three lots. These data are:

Lot	I				II				III			
Roll	1	2	3	4	5	6	7	8	9	10	11	12
	1.5	1.5	2.7	3.0	1.9	2.3	1.8	1.9	2.5	3.2	1.4	7.8
	1.7	1.6	1.9	2.4	1.5	2.4	2.9	3.5	2.9	5.5	1.5	5.2
	1.6	1.7	2.0	2.6	2.1	2.4	4.7	2.8	3.3	7.1	3.4	5.0

(a) Assuming lots are fixed and rolls are random, write down an appropriate model with assumptions for these data.

(b) Find the ANOVA table for these data.

(c) Find point estimates for the variance components σ_{ε}^2 and σ_{β}^2, where σ_{β}^2 represents component for the rolls effect.

(d) Find a formula for $Var(\bar{Y}_{i..})$, the mean for the ith lot.

(e) Based on the results of this experiment find estimates for the $Var(\bar{Y}_{i..})$ based upon the following situations:

i) 36 observations with 3 lots, 4 rolls per lot, subsamples of size 3.

ii) 30 observations with 3 lots, 5 rolls per lot, subsamples of size 2.

(Notice the anomaly that the design with the fewest observations (30 to 36) could well produce more precise estimates.)

Theoretical Exercises

18–10 Refer to the model (18.1.1) for a two–factor split–plot design in an RCBD.

(a) Find expressions for $Cov(Y_{ijk}, Y_{uvw})$.

(b) Use the E(MS) algorithm to verify the E(MS) entries in Table 18.1.2.

(c) Find expressions for the sums of squares in Table 18.1.2 and verify the computational formulas for these sums of squares given in (18.1.2).

(d) Verify that the point estimates given in (18.1.3) for the variance components are unbiased.

(e) Derive the formulas for the variances of the estimators for $\alpha_j - \alpha_\ell$ and $\beta_k - \beta_m$ given in (18.1.4).

(f) Show that $Cov[(\rho\alpha)_{ij}, (\rho\alpha)_{\ell m}] = -\sigma_{\rho\alpha}^2/a$ when $i = \ell$ and $j \neq m$. What is the value when $i \neq \ell$?

18–11 Refer to the split–plot design in a balanced CRD with ANOVA table given in Table 18.1.4. Assume factors A and B are both fixed.

(a) Write down an appropriate model for this situation assuming subsamples of size r at each condition.

(b) Find expressions for the E(MS).

(c) Describe the form of the f–tests.

(d) Describe how to carry out a Bonferroni multiple comparison for the levels of factor A.

18–12 Refer to the split–plot design in a Latin–square design with ANOVA table given in Table 18.1.4. Assume rows and columns are random, while factors A and B are fixed.

(a) Write down an appropriate model for this situation.
(b) Find expressions for the E(MS).
(c) Describe the form of the f–tests.

18–13 For each of the following variants of a split–plot design, write down the ANOVA source and degrees of freedom.

(a) Two–stage randomization in 3 blocks. Three levels of A over whole plots; treatment combinations of B and C (2 levels each) over subplots.
(b) Two–stage randomization in 3 blocks. Treatment combinations of A and B (2 levels each) over whole plots; 3 levels of C over subplots.
(c) Same as (b) except replication in a Latin–square.
(d) Two–stage randomization in 3 blocks. Two levels of A over whole plots; 3 levels of B over whole plots.

18–14 Consider the crossover design with ANOVA table as given in Table 18.2.2.

(a) Give formulas for the sums of squares in the table.
(b) Derive the formula for E(MSTR).
(c) Justify the entries for the degrees of freedom.
(d) Find formulas for point estimates for the variance components σ_β^2 and σ_ε^2.

18–15 Consider the crossover design with model (18.2.2).

(a) Find the ANOVA table.
(b) Assuming subjects are random, find the E(MS) and describe how to carry out appropriate f–tests.

18–16 Consider the second–order model (18.3.2).

(a) Explain why this model cannot be fit with data from a 2^2 factorial experiment. (Hint: Look at the entries in the **X** matrix.)

(b) Does adding a center point to a 2^k factorial enable you to fit a second–order model? Explain why.

18–17 Consider a balanced 2^2 factorial experiment with n_F observations. Add n_0 observations at the center point. Let $\bar{y}_F$ and $\bar{y}_0$ be the mean of the observations at the factorial and center points, respectively. Assume a normal linear model.

(a) Show that $Var(\bar{Y}_F - \bar{Y}_0) = \sigma^2 (n_F + n_0)/(n_F n_0)$.

(b) If s^2 is an estimate for σ^2, then find the form of a t–test for $H_0 : \mu_F = \mu_0$ vs. $H_1 : \mu_F \neq \mu_0$ (curvature).

(c) Deduce that the form of an f–test for the hypothesis in part (b) is given by $f = n_F n_0 (\bar{y}_F - \bar{y}_0)^2/[(n_F + n_0)\, s^2]$.

(d) From the f–test in part (c) obtain the formula (18.3.3) for the curvature sum of squares. How many degrees of freedom does this sum of squares have?

18–18 Another design that can be used to generate data useful for fitting a first–order model is a simplex–design. For the case of two factors the design points, when plotted, correspond to the vertices of an equilateral triangle. For three factors the design points are the vertices of a regular tetrahedron.

(a) For $k = 2$ show that the design points can be written as $(-1,0)$, $(1,0)$, $(0, \sqrt{3})$.

(b) For $k = 2$ exhibit the **X** matrix for a first–order model. Show that the x_1 and x_2 columns are orthogonal.

(c) For $k = 2$ exhibit the **X** matrix for a second–order model. Can we fit this model with data from a simplex design?

APPENDIX A MATRIX ALGEBRA

A.1 Matrices

In this appendix we present a review of matrix algebra. We emphasize only the properties and theorems concerning matrices that are used somewhere in the text.

Matrices

An $m \times n$ matrix **A** is a rectangular array of elements consisting of m rows and n columns represented by

$$\mathbf{A} = \begin{bmatrix} a_{11} & a_{12} & \cdots & a_{1n} \\ a_{21} & a_{22} & \cdots & a_{2n} \\ \cdot & \cdot & & \cdot \\ \cdot & \cdot & & \cdot \\ \cdot & \cdot & & \cdot \\ a_{m1} & a_{m2} & \cdots & a_{mn} \end{bmatrix} = [a_{ij}] \ . \qquad \text{(A.1.1.)}$$

The **order** (or **dimension**) of the matrix is $m \times n$. In this text we will assume that the elements of the matrices are real numbers.

Two $m \times n$ matrices $\mathbf{A} = [a_{ij}]$ and $\mathbf{B} = [b_{ij}]$ are said to be **equal** if and only if $a_{ij} = b_{ij}$ for all $i = 1,\cdots,m$ and $j = 1,\cdots, n$. That is, $\mathbf{A} = \mathbf{B}$ whenever each element of **A** is equal to the corresponding element in **B**.

Special Matrices

The following definitions present some special matrices that are useful in matrix algebra.

Definition A.1.1: An $n \times n$ matrix **A** is called a **square matrix** of order n. The elements $a_{11}, a_{22},\cdots,a_{nn}$, are called the **diagonal** elements of **A**.

Definition A.1.2: A **diagonal matrix D** is a square matrix whose off–diagonal elements are all zero; i.e., if $\mathbf{D} = [d_{ij}]$, then $d_{ij} = 0$ when $i \neq j$.

Definition A.1.3: An **identity matrix**, denoted by I, is a diagonal matrix in which each diagonal element is the number one.

Definition A.1.4: A matrix whose elements are all zero is called a **null** (or **zero**) **matrix** and is denoted by **0**.

Definition A.1.5: The **transpose** of an $m \times n$ matrix **A**, denoted by $\mathbf{A}'$, is the $n \times m$ matrix obtained by interchanging the rows and columns of **A**.

Definition A.1.6: A square matrix is said to be a **symmetric matrix** if $\mathbf{A}' = \mathbf{A}$; i.e., $a_{ij} = a_{ji}$.
As an illustration consider matrices **A**, **B**, and **C** given by

$$\mathbf{A} = \begin{bmatrix} 2 & 5 \\ 1 & 6 \end{bmatrix} \quad \mathbf{B} = \begin{bmatrix} 2 & 0 \\ 0 & 6 \end{bmatrix} \quad \mathbf{C} = \begin{bmatrix} 1 & 0 \\ 0 & 1 \end{bmatrix} .$$

We note that all of these matrices are square, **B** is a diagonal matrix, and **C** is an identity matrix. The transpose of matrix **A** is $\mathbf{A}' = \begin{bmatrix} 2 & 1 \\ 5 & 6 \end{bmatrix}$, and since $\mathbf{A} \neq \mathbf{A}'$ the matrix is not symmetric.

Matrix Operations

We describe how to perform matrix operations such as addition and multiplication.

1. Addition: If **A** and **B** are two $m \times n$ matrices, then the sum of the two matrices is $\mathbf{A} + \mathbf{B} = \mathbf{C}$, where $c_{ij} = a_{ij} + b_{ij}$.

2. Scalar Multiplication: If k is a scalar (i.e., a constant) and **A** is a matrix, then $k\mathbf{A} = [ka_{ij}]$. Note that $k\mathbf{A} = \mathbf{A}k$.

3. Matrix Multiplication: If **A** is an $m \times n$ matrix and **B** is an $n \times p$ matrix, then the product $\mathbf{AB} = \mathbf{C}$ is an $m \times p$ matrix such that

$$c_{ij} = \sum_{k=1}^{n} a_{ik} b_{kj} .$$

4. Trace: If **A** is a square $n \times n$ matrix, the trace of **A**, denoted

by $\text{tr}(\mathbf{A})$, is equal to the sum of the diagonal elements of $\mathbf{A}$, i.e.,

$$\text{tr}(\mathbf{A}) = \sum_{i=1}^{n} a_{ii} .$$

Using the matrices $\mathbf{A}$, $\mathbf{B}$, $\mathbf{C}$ given earlier in this section we note that

$$\mathbf{A+B} = \begin{bmatrix} 4 & 5 \\ 1 & 12 \end{bmatrix}, \quad 2\mathbf{A} = \begin{bmatrix} 4 & 10 \\ 2 & 12 \end{bmatrix},$$

$$\mathbf{AB} = \begin{bmatrix} 2(2) + 5(0) & 2(0) + 5(6) \\ 1(2) + 6(0) & 1(0) + 6(6) \end{bmatrix} = \begin{bmatrix} 4 & 30 \\ 2 & 36 \end{bmatrix},$$

and $\text{tr}(\mathbf{A}) = 2 + 6 = 8$.

An important operation for matrices involves the inverse of a matrix.

Definition A.1.7: Given a square matrix $\mathbf{A}$. If there exists a square matrix $\mathbf{A}^{-1}$ which satisfies the relation $\mathbf{A}^{-1}\mathbf{A} = \mathbf{AA}^{-1} = \mathbf{I}$, then $\mathbf{A}^{-1}$ is called the **inverse** of $\mathbf{A}$.

If a matrix $\mathbf{A}$ has an inverse, then $\mathbf{A}$ is called a **nonsingular** matrix; if not, it is called **singular**. For matrix $\mathbf{A}$ it is easy to observe that the inverse matrix is

$$\mathbf{A}^{-1} = \begin{bmatrix} 6/7 & -5/7 \\ -1/7 & 2/7 \end{bmatrix}, \text{ since}$$

$$\mathbf{A}^{-1}\mathbf{A} = \mathbf{AA}^{-1} = \begin{bmatrix} 1 & 0 \\ 0 & 1 \end{bmatrix}.$$

The inverse of a matrix can be found using procedures such as row reduction or cofactors. We present the cofactor approach later in this section. Inverses are very easy to find for diagonal matrices as noted in the following theorem.

Theorem A.1.1: If $\mathbf{D}$ is a diagonal matrix, where d_i is the ith diagonal element, and if $d_i \neq 0$ for all i, then $\mathbf{D}^{-1}$ exists, and is a diagonal matrix where the ith diagonal element of $\mathbf{D}^{-1}$ is $1/d_i$. As an illustration, the inverse of the diagonal matrix $\mathbf{B}$ above is $\begin{bmatrix} 1/2 & 0 \\ 0 & 1/6 \end{bmatrix}$.

The following definitions introduce two special classes of matrices that are useful.

Definition A.1.8: If $\mathbf{A}$ is a square matrix such that $\mathbf{A}'\mathbf{A} = \mathbf{I}$, then $\mathbf{A}$ is said to be an **orthogonal matrix**. Note that if $\mathbf{A}$ is orthogonal then $\mathbf{A}' = \mathbf{A}^{-1}$.

Definition A.1.9: A square matrix $\mathbf{A}$ is called **idempotent** if $\mathbf{A}^2 = \mathbf{A}$.

For example, matrix $\mathbf{D}$ given below is orthogonal since $\mathbf{D}'\mathbf{D} = \mathbf{I}$ and matrix $\mathbf{E}$ is idempotent since $\mathbf{E}^2 = \mathbf{E}\,\mathbf{E} = \mathbf{E}$.

$$\mathbf{D} = \begin{bmatrix} \frac{1}{\sqrt{2}} & \frac{1}{\sqrt{2}} \\ \frac{1}{\sqrt{2}} & -\frac{1}{\sqrt{2}} \end{bmatrix} \qquad \mathbf{E} = \begin{bmatrix} \frac{1}{2} & -\frac{1}{2} \\ -\frac{1}{2} & \frac{1}{2} \end{bmatrix} .$$

Some Properties of Matrix Operations

Listed below are useful properties of matrices. We assume in all cases that the dimensions of the matrices are compatible.

Theorem A.1.2: The following properties hold.

(a) $\mathbf{IA} = \mathbf{AI} = \mathbf{A}$
(b) $\mathbf{0A} = \mathbf{0}$ and $\mathbf{A0} = \mathbf{0}$
(c) $\mathbf{A} + \mathbf{B} = \mathbf{B} + \mathbf{A}$
(d) $(\mathbf{A} + \mathbf{B}) + \mathbf{C} = \mathbf{A} + (\mathbf{B} + \mathbf{C})$
(e) $k(\mathbf{A} + \mathbf{B}) = k\mathbf{A} + k\mathbf{B}$ where k is a scalar.
(f) $(\mathbf{AB})\mathbf{C} = \mathbf{A}(\mathbf{BC})$
(g) $\mathbf{A}(\mathbf{B} + \mathbf{C}) = \mathbf{AB} + \mathbf{AC}$ and $(\mathbf{A} + \mathbf{B})\mathbf{C} = \mathbf{AC} + \mathbf{BC}$
(h) $\mathbf{A}k\mathbf{B} = k\mathbf{AB}$ where k is a scalar.
(i) $(\mathbf{A}')' = \mathbf{A}$
(j) $(\mathbf{A} + \mathbf{B})' = \mathbf{A}' + \mathbf{B}'$
(k) $(\mathbf{AB})' = \mathbf{B}'\mathbf{A}'$ and $(\mathbf{ABC})' = \mathbf{C}'\mathbf{B}'\mathbf{A}'$
(l) $(\mathbf{A}^{-1})^{-1} = \mathbf{A}$

(m) $(\mathbf{AB})^{-1} = \mathbf{B}^{-1}\mathbf{A}^{-1}$ and $(\mathbf{ABC})^{-1} = \mathbf{C}^{-1}\mathbf{B}^{-1}\mathbf{A}^{-1}$

(n) $(\mathbf{A}')^{-1} = (\mathbf{A}^{-1})'$

(o) $\mathbf{A}'\mathbf{A}$ and $\mathbf{AA}'$ are symmetric matrices.

(p) If $\mathbf{A}$ is a symmetric matrix, then $\mathbf{A}^{-1}$ is also symmetric.

(q) If $\mathbf{A}'\mathbf{A} = \mathbf{0}$, then $\mathbf{A} = \mathbf{0}$.

(r) $\text{tr}(\mathbf{A} + \mathbf{B}) = \text{tr}(\mathbf{A}) + \text{tr}(\mathbf{B})$

(s) $\text{tr}(\mathbf{AB}) = \text{tr}(\mathbf{BA})$ and $\text{tr}(\mathbf{ABC}) = \text{tr}(\mathbf{CAB}) = \text{tr}(\mathbf{BCA})$.

Determinants

Another operation that can be performed on square matrices is finding the determinant of a matrix. The **determinant** of a matrix, denoted by $|\mathbf{A}|$, can be found using the process we now describe. For a 2×2 matrix $\mathbf{A} = [a_{ij}]$ the determinant is given by

$$|\mathbf{A}| = a_{11}\, a_{22} - a_{12}\, a_{21}.$$

Observe that the determinant of a matrix is a number.

For higher order $n \times n$ matrices we can find the determinant using cofactors. The **cofactor** of the element a_{ij}, denoted by $\mathbf{A}_{ij}$, is $(-1)^{i+j}$ times the determinant of the $(n - 1) \times (n - 1)$ matrix obtained by deleting the ith row and jth column. For example, in a 3×3 matrix the cofactor $\mathbf{A}_{12}$ is $(-1)^{1+2} \begin{vmatrix} a_{21} & a_{23} \\ a_{31} & a_{33} \end{vmatrix} = -(a_{21}a_{33} - a_{23}a_{31})$. We can find the determinant of an $n \times n$ matrix $\mathbf{A}$ as follows:

$$|\mathbf{A}| = \sum_{i=1}^{n} a_{ij}\mathbf{A}_{ij}, \quad \text{for any } j \text{ selected from } 1,2,\cdots,n. \qquad \text{(A.1.2)}$$

There is a more formal way of defining the determinant of a matrix. For the record we provide it in the following definition, but hasten to point out that it is rarely used as a means to calculate the value of a determinant.

Definition A.1.10: The determinant of an $n \times n$ matrix $\mathbf{A}$, is defined to be the number

$$|\mathbf{A}| = \Sigma(-1)^{i} a_{1j}a_{2k} \cdots a_{nr}$$

where j, k,···,r range over the n! permutations of the numbers 1,2,···,n, taken n at a time, and where i is the number of rearrangements of j,k,···,r from the normal order 1,2,···,n.

Determinants are related to inverses of a matrix as demonstrated in the following definition and theorem.

Definition A.1.11: The square matrix **A** is said to be **singular** if $|\mathbf{A}| = 0$, **nonsingular** if $|\mathbf{A}| \neq 0$.

Theorem A.1.3: A necessary and sufficient condition that a square matrix has an inverse is that $|\mathbf{A}| \neq 0$, that is **A** is nonsingular.

Theorem A.1.4: If $\mathbf{A} = [a_{ij}]$ is a square, nonsingular matrix, then $\mathbf{A}^{-1} = [\omega_{ij}]$, where $\omega_{ij} = \mathbf{A}_{ji}/|\mathbf{A}|$. Here $\mathbf{A}_{ji}$ is the cofactor of a_{ji}.

For the matrix $\mathbf{A} = \begin{bmatrix} 1 & 0 & 1 \\ 0 & 2 & 0 \\ 0 & 0 & 2 \end{bmatrix}$ we can use Theorem A.1.4 to find the inverse. First we find the cofactor's matrix to be $\mathbf{A}_{ij} = \begin{bmatrix} 4 & 0 & 0 \\ 0 & 2 & 0 \\ -2 & 0 & 2 \end{bmatrix}$.

Using row 2 and formula (A.1.2) we find $|\mathbf{A}| = 0 + 2\,[2-0] + 0 = 4$.

The inverse is then $\mathbf{A}^{-1} = [\omega_{ij}] = \mathbf{A}_{ji}/|\mathbf{A}| = \frac{1}{4}\begin{bmatrix} 4 & 0 & -2 \\ 0 & 2 & 0 \\ 0 & 0 & 2 \end{bmatrix} = \begin{bmatrix} 1 & 0 & -.5 \\ 0 & .5 & 0 \\ 0 & 0 & .5 \end{bmatrix}$.

Some properties of determinants are given.

Theorem A.1.5: Consider a square $n \times n$ matrix **A**.

(a) If **A** is diagonal, then $|\mathbf{A}|$ equals the product of the diagonal elements.

(b) $|\mathbf{A}'| = |\mathbf{A}|$.

(c) If **A** and **B** are $n \times n$ matrices, then $|\mathbf{AB}| = |\mathbf{A}|\,|\mathbf{B}|$.

(d) $|\mathbf{A}^{-1}| = |\mathbf{A}|^{-1}$

(e) $|k\mathbf{A}| = k^{n}|\mathbf{A}|$ where k is a scalar.

As an illustration, the determinant of the diagonal matrix

$$\mathbf{A} = \begin{bmatrix} 2 & 0 & 0 \\ 0 & 4 & 0 \\ 0 & 0 & 5 \end{bmatrix} \text{ is } |\mathbf{A}| = (2)(4)(5) = 40.$$

A.2 Vectors

A matrix with only one row or one column is called a **vector**. An $m \times 1$ matrix is called a column vector, denoted by $\mathbf{x}$. The corresponding $1 \times m$ row vector is denoted by $\mathbf{x}'$, that is $\mathbf{x}' = (x_1, x_2, \cdots, x_m)$. Since vectors are actually matrices the matrix properties presented in the previous section hold for vectors.

When there are two or more vectors it is often useful to know if these vectors are "related" to each other. The following definition provides us with one way to determine relationships among vectors.

Definition A.2.1: A set of m n–dimensional vectors $\{\mathbf{x}_1, \mathbf{x}_2, \cdots, \mathbf{x}_m\}$ is **linearly dependent** if and only if there exist some scalars c_1, $c_2, \cdots, c_m$, not all zero, such that

$$c_1\mathbf{x}_1 + c_2\mathbf{x}_2 + \cdots + c_m\mathbf{x}_m = \mathbf{0}.$$

If the set is not linearly dependent it is **linearly independent**.

As an illustration consider three vectors

$$\mathbf{x}_1' = (2,4),\ \ \mathbf{x}_2' = (1,5),\ \ \mathbf{x}_3' = (3,3).$$

We observe that $2\,\mathbf{x}_1 - \mathbf{x}_2 - \mathbf{x}_3 = \mathbf{0}$; therefore, the vectors $\mathbf{x}_1$, $\mathbf{x}_2$, and $\mathbf{x}_3$ are linearly dependent. However, it is easy to show that $\mathbf{x}_1$ and $\mathbf{x}_2$ are linearly independent vectors, that is, there are no non–zero constants c_1 and c_2 such that $c_1\mathbf{x}_1 + c_2\mathbf{x}_2 = \mathbf{0}$. We should point out that if a set of vectors is linearly independent, then any subset is also linearly independent.

Orthogonality is another important vector relationship.

Definition A.2.2: Two vectors $\mathbf{x}$ and $\mathbf{y}$ are said to be **orthogonal** if $\mathbf{x}'\mathbf{y} = 0$ (or $\mathbf{y}'\mathbf{x} = 0$).

Definition A.2.3: A set of vectors $\{\mathbf{x}_1, \mathbf{x}_2, \cdots, \mathbf{x}_m\}$ is said to be an orthonormal set if $\mathbf{x}_i'\mathbf{x}_i = 1$, $i = 1,2,\cdots,m$, and $\mathbf{x}_i'\mathbf{x}_j = 0$, $i \neq j$.

For example, vectors $\mathbf{x}' = (1,0,1)$ and $\mathbf{y}' = (-1, -1,1)$ are orthogonal since $\mathbf{x}'\mathbf{y} = 0$. The vectors $\mathbf{x}_1' = (0,-1)$ $\mathbf{x}_2' = (1,0)$ are both orthogonal and an orthonormal set. We can relate these ideas about orthogonal and orthonormal vectors to an orthogonal matrix.

Theorem A.2.1: The rows (columns) of an orthogonal matrix **A** are each of unit length (i.e., $\mathbf{x}_i'\mathbf{x}_i = 1$) and the rows (columns) of **A** are mutually orthogonal, i.e., the set of rows (columns) of an orthogonal matrix forms an orthonormal set of vectors.

A.3 Rank of a Matrix

An important concept for a matrix is its rank, which is based upon the number of linearly independent rows or columns. A useful theorem and the formal definition of rank follows.

Theorem A.3.1: The maximum number of linearly independent columns in an $m \times n$ matrix **A** is equal to the maximum number of linearly independent rows.

Definition A.3.1: The **rank** of an $m \times n$ matrix **A** is the maximum number of linearly independent rows (columns) in **A**.

As an illustration consider the matrices **A** and **B** given below. Matrix **A** has two linearly independent columns; therefore, **A** has rank two. Matrix **B** has two linearly independent rows and also has rank two.

$$\mathbf{A} = \begin{bmatrix} 1 & 0 \\ 1 & 1 \\ 0 & 1 \end{bmatrix} \qquad \mathbf{B} = \begin{bmatrix} 1 & 0 & -1 \\ 0 & 1 & 1 \\ 1 & 0 & -1 \end{bmatrix}.$$

In the previous illustration we note that the largest possible rank for the 3×2 matrix **A** is 2. Since **A** attained this largest possible rank we call **A** a **full rank** matrix.

Definition A.3.2: Suppose matrix **A** is $m \times n$. If the rank of **A** is m, **A** is said to be **full row rank**. If the rank of **A** is n, **A** is said to be **full column rank**. If the rank of **A** is the largest possible value (i.e., the smaller of m and n), then **A** is said to be a **full rank** matrix. Otherwise **A** is said to be **less than full rank**.

We note that in the previous illustration matrix **A** is full rank, while matrix **B** is less than full rank.

Several useful theorems involving the rank of a matrix are presented.

Theorem A.3.2: If **A** is an idempotent matrix, then tr(**A**) equals the rank of **A**.

Theorem A.3.3: If **A** is an $n \times n$ matrix and if $|\mathbf{A}| \neq 0$, then the rank of **A** is n. If $|\mathbf{A}| = 0$, then the rank of **A** is less than n.

Theorem A.3.4: Suppose **A** is an $m \times n$ matrix. The rank of $\mathbf{AA}'$ equals the rank of $\mathbf{A}'\mathbf{A}$ equals the rank of **A** equals the rank of $\mathbf{A}'$.

Theorem A.3.5: If matrix **A** has full row rank, $\mathbf{AA}'$ is nonsingular. Also, if matrix **B** has full column rank, $\mathbf{B}'\mathbf{B}$ is nonsingular.

Theorem A.3.6: If **BA** and **AC** exist where **B** and **C** are nonsingular matrices, then the rank of **BA** equals the rank of **AC** equals the rank of **A**.

Theorem A.3.7: The rank of the product **AB** of the two matrices **A** and **B** is less than or equal to the rank of **A** and is less than or equal to the rank of **B**.

Theorem A.3.8: If **A** is a full rank square matrix, then **A** is nonsingular and it has an inverse. If **A** is less than full rank, then **A** is singular and does not have an inverse.

Theorem A.3.9: Consider a linear system of equations defined by $\mathbf{A}\,\mathbf{x} = \mathbf{c}$, where **A** is an $m \times n$ matrix of constants, $\mathbf{x}' = (x_1, \cdots, x_n)$, and **c** is an $m \times 1$ vector of constants. The system of equations has a solution if and only if the rank of **A** is the same as the rank of the matrix $[\mathbf{A}, \mathbf{c}]$. Here $[\mathbf{A}, \mathbf{c}]$ represents the $m \times (n+1)$ matrix with **A** representing the first n columns and **c** the last column.

A.4 Linear and Quadratic Forms

Our goals in this section is to introduce the concepts of linear and quadratic forms and to present some results involving these concepts.

Definition A.4.1: If **a** is an $n \times 1$ vector of constants and **x** is an $n \times 1$ vector, then $\mathbf{a}'\mathbf{x} = \mathbf{x}'\mathbf{a} = \sum_{i=1}^{n} a_i x_i$ is said to be a **linear form** in $x_1, x_2, \cdots, x_n$. If **A** is an $m \times n$ matrix of constants and **x** is an n–dimensional vector then **Ax** is said to be a system of m linear forms in $x_1, x_2, \cdots, x_n$.

Definition A.4.2: If **A** is an n × n matrix of constants and **x** is an n–dimensional vector, then $\mathbf{x}'\mathbf{A}\mathbf{x} = \sum_{i=1}^{n} \sum_{j=1}^{n} a_{ij} x_i x_j$ is said to be a **quadratic form** in $x_1, x_2, \cdots, x_n$. Since the matrix of every quadratic form can always be chosen to be a symmetric matrix, we shall always assume that the matrix **A** associated with the quadratic form $\mathbf{x}'\mathbf{A}\mathbf{x}$ is symmetric.

If $\mathbf{a}' = (2,1,-3)$, $\mathbf{A} = \begin{bmatrix} 2 & 1 & -2 \\ 1 & 3 & -1 \\ -2 & -1 & 4 \end{bmatrix}$ and $\mathbf{x}' = (x_1, x_2, x_3)$, then the linear form $\mathbf{a}'\mathbf{x}$ is $2x_1 + x_2 - 3x_3$. Notice that a linear form $\mathbf{a}'\mathbf{x}$ is simply a **linear combination** of the entries in **x**. The quadratic form $\mathbf{x}'\,\mathbf{A}\mathbf{x}$ is $2x_1^2 + 3x_2^2 + 4x_3^2 + 2x_1x_2 - 4x_1x_3 - 2x_2x_3$.

A property of matrices that is based upon quadratic forms is positive definite.

Definition A.4.3: The quadratic form $\mathbf{x}'\mathbf{A}\mathbf{x}$ is said to be **positive definite** if and only if $\mathbf{x}'\mathbf{A}\mathbf{x} > 0$ for every **x** where $\mathbf{x} \neq \mathbf{0}$. The symmetric matrix **A** is called a **positive definite matrix** if and only if the corresponding quadratic form $\mathbf{x}'\mathbf{A}\mathbf{x}$ is positive definite.

There is a fairly simple way to check whether a matrix is positive definite.

Theorem A.4.1: The symmetric matrix $\mathbf{A} = [a_{ij}]$ is positive definite if and only if $a_{ii} > 0$ for all i, and the following is true: $a_{11} > 0$,

$$\begin{vmatrix} a_{11} & a_{12} \\ a_{21} & a_{22} \end{vmatrix} > 0, \cdots, |\mathbf{A}| > 0.$$

As an illustration consider two matrices **A** and **B** given by $\mathbf{A} = \begin{bmatrix} 2 & 2 \\ 2 & 3 \end{bmatrix}$

$\mathbf{B} = \begin{bmatrix} 2 & 0 & 1 \\ 0 & -1 & 0 \\ 1 & 0 & 3 \end{bmatrix}$. It is clear that matrix **A** is positive definite since $a_{ii} > 0$ and $|\mathbf{A}| = 2 > 0$. Matrix **B** is not positive definite since $b_{22} < 0$.

Two results are given concerning positive definite matrices.

Theorem A.4.2: Positive definite matrices are nonsingular.

Theorem A.4.3: If matrix **A** is positive definite, $\mathbf{A}^{-1}$ is also positive definite.

For a symmetric matrix **A**, if $\mathbf{x}'\mathbf{A}\mathbf{x} \geq 0$ for all vectors **x**, then **A** is called a **positive semi–definite** matrix. The following theorem involves such matrices.

Theorem A.4.4: Consider an $m \times n$ matrix **A**.

(a) $\mathbf{A}'\mathbf{A}$ is a symmetric matrix.
(b) $\mathbf{A}'\mathbf{A}$ is positive semi–definite.
(c) If **A** has rank n, then $\mathbf{A}'\mathbf{A}$ is full rank and $\mathbf{A}'\mathbf{A}$ is positive definite.

A.5 Matrix Factorizations

There are several representations or factorizations of matrices that are used in this text, and we now present some of them in this section. Some of these representations involve characteristic roots which we consider first.

Definition A.5.1: A **characteristic root** (or **eigenvalue**) of an $n \times n$ matrix **A** is a scalar λ such that $\mathbf{A}\mathbf{x} = \lambda\mathbf{x}$ for some vector $\mathbf{x} \neq \mathbf{0}$. The vector **x** is called a **characteristic vector** (or **eigenvector**) of **A**.

Theorem A.5.1: The number of nonzero characteristic roots of a matrix **A** is equal to the rank of **A**.

Theorem A.5.2: The characteristic roots of a symmetric matrix are positive if and only if the matrix is positive definite.

Theorem A.5.3: The characteristic roots of an idempotent matrix are all zero or one.

Theorem A.5.4: For an $n \times n$ matrix $\mathbf{A}$, the sum of the characteristic roots equals the trace of the matrix.

We now present some matrix factorizations.

Theorem A.5.5: For every symmetric matrix $\mathbf{A}$ there exists an orthogonal matrix $\mathbf{C}$ such that $\mathbf{C}'\mathbf{AC} = \mathbf{D}$ is a diagonal matrix whose diagonal elements are the characteristic roots of $\mathbf{A}$.

Theorem A.5.6: If $\mathbf{A}$ is a symmetric positive definite matrix, there exists a nonsingular matrix $\mathbf{B}$ such that $\mathbf{B}'\mathbf{AB} = \mathbf{I}$.

Theorem A.5.7: An $n \times n$ symmetric matrix $\mathbf{A}$ of rank r can be written $\mathbf{MM}'$ where $\mathbf{M}$ is $n \times r$ of rank r, i.e., $\mathbf{M}$ has full column rank.

Theorem A.5.8: If $\mathbf{A}$ is an idempotent, symmetric matrix of rank r, then there exists an orthogonal matrix $\mathbf{C}$ such that $\mathbf{C}'\mathbf{AC} = \begin{bmatrix} \mathbf{I}_r & \mathbf{0} \\ \mathbf{0} & \mathbf{0} \end{bmatrix}$. Here $\mathbf{I}_r$ is the $r \times r$ identity matrix and $\mathbf{0}$ is the null vector or matrix.

Theorem A.5.9: (Principal — Axis Theorem) For every quadratic form $\mathbf{x}'\mathbf{Ax}$ there exists an orthogonal $n \times n$ matrix $\mathbf{C}$ so that $\mathbf{y} = \mathbf{C}'\mathbf{x}$ and $\mathbf{x}'\mathbf{Ax} = \sum_{i=1}^{n} \omega_i y_i^2$, for some constants ω_i.

A.6 Differentiation of Vectors and Quadratic Forms

In this section we present procedures for differentiating vectors and quadratic forms. Let $f(\mathbf{x})$ represent the n–dimensional function $f(x_1, \cdots, x_n)$. The derivative of $f(\mathbf{x})$ denoted by $\frac{d}{d\mathbf{x}} f(\mathbf{x})$ is given as follows:

Definition A.6.1: $\frac{d}{d\mathbf{x}} f(\mathbf{x}) = \begin{bmatrix} \partial f(\mathbf{x})/\partial x_1 \\ \partial f(\mathbf{x})/\partial x_2 \\ \vdots \\ \partial f(\mathbf{x})/\partial x_n \end{bmatrix}$.

As an example, if $f(\mathbf{x}) = 2x_1 - 3x_2 + 4x_1x_3$, then the derivative is the 3×1 vector $\frac{d}{d\mathbf{x}} f(\mathbf{x}) = \begin{bmatrix} 2 + 4x_3 \\ -3 \\ 4x_1 \end{bmatrix}$. Some rules for differentiation are given.

Theorem A.6.1: Let $\mathbf{x}$ be an $n \times 1$ vector, $\mathbf{a}$ be an $n \times 1$ vector of constants and $\mathbf{A}$ be an $n \times n$ symmetric matrix of constants. The following are true:

(a) $\frac{d}{d\mathbf{x}} \mathbf{a}'\mathbf{x} = \mathbf{a} = \frac{d}{d\mathbf{x}} \mathbf{x}'\mathbf{a}$.

(b) $\frac{d}{d\mathbf{x}} \mathbf{x}'\mathbf{A}\,\mathbf{x} = 2\mathbf{A}\,\mathbf{x}$.

(c) $\frac{d}{d\mathbf{x}} \mathbf{x}'\mathbf{x} = 2\mathbf{x}$.

(d) $\frac{d}{d\mathbf{x}} \operatorname{tr}(\mathbf{a}'\mathbf{x}) = \mathbf{a}$.

Notice that many of these derivative rules mirror the rules for the univariate case. The same is true for finding maximum and minimum. For a function $f(\mathbf{x})$ the vector $\mathbf{x}_0$ is called a **stationary point** if $\frac{d}{d\mathbf{x}} f(\mathbf{x})$ is zero when evaluated at $\mathbf{x}_0$. To classify this point we use the second partials matrix (sometimes called the **Hessian matrix**). This matrix is defined by

$$\mathbf{H} = [h_{ij}], \text{ where } h_{ij} = \frac{d^2 f(\mathbf{x})}{dx_i dx_j}.$$

If $\mathbf{H}$ is a positive definite matrix, then $\mathbf{x}_0$ is a minimum. If $\mathbf{H}$ is a negative definite matrix, then $\mathbf{x}_0$ is a maximum. A matrix $\mathbf{H}$ is **negative definite** if $(-1)\mathbf{H}$ is positive definite.

Application Exercises

For Exercises A–1 through A–8 use

$$\mathbf{A} = \begin{bmatrix} 3 & 1 & 2 \\ 0 & -1 & 2 \\ 4 & 2 & 0 \end{bmatrix}, \mathbf{B} = \begin{bmatrix} 1 & -1 & 2 \\ 2 & 0 & 1 \\ 3 & 1 & 1 \end{bmatrix}$$

$$\mathbf{C} = \begin{bmatrix} 2 & -1 & 0 \\ 0 & 2 & 3 \end{bmatrix}, \mathbf{x} = \begin{bmatrix} x_1 \\ x_2 \\ x_3 \end{bmatrix}.$$

A–1 (a) Show that $\mathbf{C}(\mathbf{A} + \mathbf{B}) = \mathbf{CA} + \mathbf{CB}$.
(b) Show that $(\mathbf{A} + \mathbf{B})' = \mathbf{A}' + \mathbf{B}'$.
(c) Show that $(\mathbf{AB})' = \mathbf{B}'\mathbf{A}'$ and $(\mathbf{CA})' = \mathbf{A}'\mathbf{C}'$.
(d) Find $\text{tr}(\mathbf{A})$.
(e) Find $3\mathbf{A} - 4\mathbf{B}$.

A–2 (a) Find $|\mathbf{A}|$ and $|\mathbf{B}|$.
(b) Show that $|\mathbf{A}|\ |\mathbf{B}| = |\mathbf{AB}|$.

A–3 (a) Find $\mathbf{A}^{-1}$.
(b) Find $\mathbf{B}^{-1}$.
(c) Show that $(\mathbf{AB})^{-1} = \mathbf{B}^{-1}\mathbf{A}^{-1}$.

A–4 Show that $\mathbf{C}'\mathbf{C}$ and $\mathbf{CC}'$ are symmetric matrices.

A–5 (a) Find $\mathbf{x}'\mathbf{x}$ and $\mathbf{xx}'$.
(b) Find $\mathbf{x}'\mathbf{Bx}$.

A–6 Verify that $\text{tr}(\mathbf{AB}) = \text{tr}(\mathbf{BA})$.

A–7 Show that the row vectors of $\mathbf{C}$ are linearly independent while the column vectors of $\mathbf{C}$ are linearly dependent.

A–8 Find the rank of $\mathbf{C}$, $\mathbf{C}'\mathbf{C}$ and $\mathbf{CC}'$.

A–9 Show that the matrix $\mathbf{P}$ is orthogonal where

$$\mathbf{P} = \begin{bmatrix} 1/\sqrt{6} & 2/\sqrt{5} & 1/\sqrt{30} \\ -2/\sqrt{6} & 1/\sqrt{5} & -2/\sqrt{30} \\ 1/\sqrt{6} & 0 & -5/\sqrt{30} \end{bmatrix} .$$

A–10 Show that the matrix $\mathbf{Q} = \begin{bmatrix} \frac{2}{3} & -\frac{1}{3} & -\frac{1}{3} \\ -\frac{1}{3} & \frac{2}{3} & -\frac{1}{3} \\ -\frac{1}{3} & -\frac{1}{3} & \frac{2}{3} \end{bmatrix}$ is idempotent.

Show that the rank of $\mathbf{Q} = \mathrm{tr}(\mathbf{Q})$.

A–11 Consider the vectors $\mathbf{x}' = (2, -1, 1)$, $\mathbf{y}' = (0, 1, 1)$ and $\mathbf{z}' = (3, 6, 0)$.

(a) Are these vectors linearly independent?
(b) Which pairs of vectors are orthogonal?

A–12 (a) Is the matrix $\mathbf{A}$ given at the beginning of the exercises positive definite?

(b) Is the matrix $\mathbf{S} = \begin{bmatrix} 2 & 1 & -1 \\ 0 & 1 & 0 \\ -1 & 0 & 4 \end{bmatrix}$ positive definite?

A–13 Consider the matrix $\mathbf{A} = \begin{bmatrix} 4 & 2 \\ 2 & 1 \end{bmatrix}$.

(a) Show that $\lambda_1 = 5$ and $\lambda_2 = 0$ are the characteristic roots that correspond to the characteristic vectors

$$\mathbf{x}_1' = \left[\frac{2}{\sqrt{5}}, \frac{1}{\sqrt{5}}\right] \text{ and } \mathbf{x}_2' = \left[-\frac{1}{\sqrt{5}}, \frac{2}{\sqrt{5}}\right], \text{ respectively.}$$

(b) Let $\mathbf{C} = \begin{bmatrix} \frac{2}{\sqrt{5}} & -\frac{1}{\sqrt{5}} \\ \frac{1}{\sqrt{5}} & \frac{2}{\sqrt{5}} \end{bmatrix}$. Show that $\mathbf{C}$ is orthogonal.

(c) Show that $\mathbf{C}'\mathbf{AC} = \begin{bmatrix} 5 & 0 \\ 0 & 0 \end{bmatrix}$. (Note Theorem A.5.5.)

A–14 Let $f(\mathbf{x}) = x_1^2 - 4x_1 + x_2^2 + 2x_2 + 5$.

(a) Find $\frac{d}{d\mathbf{x}} f(\mathbf{x})$.

(b) Find the stationary point and use the Hessian matrix to classify it.

Theoretical Exercises

A–15 Prove Theorem A.1.2 parts m and n.

A–16 Prove Theorem A.1.2 part o.

A–17 Prove Theorem A.1.2 part q.

A–18 Prove Theorem A.1.2 part r.

A–19 Show that $\begin{bmatrix} a & b \\ c & d \end{bmatrix}^{-1} = \frac{1}{ad - bc} \begin{bmatrix} d & -b \\ -c & a \end{bmatrix}$ as long as $ad \neq bc$.

A–20 (a) Suppose $\mathbf{X}$ is an $n \times p$ matrix of rank p. Verify that the matrices $\mathbf{X}(\mathbf{X}'\mathbf{X})^{-1}\mathbf{X}'$ and $\mathbf{I} - \mathbf{X}(\mathbf{X}'\mathbf{X})^{-1}\mathbf{X}'$ are symmetric and idempotent.

(b) Show that $tr[\mathbf{X}(\mathbf{X}'\mathbf{X})^{-1}\mathbf{X}'] = p$.

A–21 Prove Theorem A.4.4. Hint: For parts (b) and (c) let $\mathbf{y} = \mathbf{Ax}$.

A–22 Refer to Theorem A.5.5. Assume $\mathbf{A}$ is positive definite so that the characteristic roots of $\mathbf{A}$ are positive. Denote $\mathbf{D}^{1/2} =$

$\text{diag}[\sqrt{d_1}, \sqrt{d_2}, \cdots, \sqrt{d_p}]$ where $\mathbf{D}$ is the diagonal matrix in the relation $\mathbf{C}'\mathbf{A}\mathbf{C} = \mathbf{D}$. Show that $(\mathbf{C}\mathbf{D}^{-1/2})'\mathbf{A}(\mathbf{C}\mathbf{D}^{-1/2}) = \mathbf{I}$.

A–23 Prove Theorem A.6.1 parts (a) and (c).

A–24 Let $\mathbf{C}$ be an $n \times p$ matrix and $\mathbf{y}$ a $n \times 1$ vector. Show that $\mathbf{C}'[\mathbf{C},\mathbf{y}] = [\mathbf{C}'\mathbf{C}, \mathbf{C}'\mathbf{y}]$.

APPENDIX B TABLES

Table B.1 F-Distribution

UPPER α POINT* OF F WITH ν_1 AND ν_2 D.F.

$\alpha = 0.10$

ν_2 \ ν_1	1	2	3	4	5	6	7	8	9
1	39.9	49.5	53.6	55.8	57.2	58.2	58.9	59.4	59.9
2	8.53	9.00	9.16	9.24	9.29	9.33	9.35	9.37	9.38
3	5.54	5.46	5.39	5.34	5.31	5.28	5.27	5.25	5.24
4	4.54	4.32	4.19	4.11	4.05	4.01	3.98	3.95	3.94
5	4.06	3.78	3.62	3.52	3.45	3.40	3.37	3.34	3.32
6	3.78	3.46	3.29	3.18	3.11	3.05	3.01	2.98	2.96
7	3.59	3.26	3.07	2.96	2.88	2.83	2.78	2.75	2.72
8	3.46	3.11	2.92	2.81	2.73	2.67	2.62	2.59	2.56
9	3.36	3.01	2.81	2.69	2.61	2.55	2.51	2.47	2.44
10	3.29	2.92	2.73	2.61	2.52	2.46	2.41	2.38	2.35
11	3.23	2.86	2.66	2.54	2.45	2.39	2.34	2.30	2.27
12	3.18	2.81	2.61	2.48	2.39	2.33	2.28	2.24	2.21
13	3.14	2.76	2.56	2.43	2.35	2.28	2.23	2.20	2.16
14	3.10	2.73	2.52	2.39	2.31	2.24	2.19	2.15	2.12
15	3.07	2.70	2.49	2.36	2.27	2.21	2.16	2.12	2.09
16	3.05	2.67	2.46	2.33	2.24	2.18	2.13	2.09	2.06
17	3.03	2.64	2.44	2.31	2.22	2.15	2.10	2.06	2.03
18	3.01	2.62	2.42	2.29	2.20	2.13	2.08	2.04	2.00
19	2.99	2.61	2.40	2.27	2.18	2.11	2.06	2.02	1.98
20	2.97	2.59	2.38	2.25	2.16	2.09	2.04	2.00	1.96
21	2.96	2.57	2.36	2.23	2.14	2.08	2.02	1.98	1.95
22	2.95	2.56	2.35	2.22	2.13	2.06	2.01	1.97	1.93
23	2.94	2.55	2.34	2.21	2.11	2.05	1.99	1.95	1.92
24	2.93	2.54	2.33	2.19	2.10	2.04	1.98	1.94	1.91
25	2.92	2.53	2.32	2.18	2.09	2.02	1.97	1.93	1.89
26	2.91	2.52	2.31	2.17	2.08	2.01	1.96	1.92	1.88
27	2.90	2.51	2.30	2.17	2.07	2.00	1.95	1.91	1.87
28	2.89	2.50	2.29	2.16	2.06	2.00	1.94	1.90	1.87
29	2.89	2.50	2.28	2.15	2.06	1.99	1.93	1.89	1.86
30	2.88	2.49	2.28	2.14	2.05	1.98	1.93	1.88	1.85
40	2.84	2.44	2.23	2.09	2.00	1.93	1.87	1.83	1.79
60	2.79	2.39	2.18	2.04	1.95	1.87	1.82	1.77	1.74
120	2.75	2.35	2.13	1.99	1.90	1.82	1.77	1.72	1.68
∞	2.71	2.30	2.08	1.94	1.85	1.77	1.72	1.67	1.63

* Rounded off to three significant figures from tables of M. Merrington and C. M. Thompson in *Biometrika*, Vol. 33, pp. 78–87, 1943. Reproduced with the kind permission of the authors and the editor.

Table B.1 (continued) F-Distribution

UPPER α POINT* OF F WITH ν_1 AND ν_2 D.F.

$\alpha = 0.10$

ν_2 \ ν_1	10	12	15	20	24	30	40	60	120	∞
1	60.2	60.7	61.2	61.7	62.0	62.3	62.5	62.8	63.1	63.3
2	9.39	9.41	9.42	9.44	9.45	9.46	9.47	9.47	9.48	9.49
3	5.23	5.22	5.20	5.18	5.18	5.17	5.16	5.15	5.14	5.13
4	3.92	3.90	3.87	3.84	3.83	3.82	3.80	3.79	3.78	3.76
5	3.30	3.27	3.24	3.21	3.19	3.17	3.16	3.14	3.12	3.10
6	2.94	2.90	2.87	2.84	2.82	2.80	2.78	2.76	2.74	2.72
7	2.70	2.67	2.63	2.59	2.58	2.56	2.54	2.51	2.49	2.47
8	2.54	2.50	2.46	2.42	2.40	2.38	2.36	2.34	2.32	2.29
9	2.42	2.38	2.34	2.30	2.28	2.25	2.23	2.21	2.18	2.16
10	2.32	2.28	2.24	2.20	2.18	2.16	2.13	2.11	2.08	2.06
11	2.25	2.21	2.17	2.12	2.10	2.08	2.05	2.03	2.00	1.97
12	2.19	2.15	2.10	2.06	2.04	2.01	1.99	1.96	1.93	1.90
13	2.14	2.10	2.05	2.01	1.98	1.96	1.93	1.90	1.88	1.85
14	2.10	2.05	2.01	1.96	1.94	1.91	1.89	1.86	1.83	1.80
15	2.06	2.02	1.97	1.92	1.90	1.87	1.85	1.82	1.79	1.76
16	2.03	1.99	1.94	1.89	1.87	1.84	1.81	1.78	1.75	1.72
17	2.00	1.96	1.91	1.86	1.84	1.81	1.78	1.75	1.72	1.69
18	1.98	1.93	1.89	1.84	1.81	1.78	1.75	1.72	1.69	1.66
19	1.96	1.91	1.86	1.81	1.79	1.76	1.73	1.70	1.67	1.63
20	1.94	1.89	1.84	1.79	1.77	1.74	1.71	1.68	1.64	1.61
21	1.92	1.88	1.83	1.78	1.75	1.72	1.69	1.66	1.62	1.59
22	1.90	1.86	1.81	1.76	1.73	1.70	1.67	1.64	1.60	1.57
23	1.89	1.84	1.80	1.74	1.72	1.69	1.66	1.62	1.59	1.55
24	1.88	1.83	1.78	1.73	1.70	1.67	1.64	1.61	1.57	1.53
25	1.87	1.82	1.77	1.72	1.69	1.66	1.63	1.59	1.56	1.52
26	1.86	1.81	1.76	1.71	1.68	1.65	1.61	1.58	1.54	1.50
27	1.85	1.80	1.75	1.70	1.67	1.64	1.60	1.57	1.53	1.49
28	1.84	1.79	1.74	1.69	1.66	1.63	1.59	1.56	1.52	1.48
29	1.83	1.78	1.73	1.68	1.65	1.62	1.58	1.55	1.51	1.47
30	1.82	1.77	1.72	1.67	1.64	1.61	1.57	1.54	1.50	1.46
40	1.76	1.71	1.66	1.61	1.57	1.54	1.51	1.47	1.42	1.38
60	1.71	1.66	1.60	1.54	1.51	1.48	1.44	1.40	1.35	1.29
120	1.65	1.60	1.55	1.48	1.45	1.41	1.37	1.32	1.26	1.19
∞	1.60	1.55	1.49	1.42	1.38	1.34	1.30	1.24	1.17	1.00

* Rounded off to three significant figures from tables of M. Merrington and C. M. Thompson in *Biometrika*, Vol. 33, pp. 78–87, 1943. Reproduced with the kind permission of the authors and the editor.

Table B.1 (continued) F-Distribution

UPPER α POINT* OF F WITH ν_1 AND ν_2 D.F.

$\alpha = 0.05$

ν_2 \ ν_1	1	2	3	4	5	6	7	8	9
1	161	200	216	225	230	234	237	239	241
2	18.5	19.0	19.2	19.2	19.3	19.3	19.4	19.4	19.4
3	10.1	9.55	9.28	9.12	9.01	8.94	8.89	8.85	8.81
4	7.71	6.94	6.59	6.39	6.26	6.16	6.09	6.04	6.00
5	6.61	5.79	5.41	5.19	5.05	4.95	4.88	4.82	4.77
6	5.99	5.14	4.76	4.53	4.39	4.28	4.21	4.15	4.10
7	5.59	4.74	4.35	4.12	3.97	3.87	3.79	3.73	3.68
8	5.32	4.46	4.07	3.84	3.69	3.58	3.50	3.44	3.39
9	5.12	4.26	3.86	3.63	3.48	3.37	3.29	3.23	3.18
10	4.96	4.10	3.71	3.48	3.33	3.22	3.14	3.07	3.02
11	4.84	3.98	3.59	3.36	3.20	3.09	3.01	2.95	2.90
12	4.75	3.89	3.49	3.26	3.11	3.00	2.91	2.85	2.80
13	4.67	3.81	3.41	3.18	3.03	2.92	2.83	2.77	2.71
14	4.60	3.74	3.34	3.11	2.96	2.85	2.76	2.70	2.65
15	4.54	3.68	3.29	3.06	2.90	2.79	2.71	2.64	2.59
16	4.49	3.63	3.24	3.01	2.85	2.74	2.66	2.59	2.54
17	4.45	3.59	3.20	2.96	2.81	2.70	2.61	2.55	2.49
18	4.41	3.55	3.16	2.93	2.77	2.66	2.58	2.51	2.46
19	4.38	3.52	3.13	2.90	2.74	2.63	2.54	2.48	2.42
20	4.35	3 49	3.10	2.87	2.71	2.60	2.51	2.45	2.39
21	4.32	3.47	3.07	2.84	2.68	2.57	2.49	2.42	2.37
22	4.30	3.44	3.05	2.82	2.66	2.55	2.46	2.40	2.34
23	4.28	3.42	3.03	2.80	2.64	2.53	2.44	2.37	2.32
24	4.26	3.40	3.01	2.78	2.62	2.51	2.42	2.36	2.30
25	4.24	3.39	2.99	2.76	2.60	2.49	2.40	2.34	2.28
26	4.23	3.37	2.98	2.74	2.59	2.47	2.39	2.32	2.27
27	4.21	3.35	2.96	2.73	2.57	2.46	2.37	2.31	2.25
28	4.20	3.34	2.95	2.71	2.56	2.45	2.36	2.29	2.24
29	4.18	3.33	2.93	2.70	2.55	2.43	2.35	2.28	2.22
30	4.17	3.32	2.92	2.69	2.53	2.42	2.33	2.27	2.21
40	4.08	3.23	2.84	2.61	2.45	2.34	2.25	2.18	2.12
60	4.00	3.15	2.76	2.53	2.37	2.25	2.17	2.10	2.04
120	3.92	3.07	2.68	2.45	2.29	2.17	2.09	2.02	1.96
∞	3.84	3.00	2.60	2.37	2.21	2.10	2.01	1.94	1.88

* Rounded off to three significant figures from tables of M. Merrington and C. M. Thompson in *Biometrika*, Vol. 33, pp. 78–87, 1943. Reproduced with the kind permission of the authors and the editor.

Table B.1 (continued) F-Distribution

UPPER α POINT* OF F WITH ν_1 AND ν_2 D.F.

$\alpha = 0.05$

ν_2 \ ν_1	10	12	15	20	24	30	40	60	120	∞
1	242	244	246	248	249	250	251	252	253	254
2	19.4	19.4	19.4	19.4	19.5	19.5	19.5	19.5	19.5	19.5
3	8.79	8.74	8.70	8.66	8.64	8.62	8.59	8.57	8.55	8.53
4	5.96	5.91	5.86	5.80	5.77	5.75	5.72	5.69	5.66	5.63
5	4.74	4.68	4.62	4.56	4.53	4.50	4.46	4.43	4.40	4.36
6	4.06	4.00	3.94	3.87	3.84	3.81	3.77	3.74	3.70	3.67
7	3.64	3.57	3.51	3.44	3.41	3.38	3.34	3.30	3.27	3.23
8	3.35	3.28	3.22	3.15	3.12	3.08	3.04	3.01	2.97	2.93
9	3.14	3.07	3.01	2.94	2.90	2.86	2.83	2.79	2.75	2.71
10	2.98	2.91	2.85	2.77	2.74	2.70	2.66	2.62	2.58	2.54
11	2.85	2.79	2.72	2.65	2.61	2.57	2.53	2.49	2.45	2.40
12	2.75	2.69	2.62	2.54	2.51	2.47	2.43	2.38	2.34	2.30
13	2.67	2.60	2.53	2.46	2.42	2.38	2.34	2.30	2.25	2.21
14	2.60	2.53	2.46	2.39	2.35	2.31	2.27	2.22	2.18	2.13
15	2.54	2.48	2.40	2.33	2.29	2.25	2.20	2.16	2.11	2.07
16	2.49	2.42	2.35	2.28	2.24	2.19	2.15	2.11	2.06	2.01
17	2.45	2.38	2.31	2.23	2.19	2.15	2.10	2.06	2.01	1.96
18	2.41	2.34	2.27	2.19	2.15	2.11	2.06	2.02	1.97	1.92
19	2.38	2.31	2.23	2.16	2.11	2.07	2.03	1.98	1.93	1.88
20	2.35	2.28	2.20	2.12	2.08	2.04	1.99	1.95	1.90	1.84
21	2.32	2.25	2.18	2.10	2.05	2.01	1.96	1.92	1.87	1.81
22	2.30	2.23	2.15	2.07	2.03	1.98	1.94	1.89	1.84	1.78
23	2.27	2.20	2.13	2.05	2.01	1.96	1.91	1.86	1.81	1.76
24	2.25	2.18	2.11	2.03	1.98	1.94	1.89	1.84	1.79	1.73
25	2.24	2.16	2.09	2.01	1.96	1.92	1.87	1.82	1.77	1.71
26	2.22	2.15	2.07	1.99	1.95	1.90	1.85	1.80	1.75	1.69
27	2.20	2.13	2.06	1.97	1.93	1.88	1.84	1.79	1.73	1.67
28	2.19	2.12	2.04	1.96	1.91	1.87	1.82	1.77	1.71	1.65
29	2.18	2.10	2.03	1.94	1.90	1.85	1.81	1.75	1.70	1.64
30	2.16	2.09	2.01	1.93	1.89	1.84	1.79	1.74	1.68	1.62
40	2.08	2.00	1.92	1.84	1.79	1.74	1.69	1.64	1.58	1.51
60	1.99	1.92	1.84	1.75	1.70	1.65	1.59	1.53	1.47	1.39
120	1.91	1.83	1.75	1.66	1.61	1.55	1.50	1.43	1.35	1.25
∞	1.83	1.75	1.67	1.57	1.52	1.46	1.39	1.32	1.22	1.00

* Rounded off to three significant figures from tables of M. Merrington and C. M. Thomspon in *Biometrika*, Vol. 33, pp. 78–87, 1943. Reproduced with the kind permission of the authors and the editor.

Table B.1 (continued) F-Distribution

UPPER α POINT* OF F WITH ν_1 AND ν_2 D.F.

$\alpha = 0.01$

ν_2 \ ν_1	1	2	3	4	5	6	7	8	9
1	4050	5000	5400	5620	5760	5860	5930	5980	6020
2	98.5	99.0	99.2	99.2	99.3	99.3	99.4	99.4	99.4
3	34.1	30.8	29.5	28.7	28.2	27.9	27.7	27.5	27.3
4	21.2	18.0	16.7	16.0	15.5	15.2	15.0	14.8	14.7
5	16.3	13.3	12.1	11.4	11.0	10.7	10.5	10.3	10.2
6	13.7	10.9	9.78	9.15	8.75	8.47	8.26	8.10	7.98
7	12.2	9.55	8.45	7.85	7.46	7.19	6.99	6.84	6.72
8	11.3	8.65	7.59	7.01	6.63	6.37	6.18	6.03	5.91
9	10.6	8.02	6.99	6.42	6.06	5.80	5.61	5.47	5.35
10	10.0	7.56	6.55	5.99	5.64	5.39	5.20	5.06	4.94
11	9.65	7.21	6.22	5.67	5.32	5.07	4.89	4.74	4.63
12	9.33	6.93	5.95	5.41	5.06	4.82	4.64	4.50	4.39
13	9.07	6.70	5.74	5.21	4.86	4.62	4.44	4.30	4.19
14	8.86	6.51	5.56	5.04	4.69	4.46	4.28	4.14	4.03
15	8.68	6.36	5.42	4.89	4.56	4.32	4.14	4.00	3.89
16	8.53	6.23	5.29	4.77	4.44	4.20	4.03	3.89	3.78
17	8.40	6.11	5.18	4.67	4.34	4.10	3.93	3.79	3.68
18	8.29	6.01	5.09	4.58	4.25	4.01	3.84	3.71	3.60
19	8.18	5.93	5.01	4.50	4.17	3.94	3.77	3.63	3.52
20	8.10	5.85	4.94	4.43	4.10	3.87	3.70	3.56	3.46
21	8.02	5.78	4.87	4.37	4.04	3.81	3.64	3.51	3.40
22	7.95	5.72	4.82	4.31	3.99	3.76	3.59	3.45	3.35
23	7.88	5.66	4.76	4.26	3.94	3.71	3.54	3.41	3.30
24	7.82	5.61	4.72	4.22	3.90	3.67	3.50	3.36	3.26
25	7.77	5.57	4.68	4.18	3.85	3.63	3.46	3.32	3.22
26	7.72	5.53	4.64	4.14	3.82	3.59	3.42	3.29	3.18
27	7.68	5.49	4.60	4.11	3.78	3.56	3.39	3.26	3.15
28	7.64	5.45	4.57	4.07	3.75	3.53	3.36	3.23	3.12
29	7.60	5.42	4.54	4.04	3.73	3.50	3.33	3.20	3.09
30	7.56	5.39	4.51	4.02	3.70	3.47	3.30	3.17	3.07
40	7.31	5.18	4.31	3.83	3.51	3.29	3.12	2.99	2.89
60	7.08	4.98	4.13	3.65	3.34	3.12	2.95	2.82	2.72
120	6.85	4.79	3.95	3.48	3.17	2.96	2.79	2.66	2.56
∞	6.63	4.61	3.78	3.32	3.02	2.80	2.64	2.51	2.41

* Rounded off to three significant figures from tables of M. Merrington and C. M. Thompson in *Biometrika*, Vol. 33, pp. 78–87, 1943. Reproduced with the kind permission of the authors and the editor.

Table B.1 (continued) F-Distribution

UPPER α POINT* OF F WITH ν_1 AND ν_2 D.F.

$\alpha = 0.01$

ν_2 \ ν_1	10	12	15	20	24	30	40	60	120	∞
1	6060	6110	6160	6210	6230	6260	6290	6310	6340	6370
2	99.4	99.4	99.4	99.4	99.5	99.5	99.5	99.5	99.5	99.5
3	27.2	27.1	26.9	26.7	26.6	26.5	26.4	26.3	26.2	26.1
4	14.5	14.4	14.2	14.0	13.9	13.8	13.7	13.7	13.6	13.5
5	10.1	9.89	9.72	9.55	9.47	9.38	9.29	9.20	9.11	9.02
6	7.87	7.72	7.56	7.40	7.31	7.23	7.14	7.06	6.97	6.88
7	6.62	6.47	6.31	6.16	6.07	5.99	5.91	5.82	5.74	5.65
8	5.81	5.67	5.52	5.36	5.28	5.20	5.12	5.03	4.95	4.86
9	5.26	5.11	4.96	4.81	4.73	4.65	4.57	4.48	4.40	4.31
10	4.85	4.71	4.56	4.41	4.33	4.25	4.17	4.08	4.00	3.91
11	4.54	4.40	4.25	4.10	4.02	3.94	3.86	3.78	3.69	3.60
12	4.30	4.16	4.01	3.86	3.78	3.70	3.62	3.54	3.45	3.36
13	4.10	3.96	3.82	3.66	3.59	3.51	3.43	3.34	3.25	3.17
14	3.94	3.80	3.66	3.51	3.43	3.35	3.27	3.18	3.09	3.00
15	3.80	3.67	3.52	3.37	3.29	3.21	3.13	3.05	2.96	2.87
16	3.69	3.55	3.41	3.26	3.18	3.10	3.02	2.93	2.84	2.75
17	3.59	3.46	3.31	3.16	3.08	3.00	2.92	2.83	2.75	2.65
18	3.51	3.37	3.23	3.08	3.00	2.92	2.84	2.75	2.66	2.57
19	3.43	3.30	3.15	3.00	2.92	2.84	2.76	2.67	2.58	2.49
20	3.37	3.23	3.09	2.94	2.86	2.78	2.69	2.61	2.52	2.42
21	3.31	3.17	3.03	2.88	2.80	2.72	2.64	2.55	2.46	2.36
22	3.26	3.12	2.98	2.83	2.75	2.67	2.58	2.50	2.40	2.31
23	3.21	3.07	2.93	2.78	2.70	2.62	2.54	2.45	2.35	2.26
24	3.17	3.03	2.89	2.74	2.66	2.58	2.49	2.40	2.31	2.21
25	3.13	2.99	2.85	2.70	2.62	2.54	2.45	2.36	2.27	2.17
26	3.09	2.96	2.81	2.66	2.58	2.50	2.42	2.33	2.23	2.13
27	3.06	2.93	2.78	2.63	2.55	2.47	2.38	2.29	2.20	2.10
28	3.03	2.90	2.75	2.60	2.52	2.44	2.35	2.26	2.17	2.06
29	3.00	2.87	2.73	2.57	2.49	2.41	2.33	2.23	2.14	2.03
30	2.98	2.84	2.70	2.55	2.47	2.39	2.30	2.21	2.11	2.01
40	2.80	2.66	2.52	2.37	2.29	2.20	2.11	2.02	1.92	1.80
60	2.63	2.50	2.35	2.20	2.12	2.03	1.94	1.84	1.73	1.60
120	2.47	2.34	2.19	2.03	1.95	1.86	1.76	1.66	1.53	1.38
∞	2.32	2.18	2.04	1.88	1.79	1.70	1.59	1.47	1.32	1.00

* Rounded off to three significant figures from tables of M. Merrington and C. M. Thompson in *Biometrika*, Vol. 33, pp. 78–87, 1943. Reproduced with the kind permission of the authors and the editor.

Table B.2 t-Distribution

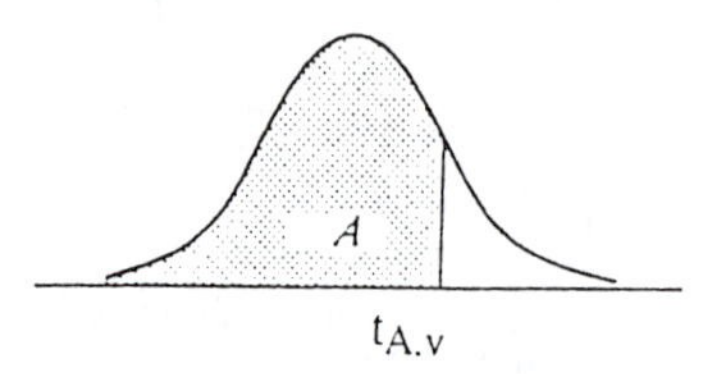

ν	A .60	.70	.80	.85	.90	.95	.975
1	0.325	0.727	1.376	1.963	3.078	6.314	12.706
2	0.289	0.617	1.061	1.386	1.886	2.920	4.303
3	0.277	0.584	0.978	1.250	1.638	2.353	3.182
4	0.271	0.569	0.941	1.190	1.533	2.132	2.776
5	0.267	0.559	0.920	1.156	1.476	2.015	2.571
6	0.265	0.553	0.906	1.134	1.440	1.943	2.447
7	0.263	0.549	0.896	1.119	1.415	1.895	2.365
8	0.262	0.546	0.889	1.108	1.397	1.860	2.306
9	0.261	0.543	0.883	1.100	1.383	1.833	2.262
10	0.260	0.542	0.879	1.093	1.372	1.812	2.228
11	0.260	0.540	0.876	1.088	1.363	1.796	2.201
12	0.259	0.539	0.873	1.083	1.356	1.782	2.179
13	0.259	0.537	0.870	1.079	1.350	1.771	2.160
14	0.258	0.537	0.868	1.076	1.345	1.761	2.145
15	0.258	0.536	0.866	1.074	1.341	1.753	2.131
16	0.258	0.535	0.865	1.071	1.337	1.746	2.120
17	0.257	0.534	0.863	1.069	1.333	1.740	2.110
18	0.257	0.534	0.862	1.067	1.330	1.734	2.101
19	0.257	0.533	0.861	1.066	1.328	1.729	2.093
20	0.257	0.533	0.860	1.064	1.325	1.725	2.086
21	0.257	0.532	0.859	1.063	1.323	1.721	2.080
22	0.256	0.532	0.858	1.061	1.321	1.717	2.074
23	0.256	0.532	0.858	1.060	1.319	1.714	2.069
24	0.256	0.531	0.857	1.059	1.318	1.711	2.064
25	0.256	0.531	0.856	1.058	1.316	1.708	2.060
26	0.256	0.531	0.856	1.058	1.315	1.706	2.056
27	0.256	0.531	0.855	1.057	1.314	1.703	2.052
28	0.256	0.530	0.855	1.056	1.313	1.701	2.048
29	0.256	0.530	0.854	1.055	1.311	1.699	2.045
30	0.256	0.530	0.854	1.055	1.310	1.697	2.042
40	0.255	0.529	0.851	1.050	1.303	1.684	2.021
60	0.254	0.527	0.848	1.045	1.296	1.671	2.000
120	0.254	0.526	0.845	1.041	1.289	1.658	1.980
∞	0.253	0.524	0.842	1.036	1.282	1.645	1.960

Table B.2 (continued) t-Distribution

	A						
ν	.98	.985	.99	.9925	.995	.9975	.9995
1	15.895	21.205	31.821	42.434	63.657	127.322	636.590
2	4.849	5.643	6.965	8.073	9.925	14.089	31.598
3	3.482	3.896	4.541	5.047	5.841	7.453	12.924
4	2.999	3.298	3.747	4.088	4.604	5.598	8.610
5	2.757	3.003	3.365	3.634	4.032	4.773	6.869
6	2.612	2.829	3.143	3.372	3.707	4.317	5.959
7	2.517	2.715	2.998	3.203	3.499	4.029	5.408
8	2.449	2.634	2.896	3.085	3.355	3.833	5.041
9	2.398	2.574	2.821	2.998	3.250	3.690	4.781
10	2.359	2.527	2.764	2.932	3.169	3.581	4.587
11	2.328	2.491	2.718	2.879	3.106	3.497	4.437
12	2.303	2.461	2.681	2.836	3.055	3.428	4.318
13	2.282	2.436	2.650	2.801	3.012	3.372	4.221
14	2.264	2.415	2.624	2.771	2.977	3.326	4.140
15	2.249	2.397	2.602	2.746	2.947	3.286	4.073
16	2.235	2.382	2.583	2.724	2.921	3.252	4.015
17	2.224	2.368	2.567	2.706	2.898	3.222	3.965
18	2.214	2.356	2.552	2.689	2.878	3.197	3.922
19	2.205	2.346	2.539	2.674	2.861	3.174	3.883
20	2.197	2.336	2.528	2.661	2.845	3.153	3.849
21	2.189	2.328	2.518	2.649	2.831	3.135	3.819
22	2.183	2.320	2.508	2.639	2.819	3.119	3.792
23	2.177	2.313	2.500	2.629	2.807	3.104	3.768
24	2.172	2.307	2.492	2.620	2.797	3.091	3.745
25	2.167	2.301	2.485	2.612	2.787	3.078	3.725
26	2.162	2.296	2.479	2.605	2.779	3.067	3.707
27	2.158	2.291	2.473	2.598	2.771	3.057	3.690
28	2.154	2.286	2.467	2.592	2.763	3.047	3.674
29	2.150	2.282	2.462	2.586	2.756	3.038	3.659
30	2.147	2.278	2.457	2.581	2.750	3.030	3.646
40	2.123	2.250	2.423	2.542	2.704	2.971	3.551
60	2.099	2.223	2.390	2.504	2.660	2.915	3.460
120	2.076	2.196	2.358	2.468	2.617	2.860	3.373
∞	2.054	2.170	2.326	2.432	2.576	2.807	3.291

Table B.3 Hartley Tables

α = .05

	p										
df	2	3	4	5	6	7	8	9	10	11	12
2	39.0	87.5	142	202	266	333	403	475	550	626	704
3	15.4	27.8	39.2	50.7	62.0	72.9	83.5	93.9	104	114	124
4	9.60	15.5	20.6	25.2	29.5	33.6	37.5	41.1	44.6	48.0	51.4
5	7.15	10.8	13.7	16.3	18.7	20.8	22.9	24.7	26.5	28.2	29.9
6	5.82	8.38	10.4	12.1	13.7	15.0	16.3	17.5	18.6	19.7	20.7
7	4.99	6.94	8.44	9.70	10.8	11.8	12.7	13.5	14.3	15.1	15.8
8	4.43	6.00	7.18	8.12	9.03	9.78	10.5	11.1	11.7	12.2	12.7
9	4.03	5.34	6.31	7.11	7.80	8.41	8.95	9.45	9.91	10.3	10.7
10	3.72	4.85	5.67	6.34	6.92	7.42	7.87	8.28	8.66	9.01	9.34
12	3.28	4.16	4.79	5.30	5.72	6.09	6.42	6.72	7.00	7.25	7.48
15	2.86	3.54	4.01	4.37	4.68	4.95	5.19	5.40	5.59	5.77	5.93
20	2.46	2.95	3.29	3.54	3.76	3.94	4.10	4.24	4.37	4.49	4.59
30	2.07	2.40	2.61	2.78	2.91	3.02	3.12	3.21	3.29	3.36	3.39
60	1.67	1.85	1.96	2.04	2.11	2.17	2.22	2.26	2.30	2.33	2.36
∞	1.00	1.00	1.00	1.00	1.00	1.00	1.00	1.00	1.00	1.00	1.00

α = .01

	p										
df	2	3	4	5	6	7	8	9	10	11	12
2	199	448	729	1,036	1,362	1,705	2,063	2,432	2,813	3,204	3,605
3	47.5	85	120	151	184	216	249	281	310	337	361
4	23.2	37	49	59	69	79	89	97	106	113	120
5	14.9	22	28	33	38	42	46	50	54	57	60
6	11.1	15.5	19.1	22	25	27	30	32	34	36	37
7	8.89	12.1	14.5	16.5	18.4	20	22	23	24	26	27
8	7.50	9.9	11.7	13.2	14.5	15.8	16.9	17.9	18.9	19.8	21
9	6.54	8.5	9.9	11.1	12.1	13.1	13.9	14.7	15.3	16.0	16.6
10	5.85	7.4	8.6	9.6	10.4	11.1	11.8	12.4	12.9	13.4	13.9
12	4.91	6.1	6.9	7.6	8.2	8.7	9.1	9.5	9.9	10.2	10.6
15	4.07	4.9	5.5	6.0	6.4	6.7	7.1	7.3	7.5	7.8	8.0
20	3.32	3.8	4.3	4.6	4.9	5.1	5.3	5.5	5.6	5.8	5.9
30	2.63	3.0	3.3	3.4	3.6	3.7	3.8	3.9	4.0	4.1	4.2
60	1.96	2.2	2.3	2.4	2.4	2.5	2.5	2.6	2.6	2.7	2.7
∞	1.00	1.0	1.0	1.0	1.0	1.0	1.0	1.0	1.0	1.0	1.0

Source: Reprinted, with permission, from H. A. David, "Upper 5 and 1% Points of the Maximum *F*-Ratio," *Biometrika* 39 (1952), pp. 422–24.

Table B.4 Coefficients for Orthogonal Polynomials for p Equally Spaced Treatments

p	Polynomial	treatment totals 1	2	3	4	5	6	7	8	9	10
3	Linear	−1	0	1							
	Quadratic	1	−2	1							
	Linear	−3	−1	1	3						
4	Quadratic	1	−1	−1	1						
	Cubic	−1	3	−3	1						
	Linear	−2	−1	0	1	2					
5	Quadratic	2	−1	−2	−1	2					
	Cubic	−1	2	0	−2	1					
	Quartic	1	−4	6	−4	1					
	Linear	−5	−3	−1	1	3	5				
6	Quadratic	5	−1	−4	−4	−1	5				
	Cubic	−5	7	4	−4	−7	5				
	Quartic	1	−3	2	2	−3	1				
	Linear	−3	−2	−1	0	1	2	3			
7	Quadratic	5	0	−3	−4	−3	0	5			
	Cubic	−1	1	1	0	−1	−1	1			
	Quartic	3	−7	1	6	1	−7	3			
	Linear	−7	−5	−3	−1	1	3	5	7		
	Quadratic	7	1	−3	−5	−5	−3	1	7		
8	Cubic	−7	5	7	3	−3	−7	−5	7		
	Quartic	7	−13	−3	9	9	−3	−13	7		
	Quintic	−7	23	−17	−15	15	17	−23	7		
	Linear	−4	−3	−2	−1	0	1	2	3	4	
	Quadratic	28	7	−8	−17	−20	−17	−8	7	28	
9	Cubic	−14	7	13	9	0	−9	−13	−7	14	
	Quartic	14	−21	−11	9	18	9	−11	−21	14	
	Quintic	−4	11	−4	−9	0	9	4	−11	4	
	Linear	−9	−7	−5	−3	−1	1	3	5	7	9
	Quadratic	6	2	−1	−3	−4	−4	−3	−1	2	6
10	Cubic	−42	14	35	31	12	−12	−31	−35	−14	42
	Quartic	18	−22	−17	3	18	18	3	−17	−22	18
	Quintic	−6	14	−1	−11	−6	6	11	1	−14	6

Table B.5 Upper Percentage Points for the Bonferroni
(k = number of comparisons, υ = df for error)

$\alpha = .05$

υ \ k	2	3	4	5	6	7	8	9	10	15	20	25	30	35	40	45	50
5	3.17	3.54	3.81	4.04	4.22	4.38	4.53	4.66	4.78	5.25	5.60	5.89	6.15	6.36	6.56	6.70	6.86
7	2.84	3.13	3.34	3.50	3.64	3.76	3.86	3.95	4.03	4.36	4.59	4.78	4.95	5.09	5.21	5.31	5.40
10	2.64	2.87	3.04	3.17	3.28	3.37	3.45	3.52	3.58	3.83	4.01	4.15	4.27	4.37	4.45	4.53	4.59
12	2.56	2.78	2.94	3.06	3.15	3.24	3.31	3.37	3.43	3.65	3.80	3.93	4.04	4.13	4.20	4.26	4.32
15	2.49	2.69	2.84	2.95	3.04	3.11	3.18	3.24	3.29	3.48	3.62	3.74	3.82	3.90	3.97	4.02	4.07
20	2.42	2.61	2.75	2.85	2.93	3.00	3.06	3.11	3.16	3.33	3.46	3.55	3.63	3.70	3.76	3.80	3.85
24	2.39	2.58	2.70	2.80	2.88	2.94	3.00	3.05	3.09	3.26	3.38	3.47	3.54	3.61	3.66	3.70	3.74
30	2.36	2.54	2.66	2.75	2.83	2.89	2.94	2.99	3.03	3.19	3.30	3.39	3.46	3.52	3.57	3.61	3.65
40	2.33	2.50	2.62	2.71	2.78	2.84	2.89	2.93	2.97	3.12	3.23	3.31	3.38	3.43	3.48	3.51	3.55
60	2.30	2.47	2.58	2.66	2.73	2.79	2.84	2.88	2.92	3.06	3.16	3.24	3.30	3.34	3.39	3.42	3.46
120	2.27	2.43	2.54	2.62	2.68	2.74	2.79	2.83	2.86	2.99	3.09	3.16	3.22	3.27	3.31	3.34	3.37
∞	2.24	2.39	2.50	2.58	2.64	2.69	2.74	2.77	2.81	2.94	3.02	3.09	3.15	3.19	3.23	3.26	3.29

Table B.6 Upper Percentage Points for the Studentized Range ($\alpha = .05$, p = number of treatments, υ = df for error)

$\alpha = 0.05$

ν \ p	2	3	4	5	6	7	8	9	10	11	12	13	14	15	16	17	18	19	20
1	18.0	27.0	32.8	37.1	40.4	43.1	45.4	47.4	49.1	50.6	52.0	53.2	54.3	55.4	56.3	57.2	58.0	58.8	59.6
2	6.08	8.33	9.80	10.9	11.7	12.4	13.0	13.5	14.0	14.4	14.7	15.1	15.4	15.7	15.9	16.1	16.4	16.6	16.8
3	4.50	5.91	6.82	7.50	8.04	8.48	8.85	9.18	9.46	9.72	9.95	10.2	10.3	10.5	10.7	10.8	11.0	11.1	11.2
4	3.93	5.04	5.76	6.29	6.71	7.05	7.35	7.60	7.83	8.03	8.21	8.37	8.52	8.66	8.79	8.91	9.03	9.13	9.23
5	3.64	4.60	5.22	5.67	6.03	6.33	6.58	6.80	6.99	7.17	7.32	7.47	7.60	7.72	7.83	7.93	8.03	8.12	8.21
6	3.46	4.34	4.90	5.30	5.63	5.90	6.12	6.32	6.49	6.65	6.79	6.92	7.03	7.14	7.24	7.34	7.43	7.51	7.59
7	3.34	4.16	4.68	5.06	5.36	5.61	5.82	6.00	6.16	6.30	6.43	6.55	6.66	6.76	6.85	6.94	7.02	7.10	7.17
8	3.26	4.04	4.53	4.89	5.17	5.40	5.60	5.77	5.92	6.05	6.18	6.29	6.39	6.48	6.57	6.65	6.73	6.80	6.87
9	3.20	3.95	4.41	4.76	5.02	5.24	5.43	5.59	5.74	5.87	5.98	6.09	6.19	6.28	6.36	6.44	6.51	6.58	6.64
10	3.15	3.88	4.33	4.65	4.91	5.12	5.30	5.46	5.60	5.72	5.83	5.93	6.03	6.11	6.19	6.27	6.34	6.40	6.47
11	3.11	3.82	4.26	4.57	4.82	5.03	5.20	5.35	5.49	5.61	5.71	5.81	5.90	5.98	6.06	6.13	6.20	6.27	6.33
12	3.08	3.77	4.20	4.51	4.75	4.95	5.12	5.27	5.39	5.51	5.61	5.71	5.80	5.88	5.95	6.02	6.09	6.15	6.21
13	3.06	3.73	4.15	4.45	4.69	4.88	5.05	5.19	5.32	5.43	5.53	5.63	5.71	5.79	5.86	5.93	5.99	6.05	6.11
14	3.03	3.70	4.11	4.41	4.64	4.83	4.99	5.13	5.25	5.36	5.46	5.55	5.64	5.71	5.79	5.85	5.91	5.97	6.03
15	3.01	3.67	4.08	4.37	4.59	4.78	4.94	5.08	5.20	5.31	5.40	5.49	5.57	5.65	5.72	5.78	5.85	5.90	5.96
16	3.00	3.65	4.05	4.33	4.56	4.74	4.90	5.03	5.15	5.26	5.35	5.44	5.52	5.59	5.66	5.73	5.79	5.84	5.90
17	2.98	3.63	4.02	4.30	4.52	4.70	4.86	4.99	5.11	5.21	5.31	5.39	5.47	5.54	5.61	5.67	5.73	5.79	5.84
18	2.97	3.61	4.00	4.28	4.49	4.67	4.82	4.96	5.07	5.17	5.27	5.35	5.43	5.50	5.57	5.63	5.69	5.74	5.79
19	2.96	3.59	3.98	4.25	4.47	4.65	4.79	4.92	5.04	5.14	5.23	5.31	5.39	5.46	5.53	5.59	5.65	5.70	5.75
20	2.95	3.58	3.96	4.23	4.45	4.62	4.77	4.90	5.01	5.11	5.20	5.28	5.36	5.43	5.49	5.55	5.61	5.66	5.71
24	2.92	3.53	3.90	4.17	4.37	4.54	4.68	4.81	4.92	5.01	5.10	5.18	5.25	5.32	5.38	5.44	5.49	5.55	5.59
30	2.89	3.49	3.85	4.10	4.30	4.46	4.60	4.72	4.82	4.92	5.00	5.08	5.15	5.21	5.27	5.33	5.38	5.43	5.47
40	2.86	3.44	3.79	4.04	4.23	4.39	4.52	4.63	4.73	4.82	4.90	4.98	5.04	5.11	5.16	5.22	5.27	5.31	5.36
60	2.83	3.40	3.74	3.98	4.16	4.31	4.44	4.55	4.65	4.73	4.81	4.88	4.94	5.00	5.06	5.11	5.15	5.20	5.24
120	2.80	3.36	3.68	3.92	4.10	4.24	4.36	4.47	4.56	4.64	4.71	4.78	4.84	4.90	4.95	5.00	5.04	5.09	5.13
∞	2.77	3.31	3.63	3.86	4.03	4.17	4.29	4.39	4.47	4.55	4.62	4.68	4.74	4.80	4.85	4.89	4.93	4.97	5.01

Table B.6 (continued) Upper Percentage Points for the Studentized Range
($\alpha = .10$, p = number of treatments, υ = df for error)

$\alpha = 0.10$

ν \ p	2	3	4	5	6	7	8	9	10	11	12	13	14	15	16	17	18	19	20
1	8.93	13.4	16.4	18.5	20.2	21.5	22.6	23.6	24.5	25.2	25.9	26.5	27.1	27.6	28.1	28.5	29.0	29.3	29.7
2	4.13	5.73	6.77	7.54	8.14	8.63	9.05	9.41	9.72	10.0	10.3	10.5	10.7	10.9	11.1	11.2	11.4	11.5	11.7
3	3.33	4.47	5.20	5.74	6.16	6.51	6.81	7.06	7.29	7.49	7.67	7.83	7.98	8.12	8.25	8.37	8.48	8.58	8.68
4	3.01	3.98	4.59	5.03	5.39	5.68	5.93	6.14	6.33	6.49	6.65	6.78	6.91	7.02	7.13	7.23	7.33	7.41	7.50
5	2.85	3.72	4.26	4.66	4.98	5.24	5.46	5.65	5.82	5.97	6.10	6.22	6.34	6.44	6.54	6.63	6.71	6.79	6.86
6	2.75	3.56	4.07	4.44	4.73	4.97	5.17	5.34	5.50	5.64	5.76	5.87	5.98	6.07	6.16	6.25	6.32	6.40	6.47
7	2.68	3.45	3.93	4.28	4.55	4.78	4.97	5.14	5.28	5.41	5.53	5.64	5.74	5.83	5.91	5.99	6.06	6.13	6.19
8	2.63	3.37	3.83	4.17	4.43	4.65	4.83	4.99	5.13	5.25	5.36	5.46	5.56	5.64	5.72	5.80	5.87	5.93	6.00
9	2.59	3.32	3.76	4.08	4.34	4.54	4.72	4.87	5.01	5.13	5.23	5.33	5.42	5.51	5.58	5.66	5.72	5.79	5.85
10	2.56	3.27	3.70	4.02	4.26	4.47	4.64	4.78	4.91	5.03	5.13	5.23	5.32	5.40	5.47	5.54	5.61	5.67	5.73
11	2.54	3.23	3.66	3.96	4.20	4.40	4.57	4.71	4.84	4.95	5.05	5.15	5.23	5.31	5.38	5.45	5.51	5.57	5.63
12	2.52	3.20	3.62	3.92	4.16	4.35	4.51	4.65	4.78	4.89	4.99	5.08	5.16	5.24	5.31	5.37	5.44	5.49	5.55
13	2.50	3.18	3.59	3.88	4.12	4.30	4.46	4.60	4.72	4.83	4.93	5.02	5.10	5.18	5.25	5.31	5.37	5.43	5.48
14	2.49	3.16	3.56	3.85	4.08	4.27	4.42	4.56	4.68	4.79	4.88	4.97	5.05	5.12	5.19	5.26	5.32	5.37	5.43
15	2.48	3.14	3.54	3.83	4.05	4.23	4.39	4.52	4.64	4.75	4.84	4.93	5.01	5.08	5.15	5.21	5.27	5.32	5.38
16	2.47	3.12	3.52	3.80	4.03	4.21	4.36	4.49	4.61	4.71	4.81	4.89	4.97	5.04	5.11	5.17	5.23	5.28	5.33
17	2.46	3.11	3.50	3.78	4.00	4.18	4.33	4.46	4.58	4.68	4.77	4.86	4.93	5.01	5.07	5.13	5.19	5.24	5.30
18	2.45	3.10	3.49	3.77	3.98	4.16	4.31	4.44	4.55	4.65	4.75	4.83	4.90	4.98	5.04	5.10	5.16	5.21	5.26
19	2.45	3.09	3.47	3.75	3.97	4.14	4.29	4.42	4.53	4.63	4.72	4.80	4.88	4.95	5.01	5.07	5.13	5.18	5.23
20	2.44	3.08	3.46	3.74	3.95	4.12	4.27	4.40	4.51	4.61	4.70	4.78	4.85	4.92	4.99	5.05	5.10	5.16	5.20
24	2.42	3.05	3.42	3.69	3.90	4.07	4.21	4.34	4.44	4.54	4.63	4.71	4.78	4.85	4.91	4.97	5.02	5.07	5.12
30	2.40	3.02	3.39	3.65	3.85	4.02	4.16	4.28	4.38	4.47	4.56	4.64	4.71	4.77	4.83	4.89	4.94	4.99	5.03
40	2.38	2.99	3.35	3.60	3.80	3.96	4.10	4.21	4.32	4.41	4.49	4.56	4.63	4.69	4.75	4.81	4.86	4.90	4.95
60	2.36	2.96	3.31	3.56	3.75	3.91	4.04	4.16	4.25	4.34	4.42	4.49	4.56	4.62	4.67	4.73	4.78	4.82	4.86
120	2.34	2.93	3.28	3.52	3.71	3.86	3.99	4.10	4.19	4.28	4.35	4.42	4.48	4.54	4.60	4.65	4.69	4.74	4.78
∞	2.33	2.90	3.24	3.48	3.66	3.81	3.93	4.04	4.13	4.21	4.28	4.35	4.41	4.47	4.52	4.57	4.61	4.65	4.69

Table B.7 Pearson – Hartley Tables
(υ_1 = df for numerator, υ_2 = df for denominator)

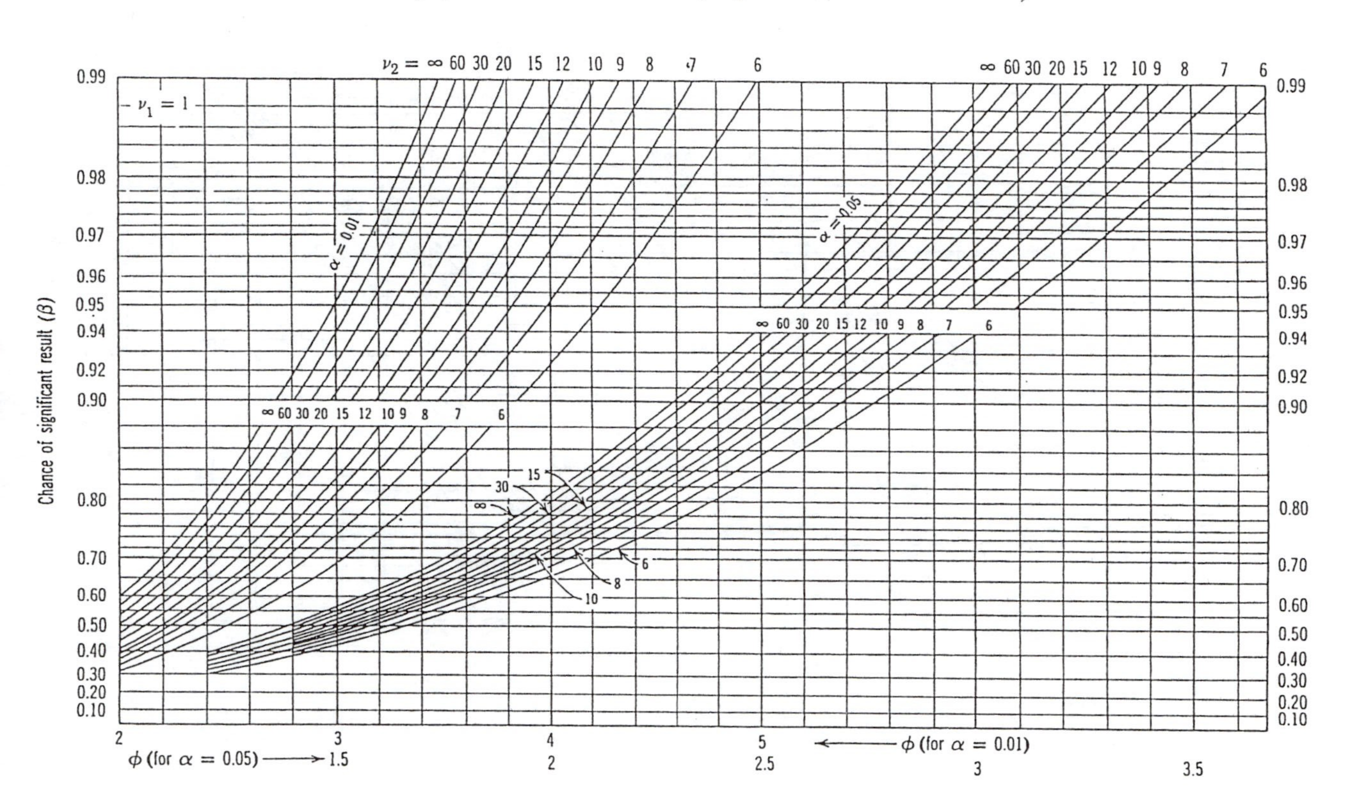

Table B.7 (continued) Pearson – Hartley Tables
(υ_1 = df for numerator, υ_2 = df for denominator)

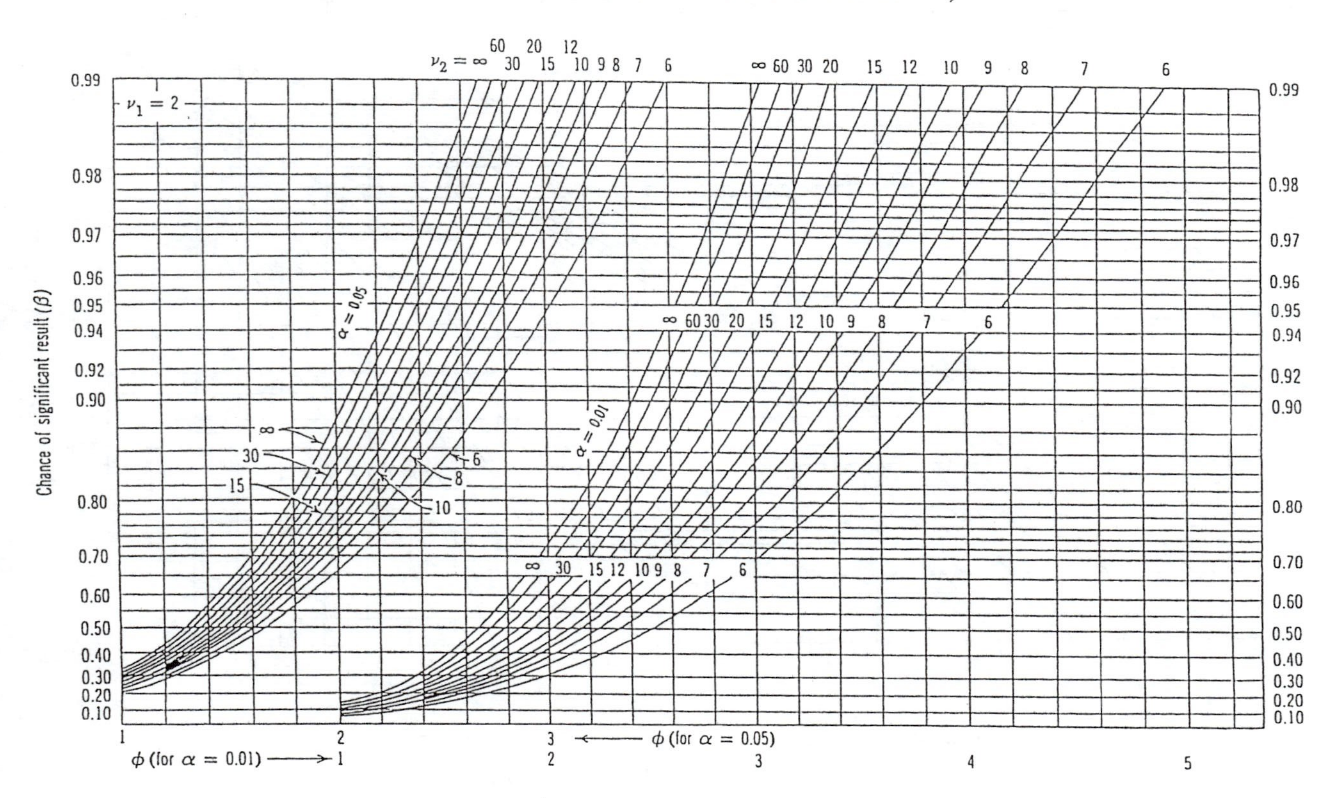

Table B.7 (continued) Pearson – Hartley Tables
(υ_1 = df for numerator, υ_2 = df for denominator)

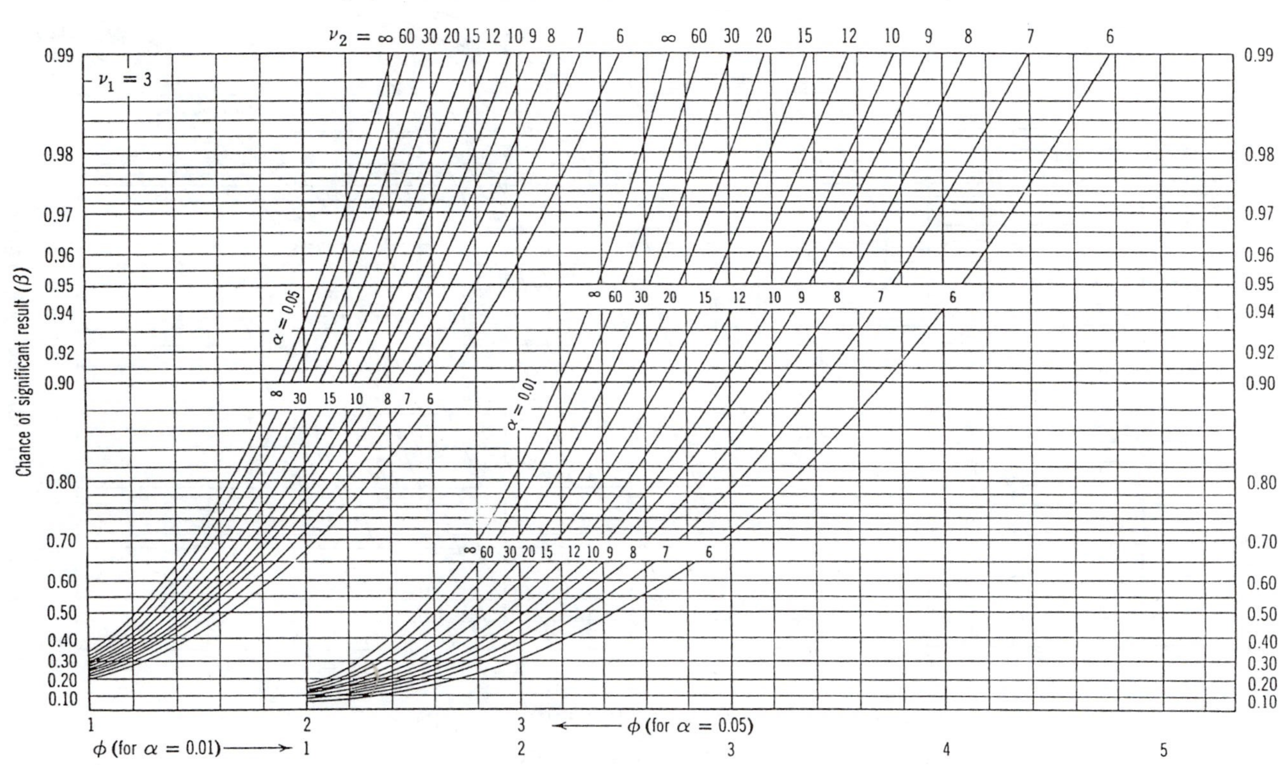

Table B.7 (continued) Pearson – Hartley Tables

(υ_1 = df for numerator, υ_2 = df for denominator)

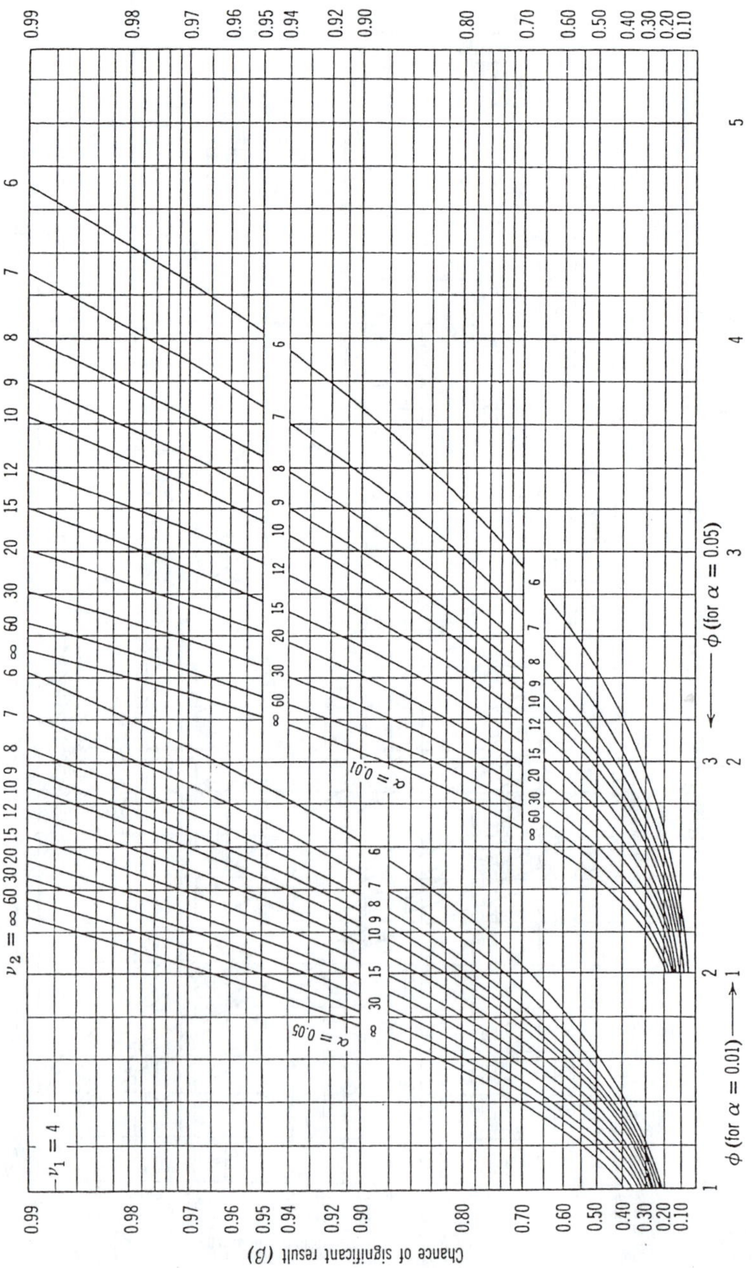

$\nu_1 = 4$

$\nu_2 = \infty$ 60 30 20 15 12 10 9 8 7 6 ∞ 60 30 20 15 12 10 9 8 7 6

$\alpha = 0.05$

∞ 30 15 10 9 8 7 6

$\alpha = 0.01$

∞ 60 30 20 15 12 10 9 8 7 6

∞ 60 30 20 15 12 10 9 8 7 6

Chance of significant result (β)

0.99 0.98 0.97 0.96 0.95 0.94 0.92 0.90 0.80 0.70 0.60 0.50 0.40 0.30 0.20 0.10

ϕ (for $\alpha = 0.01$) ⟶ 1 2 3

⟵ ϕ (for $\alpha = 0.05$) 1 2 3 4 5

Table B.7 (continued) Pearson – Hartley Tables
(υ_1 = df for numerator, υ_2 = df for denominator)

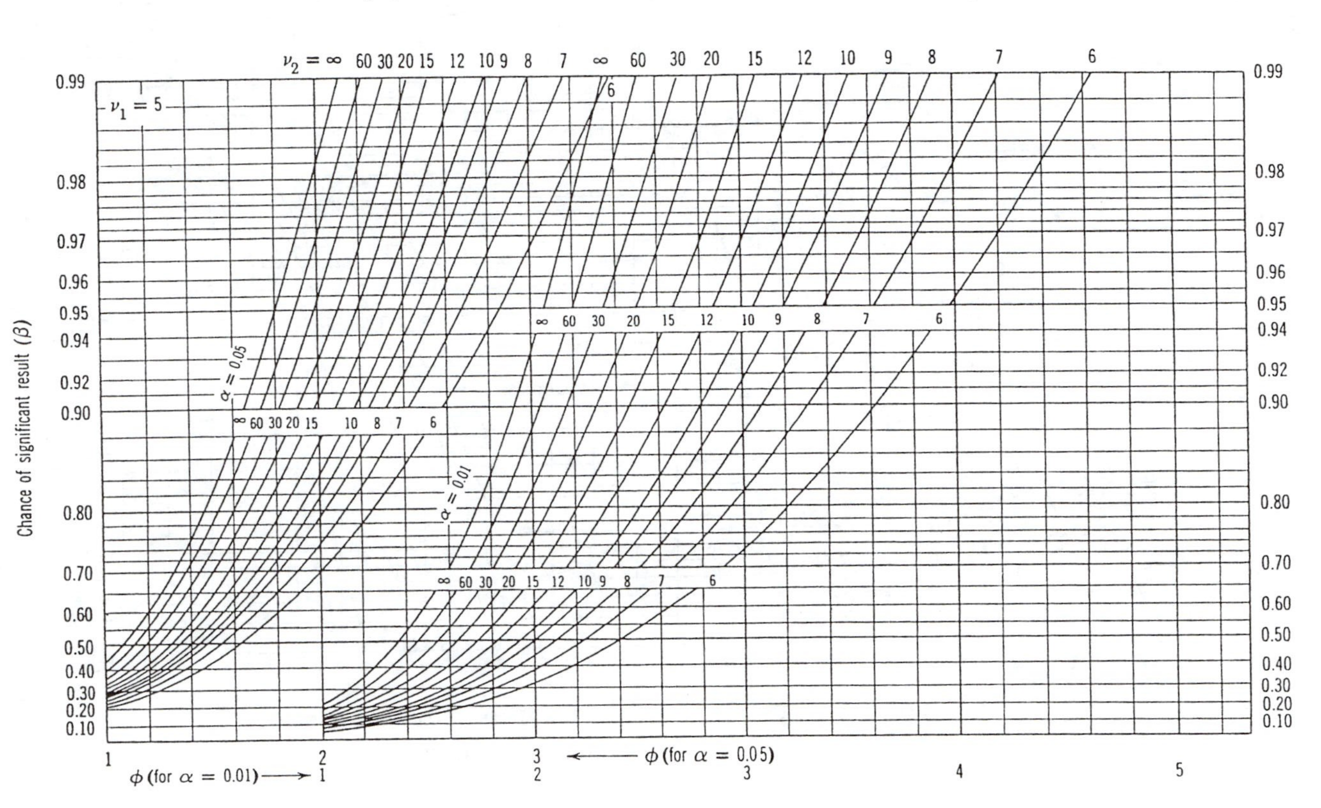

Table B.7 (continued) Pearson – Hartley Tables
(υ_1 = df for numerator, υ_2 = df for denominator)

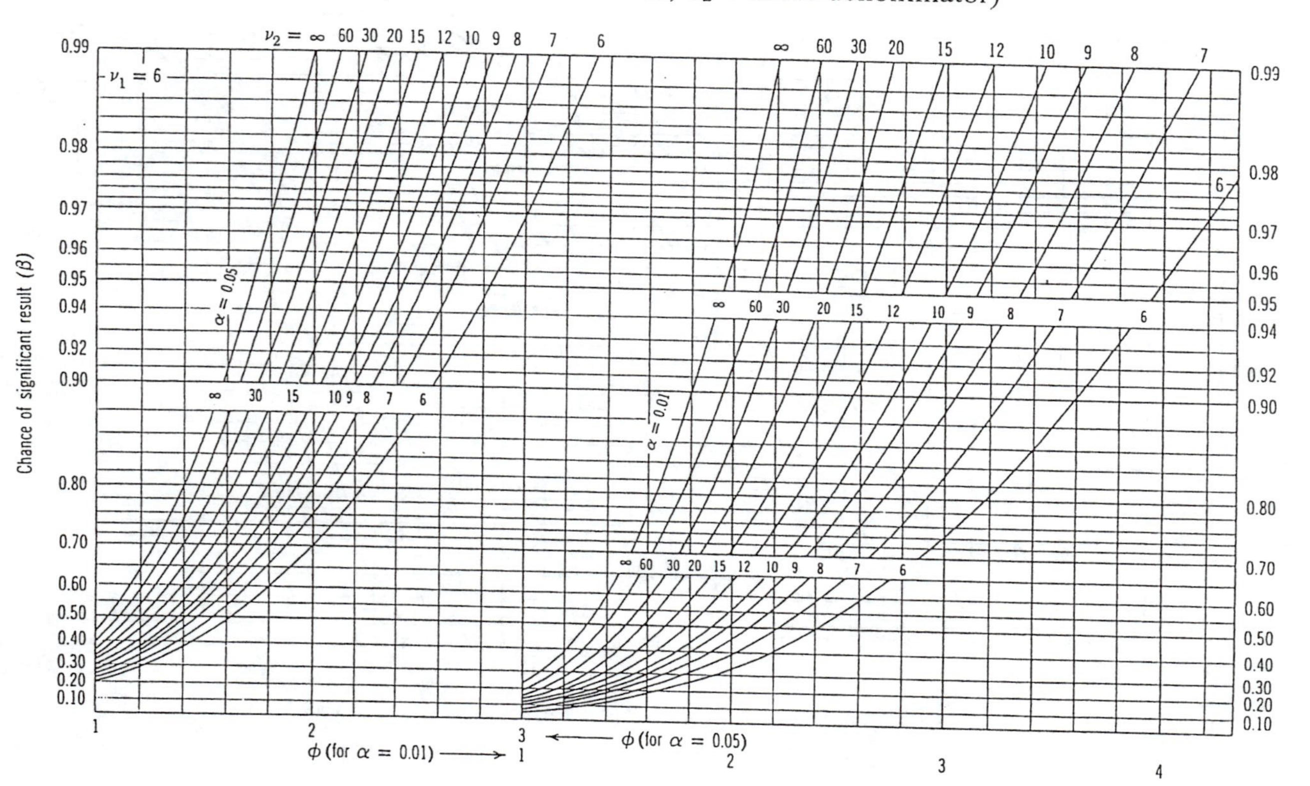

Table B.7 (continued) Pearson – Hartley Tables
(υ_1 = df for numerator, υ_2 = df for denominator)

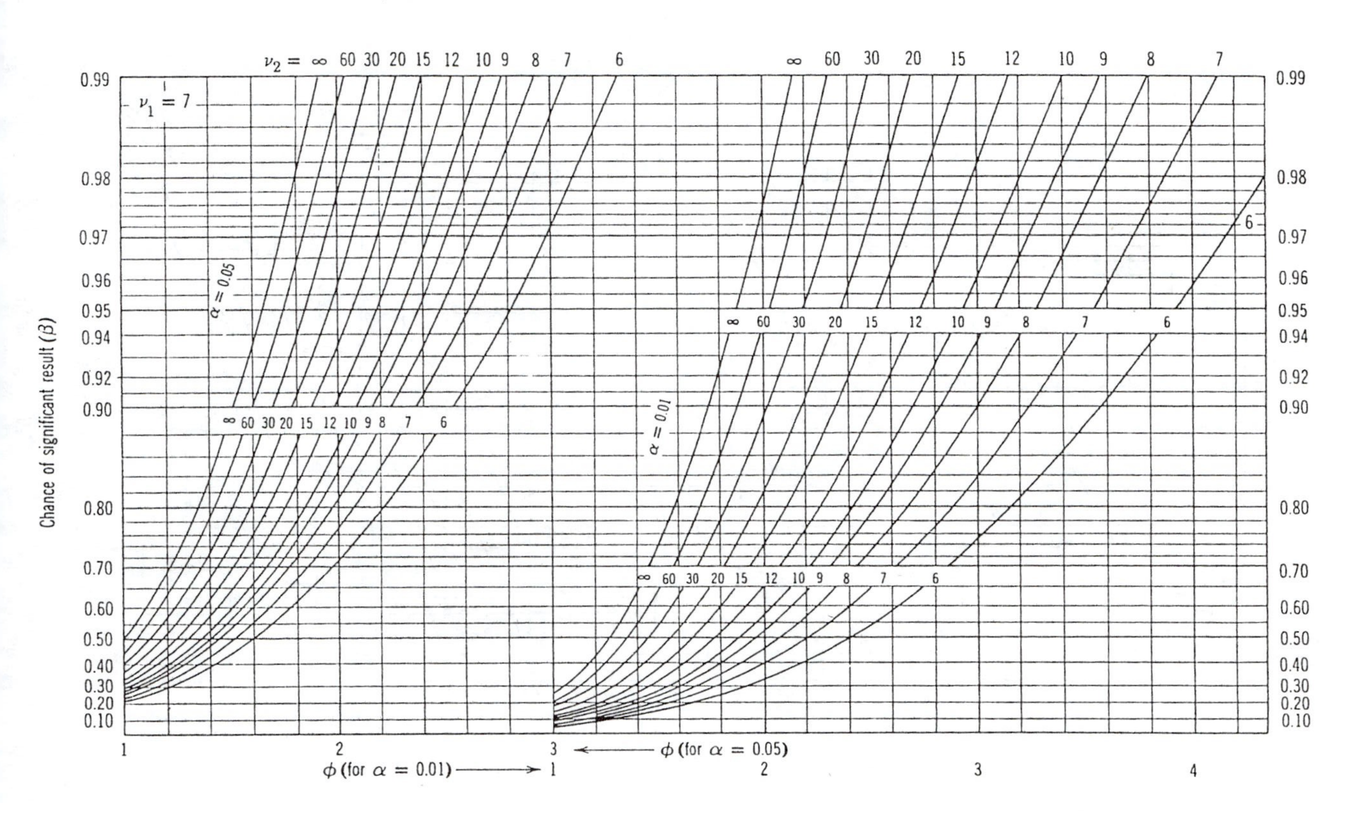

Table B.7 (continued) Pearson – Hartley Tables
(υ_1 = df for numerator, υ_2 = df for denominator)

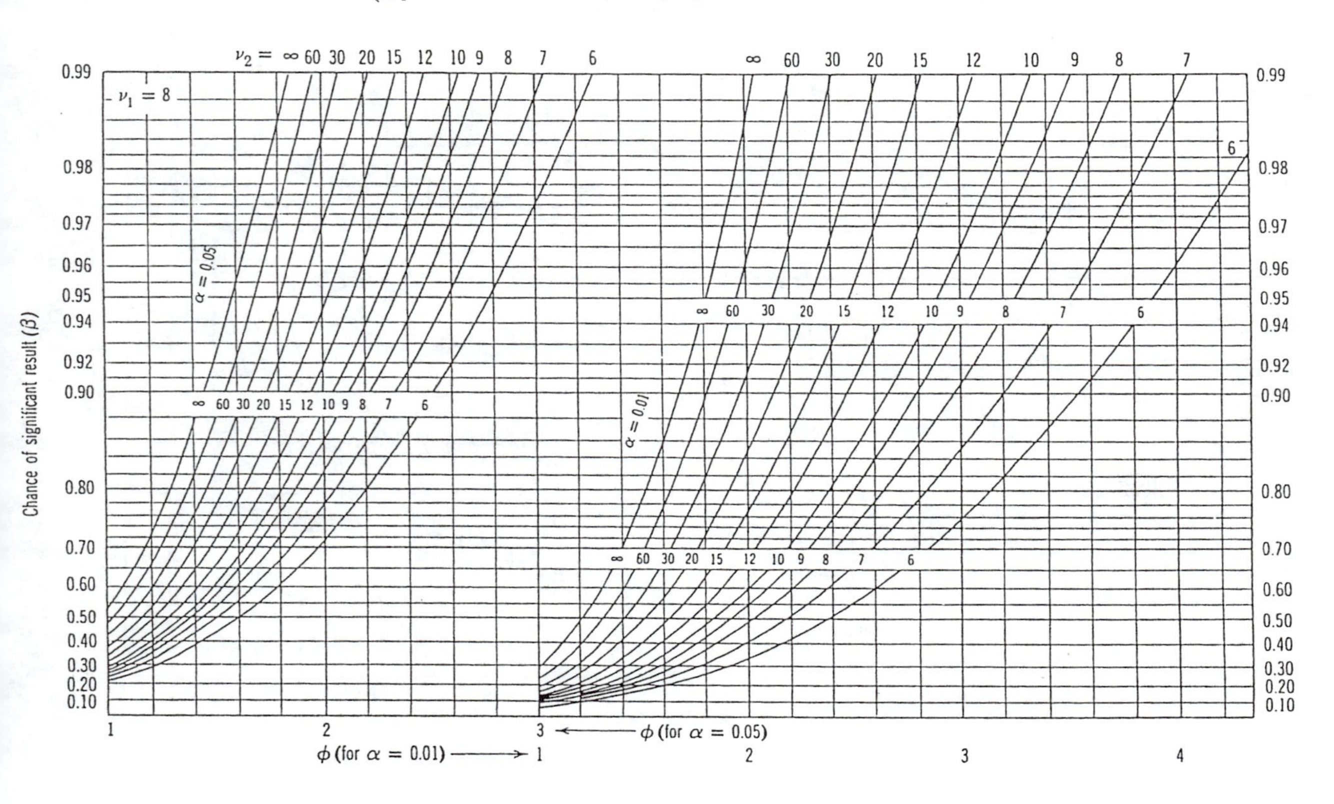

Table B.8 Values of $d_{.05,k,\upsilon}$ for Duncan's Multiple Range Test
(p = number of treatments, υ = df for error)

α = .05

ν \ p	2	3	4	5	6	7	8	9	10	11	12	13	14	15	16	17	18	19
1	17.97	17.97	17.97	17.97	17.97	17.97	17.97	17.97	17.97	17.97	17.97	17.97	17.97	17.97	17.97	17.97	17.97	17.97
2	6.085	6.085	6.085	6.085	6.085	6.085	6.085	6.085	6.085	6.085	6.085	6.085	6.085	6.085	6.085	6.085	6.085	6.085
3	4.501	4.516	4.516	4.516	4.516	4.516	4.516	4.516	4.516	4.516	4.516	4.516	4.516	4.516	4.516	4.516	4.516	4.516
4	3.927	4.013	4.033	4.033	4.033	4.033	4.033	4.033	4.033	4.033	4.033	4.033	4.033	4.033	4.033	4.033	4.033	4.033
5	3.635	3.749	3.797	3.814	3.814	3.814	3.814	3.814	3.814	3.814	3.814	3.814	3.814	3.814	3.814	3.814	3.814	3.814
6	3.461	3.587	3.649	3.680	3.694	3.697	3.697	3.697	3.697	3.697	3.697	3.697	3.697	3.697	3.697	3.697	3.697	3.697
7	3.344	3.477	3.548	3.588	3.611	3.622	3.626	3.626	3.626	3.626	3.626	3.626	3.626	3.626	3.626	3.626	3.626	3.626
8	3.261	3.399	3.475	3.521	3.549	3.566	3.575	3.579	3.579	3.579	3.579	3.579	3.579	3.579	3.579	3.579	3.579	3.579
9	3.199	3.339	3.420	3.470	3.502	3.523	3.536	3.544	3.547	3.547	3.547	3.547	3.547	3.547	3.547	3.547	3.547	3.547
10	3.151	3.293	3.376	3.430	3.465	3.489	3.505	3.516	3.522	3.525	3.526	3.526	3.526	3.526	3.526	3.526	3.526	3.526
11	3.113	3.256	3.342	3.397	3.435	3.462	3.480	3.493	3.501	3.506	3.509	3.510	3.510	3.510	3.510	3.510	3.510	3.510
12	3.082	3.225	3.313	3.370	3.410	3.439	3.459	3.474	3.484	3.491	3.496	3.498	3.499	3.499	3.499	3.499	3.499	3.499
13	3.055	3.200	3.289	3.348	3.389	3.419	3.442	3.458	3.470	3.478	3.484	3.488	3.490	3.490	3.490	3.490	3.490	3.490
14	3.033	3.178	3.268	3.329	3.372	3.403	3.426	3.444	3.457	3.467	3.474	3.479	3.482	3.484	3.484	3.485	3.485	3.485
15	3.014	3.160	3.250	3.312	3.356	3.389	3.413	3.432	3.446	3.457	3.465	3.471	3.476	3.478	3.480	3.481	3.481	3.481
16	2.998	3.144	3.235	3.298	3.343	3.376	3.402	3.422	3.437	3.449	3.458	3.465	3.470	3.473	3.477	3.478	3.478	3.478
17	2.984	3.130	3.222	3.285	3.331	3.366	3.392	3.412	3.429	3.441	3.451	3.459	3.465	3.469	3.473	3.475	3.476	3.476
18	2.971	3.118	3.210	3.274	3.321	3.356	3.383	3.405	3.421	3.435	3.445	3.454	3.460	3.465	3.470	3.472	3.474	3.474
19	2.960	3.107	3.199	3.264	3.311	3.347	3.375	3.397	3.415	3.429	3.440	3.449	3.456	3.462	3.467	3.470	3.472	3.473
20	2.950	3.097	3.190	3.255	3.303	3.339	3.368	3.391	3.409	3.424	3.436	3.445	3.453	3.459	3.464	3.467	3.470	3.472
24	2.919	3.066	3.160	3.226	3.276	3.315	3.345	3.370	3.390	3.406	3.420	3.432	3.441	3.449	3.456	3.461	3.465	3.469
30	2.888	3.035	3.131	3.199	3.250	3.290	3.322	3.349	3.371	3.389	3.405	3.418	3.430	3.439	3.447	3.454	3.460	3.466
40	2.858	3.006	3.102	3.171	3.224	3.266	3.300	3.328	3.352	3.373	3.390	3.405	3.418	3.429	3.439	3.448	3.456	3.463
60	2.829	2.976	3.073	3.143	3.198	3.241	3.277	3.307	3.333	3.355	3.374	3.391	3.406	3.419	3.431	3.442	3.451	3.460
120	2.800	2.947	3.045	3.116	3.172	3.217	3.254	3.287	3.314	3.337	3.359	3.377	3.394	3.409	3.423	3.435	3.446	3.457
∞	2.772	2.918	3.017	3.089	3.146	3.193	3.232	3.265	3.294	3.320	3.343	3.363	3.382	3.399	3.414	3.428	3.442	3.454

REFERENCES

Anderson, T.W., *An Introduction to Multivariate Statistical Analysis*, 2d ed, Wiley, New York, 1984.

Bartlett, M.S., Properties of sufficiency and statistical tests, *Proceedings of the Royal Society A*, 160, 268-282, 1937.

Beck, A. R., Rial, W. and Rickels, K., Short form of depression inventory: cross validation, *Psychology Reports*, 34, 1184-1185, 1974.

Bennett, C.A. and Franklin, N. L., *Statistical Analysis in Chemistry and the Chemical Industry,* Wiley, New York, 1954.

Box, G.E.P., Nonnormality and tests on variances, *Biometrika*, 40, 318-335, 1953.

Box, G.E.P., Some theorems on quadratic forms applied in the study of analysis of variance problems, I. Effect of inequality of variance in the one-way classification, *Annals of Mathematical Statistics*, 25, 290-302, 1954a.

Box, G.E.P., Some theorems on quadratic forms applied in the study of analysis of variance problems, II. Effect of inequality of variance and correlation between errors in the two-way classification, *Annals of Mathematical Statistics*, 25, 484-498, 1954b.

Box, G.E.P. and Cox, D.R., An analysis of transformations, *Journal of the Royal Statistical Society*, B 26, 211-243, 1964.

Bradu, D. and Gabriel, K.R., The biplot as a diagnostic tool for models of two-way tables, *Technometrics*, 20, 47-68, 1978.

Carmer, S.G. and Swanson, M.R., An evaluation of ten pairwise multiple comparison procedures by Monte Carlo methods, *Journal of the American Statistical Association*, 68, 66-74, 1973.

Cochran, W.G., The distribution of the largest of a set of estimated variances as a fraction of their total, *Annals of Eugenics*, 11, 47-52, 1941.

Cochran, W.G., Some consequences when the assumptions for the analysis of variances are not satisfied, *Biometrics*, 3, 22-38, 1947.

Cochran, W.G. and Cox, G.M., *Experimental Designs*, 2d ed, Wiley, New York, 1957.

Coleman, D.E. and Montogomery, D.C., A systematic approach to planning for a designed industrial experiment, *Technometrics*, 35, 1-12, 1993.

Conover, W.J., *Practical Nonparametric Statistics*, 2d ed, Wiley, New York, 1980.

Cornfield, J. and Tukey, J.W., Average values of mean squares in factorials, *Annals of Mathematical Statistics*, 27, 907-949, 1956.

Daniel, W.W., *Applied Nonparametric Statistics*, PWS-Kent, Boston, 1990.

Davis, J., Performance of young hearing-impaired children on a test of basic Concepts, *Journal of Speech Hearing Research*, 17, 342-351, 1974.

Draper, N.R. and Smith, H., *Applied Regression Analysis*, 2d ed, Wiley, New York, 1981.

Duncan, D.B., Multiple range and multiple F tests, *Biometrics*, 11, 1-42, 1955.

Dunnett,C.W., A multiple comparison procedure for comparing several treatments with a control, *Journal of the American Statistical Association*, 509, 1096-1121, 1955.

Einot, I. and Gabriel, K.R., A study of the powers of several multiple comparisons, *Journal of the American Statistical Association*, 70, 574-583, 1975.

Graybill, F.A., *Theory and Application of the Linear Model*. Duxbury, Belmont CA, 1976.

Hartley, H.O., Testing the homogeneity of a set of variances, *Biometrika*, 31, 249-255, 1940.

Hartley, H.O., The maximum F-ratio as a short-cut test for heterogeneity of variance, *Biometrika*, 37, 308-312, 1950.

Hayter, A.J., A proof of the conjecture that the Tukey-Kramer multiple comparison procedure is conservative, *Annals of Statistics*, 12, 61-75, 1984.

Hicks, C.R., *Fundamental Concepts in the Design of Experiments*, 2d ed, Holt, Rinehart & Winston, New York, 1973.

Hochbery, Y. and Tamhane, A.C., *Multiple Comparison Procedures*, Wiley, New York, 1987.

Hollander, M. and Wolfe, D.A., *Nonparametric Statistical Methods*, Wiley, New York, 1973.

John, P.W.M., *Statistical Design and Analysis of Experiments*, Macmillan, New York, 1971.

Johnson, N.R. and Leone, F.C., *Statistics and Experimental Designs in Engineering and the Physical Sciences*, Vol 2, Wiley, New York, 1977.

Keuhl, R.O., *Statistical Principles of Research Design and Analysis*, Duxbury, Belmont, CA, 1994.

Keuls, M., The use of the studentized range in conjunction with an analysis of variance, *Euphytica*, 1, 112-122, 1952.

Montgomery, D.C., *Design and Analysis of Experiments*, 4th ed, Wiley, New York, 1997.

Meyer, S.J., The relation of parental conflict, dominance, and inconsistency of love to anorexia nervosa and bulimia in adult women, Master's Thesis, Miami University, Oxford, Ohio, 1986.

Myers, R.H., *Classical and Modern Regression with Applications*, Duxbury, Boston, 1990.

Myer, R.H. and Minton, J.S., *A First Course in the Theory of Linear Statistical Models*, Duxbury, Boston, 1991.

Neter, J, Wasserman, W. and Kutner, M.H., *Applied Linear Statistical Models*, 3d ed, Irwin, Homewood, IL, 1988.

Newman, D., The distribution of the range in samples from a normal population, expressed in terms of an independent estimate of standard deviation, *Biometrika*, 31, 20-30, 1939.

Pearson, E.S., The analysis of variance in cases of non-normal variation, *Biometrika*, 23, 114-133, 1931.

Pearson, E.S. and Hartely, H.O., Charts of the power function for analysis-of-variance tests, derived from non-central F distribution, *Biometrika*, 38, 112-130, 1951.

Pearson, E.S. and Hartely, H.O., *Biometrika Tables for Statisticians*, 3d ed, Cambridge, New York, 1966.

Satterthwaite, F.E., An approximate distribution of estimates of variance components, *Biometrics*, 2, 110-114, 1946.

Scheaffer, R.L. and McClave, J.T., *Probability and Statistics for Engineers*, 2d ed, Duxbury, Boston, 1982.

Scheffé, H., *The Analysis of Variance*, Wiley, New York, 1959.

Searle, S.R., *Linear Models*, Wiley, New York, 1971.

Searle, S.R., *Linear Models for Unbalanced Data*, Wiley, New York, 1987.

Seber, G.A.F., *Linear Regression Analysis*, Wiley, New York, 1977.

Sincich, T., *Statistics by Example*, Dellen, San Francisco, 1982.

Tukey, J.W., One degree of freedom for non-additivity, *Biometrics*, 5, 232-242, 1949.

Walpole, R.E. and Myers, R.H., *Probability and Statistics for Engineers and Scientists*, Macmillan, New York, 1993.

Williams, E.J., Experimental designs balanced for the estimation of residual effects of treatments, *Australian Journal of Scientific Research*, 2, 149-168, 1949.

SOLUTIONS FOR SELECTED ODD-NUMBERED EXERCISES

Chapter 2

2–1 (a) No (b) Yes (c) No (d) Yes (e) No (f) Yes (g) No (h) No (i) No (j) Yes (k) Yes (l) No (m) Yes (n) Yes

2–3 (b) $\hat{y} = 29.682 - 1.009x$ (c) 22.114

2–5 (b)

$$\mathbf{y} = \begin{bmatrix} 30 \\ 50 \\ 40 \\ 60 \\ 20 \\ 30 \\ 40 \\ 50 \end{bmatrix} \quad \mathbf{x} = \begin{bmatrix} 1 & -1 & -1 & 1 \\ 1 & -1 & 1 & -1 \\ 1 & 1 & -1 & -1 \\ 1 & 1 & 1 & 1 \\ 1 & -1 & -1 & 1 \\ 1 & -1 & 1 & -1 \\ 1 & 1 & -1 & -1 \\ 1 & 1 & 1 & 1 \end{bmatrix} \quad \mathbf{b} = \begin{bmatrix} 40 \\ 7.5 \\ 7.5 \\ 0 \end{bmatrix}$$

(c) Rank is 3.

2–7 (a) $Y_{ij} = \mu + \tau_i + \varepsilon_{ij}$; $i = 1,2,3$; $j = 1,2,3,4,5$.

(c) Rank is 3.

2–9 Place the expressions for b_0 and b_1 given in (2.3.3) in Equation (2.3.2) and simplify.

2–11 Use

$$\mathbf{y} = \begin{bmatrix} y_1 \\ \vdots \\ y_n \end{bmatrix} \quad \mathbf{x} = \begin{bmatrix} 1 & x_1 \\ \vdots & \vdots \\ 1 & x_n \end{bmatrix} \quad \mathbf{b} = \begin{bmatrix} b_0 \\ b_1 \end{bmatrix}$$

2–13 (a) $a = \bar{y}$ $b = \Sigma y_i(x_i - \bar{x})/\Sigma(x_i - \bar{x})^2$

(b) $\hat{y} = 3 + 2(x-2)$ (c) The models are the same since $\beta_0 = \alpha - \beta\bar{x}$ and $\beta_1 = \beta$.

2–15 (a) Let $\mathbf{Y}' = [Y_{11}\ Y_{12}\ Y_{13}\ Y_{14}\ Y_{21}\ Y_{22}\ Y_{23}\ Y_{31}\ Y_{32}]$, $\boldsymbol{\beta}' = (\mu_1\ \mu_2\ \mu_3)$ (b) Rank is 3, which is full rank.

Chapter 3

3–1 (a) $\mathbf{b}' = (2\ \ .6\ \ 1)$ (b) 2.6 (c) .4

3–3 (a) $\hat{y} = 40 + 7.5x_1 + 7.5x_2$ (b) 46 (c) 300

3–5 (a) $\hat{y} = 10 - 2x$ (b) i) 10 ii) 8 (c) 18

3–9 One solution is a matrix with the following four rows: –.75 0 0 .5, 0 0 0 0, 0 0 0 0, .5 0 0 –.25.

3–11 (a) i) $\mathbf{b}' = (\bar{y}_{\cdot\cdot}, \bar{y}_{1\cdot} - \bar{y}_{\cdot\cdot}, \bar{y}_{2\cdot} - \bar{y}_{\cdot\cdot})$

ii) $\mathbf{b}' = (.5(\bar{y}_{1\cdot} + \bar{y}_{2\cdot}), .5(\bar{y}_{1\cdot} - \bar{y}_{2\cdot}), .5\,(\bar{y}_{2\cdot} - \bar{y}_{1\cdot}))$

iii) $\mathbf{b}' = (\bar{y}_{1\cdot}, 0, \bar{y}_{2\cdot} - \bar{y}_{1\cdot})$ (b) $\bar{y}_{i\cdot}$ for all three solutions.

(c) $4\bar{y}^2_{1\cdot} + 2\bar{y}^2_{2\cdot}$

3–13 38.819

3–15 (b) i) $\bar{y}_{2\cdot} - \bar{y}_{1\cdot}$ ii) $\bar{y}_{1\cdot} - 2\bar{y}_{2\cdot} + \bar{y}_{3\cdot}$

3–19 (a) i) $\hat{y} = 10.1 + 1.15x$ ii) $\hat{y} = 14.7 + 1.15x_1$ (b) 13.55 for both models (c) 5.65 for both models.

3–29 Disprove with a counterexample.

3–33 (SSe based upon $W_i = aY_i$) = (SSe based upon $W_i = aY_i + c$) = a^2 (SSe based on Y_i).

Chapter 4

4–1 (a) $\sigma^2/2$, $\sigma^2/12$, $-\sigma^2/6$ (b) 4

4–3 (a) 4.67 (b) 59.83 (c) σ^2, $\sigma^2/6$ (d) $N(\mu + \tau_3, \sigma^2/6)$

4–5 $\sigma^2\mathbf{B}$, where $\mathbf{B}$ has rows 1/5, 0, 0; 0, 1/10, 0; 0, 0, 1/14.

4–7 (a) $\sigma^2/8$ (b) 46 (c) $.17\sigma^2$ (d) $N(\beta_1, \sigma^2/8)$

4–9 (a) Yes (b) N(19, 28) (c) $MN(\boldsymbol{\mu},\Sigma)$, where $\boldsymbol{\mu}' = (20, 12, 17)$ and Σ has rows 8, –4, –6; –4, 25, 34; –6, 34, 47 (d) $MN(\boldsymbol{\mu},\Sigma)$, where $\boldsymbol{\mu}' = (8,5)$ and Σ has rows 7, 4; 4, 8.

4–13 $\text{Var}(\hat{\beta}_0) = \sigma^2\Sigma x_i^2/(n\Sigma(x_i - \bar{x})^2)$, $\text{Var}(\hat{\beta}_1) = \sigma^2/\Sigma(x_i - \bar{x})^2$

$\text{Cov}(\hat{\beta}_0, \hat{\beta}_1) = -\sigma^2\Sigma x_i/(n\Sigma(x_i - \bar{x})^2)$

4–23 (b) Yes (c) Every linear combination of the elements of $\boldsymbol{\beta}$.

(d) $\bar{Y}_{1\cdot} - \bar{Y}_{2\cdot}$ (e) $7\sigma^2/12$

4–25 (a) The X matrix has rows 0, 1, 0; 0, 0, 1; 1, 1, 0; 1, 0, 1.
(b) Both are estimable. (c) The three normal equations are $2a + t_1 + t_2 = y_{2\cdot}$, $a + 2t_1 = y_{\cdot 1}$, $a + 2t_2 = y_{\cdot 2}$. (d) $\bar{y}_{\cdot 1} - \bar{y}_{\cdot 2}$

(e) i) $c' = (.5, -.5, .5, -.5)$ ii) σ^2 (f) ii) $c' = (1, -1, 0, 0)$

iii) Not BLUE (g) i) 0 ii) 28 (h) $19\sigma^2$

4–29 $N(\mu, \sigma^2/n)$

4–31 (a) $\rho = 0,\ \mu_1 = 5,\ \mu_2 = -3,\ \sigma_1 = 1,\ \sigma_2 = 1.$
(b) $\rho = 0,\ \mu_1 = -1,\ \mu_2 = 3,\ \sigma_1 = 1,\ \sigma_2 = 1.$
(c) $\rho = .5,\ \mu_1 = 5,\ \mu_2 = 8,\ \sigma_1 = 1,\ \sigma_2 = 2.$

Chapter 5

5–1 .2

5–3 $56 + 6\sigma^2$

5–5 (a) .282 (b) (–1.102, –.916) (c) (21.74, 22.49)

5–7 (–4.997, .997), Diets are not significantly different.

5–9 (–6.42, –3.78), Measures the difference between the average of the means for locations 1 and 2 versus location 3.

5–11 $t = 0$, Fail to reject H_0.

5–13 (a) (41.6, 53.1) (b) $f = 14.5$, Reject H_0 (c) (4.39, 20.61)

5–15 (a) i) $t = 6.79$, Reject H_0 ii) $f = 46.1$, Reject H_0 (b) i) $f = 24.2$, Reject H_0 ii) $f = 26.6$, Reject H_0 iii) $f = 0$, Fail to reject H_0

5–19 Chi–square with $n - p$ degrees of freedom, $E(Q) = n - p$.

5–23 $b_0 \pm t_{\frac{\alpha}{2},n-2}\sqrt{s^2\Sigma x_i^2/(n\Sigma(x_i - \bar{x})^2)}$, $b_1 \pm t_{\frac{\alpha}{2},n-2}\sqrt{s^2/\Sigma(x_i - \bar{x})^2}$

5–25 $t = b_1/\sqrt{s^2/\Sigma(x_i - \bar{x})^2}$

5–27 $t = b_j/\sqrt{s^2 c_{jj}}$, c_{jj} is the jth entry in $(\mathbf{X}'\mathbf{X})^{-1}$, $j = 0,1,\cdots,k$.

5–31 No.

Chapter 6

6–1 (a) 1028.814 (b) 953.687 (c) 67.985 (d) 1.684 (e) –0.131 (f) 0.796

6–3 Verify results given in Example 6.6.1.

6–5 (–5.776, –0.224)

6–7 $y = 1 - 3.5x + 0.5x^2$

6–9 (i) $\hat{y} = 110.000 + 0.287x + 0.333x^2$

(ii) $\hat{y} = 146.125 - 36.125x_1 - 26.375x_2$ for given x_1, x_2 coding.

(iii) $\hat{y} = 125.292 + 18.063x_1 + 2.701x_2$ for given x_1, x_2 coding.

(a) Identical (b) 146.125 (c) SSr = 5220.062 + 368.521, Linear in nature.

6–11 16.375, 22.675, 19.000 Average yield is parabolic in nature, yet increases with temperature.

6–13 (a) and (b) Hint: Use the fact that SSt = SSr + SSe.

6–15 (a) For A: –15.973 + 50.326x; for B: –12.654 + 54.339x.
(b) t = 0.271, no statistical difference in effective rates.
(c) f = 6.046, a statistical difference in potency exists.

6–17 b_1 is not the leading coefficient in the fitted function.

6–19 $\mathbf{X'y} = \begin{bmatrix} 3007 \\ 289 \\ 133 \end{bmatrix}$

Chapter 7

7–1 (b) 51 (c) (49.69, 52.31) (d) SStr = 51.568, SSe = 33.828, SSt = 85.396, f = 11.43, (e) f = 11.43 Reject H_0 (f) H = 6.51, Fail to reject H_0.

7–3 (a) SStr = 2017.0556, SSe = 1415.1667, SSt = 3432.222, f = 23.52 (b) f = 23.52, Reject H_0 (c) (18.14, 25.86) (d) (9.37, 20.29)
(e) H = 9.21, Reject H_0 (f) SStr = 17.041, SSe = 9.322, f = 30.16, Reject $H_0 : \tau_1 = \tau_2 = \tau_3$, interval for μ_2 is (2.70, 3.32), interval for $\mu_1 - \mu_3$ is (.97, 1.85), H = 2.19 so fail to reject H_0.

7–5 (a) SStr = 220.9167, SSe = 1658.4167, SSt = 1879.333, f = 1.4 (b) f = 1.4, Fail to reject H_0 (c) (–1.82, 16.66) (d) No
(e) t = –.19, No significant difference.

7–7 (a) $Y = \alpha_0 + \alpha_1 x_1 + \alpha_2 x_2 + \varepsilon$ (b) SStr = 1484.9, SSe = 459.0, SSt = 1943.9, f = 38.82 (c) f = 38.82, Reject H_0 (d) (–1.96, 4.56)

7–9 (a) H = 3.87, Fail to reject H_0 (b) No (c) Reasonable assumptions

7–13 b = .016, Fail to reject H_0.

7–15 (a) $\Sigma\Sigma(y_{ij} - \bar{y}_{i\cdot})^2$ (b) f = MStr/MSe (c) same

7–17 (ii) and (iv) are estimable, (i) and (iii) are not

7–23 (e) $(\bar{y}_{i\cdot} - \bar{y}_{\cdot\cdot}) \pm t_{\frac{\alpha}{2},n-p} \sqrt{s^2(1/n_i - 1/n)}$ (f) Interval for τ_A is (1.161, 3.295), Interval for μ_A is (49.69, 52.31).

7–27 (b) .35, power increases as sample size increases (c) .75, power decreases as variance increases

Chapter 8

8–1 f = 2.55 (no significant difference between types of blondes), f = 1.08 (no significant difference between types of brunettes), f = 18.45 (significant difference between blondes and brunetttes).

8–3 $\hat{y} = 117.8 + .804167x - .003812x^2 + .000005x^3$

8–5 Since it is unbalanced you cannot partition the treatment sum of squares using the contrasts $\tau_3 - \tau_1$ and $\tau_1 - 2\tau_2 + \tau_3$. Use polynomial regression to find the linear sum of squares to be 3209.7 and the quadratic sum of squares is 88.2.

8–7 Width of confidence interval is 34.29.

8–9 Treatment 3 is significantly different from treatments 1 and 2.

Chapter 9

9–1 (a) $\underline{2 \quad 3}$ 1 (b) Intervals for (1,2), (1,3), (2,3) are: (1.85, 6.35), (.33, 4.84), (–.74, 3.77)

9–3 Critical values are: LSD (10.61), Tukey (14.32), Bonferroni (15.06), Scheffe (15.60)

9–5 (a) (2.55, 17.45), (1.72, 16.62), (–5.35, 5.19) (b) (1.30, 18.70), (.47, 17.87), (–6.24, 6.08) (c) Bonferroni intervals are shorter.

9–7 (a) Critical values are: 21.04, 19.84, outcome 3 $\underline{2 \quad 1}$.

(b) Critical value is 20.64, outcome 3 $\underline{2 \quad 1}$.

Chapter 10

10–1 (a) SStr = 2.181 SSbl = 8.424 SSe = .132 f = 82.48 (Reject H_0 for treatments), f = 127.43 (Reject H_0 for blocks) (b) (.535, .831) (c) Critical value is .18, graphical outcome $\underline{2 \quad 0}$ 1 (d) f = 1.91, Fail to reject H_0 (e) 43.1

10–3 (a) SStr = 501.75 SSbl = 398.00 SSe = 58.00 f = 34.6 (Reject H_0 for treatments), f = 20.59 (Reject H_0 for blocks) (b) Critical value is 4.128, graphical outcome is 4 <u>1 2</u> 3 (c) No problems.

10–5 (a) t = 5.761, Significant difference between before and after (b) SStr = 256.0 SSbl = 5142.0 SSe = 54.0 f = 33.19, $f = t^2$ (same test) (c) f = .69, Fail to reject H_0, test less powerful (d) For RCBD: (4.72, 11.28), for CRD: (–12.66, 28.66).

10–7 (12.45, 20.55)

10–9 Critical value is .9, treatment 1 is significantly different from treatments 2,3,5.

10–11 (b) Confidence interval is

$$(\bar{y}_{i\cdot} + \bar{y}_{\cdot j} - \bar{y}_{\cdot\cdot}) \pm t_{\frac{\alpha}{2},(p-1)(r-1)} \sqrt{s^2 \ (p+r-1)/pr}.$$

10–13 Solution is $m = \bar{y}_{p\cdot} + \bar{y}_{\cdot r} - \bar{y}_{\cdot\cdot}$, $t_i = \bar{y}_{i\cdot} - \bar{y}_{p\cdot}$, $b_j = \bar{y}_{\cdot j} - \bar{y}_{\cdot\cdot}$, contrast estimate same as in (10.3.5).

10–17 Not equivalent for one–sided.

10–23 (a) 5 (b) 9

Chapter 11

11–1 (a) f = 2.38, Fail to reject H_0 (b) Outlier increases s^2 and decreases power of f–test (c) i) f = 58.88, Reject H_0 ii) Decreased s^2 and more powerful f–test iii) Assumptions seem reasonable iv) <u>2 4</u> 1 3, brands 2 and 4 seem best v) Tukey on SAS is based upon means, which does not provide the proper estimate in some cases.

11–3 (a) i) For operators: f = 9.38, Reject H_0; For days: f = 2.03, Fail to reject H_0 ii) A B C D (A B C underlined; C D underlined) iii) t = –4.01, Reject H_0 iv) Assumptions seem reasonable (b) Interval for μ_{22} is (71.56, 92.78) (c) ANOVA Table has unadjusted sum of square for days (blocks).

11–5 (a) SStr = 11664.69, SSbl = 706.36. Inverse matrix is not block diagonal which implies that treatment and blocks are not orthogonal (b) Model can be used to find the ANOVA Table, for example, SStr = SS $(\alpha_1, \alpha_2, \alpha_3 | \alpha_0, \alpha_4, \alpha_5, \alpha_6, \alpha_7)$ (c) Use x_1 = 1 for 200, 2 for 300, 3 for 400, 5 for 600.

11–7 (a) $\tau_2 - \tau_1 + (\beta_1 + \beta_2 + \beta_3 - 3\beta_4)/12$ (b) Difference is not contaminated by block effects. (c) $\sigma^2 + 27\tau_1^2/7 + 3\tau_2^2 + 4\beta_4^2/21 + 2\beta_4(\tau_1 - 7\tau_2)/7$ (d) $\sigma^2 + 3(\tau_1 - \tau_2)^2/2$ (e) It is free of block effects.

Chapter 12

12–3 (a) SS(field) = 107.1875 SS(method) = 303.6875 SS(variety) = 29.6875 SSe = 10.875 (b) Field: f = 19.71, Reject H_0; Method: f = 55.85, Reject H_0; Variety: f = 5.46, Reject H_0. (c) No significant differences among varieties using a Tukey procedure. (d) Methods 1 2 3 4, Weed spray is significantly better than pesticide. (e) No serious problems indicated by the residuals.

12–5 $SS(\mu,\tau) = SS(\tau|\mu,\rho,\gamma) = SS(\tau|\mu,\rho) = 185.1875$.

12–7 Adjusted sum of squares: SS(tech) = 4.167, f = .16; SS(week) = 18.0, f = 1.02; SS(schedule) = 121.5, f = 4.58; None of the f–tests yield significant results; Unadjusted sum of squares: SS(tech) = 24.25, SS(week) = 18.0, SS(schedule) = 141.583.

12–11 Estimates for μ, τ_A, ρ_I and γ_1 are: 2.824, .596, –.024 –.104.

12–13 (b) $y_{ij(k)}$ (c) SSe = 0 (d) No, no error term.

12–17 (b) SS(row) = 85.25 f = 7.93, SS(col) = 25.25 f = 2.35, SS(trt) = 4.25 f = .4, SS(layer) = 26.25 f = 2.44, None of the f–values are significant, all f–values have 3 and 3 degrees of freedom.

Chapter 13

13–1 (a) SS(Supp) = .5104 SS(whey) = 6.6912 SS(int) = 3.7246 SSe = .48 (b) Interaction f = 41.38, Significant interaction; Both main effects are also significant (f = 17.01 for supplement and f = 74.35 for whey). (c) No supplement and high whey yields the largest mean; however, it is not significantly different from no supplement and medium whey. (d) (.33, .93) (e) No problems suggested by the residuals.

13–3 (a) SS(var) = 498.778 SS(pest) = 969.389 SS(int) = 8.889 SSe = 314 (b) Interaction f = .06, not significant; Both main effects are significant (f = 7.15 for variety and f = 27.79 for pesticide) (c) Critical value is 9.5214, outcome is: Var 2 3 1. (d) No problems suggested by residuals.

13–5 ANOVA Table: SS(humid) = 30.375 SS(temp) = 385.125 SS(int) = 17.458 SSe = 170.0 Effects: linear (SSQ = 39.001, f = 3.08, not significant); quadratic (SSQ = 332.160, f= 33.67, significant); cubic (SSQ = 13.964, f = 1.42, not significant); linear is unadjusted, quadratic is adjusted for linear, cubic is adjusted for both.

13–7 (a) f = .14, not significant (b) SSa = 385.20 SSb = 176.817 SSe = 104.6 (c) Test for A: f = 9.21, Reject H_0; Test for B : f = 8.45, Reject H_0. (d) Point estimate is 22, Interval is (13.7, 30.3). (e) A 3 $\underline{1 \quad 2}$ (f) t = –2.29, not significant (g) Cannot estimate $(\alpha\beta)_{22}$.

13–11 (a) Price H $\underline{\text{M} \quad \text{L}}$ (b) Price H $\underline{\text{M} \quad \text{L}}$

13–15 For setting I : $\mu = 14$, $\alpha_1 = -2$, $\alpha_2 = 2$, $\beta_1 = 0$, $\beta_2 = 2$, $\beta_3 = -2$, all $(\alpha\beta)_{ij} = 0$; For setting II: $\mu = 14$, $\alpha_1 = 1$, $\alpha_2 = -1$, $\beta_1 = 2$, $\beta_2 = 1$, $\beta_3 = 3$, $(\alpha\beta)_{11} = -2$, $(\alpha\beta)_{21} = 2$, $(\alpha\beta)_{12} = 3$, $(\alpha\beta)_{22} = -3$, $(\alpha\beta)_{13} = -1$, $(\alpha\beta)_{23} = 1$.

13–21 $\Sigma\Sigma c_{ij} \bar{y}_{ij\cdot} \pm t_{\frac{\alpha}{2}, ab(r-1)} \sqrt{s^2 \Sigma\, \Sigma c_{ij}^2 / r}$

13–23 Let A1 = 1,0, –1 if A = 1,2,3, respectively. Let A2 = 1,0,–1 if A = 2,1,3, respectively. Define B1 and B2 similarly for factor B. The reparametrized model is $Y = \gamma_0 + \gamma_1$ A1 + γ_2 A2 + γ_3 B1 + γ_4 B4 + γ_5 (A1 * B1) + γ_6 (A1 * B2) + γ_7 (A2 * B1) + γ_8 (A2 * B2) + ε. Then $\gamma_0 = \mu$, $\gamma_1 = \alpha_1$, $\gamma_2 = \alpha_2$, $\gamma_3 = \beta_1$, $\gamma_4 = \beta_2$, $\gamma_5 = (\alpha\beta)_{11}$, $\gamma_6 = (\alpha\beta)_{12}$, $\gamma_7 = (\alpha\beta)_{21}$, $\gamma_8 = (\alpha\beta)_{22}$.

13–27 Neither function is estimable.

13–29 (a) r = 5 (b) r = 4

Chapter 14

14–1 All three and two–factor interactions are not significant, Format (SSQ = 185.28, f = 1.12, not significant), Review (SSQ = 536.28, f = 3.24, not significant), Homework (SSQ = 2064.03, f = 12.47, significant) (b) (6.67, 25.45) (c) Normality assumption is questionable.

14–3 (a) Spray * Variety (SSQ = 16.78, f = 1.56, not significant), Spray (SSQ = 1701.39, f = 315.72, significant), Variety (SSQ = 1007.44, f = 93.47, significant), Field (SSQ = 304.78, f = 28.28, significant) (b) All variety pairs are significantly different.

14–5 A = format, B = review, C = homework; sum of squares are: SSa, SSb, SSc (see 14–1), SSab = 3.78, SSac = 1.53, SSbc = 7.03, SSabc = 13.78.

14–7 (a) A (SSQ = 90312.5, f = 4.27), B(SSQ = 15312.5, f = .72), C(SSQ = 2812.5, f = .13), D(SSQ = 37812.5, f = 1.79), none of the tests are significant, A has the most effect (b) Alias pairs are: (A, BCD), (B, ACD), (C, ABD), (D, ABC), (AB,CD), (AC, BD), (AD, BC), (I, ABCD) (c) SSa, SSb, SSc, SSd are the same as in part (a), SSab = 312.5, SSac = 37812.5, SSad = 25312.5.

14–9 Design points are: (1), a, ab, ac, ad ,bc, bd, cd, abcd; same design points used in 14–7.

14–11 A $\underline{1 \quad 3}$ 2

14–13 (c) $t = \Sigma\, c_j \bar{y}_{\cdot j \cdot \cdot} / \sqrt{s^2 \Sigma c_j/(rac)}$ (d) Decide $\mu_{ijk} \pm \mu_{def}$ if $|\bar{y}_{ijk\cdot} - \bar{y}_{def\cdot}| \geq q_{\alpha}$, abc, abc(r–1) $\sqrt{s^2/r}$.

14–15 (b) Rank = 7 (d) $m = \bar{y}_{\cdot\cdot\cdot}$ $a_i = \bar{y}_{i\cdot\cdot} - \bar{y}_{\cdot\cdot\cdot}$, $b_j = \bar{y}_{\cdot j\cdot} - \bar{y}_{\cdot\cdot\cdot}$, $(ab)_{ij} = \bar{y}_{ij\cdot} - \bar{y}_{i\cdot\cdot} - \bar{y}_{\cdot j\cdot} + \bar{y}_{\cdot\cdot\cdot}$, $r_k = \bar{y}_{\cdot\cdot k} - \bar{y}_{\cdot\cdot\cdot}$ (e) $SSe = \Sigma\Sigma\Sigma(y_{ijk} - \bar{y}_{ij\cdot} - \bar{y}_{\cdot\cdot k} + \bar{y}_{\cdot\cdot\cdot})^2$ (f) $SSblock = ab\,\Sigma\,(\bar{y}_{\cdot\cdot k} - \bar{y}_{\cdot\cdot\cdot})^2$.

14–17 (c) $(\bar{y}_{i\cdot\cdot\cdot} - \bar{y}_{t\cdot\cdot\cdot}) \pm t_{\frac{\alpha}{2},v} \sqrt{2s^2/(ab^2)}$, $v = a^2b^2 - 3ab + 2$

(d) Decide $\alpha_i \pm \alpha_t$ if $|\bar{y}_{i\cdot\cdot\cdot} - \bar{y}_{t\cdot\cdot\cdot}| \geq q_{\alpha,a,v}\sqrt{s^2/(ab^2)}$

14–19 Effect = .5 (least squares estimate for $\alpha_H - \alpha_L$).

Chapter 15

15–1 (a) f = .51, No significant difference between slopes (b) f = 15.43, Significant treatment effect (c) (1.23, 1.55) (d) 2 $\underline{1 \quad 3}$

15–3 (a) f = .68, No significant difference between slopes for three treatments (b) SSe = 75.061, treatments (SSQ = 57.377, f = 3.44, not significant), blocks (SSQ = 29.409, f = .71, not significant), covariate (SSQ = 43.495, f = 5.22, significant)

(c) No significant differences (d) treatments (SSQ = 22.056, f = 1.86), SSe = 118.556, larger error term with RCBD and smaller f

15–5 (6.87, 15.09)

15–7 (–5.01, 12.01)

15–13 $\alpha_0 = \mu + \tau_3$, $\alpha_3 = \beta_3$, $\alpha_1 = \tau_1 - \tau_3$, $\alpha_2 = \tau_2 - \tau_3$, $\alpha_{13} = \beta_1 - \beta_3$, $\alpha_{23} = \beta_2 - \beta_3$.

15–17 (a) $Y = \alpha_0 + \alpha_1 x_1 + \alpha_2 x_2 + \alpha_{22} x_2^2 + \alpha_{12} x_1 x_2 + \alpha_{122} x_1 x_2^2 + \varepsilon$; $x_1 = 1$ if treatment 1, 0 otherwise; x_2 = covariate.

Chapter 16

16–1 (a) Same model and assumptions as in (16.4.1) (b and c) Creek (SSQ = 2.678, f = 48.69, significant), Sample (SSQ = .33, f = 2.95, significant), SSe = .14 (d) All levels of creek are significantly different.

16–3 (a) Same model and assumptions as in (16.6.1) with oven random and temperature fixed, (b and c) Interaction (SSQ = 6.75, f = .69, not significant), Oven (SSQ = 36.125, f = 7.41, significant), Temperature (SSQ = 108.25, f = 48.11, significant) (d) (3.327, 5.923) (e) $\underline{3 \quad 1} \quad 2$ (f) 1.736

16–5 (a) (.85, 126) (b) (11.18, 20.32)

16–7 (b) 0

16–13 $\Sigma c_i \bar{y}_{i\cdot\cdot} \pm t_{\frac{\alpha}{2}, v} \sqrt{MSb(a)\Sigma c_i^2/bc}$, $v = a(b-1)$

16–15 (d) $\bar{y}_{i\cdot} \pm t_{\frac{\alpha}{2}, v} \sqrt{[(p-1)MSe+MSbl]/(rp)}$, v found using (16.8.3).

16–17 (e) $\hat{\sigma}^2_\beta = (MSB - MSAB)/(ar)$, $\hat{\sigma}^2_{\alpha\beta} = (MSAB - MSE)/r$, $\hat{\sigma}^2_\varepsilon = MSE$, (f) $(vs^2_\beta/\chi^2_{\frac{\alpha}{2}, v},\ vs^2_\beta/\chi^2_{1-\frac{\alpha}{2}, v})$, v given by (16.8.3).

Chapter 17

17–1 (a) Same model and assumptions as in (17.1.1) (b) States: f = 8.24, significant (c) State one: (62.372, 80.378), State two: (47.435, 65.441)

17–5 The estimated variances are i) .0715, ii) .0998. Design i) is more efficient.

17–7 (a) In the model, clinic and subject are random, gender (g) and time (t) are fixed factors; subject is nested within gender and clinic. (c) SSc = 10.667, SSg = 42.667, SSt = 64.083, SSsub = 8.333, SScg = 4.167, SSct = 1.083, SSgt = .583 (d) For time: f = MSt/MSct = 59.15, which is significant.

17–9 (a) $s_\alpha^2 = [MSa - MSb(a)]/(bcr)$, $s_\beta^2 = [MSb(a) - MSc(b,a)]/(cr)$ $s_\gamma^2 = [MSc(b,a) - MSe]/r$, $s_\varepsilon^2 = MSe$.

17–15 $Y_{ijk\ell} = \mu + \alpha_i + \beta_j + (\alpha\beta)_{ij} + \gamma_{k(ij)} + \varepsilon_{ijk\ell}$, i = 1,2,3,4; j = 1,...,5; k = 1,2,3, ℓ = 1,2. The degrees of freedom for the sum of squares are: A (3), B(4), AB (12), C(A,B) (40), error (60), total (119).

Chapter 18

18–1 (a) Variance components are: farm (75.14), farm*spray (1.04), error (5.5) (b) (1.02, 9.64)

18–3 (a) Same model and assumptions as (18.2.1). (b) SSd = 39.2, SSseq = .8, SSsub(seq) = 3.4, SSperiod = .8, SSe = 3.0 (c) Drug (f = 104.53), significant), Sequence (f = 1.88, not significant), period (f = 2.13, not significant) (d) No.

18–5 $\hat{y} = 12.583 + 1.583x_1 - 4.75x_2 - 2.083x_1x_2$

18–7 81 versus 25.

18–9 (a) Same model and assumptions as (16.4.1) (b) SSlot = 27.42, SSroll(lot) = 36.38, SSe = 21.8 (c) Roll (1.045), Error (.908) (d) $(\sigma_\varepsilon^2 + c\sigma_\beta^2)/(bc)$, β represents roll, c = 3, b = 4 (e) Estimated variances are for i) .337 for ii) .2998.

INDEX